AF271363

The Mathematics of Signal Processing

Arising from courses taught by the authors, this largely self-contained treatment is ideal for mathematicians who are interested in applications or for students from applied fields who want to understand the mathematics behind their subject.

Early chapters cover Fourier analysis, functional analysis, probability and linear algebra, all of which have been chosen to prepare the reader for the applications to come. The book includes rigorous proofs of core results in compressive sensing and wavelet convergence. Fundamental is the treatment of the linear system $y = \Phi x$ in both finite and infinite dimensions. There are three possibilities: the system is determined, overdetermined or underdetermined, each with different aspects.

The authors assume only basic familiarity with advanced calculus, linear algebra and matrix theory, and modest familiarity with signal processing, so the book is accessible to students from the advanced undergraduate level. Many exercises are also included.

STEVEN B. DAMELIN is Full Professor of Mathematics and Director for the Unit of Advances in Mathematics and its Applications, USA.

WILLARD MILLER, JR. is Professor Emeritus in the School of Mathematics at the University of Minnesota.

Cambridge Texts in Applied Mathematics

All titles listed below can be obtained from good booksellers or from Cambridge University Press. For a complete series listing, visit www.cambridge.org/mathematics.

The Mathematics of Signal Processing

STEVEN B. DAMELIN
Unit of Advances in Mathematics and its Applications

WILLARD MILLER, JR.
University of Minnesota

CAMBRIDGE
UNIVERSITY PRESS

CAMBRIDGE
UNIVERSITY PRESS

Shaftesbury Road, Cambridge CB2 8EA, United Kingdom

One Liberty Plaza, 20th Floor, New York, NY 10006, USA

477 Williamstown Road, Port Melbourne, VIC 3207, Australia

314–321, 3rd Floor, Plot 3, Splendor Forum, Jasola District Centre, New Delhi – 110025, India

103 Penang Road, #05–06/07, Visioncrest Commercial, Singapore 238467

Cambridge University Press is part of Cambridge University Press & Assessment,
a department of the University of Cambridge.

We share the University's mission to contribute to society through the pursuit of
education, learning and research at the highest international levels of excellence.

www.cambridge.org
Information on this title: www.cambridge.org/9781107601048

First published 2012

A catalogue record for this publication is available from the British Library

ISBN   978-1-107-01322-3   Hardback
ISBN   978-1-107-60104-8   Paperback

Dedicated to our parents, children and partners

Contents

Preface

Basically, this is a book about mathematics, pitched at the advanced undergraduate/beginning graduate level, where ideas from signal processing are used to motivate much of the material, and applications of the theory to signal processing are featured. It is meant for math students who are interested in potential applications of mathematical structures and for students from the fields of application who want to understand the mathematical foundations of their subject. The first few chapters cover rather standard material in Fourier analysis, functional analysis, probability theory and linear algebra, but the topics are carefully chosen to prepare the student for the more technical applications to come. The mathematical core is the treatment of the linear system $y = \Phi x$ in both finite-dimensional and infinite-dimensional cases. This breaks up naturally into three categories in which the system is determined, overdetermined or underdetermined. Each has different mathematical aspects and leads to different types of application. There are a number of books with some overlap in coverage with this volume, e.g., [11, 15, 17, 19, 53, 69, 71, 72, 73, 82, 84, 95, 99, 101], and we have profited from them. However, our text has a number of features, including its coverage of subject matter, that together make it unique. An important aspect of this book on the interface between fields is that it is largely self-contained. Many such books continually refer the reader elsewhere for essential background material. We have tried to avoid this. We assume the reader has a basic familiarity with advanced calculus and with linear algebra and matrix theory up through the diagonalization of symmetric or self-adjoint matrices. Most of the remaining development of topics is self-contained. When we do need to call on technical results not proved in the text, we try to be specific. Little in the way of formal knowledge about signal processing is assumed. Thus while this means that many interesting topics cannot be covered in a text of modest size, the topics that are treated have a logical coherence, and the reader is not continually distracted by appeals to other books and papers. There are many exercises. In most of the volume the logic of the mathematical topics predominates, but in a few chapters, particularly for compressive sensing and for parsimonious representation of data, the issues in the area of application predominate and mathematical topics are introduced as appropriate to tackle the applied problems. Some of the sections, designated by "Digging deeper" are more technical and can be mostly skipped on a first reading. We usually give a nontechnical description of the principal results of these sections. The book is sufficiently flexible to provide relatively easy access to new ideas for students or instructors who wish to skip around, while filling in the background details for others. We include a large list of references for the reader who wants to "dig deeper." In particular, this is the case in the chapter on the parsimonious representation of data.

This book arose from courses we have both taught and from ongoing research. The idea of writing the book originated while the first author was a New Directions Professor of Imaging at the Institute for Mathematics and its Applications, The University of Minnesota during the 05–06 academic year. The authors acknowledge support from the National Science Foundation; the Centre for High Performance Computing, Cape Town; the Institute for Mathematics and its Applications, University of Minnesota; the School of Computational and Applied Mathematics, the University of the Witwatersrand, Johannesburg; Georgia Southern University; and the United States Office of Airforce Research. We are indebted to a large number of colleagues and students who have provided valuable feedback on this project, particularly Li Lin and Peter Mueller who tested the compressive sensing algorithms. All figures in this book were generated by us from open source programs such as CVX, Maple or MATLAB, or from licensed MATLAB wavelet and signal processing toolboxes.

In closing, we thank the staff at Cambridge University Press, especially David Tranah and Jon Billam, for their support and cooperation during the preparation of this volume and we look forward to working with them on future projects.

Introduction

Consider a linear system $y = \Phi x$ where Φ can be taken as an $m \times n$ matrix acting on Euclidean space or more generally, a linear operator on a Hilbert space. We call the vector x a signal or input, Φ the transform–sample matrix–filter and the vector y the sample or output. The problem is to reconstruct x from y, or more generally, to reconstruct an altered version of x from an altered y. For example, we might analyze the signal x in terms of frequency components and various combinations of time and frequency components y. Once we have analyzed the signal we may alter some of the component parts to eliminate undesirable features or to compress the signal for more efficient transmission and storage. Finally, we reconstitute the signal from its component parts.

The three typical steps in this process are:

- **Analysis.** Decompose the signal into basic components. This is called analysis. We will think of the signal space as a vector space and break it up into a sum of subspaces, each of which captures a special feature of a signal.
- **Processing.** Modify some of the basic components of the signal that were obtained through the analysis. This is called processing.
- **Synthesis.** Reconstitute the signal from its (altered) component parts. This is called synthesis. Sometimes, we will want perfect reconstruction. Sometimes only perfect reconstruction with high probability. If we don't alter the component parts, we usually want the synthesized signal to agree exactly with the original signal. We will also be interested in the convergence properties of an altered signal with respect to the original signal, e.g., how well a reconstituted signal, from which some information may have been dropped, approximates the original signal. Finally we look at problems where the "signal" lies in some

high dimensional Euclidean space in the form of discrete data and where the "filter" is not necessarily linear.

We will look at several methods for signal analysis. We will cover:

- Fourier series and Fourier integrals, infinite products.
- Windowed Fourier transforms.
- Continuous wavelet transforms.
- Filter banks, bases, frames.
- Discrete transforms, Z transforms, Haar wavelets and Daubechies wavelets, singular value decomposition.
- Compressive sampling/compressive sensing.
- Topics in the the parsimonious representation of data.

We break up our treatment into several cases, both theoretical and applied: (1) The system is invertible (Fourier series, Fourier integrals, finite Fourier transform, Z transform, Riesz basis, discrete wavelets, etc.). (2) The system is underdetermined, so that a unique solution can be obtained only if x is restricted (compressive sensing). (3) The system is overdetermined (bandlimited functions, windowed Fourier transform, continuous wavelet transform, frames). In the last case one can throw away some information from y and still recover x. This is the motivation of frame theory, discrete wavelets from continuous wavelets, Shannon sampling, filterbanks, etc. (4) The signal space is a collection of data in some containing Euclidean space.

Each of these cases has its own mathematical peculiarities and opportunity for application. Taken together, they form a logically coherent whole.

1

Normed vector spaces

The purpose of this chapter is to introduce key structural concepts that are needed for theoretical transform analysis and are part of the common language of modern signal processing and computer vision. One of the great insights of this approach is the recognition that natural abstractions which occur in analysis, algebra and geometry help to unify the study of the principal objects which occur in modern signal processing. Everything in this book takes place in a *vector space*, a linear space of objects closed under associative, distributive and commutative laws. The vector spaces we study include vectors in Euclidean and complex space and spaces of functions such as polynomials, integrable functions, approximation spaces such as wavelets and images, spaces of bounded linear operators and compression operators (infinite dimensional). We also need geometrical concepts such as distance and shortest (perpendicular) distance, and sparsity. This chapter first introduces important concepts of vector space and subspace which allow for general ideas of linear independence, span and basis to be defined. Span tells us for example, that a linear space may be generated from a smaller collection of its members by linear combinations. Thereafter, we discuss Riemann integrals and introduce the notion of a normed linear space and metric space. Metric spaces are spaces, nonlinear in general, where a notion of distance and hence limit makes sense. Normed spaces are generalizations of "absolute value" spaces. All normed spaces are metric spaces. The geometry of Euclidean space is founded on the familiar properties of length and angle. In Euclidean geometry, the angle between two vectors is specified in terms of the dot product, which itself is formalized by the notion of inner product. In this chapter, we introduce inner product space, completeness and Hilbert space with important examples. An inner product space is a generalization of a dot product space which preserves the concept of "perpendicular/orthonormal" or "shortest distance." It is a normed linear space satisfying a parallelogram law. Completeness means, roughly, closed under limiting processes and the most general function space admitting an inner product structure is a Hilbert Space. Hilbert spaces lie at the foundation of much of modern analysis, function theory and Fourier analysis, and provide the theoretical setting for modern signal processing. A Hilbert space is a complete inner product space. A basis is a spanning set which is linearly independent. We introduce orthonormal bases in finite- and infinite-dimensional Hilbert spaces and study bounded linear operators on Hilbert spaces. The characterizations of inner products on Euclidean space allows us to study least square and minimization approximations, singular values of matrices and ℓ_1 optimization. An important idea developed is of natural examples motivating an abstract theory which in turn leads to the ability to understand more complex objects but with the same underlying features.

1.1 Definitions

The most basic object in this text is a vector space V over a field $\mathbb{F}$, the latter being the field of real numbers $\mathbb{R}$ or the field of complex numbers $\mathbb{C}$. The elements of $\mathbb{F}$ are called scalars. Vector spaces V or (linear spaces) over fields $\mathbb{F}$ capture the essential properties of $n \geq 1$ Euclidean space $\mathbb{V}^n$ which is the space of all real ($\mathbb{R}^n$) or complex ($\mathbb{C}^n$) column vectors with n entries closed under addition and scalar multiplication.

Definition 1.1 A vector space V over $\mathbb{F}$ is a collection of elements (vectors) with the following properties:[1]

- For every pair $u, v \in V$ there is defined a unique vector $w = u + v \in V$ (the sum of u and v)
- For every $\alpha \in \mathbb{F}$, $u \in V$ there is defined a unique vector $z = \alpha u \in V$ (product of α and u)
- Commutative, Associative and Distributive laws

  1. $u + v = v + u$
  2. $(u + v) + w = u + (v + w)$
  3. There exists a vector $\Theta \in V$ such that $u + \Theta = u$ for all $u \in V$

---

[1] We should strictly write $(V, \mathbb{F})$ since V depends on the field over which is defined. As this will be clear always, we suppress this notation.

4. For every $u \in V$ there is a $-u \in V$ such that $u + (-u) = \Theta$
5. $1u = u$ for all $u \in V$
6. $\alpha(\beta u) = (\alpha\beta)u$ for all $\alpha, \beta \in \mathbb{F}$
7. $(\alpha + \beta)u = \alpha u + \beta u$
8. $\alpha(u + v) = \alpha u + \alpha v$

Vector spaces are often called "linear" spaces. Given any two elements $u, v \in V$, by a linear combination of u, v we mean the sum $\alpha u + \beta v$ for any scalars α, β. Since V is a vector space, $\alpha u + \beta v \in V$ and is well defined. Θ is called the zero vector.

Examples 1.2 (a) As we have noted above, for $n \geq 1$, the space V_n is a vector space if given $u := (u_1, ..., u_n)$ and $v := (v_1, ..., v_n)$ in V_n we define $u + v := (u_1 + v_1, ..., u_n + v_n)$ and $cu := (cu_1, ..., cu_n)$ for any $c \in \mathbb{F}$.

(b) Let $n, m \geq 1$ and let $\mathcal{M}_{m \times n}$ denote the space of all real matrices of size $m \times n$. Then $\mathcal{M}_{m \times n}$ forms a real vector space under the laws of matrix addition and scalar multiplication. The Θ element is the matrix with all zero entries.

(c) Consider the space Π_n, $n \geq 1$ of real polynomials with degree $\leq n$. Then, with addition and scalar multiplication defined pointwise, Π_n becomes a real vector space. Note that the space of polynomials of degree equal to n is not closed under addition and so is not a vector space.[2]

(d) Let J be an arbitrary set and consider the space of functions $\mathcal{F}_V(J)$ as the space of all $f : J \to V$. Then defining addition and scalar multiplication by $(f + g)(x) := f(x) + g(x)$ and $(cf)(x) := cf(x)$ for $f, g \in \mathcal{F}_V$ and $x \in J$, $\mathcal{F}_V(J)$ is a vector space over the same field as V.

Sometimes, we are given a vector space V and a nonempty subset W of V that we need to study. It may happen that W is not closed under linear combinations. A nonempty subset of V will be called a subspace of V if it is closed under linear combinations of its elements with respect to the same field as V. More precisely we have

Definition 1.3 A nonempty set W in V is a subspace of V if $\alpha u + \beta v \in W$ for all $\alpha, \beta \in \mathbb{F}$ and $u, v \in W$.

---

[2] Indeed, consider $(x^3 + 100) + (-x^3 - x) = 100 - x$.

It's easy to prove that W is itself a vector space over $\mathbb{F}$ and contains in particular the zero element of of W. Here, V and the set $\{\Theta\}$ are called the trivial subspaces of V.

We now list some examples of subspaces and non subspaces, some of which are important in what follows, in order to remind the reader of this idea.

Examples 1.4 (a) Let $V_n = \mathbb{R}^n$ and scalars c_i, $1 \le i \le n$ be given. Then the half space consisting of all n-tuples $(u_1, \ldots, u_{n-1}, 0)$ with $u_i \in V_n$, $1 \le i \le n-1$ and the set of solutions $(v_1, \ldots, v_n) \in V_n$ to the homogeneous linear equation

$$c_1 x_1 + \cdots + c_n x_n = 0$$

are each nontrivial subspaces.

(b) $C^{(n)}[a,b]$: The space of all complex-valued functions with continuous derivatives of orders $0, 1, 2, \ldots n$ on the closed, bounded interval $[a,b]$ of the real line. Let $t \in [a,b]$, i.e., $a \le t \le b$ with $a < b$. Vector addition and scalar multiplication of functions $u, v \in C^{(n)}[a,b]$ are defined by

$$[u+v](t) = u(t) + v(t) \qquad [\alpha u](t) = \alpha u(t).$$

The zero vector is the function $\Theta(t) \equiv 0$.

(c) $S(I)$: The space of all complex-valued step functions on the (bounded or unbounded) interval I on the real line.[3] s is a step function on J if there are a finite number of non-intersecting bounded intervals $I_1, \ldots, I_m$ and complex numbers $c_1, \ldots, c_m$ such that $s(t) = c_k$ for $t \in I_k$, $k = 1, \ldots, m$ and $s(t) = 0$ for $t \in I - \cup_{k=1}^m I_k$. Vector addition and scalar multiplication of step functions $s_1, s_2 \in S(I)$ are defined by

$$[s_1 + s_2](t) = s_1(t) + s_2(t) \qquad [\alpha s_1](t) = \alpha s_1(t).$$

(One needs to check that $s_1 + s_2$ and αs_1 are step functions.) The zero vector is the function $\Theta(t) \equiv 0$.

---

[3] Intervals I are the only connected subsets of $\mathbb{R}$ of the form:

– closed, meaning $[a,b] := \{x \in \mathbb{R} : a \le x \le b\}$
– open, meaning $(a,b) := \{x \in \mathbb{R} : a < x < b\}$
– half open, meaning $[a,b)$ or $(a,b]$ where

$$[a,b) := \{x \in \mathbb{R} : a \le x < b\}$$

and $(a,b]$ is similarly defined. If either a or b is $\pm\infty$, then J is open at a or b and J is unbounded. Otherwise, it is bounded.

(d) $A(I)$: The space of all analytic functions on an open interval I. Here, we recall that f is analytic on I if all its $n \geq 1$ orders derivatives exist and are finite on I and given any fixed $a \in I$, the series $\sum_{n=0}^{\infty} f^{(n)}(a)$ $(x-a)^n/n!$ converges to $f(x)$ for all x close enough to a.

(e) For clarity, we add a few examples of sets which are not subspaces. We consider V_3 for simplicity as our underlying vector space.

- The set of all vectors of the form $(u_1, u_2, 1) \in \mathbb{R}^3$. Note that $(0, 0, 0)$ is not in this set.
- The positive octant $\{(u_1, u_2, u_3) : u_i \geq 0, 1 \leq i \leq 3\}$. Note that this set is not closed under multiplication by negative scalars. [4]

We now show that given any finite collection of vectors say $u^{(1)}, u^{(2)}, \ldots, u^{(m)}$ in V for some $m \geq 1$, it is always possible to construct a subspace of V containing $u^{(1)}, u^{(2)}, \ldots, u^{(m)}$ and, moreover, this is the smallest (nontrivial) subspace of V containing all these vectors. Indeed, we have

Lemma 1.5 *Let $u^{(1)}, u^{(2)}, \ldots, u^{(m)}$ be a set of vectors in the vector space V. Denote by $[u^{(1)}, u^{(2)}, \ldots, u^{(m)}]$ the set of all vectors of the form $\alpha_1 u^{(1)} + \alpha_2 u^{(2)} + \cdots + \alpha_m u^{(m)}$ for $\alpha_i \in \mathbb{F}$. The set $[u^{(1)}, u^{(2)}, \ldots, u^{(m)}]$, called the span of the set $\{u^{(1)}, \ldots, u^{(m)}\}$, is the smallest subspace of V containing $u^{(1)}, u^{(2)}, \ldots, u^{(m)}$.*

Proof Let $u, v \in [u^{(1)}, u^{(2)}, \ldots, u^{(m)}]$. Thus,

$$u = \sum_{i=1}^{m} \alpha_i u^{(i)}, \qquad v = \sum_{i=1}^{m} \beta_i u^{(i)}$$

so

$$\alpha u + \beta v = \sum_{i=1}^{m} (\alpha \alpha_i + \beta \beta_i) u^{(i)} \in [u^{(1)}, u^{(2)}, \ldots, u^{(m)}].$$

Clearly any subspace of V, containing $u^{(1)}, u^{(2)}, \ldots, u^{(m)}$ will contain $[u^{(1)}, u^{(2)}, \ldots, u^{(m)}]$. $\square$

Note that spanning sets in vector spaces generalize the geometric notion of two vectors spanning a plane in $\mathbb{R}^3$. We now present three definitions of linear independence, dimensionality and basis and a useful characterization of a basis. We begin with the idea of linear independence. Often, all the vectors used to form a spanning set are essential.

---

[4] In fact, see Exercise 1.7, it is instructive to show that there are only two nontrivial subspaces of $\mathbb{R}^3$. (1) A plane passing through the origin and (2) a line passing through the origin.

For example, if we wish to span a plane in $\mathbb{R}^3$, we cannot use fewer than two vectors since the span of one vector is a line and thus not a plane. In some problems, however, some elements of a spanning set may be redundant. For example, we only need one vector to describe a line in $\mathbb{R}^3$ but if we are given two vectors which are parallel then their span is a line and so we really only need one of these to prescribe the line. The idea of removing superfluous elements from a spanning set leads to the idea of *linear dependence* which is given below.

Definition 1.6 The elements $u^{(1)}, u^{(2)}, \ldots, u^{(p)}$ of V are *linearly independent* if the relation $\alpha_1 u^{(1)} + \alpha_2 u^{(2)} + \cdots + \alpha_p u^{(p)} = \Theta$ for $\alpha_i \in \mathbb{F}$ holds *only* for $\alpha_1 = \alpha_2 = \cdots = \alpha_p = 0$. Otherwise $u^{(1)}, \ldots, u^{(p)}$ are *linearly dependent*.

Examples 1.7 (a) Any collection of vectors which includes the zero vector is linearly dependent.

(b) Two vectors are linearly dependent iff they are parallel. Indeed, if $u^{(1)} = c u^{(2)}$ for some $c \in \mathbb{F}$ and vectors $u^{(1)}, u^{(2)} \in V$, then $u^{(1)} - c u^{(2)} = \Theta$ is a nontrivial linear combination of $u^{(1)}$ and $u^{(2)}$ summing to Θ. Conversely, if $c_1 u^{(1)} + c_2 u^{(2)} = \Theta$ and $c_1 \neq 0$, then $u^{(1)} = -(c_2/c_1) u^{(2)}$ and if $c_1 = 0$ and $c_2 \neq 0$, then $u^{(2)} = \Theta$.

(c) The basic monomials $\{1, x, x^2, x^3, \ldots, x^n\}$ are linearly independent. See Exercise 1.9.

(d) The set of quadratic trigonometric functions

$$\{1, \cos x, \sin x, \cos^2 x, \cos x \sin x, \sin^2 x\}$$

is linearly dependent. Hint: use the fact that $\cos^2 x + \sin^2 x = 1$.

Next, we have

Definition 1.8 V is n-dimensional if there exist n linearly independent vectors in V and any $n + 1$ vectors in V are linearly dependent.

Definition 1.9 V is finite dimensional if V is n-dimensional for some integer n. Otherwise V is infinite dimensional.

For example, V_n is finite dimensional and all of the spaces $C^{(n)}[a, b]$, $S(I)$ and $A(I)$ are infinite dimensional. Next, we define the concept of a basis. In order to span a vector space or subspace, we know that we must use a sufficient number of distinct elements. On the other hand, we also know that including too many elements in a spanning set causes problems with linear independence. Optimal spanning sets are called bases. *Bases* are fundamental in signal processing, linear algebra, data compression, imaging, control and many other areas of research. We have

Definition 1.10 If there exist vectors $u^{(1)}, \ldots, u^{(n)}$, linearly independent in V and such that every vector $u \in V$ can be written in the form

$$u = \alpha_1 u^{(1)} + \alpha_2 u^{(2)} + \cdots + \alpha_n u^{(n)}, \qquad \alpha_i \in \mathbb{F},$$

($\{u^{(1)}, \ldots, u^{(n)}\}$ *spans* V), then V is n-dimensional. Such a set $\{u^{(1)}, \ldots, u^{(n)}\}$ is called a basis for V.

The following theorem gives a useful characterization of a basis.

Theorem 1.11 *Let V be an n-dimensional vector space and $u^{(1)}, \ldots, u^{(n)}$ a linearly independent set in V. Then $u^{(1)}, \ldots, u^{(n)}$ is a basis for V and every $u \in V$ can be written uniquely in the form*

$$u = \beta_1 u^{(1)} + \beta_2 u^{(2)} + \cdots + \beta_n u^{(n)}.$$

Proof Let $u \in V$. Then the set $u^{(1)}, \ldots, u^{(n)}, u$ is linearly dependent. Thus there exist $\alpha_1, \cdots, \alpha_{n+1} \in \mathbb{F}$, not all zero, such that

$$\alpha_1 u^{(1)} + \alpha_2 u^{(2)} + \cdots + \alpha_n u^{(n)} + \alpha_{n+1} u = \Theta.$$

If $\alpha_{n+1} = 0$ then $\alpha_1 = \cdots = \alpha_n = 0$. But this cannot happen, so $\alpha_{n+1} \neq 0$ and hence

$$u = \beta_1 u^{(1)} + \beta_2 u^{(2)} + \cdots + \beta_n u^{(n)}, \qquad \beta_i = -\frac{\alpha_i}{\alpha_{n+1}}.$$

Now suppose

$$u = \beta_1 u^{(1)} + \beta_2 u^{(2)} + \cdots + \beta_n u^{(n)} = \gamma_1 u^{(1)} + \gamma_2 u^{(2)} + \cdots + \gamma_n u^{(n)}.$$

Then

$$(\beta_1 - \gamma_1) u^{(1)} + \cdots + (\beta_n - \gamma_n) u^{(n)} = \Theta.$$

But the u_i form a linearly independent set, so $\beta_1 - \gamma_1 = 0, \ldots, \beta_n - \gamma_n = 0$. □

Examples 1.12 • V_n: A standard basis is:

$$e^{(1)} = (1, 0, \ldots, 0), \ e^{(2)} = (0, 1, 0, \ldots, 0), \ \ldots, \ e^{(n)} = (0, 0, \ldots, 1).$$

Proof

$$(\alpha_1, \ldots, \alpha_n) = \alpha_1 e^{(1)} + \cdots + \alpha_n e^{(n)},$$

so the vectors span. They are linearly independent because

$$(\beta_1, \cdots, \beta_n) = \beta_1 e^{(1)} + \cdots + \beta_n e^{(n)} = \Theta = (0, \cdots, 0)$$

if and only if $\beta_1 = \cdots = \beta_n = 0$. □

- V_∞, the space of all (real or complex) infinity-tuples

$$(\alpha_1, \alpha_2, \ldots, \alpha_n, \cdots).$$

1.2 Inner products and norms

The geometry of Euclidean space is built on properties of length and angle. The abstract concept of a norm on a vector space formalizes the geometrical notion of the length of a vector. In Euclidean geometry, the angle between two vectors is specified by their dot product which is itself formalized by the concept of inner product. As we shall see, norms and inner products are basic in signal processing. As a warm up, in this section, we will prove one of the most important inequalities in the theory, namely the Cauchy–Schwarz inequality, valid in any inner product space. The more familiar triangle inequality, follows from the definition of a norm. Complete inner product spaces, Hilbert spaces, are fundamental in what follows.

Definition 1.13 A vector space $\mathcal{N}$ over $\mathbb{F}$ is a normed linear space (pre Banach space) if to every $u \in \mathcal{N}$ there corresponds a real scalar $||u||$ (called the norm of u) such that

1. $||u|| \geq 0$ and $||u|| = 0$ if and only if $u = 0$.
2. $||\alpha u|| = |\alpha| \, ||u||$ for all $\alpha \in \mathbb{F}$.
3. Triangle inequality. $||u + v|| \leq ||u|| + ||v||$ for all $u, v \in \mathcal{N}$.

Assumption 3 is a generalization of the familiar triangle inequality in $\mathbb{R}^2$ that the length of any side of a triangle is bounded by the sum of the lengths of the other sides. This fact is actually an immediate consequence of the *Cauchy–Schwarz* inequality which we will state and prove later in this chapter.

Examples 1.14 • $C^{(n)}[a, b]$: Set of all complex-valued functions with continuous derivatives of orders $0, 1, 2, \ldots, n$ on the closed interval $[a, b]$ of the real line. Let $t \in [a, b]$, i.e., $a \leq t \leq b$ with $a < b$. Vector addition and scalar multiplication of functions $u, v \in C^{(n)}[a, b]$ are defined by

$$[u + v](t) = u(t) + v(t), \qquad [\alpha u](t) = \alpha u(t).$$

The zero vector is the function $\Theta(t) \equiv 0$. We defined this space earlier, but now we provide it with a norm defined by $||u|| = \int_a^b |u(t)| \, dt$.

- $S^1(I)$: Set of all complex-valued step functions on the (bounded or unbounded) interval I on the real line with norm to be defined. This space was already introduced as (c) in Examples 1.4, but here we add a norm. The space is infinite-dimensional. We define the integral of a step function as the "area under the curve," i.e., $\int_I s(t)dt \equiv \sum_{k=1}^m c_k \ell(I_k)$ where $\ell(I_k) =$ length of $I_k = b - a$ if $I_k = [a,b]$ or $[a,b)$, or $(a,b]$ or (a,b). Note that

1. $s \in S(I) \implies |s| \in S(I)$.
2. $|\int_I s(t)dt| \leq \int_I |s(t)|dt$.
3. $s_1, s_2 \in S(I) \implies \alpha_1 s_1 + \alpha_2 s_2 \in S(I)$ and $\int_I (\alpha_1 s_1 + \alpha_2 s_2)(t)dt = \alpha_1 \int_I s_1(t)dt + \alpha_2 \int_I s_2(t)dt$.

Now we define the norm by $||s|| = \int_I |s(t)|dt$. Finally, we adopt the rule that we identify $s_1, s_2 \in S(I)$, i.e., $s_1 \sim s_2$, if $s_1(t) = s_2(t)$ except at a finite number of points. (This is needed to satisfy property 1 of the norm.) Now we let $S^1(I)$ be the space of equivalence classes of step functions in $S(I)$. Then $S^1(I)$ is a normed linear space with norm $||\cdot||$.

Definition 1.15 A vector space $\mathcal{H}$ over $\mathbb{F}$ is an inner product space or pre Hilbert space if to every ordered pair $u, v \in \mathcal{H}$ there corresponds a scalar $(u,v) \in F$ such that

Case 1: $\mathbb{F} = \mathbb{C}$, Complex field

1. $(u,v) = \overline{(v,u)}$ [symmetry]
2. $(u+v,w) = (u,w) + (v,w)$ [linearity]
3. $(\alpha u, v) = \alpha(u,v)$, for all $\alpha \in \mathbb{C}$ [homogeneity]
4. $(u,u) \geq 0$, and $(u,u) = 0$ if and only if $u = \Theta$ [positivity]

Note: $(u, \alpha v) = \bar{\alpha}(u,v)$.

Case 2: $\mathbb{F} = \mathbb{R}$, Real field

1. $(u,v) = (v,u)$ [symmetry]
2. $(u+v,w) = (u,w) + (v,w)$ [linearity]
3. $(\alpha u, v) = \alpha(u,v)$, for all $\alpha \in R$ [homogeneity]
4. $(u,u) \geq 0$, and $(u,u) = 0$ if and only if $u = 0$ [positivity]

Note: $(u, \alpha v) = \alpha(u,v)$.

Unless stated otherwise, we will consider complex inner product spaces from now on. The real case is usually an obvious restriction.

Definition 1.16 Let $\mathcal{H}$ be an inner product space with inner product (u, v). The norm $||u||$ of $u \in \mathcal{H}$ is the nonnegative number $||u|| = \sqrt{(u, u)}$.

We now introduce a fundamental inequality which is valid for *any* inner product on any vector space. This inequality is inspired by the geometric interpretation of the dot product on Euclidean space in terms of the angle between vectors. It is named after two of the founders of modern analysis, Augustin Cauchy and Herman Schwarz who first established it in the case of the L_2 inner product.[5] In Euclidean geometry, the dot product between any two vectors can be characterized geometrically by the equation

$$u \cdot w = ||u|| \, ||w|| \cos \theta$$

where θ measures the angle between the vectors u and w. As $\cos \theta$ is bounded by 1 for any argument, we have that the absolute value of the dot product is bounded by the product of the lengths of the vectors

$$|u \cdot w| \leq ||u|| \, ||w||.$$

This is the simplest form of the *Cauchy–Schwarz* inequality which we state below.

Theorem 1.17 *Schwarz inequality. Let $\mathcal{H}$ be an inner product space and $u, v \in \mathcal{H}$. Then*

$$|(u, v)| \leq ||u|| \, ||v||.$$

Equality holds if and only if u, v are linearly dependent.

Proof We can suppose $u, v \neq \Theta$. Set $w = u + \alpha v$, for $\alpha \in \mathbb{C}$. Then $(w, w) \geq 0$ and $= 0$ if and only if $u + \alpha v = \Theta$. Hence

$$(w, w) = (u + \alpha v, u + \alpha v) = ||u||^2 + |\alpha|^2 \, ||v||^2 + \alpha(v, u) + \bar{\alpha}(u, v) \geq 0.$$

Set $\alpha = -(u, v)/||v||^2$. Then

$$||u||^2 + \frac{|(u, v)|^2}{||v||^2} - 2\frac{|(u, v)|^2}{||v||^2} \geq 0.$$

Thus $|(u, v)|^2 \leq ||u||^2 \, ||v||^2$. $\qquad\qquad\square$

Theorem 1.18 *Properties of the norm. Let $\mathcal{H}$ be an inner product space with inner product (u, v). Then*

[5] Some mathematicians give credit for this inequality to the Russian mathematician Viktor Bunyakovskii.

- $||u|| \geq 0$ *and* $||u|| = 0$ *if and only if* $u = \Theta$.
- $||\alpha u|| = |\alpha| \, ||u||$.
- *Triangle inequality.* $||u + v|| \leq ||u|| + ||v||$.

Proof

$$||u + v||^2 = (u + v, u + v) = ||u||^2 + (u, v) + (v, u) + ||v||^2$$

$$\leq ||u||^2 + 2||u|| \, ||v|| + ||v||^2 = (||u|| + ||v||)^2.$$

$\square$

Examples 1.19 • $\mathcal{H}_n$. This is the space of complex n-tuples V_n with inner product

$$(u, v) = \sum_{i=1}^{n} u_i \bar{v}_i$$

for vectors

$$u = (u_1, \ldots, u_n), \qquad v = (v_1, \ldots, v_n), \qquad u_i, v_i \in \mathbb{C}.$$

• $\mathbb{R}^n$ This is the space of real n-tuples V_n with inner product

$$(u, v) = \sum_{i=1}^{n} u_i v_i$$

for vectors

$$u = (u_1, \ldots, u_n), \qquad v = (v_1, \ldots, v_n), \qquad u_i, v_i \in \mathbb{R}.$$

Note that (u, v) is just the dot product. In particular for $\mathbb{R}^3$ (Euclidean 3-space) $(u, v) = ||u|| \, ||v|| \cos \phi$ where $||u|| = \sqrt{u_1^2 + u_2^2 + u_3^2}$ (the length of u), and $\cos \phi$ is the cosine of the angle between vectors u and v. The triangle inequality $||u + v|| \leq ||u|| + ||v||$ says in this case that the length of one side of a triangle is less than or equal to the sum of the lengths of the other two sides.

• $\hat{\mathcal{H}}_\infty$, the space of all complex infinity-tuples

$$u = (u_1, u_2, \ldots, u_n, \ldots)$$

such that only a finite number of the u_i are nonzero. $(u, v) = \sum_{i=1}^{\infty} u_i \bar{v}_i$.

• $\mathcal{H}_\infty$, the space of all complex infinity-tuples

$$u = (u_1, u_2, \ldots, u_n, \ldots)$$

such that $\sum_{i=1}^{\infty} |u_i|^2 < \infty$. Here, $(u, v) = \sum_{i=1}^{\infty} u_i \bar{v}_i$. (Later, we will verify that this is a vector space.)

- ℓ_2, the space of all complex infinity-tuples

$$u = (\ldots, u_{-1}, u_0, u_1, \ldots, u_n, \ldots).$$

such that $\sum_{i=-\infty}^{\infty} |u_i|^2 < \infty$. Here, $(u, v) = \sum_{i=-\infty}^{\infty} u_i \bar{v}_i$. (Later, we will verify that this is a vector space.)

- $C_2^{(n)}[a, b]$: The set of all complex-valued functions $u(t)$ with continuous derivatives of orders $0, 1, 2, \ldots n$ on the closed interval $[a, b]$ of the real line. We define an inner product by

$$(u, v) = \int_a^b u(t)\bar{v}(t) \, dt, \qquad u, v \in C_2^{(n)}[a, b].$$

- $C_2^{(n)}(a, b)$: Set of all complex-valued functions $u(t)$ with continuous derivatives of orders $0, 1, 2, \cdots n$ on the open interval (a, b) of the real line, such that $\int_a^b |u(t)|^2 \, dt < \infty$, (Riemann integral). We define an inner product by

$$(u, v) = \int_a^b u(t)\bar{v}(t) \, dt, \qquad u, v \in C_2^{(n)}(a, b).$$

Note: $u(t) = t^{-1/3} \in C_2^{(2)}(0, 1)$, but $v(t) = t^{-1}$ doesn't belong to this space.

- $S^2(I)$: Space of all complex-valued step functions on the (bounded or unbounded) interval I on the real line. This space was already introduced as (c) in Examples 1.4, but now we equip it with an inner product. Note that the product of step functions, defined by $s_1 s_2(t) \equiv s_1(t)s_2(t)$ is a step function, as are $|s_1|$ and $\bar{s}_1$. We earlier defined the integral of a step function as $\int_I s(t)dt \equiv \sum_{k=1}^m c_k \ell(I_k)$ where $\ell(I_k) =$ length of $I_k = b - a$ if $I_k = [a, b]$ or $[a, b)$, or $(a, b]$ or (a, b). Now we define the inner product by $(s_1, s_2) = \int_I s_1(t)\overline{s_2(t)}dt$. Finally, we adopt the rule that $s_1, s_2 \in S(I)$ are identified, $s_1 \sim s_2$ if $s_1(t) = s_2(t)$ except at a finite number of points. (This is needed to satisfy property 4 of the inner product, Definition 8.) Now we let $S^2(I)$ be the space of equivalence classes of step functions in $S(I)$. Then $S^2(I)$ is an inner product space.

1.3 Finite-dimensional ℓ_p spaces

In the mathematics of signal processing it is conceptually very appealing to work with general signals defined in infinite-dimensional spaces, such as the space of square integrable functions on the real line, so that we can apply the tools of calculus. On the other hand, the signals that are received, processed and transmitted in practice are almost always finite and discrete. For example, a digital photograph is determined completely by a vector of n pixels where each pixel is described by an integer that denotes its color and intensity. Cell phone conversations are sampled and digitized for transmission and reconstruction. Thus, while much of our analysis will be carried out in the infinite-dimensional arena, to connect with practical signal processing we will ultimately need to confront the reality of finite signals. In preparation for this confrontation we will introduce and study a family of norms on the vector space $\mathbb{C}^n$ of complex n-tuples

$$x = (x_1, x_2, \ldots, x_n).$$

(However, virtually all our results restrict naturally to norms on the space $\mathbb{R}^n$ of real n-tuples.) Intuitively, we will consider our signals as vectors $x, y \in \mathbb{C}^n$ for some fixed n and use $||x - y||$ to measure how closely x approximates y with respect to the norm $|| \cdot ||$.

Let $p \geq 1$ be a real number. We define the ℓ_p norm of the vector $x \in \mathbb{C}^n$ by

$$||x||_p = \left[\sum_{j=1}^n |x_j|^p \right]^{\frac{1}{p}}. \tag{1.1}$$

To justify the name, we need to verify that $||x||_p$ is in fact a norm, that it satisfies the triangle inequality $||x + y||_p \leq ||x||_p + ||y||_p$ for each $p \geq 1$. For $p = 2$ this follows immediately from the Cauchy–Schwarz inequality, while for $p = 1$ it is a simple consequence of the inequality $|\alpha + \beta| \leq |\alpha| + |\beta|$ for the absolute values of complex numbers. For general $p > 1$ the proof is more involved.

Given $p > 1$ there is a unique $q > 1$ such that $1/p + 1/q = 1$. Indeed, $q = p/(p - 1)$. A fundamental result is the following elementary inequality.

Lemma 1.20 *Let $\alpha, \beta, \gamma > 0$. Then*

$$\alpha\beta \leq \frac{\gamma^p \alpha^p}{p} + \frac{\gamma^{-q} \beta^q}{q}.$$

Proof The verification is via basic calculus. Consider the function

$$f(\gamma) = \frac{\gamma^p \alpha^p}{p} + \frac{\gamma^{-q} \beta^q}{q}$$

over the domain $\gamma > 0$. We wish to find the absolute minimum value of this function. It is easy to verify that this absolute minimum must occur at a critical point of f, i.e., at a value γ such that $f'(\gamma) = \gamma^{p-1}\alpha^p - \beta^q/\gamma^{q+1} = 0$. Thus the minimum occurs at $\gamma_0 = \beta^{1/p}/\alpha^{1/q}$, and

$$f(\gamma_0) = \frac{\gamma_0^p \alpha^p}{p} + \frac{\gamma_0^{-q}\beta^q}{q} = (\frac{\beta^{\frac{1}{p}}}{\alpha^{\frac{1}{q}}})^p \frac{\alpha^p}{p} + (\frac{\alpha^{\frac{1}{q}}}{\beta^{\frac{1}{p}}})^q \frac{\beta^q}{q} = \alpha\beta.$$

$\square$

This result leads directly to Hölder's inequality.

Theorem 1.21 *Let $p, q > 1$ such that $1/p + 1/q = 1$ and $x, y \in \mathbb{C}^n$. Then*

$$|\sum_{j=1}^n x_j y_j| \le ||x||_p \cdot ||y||_q. \tag{1.2}$$

Proof Applying Lemma 1.20 for each j we have

$$|\sum_{j=1}^n x_j y_j| \le \sum_{j=1}^n |x_j| \cdot |y_j| \le \sum_{j=1}^n \left(\frac{\lambda^p}{p}|x_j|^p + \frac{\lambda^{-q}}{q}|y_j|^q \right)$$
$$= \frac{\lambda^p}{p}||x||_p^p + \frac{\lambda^{-q}}{q}||y||_q^q, \tag{1.3}$$

for any $\lambda > 0$. To get the strongest inequality we choose λ to minimize the right-hand side of (1.3). A simple calculus argument shows that the minimum is achieved for $\lambda_0 = ||y||_q^{1/p}/||x||_p^{1/q}$. Substituting $\lambda = \lambda_0$ in (1.3), we obtain Hölder's inequality. $\square$

Note that the special case $p = q = 2$ of Hölder's inequality is the Cauchy–Schwarz inequality for real vector spaces.

Corollary 1.22 $||\cdot||_p$ *is a norm on $\mathbb{C}^n$ for $p \ge 1$.*

Proof It is evident from the explicit expression (1.1) that $||cx||_p = |c|\,||x||_p$ for any constant c and any $x \in \mathbb{C}^n$. Further $||x||_p = 0 \iff x = \Theta$. The only nontrivial fact to prove is that the triangle inequality

$$||x + y||_p \le ||x||_p + ||y||_p \tag{1.4}$$

is satisfied for all $x, y \in \mathbb{C}^n$. This is evident for $p = 1$. For $p > 1$ we employ the Hölder inequality. For fixed j we have

$$|x_j + y_j|^p = |x_j + y_j| \cdot |x_j + y_j|^{p-1} \le |x_j|\,|x_j + y_j|^{p-1} + |y_j|\,|x_j + y_j|^{p-1}.$$

Summing on j and applying the Hölder inequality to each of the sums on the right-hand side, we obtain

$$||x + y||_p^p \le (||x||_p + ||y||_p)\,||z||_q$$

where $||z||_q = ||x||_p^{p/q} = ||x||_p^{p-1}$. Dividing both sides of the inequality by $||z||_q$ we obtain the desired result. □

We have introduced a family of norms on $\mathbb{C}^n$, indexed by the real number $p \geq 1$. The space with this norm is called ℓ_p. One extreme case of this family is $p = 1$, which we can characterize as the limit $||x||_1 = \lim_{p \to 1+} ||x||_p$. On the other hand we can let p grow without bound and obtain a new norm $|| \cdot ||_\infty$ in the limit. This norm is defined by

$$||x||_\infty = \max_{j=1,\dots,n} |x_j|, \tag{1.5}$$

i.e., the maximum absolute value of the components x_j of x. The proof of the following result is elementary and is left to the exercises.

Theorem 1.23 $|| \cdot ||_\infty$ *is a norm on $\mathbb{C}^n$ and satisfies the properties* $||x||_\infty = \lim_{p \to \infty} ||x||_p$ *and* $||x||_\infty \leq ||x||_p$ *for any $p \geq 1$.*

Now we show that $||x||_p$ is a monotonically decreasing function of p. The proof is elementary, but takes several steps.

Lemma 1.24 *If $p \geq 1$ then $(|y_1| + |y_2|)^p \geq |y_1|^p + |y_2|^p$.*

Proof It is a simple exercise to see that the statement of the lemma is equivalent to showing that the function

$$f(x) = (1 + x)^p - 1 - x^p \geq 0$$

for all $x \geq 0$. Note that $f(0) = 0$ and $f'(0+) = p > 0$ so that the function is monotonically increasing from 0 for small x. The derivative $f'(x) = p[(1 + x)^{p-1} - x^{p-1}]$ is continuous for all positive x and never vanishes. Thus always $f'(x) > 0$ so $f(x)$ is a monotonically increasing function that is strictly positive for all $x > 0$. □

Lemma 1.25 *If $p \geq 1$ then*

$$(|y_1| + |y_2| + \cdots + |y_n|)^p \geq |y_1|^p + |y_2|^p + \cdots + |y_n|^p. \tag{1.6}$$

This result is true for $n = 2$ by the previous lemma. The result is then established for general n by mathematical induction.

Theorem 1.26 *If $s > r \geq 1$ then $||x||_r \geq ||x||_s$ for all $x \in \mathbb{C}^n$.*

Proof Set $y_j = |x_j|^r$, $p = s/r$ in (1.6). Thus we have

$$(|x_1|^r + |x_2|^r + \cdots + |x_n|^r)^{s/r} \geq |x_1|^s + |x_2|^s + \cdots + |x_n|^s.$$

Taking the sth root of both sides of the inequality we obtain $||x||_r \geq ||x||_s$. □

We conclude that $||x||_1 \geq ||x||_p \geq ||x||_\infty$ for all $p > 1$ and that $||x||_p$ is a monotonically decreasing function of r. We can illustrate graphically the relation between the ℓ_p norms. The unit ball in ℓ_p is the set of all $x \in \mathbb{C}^n$ such that $||x||_p \leq 1$. For $n = 2$ and real vectors $x \in \mathbb{R}^2$ we can construct the boundaries of the unit ball for various norms. Consider the boundary $||x|| = |x_1| + |x_2| = 1$ of the ℓ_1-ball. In the first quadrant this curve is the line segment $x_1 + x_2 = 1$, $x_1, x_2 \geq 0$, whereas in the second quadrant it is the line segment $-x_1 + x_2 = 1$, $x_2 \geq 0$, $x_1 \leq 0$. Continuing in this way we observe that the ℓ_1 unit ball is the interior and boundary of the square with vertices at the points $(\pm 1, 0)$, $(0, \pm 1)$. Even simpler, the unit ℓ_2-ball is the interior and boundary of the circle centered at $(0,0)$ and with radius 1. The unit ℓ_∞ ball is the interior and boundary of the square with vertices at $(\pm 1, \pm 1)$. Figure 1.1 shows the relationships between these unit balls. The ℓ_p balls for general $p > 1$ have an oval shape and all balls go through the vertices $(\pm 1, 0)$, $(0, \pm 1)$. For $n > 2$ the situation is similar. Only the ℓ_1 and ℓ_∞-balls have hyperplanes for boundaries; the rest have smooth curved boundaries. The 2^n intersection points of the coordinate axes with the unit ℓ_2-ball lie on all ℓ_p-balls.

As is true for all norms on finite-dimensional spaces, every ℓ_r norm is commensurate to every ℓ_s norm. That is, there are positive constants $c_{r,s}, d_{r,s}$ such that

$$c_{r,s}||x||_s \leq ||x||_r \leq d_{r,s}||x||_s$$

for all $x \in \mathbb{C}^n$. For example $||x||_\infty \leq ||x||_r \leq n^{1/r}||x||_\infty$ as the reader can verify. An important inequality that we shall employ later is

$$||x||_2 \leq ||x||_1 \leq \sqrt{n}||x||_2. \tag{1.7}$$

The right-hand side of this expression follows from the Cauchy–Schwarz inequality for the dot product of x and $e = (1, 1, \ldots, 1) \in \mathbb{C}^n$:

$$||x||_1 = \sum_{j=1}^n 1|x_j| \leq ||e||_2||x||_2 = \sqrt{n}||x||_2.$$

The only ℓ_p normed space with a norm inherited from an inner product is ℓ_2. In view of the fact that calculations are generally much easier for inner product spaces than for spaces that are merely normed, one might question the need for introducing the other norms in this book. However, it turns out that some of these other norms are extremely useful in signal

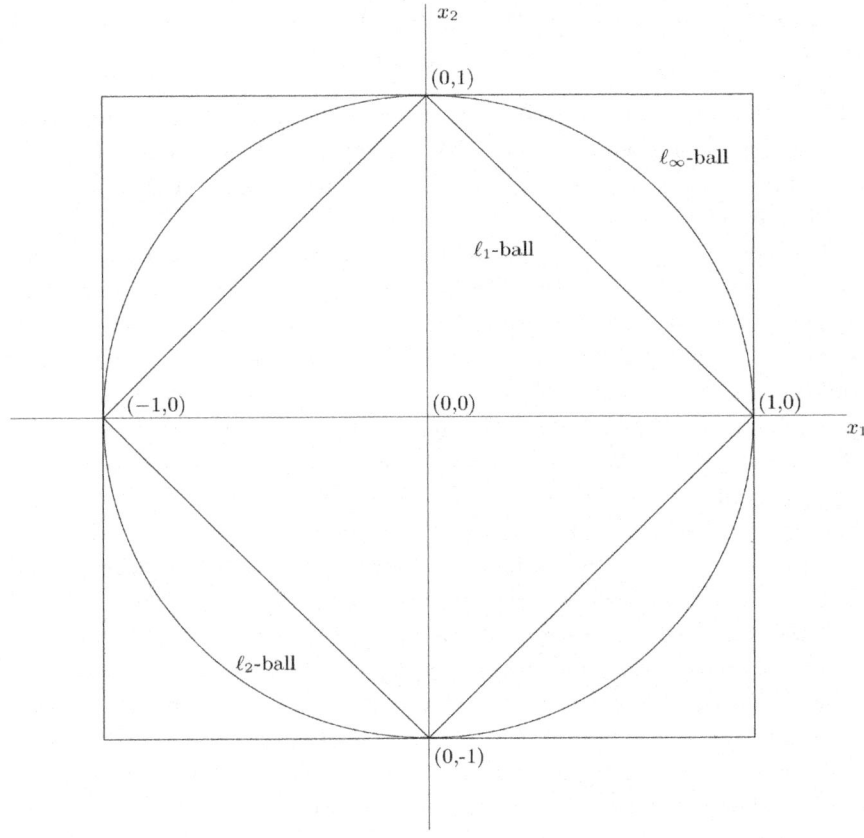

Figure 1.1 Unit balls for ℓ_p spaces with $n = 2$.

processing applications. In particular the ℓ_1 norm will play a critical role in the sensing of sparse signals.

Exercise 1.1 Verify explicitly that $\| \cdot \|_1$ is a norm on $\mathbb{C}^n$.

Exercise 1.2 Verify explicitly the steps in the proof of the Hölder inequality.

Exercise 1.3 Prove Theorem 1.23.

Exercise 1.4 Prove Lemma 1.25.

1.4 Digging deeper: completion of inner product spaces

In practice we are often presented with a vector space V which may not be complete, in a sense that we will make precise below. We describe a method to embed V in a larger space which is complete. This process is called *completion*.

Consider the space $\mathbb{R}$ of the real numbers which can be constructed from the more basic space $\mathbb{R}'$ of rational numbers. The norm of a rational number r is just the absolute value $|r|$. Every rational number can be expressed as a ratio of integers $r = n/m$. The rationals are closed under addition, subtraction, multiplication and division by nonzero numbers. Why don't we stick with the rationals and not bother with real numbers? The basic problem is that we cannot do analysis (calculus, etc.) with the rationals because they are not closed under limiting processes. For example $\sqrt{2}$ wouldn't exist. The Cauchy sequence $1, 1.4, 1.41, 1.414, \ldots$ wouldn't diverge, but would fail to converge to a rational number. There is a "hole" in the field of rational numbers and we label this hole by $\sqrt{2}$. We say that the Cauchy sequence above and all other sequences approaching the same hole are converging to $\sqrt{2}$. Each hole can be identified with the equivalence class of Cauchy sequences approaching the hole. The reals are just the space of equivalence classes of these sequences with appropriate definitions for addition and multiplication. Each rational number r corresponds to a constant Cauchy sequence $r, r, r, \ldots$ so the rational numbers can be embedded as a subset of the reals. Then one can show that the reals are *closed*: every Cauchy sequence of real numbers converges to a real number. We have filled in all of the holes between the rationals. The reals are the *closure* of the rationals.

Exercise 1.5 Show that there are "holes" in the rationals by demonstrating that $\sqrt{p}$ cannot be rational for a prime integer p. Hint: if $\sqrt{p}$ is rational then there are relatively prime integers m, n such that $\sqrt{p} = m/n$, so $m^2 = pn^2$.

The same idea works for inner product spaces and it also underlies the relation between the Riemann integral and the Lebesgue integral. To develop this idea further, it is convenient to introduce the simple but general concept of a metric space. We will carry out the basic closure construction for metric spaces and then specialize to inner product and normed spaces. It is not essential to follow all of the details of these constructions, particularly on a first reading. However it is important to grasp the essence of the argument, which is one of the great ideas in mathematics. You will note that, although the details are complicated, the same basic ideas are employed at each step of the construction.

Definition 1.27 A set $\mathcal{M}$ is called a *metric space* if for each $u, v \in \mathcal{M}$ there is a real number $\rho(u, v)$ (the metric) such that

1. $\rho(u, v) \geq 0$, and $\rho(u, v) = 0$ if and only if $u = v$
2. $\rho(u, v) = \rho(v, u)$
3. $\rho(u, w) \leq \rho(u, v) + \rho(v, w)$ (triangle inequality).

Remark: Normed spaces are metric spaces: $\rho(u, v) = ||u - v||$.

Definition 1.28 A sequence $u^{(1)}, u^{(2)}, \ldots$ in $\mathcal{M}$ is called a *Cauchy sequence* if for every $\epsilon > 0$ there exists an integer $N(\epsilon)$ such that $\rho(u^{(n)}, u^{(m)}) < \epsilon$ whenever $n, m > N(\epsilon)$.

Definition 1.29 A sequence $u^{(1)}, u^{(2)}, \ldots$ in $\mathcal{M}$ is *convergent* if for every $\epsilon > 0$ there exists an integer $M(\epsilon)$ such that $\rho(u^{(n)}, u) < \epsilon$ whenever $n, m > M(\epsilon)$. Here u is the *limit* of the sequence, and we write $u = \lim_{n \to \infty} u^{(n)}$.

Lemma 1.30 (1) *The limit of a convergent sequence is unique.*
(2) *Every convergent sequence is Cauchy.*

Proof (1) Suppose $u = \lim_{n \to \infty} u^{(n)}$, $v = \lim_{n \to \infty} u^{(n)}$. Then $\rho(u, v) \leq \rho(u, u^{(n)}) + \rho(u^{(n)}, v) \to 0$ as $n \to \infty$. Therefore $\rho(u, v) = 0$, so $u = v$. (2) $\{u^{(n)}\}$ converges to u implies $\rho(u^{(n)}, u^{(m)}) \leq \rho(u^{(n)}, u) + \rho(u^{(m)}, u) \to 0$ as $n, m \to \infty$. □

Definition 1.31 A metric space $\mathcal{M}$ is *complete* if every Cauchy sequence in $\mathcal{M}$ converges in $\mathcal{M}$.

Examples 1.32 Some examples of metric spaces:

- Any normed space. $\rho(u, v) = ||u - v||$. Finite-dimensional inner product spaces are complete.
- The space $\mathbb{C}$ of complex numbers is complete under the absolute value metric $\rho(u, v) = |u - v|$.
- $\mathcal{M}$ as the set of all rationals on the real line. $\rho(u, v) = |u - v|$ for rational numbers u, v (absolute value). Here $\mathcal{M}$ is not complete.

Definition 1.33 A subset $\mathcal{M}'$ of the metric space $\mathcal{M}$ is dense in $\mathcal{M}$ if for every $u \in \mathcal{M}$ there exists a sequence $\{u^{(n)}\} \subset \mathcal{M}$ such that $u = \lim_{n \to \infty} u^{(n)}$.

Definition 1.34 Two metric spaces $\mathcal{M}_1, \mathcal{M}_2$ are isometric if there is a 1-1 onto map $f : \mathcal{M}_1 \to \mathcal{M}_2$ such that $\rho_2(f(u), f(v)) = \rho_1(u, v)$ for all $u, v \in \mathcal{M}_1$.

Remark: We identify isometric spaces.

Theorem 1.35 *Given an incomplete metric space $\mathcal{M}$ we can extend it to a complete metric space $\overline{\mathcal{M}}$ (the completion of $\mathcal{M}$) such that* **(1)** $\mathcal{M}$ *is dense in $\overline{\mathcal{M}}$.* **(2)** *Any two such completions $\overline{\mathcal{M}}', \overline{\mathcal{M}}''$ are isometric.*

Proof (divided into parts)

1.

Definition 1.36 Two Cauchy sequences $\{u^{(n)}\}, \{\tilde{u}^{(n)}\}$ in $\mathcal{M}$ are equivalent ($\{u^{(n)}\} \sim \{\tilde{u}^{(n)}\}$) if $\rho(u^{(n)}, \tilde{u}^{(n)}) \to 0$ as $n \to \infty$.

Clearly $\sim$ is an equivalence relation, i.e.,

1. $\{u^{(n)}\} \sim \{u^{(n)}\}$, reflexive
2. If $\{u^{(n)}\} \sim \{v^{(n)}\}$ then $\{v^{(n)}\} \sim \{u^{(n)}\}$, symmetric
3. If $\{u^{(n)}\} \sim \{v^{(n)}\}$ and $\{v^{(n)}\} \sim \{w^{(n)}\}$ then $\{u^{(n)}\} \sim \{w^{(n)}\}$, transitive.

Let $\overline{\mathcal{M}}$ be the set of all equivalence classes of Cauchy sequences. An equivalence class $\overline{u}$ consists of all Cauchy sequences equivalent to a given $\{u^{(n)}\}$.

2. $\overline{\mathcal{M}}$ is a metric space. Define $\overline{\rho}(\overline{u}, \overline{v}) = \lim_{n \to \infty} \rho(u^{(n)}, v^{(n)})$, where $\{u^{(n)}\} \in \overline{u}, \{v^{(n)}\} \in \overline{v}$.

1. $\overline{\rho}(\overline{u}, \overline{v})$ exists.
   Proof
   $$\rho(u^{(n)}, v^{(n)}) \leq \rho(u^{(n)}, u^{(m)}) + \rho(u^{(m)}, v^{(m)}) + \rho(v^{(m)}, v^{(n)}),$$
   so
   $$\rho(u^{(n)}, v^{(n)}) - \rho(u^{(m)}, v^{(m)}) \leq \rho(u^{(n)}, u^{(m)}) + \rho(v^{(m)}, v^{(n)}),$$
   and
   $$|\rho(u^{(n)}, v^{(n)}) - \rho(u^{(m)}, v^{(m)})| \leq \rho(u^{(n)}, u^{(m)}) + \rho(v^{(m)}, v^{(n)}) \to 0$$
   as $n, m \to \infty$. $\qquad\square$

2. $\overline{\rho}(\overline{u}, \overline{v})$ is well defined.

Proof Let $\{u^{(n)}\}, \{u^{(n')}\} \in \overline{u}, \{v^{(n)}\}, \{v^{(n')}\} \in \overline{v}$. Does $\lim_{n \to \infty} \rho(u^{(n)}, v^{(n)}) = \lim_{n \to \infty} \rho(u^{(n')}, v^{(n')})$? Yes, because

$$\rho(u^{(n)}, v^{(n)}) \leq \rho(u^{(n)}, u^{(n')}) + \rho(u^{(n')}, v^{(n')}) + \rho(v^{(n')}, v^{(n)}),$$

so

$$|\rho(u^{(n)}, v^{(n)}) - \rho(u^{(n')}, v^{(n')})| \leq \rho(u^{(n)}, u^{(n')}) + \rho(v^{(n')}, v^{(n)}) \to 0$$

as $n \to \infty$.   □

3. $\overline{\rho}$ is a metric on $\overline{\mathcal{M}}$, i.e.

1. $\overline{\rho}(\overline{u}, \overline{v}) \geq 0$, and $= 0$ if and only if $\overline{u} = \overline{v}$

Proof $\overline{\rho}(\overline{u}, \overline{v}) = \lim_{n \to \infty} \rho(u^{(n)}, v^{(n)}) \geq 0$ and $= 0$ if and only if $\{u^{(n)}\} \sim \{v^{(n)}\}$, i.e., if and only if $\overline{u} = \overline{v}$.   □

2. $\overline{\rho}(\overline{u}, \overline{v}) = \overline{\rho}(\overline{v}, \overline{u})$, obvious

3. $\overline{\rho}(\overline{u}, \overline{v}) \leq \overline{\rho}(\overline{u}, \overline{w}) + \overline{\rho}(\overline{w}, \overline{v})$, easy

4. $\mathcal{M}$ is isometric to a metric subset $\overline{\mathcal{S}}$ of $\overline{\mathcal{M}}$.

Proof Consider the set $\mathcal{S}$ of equivalence classes $\overline{u}$ all of whose Cauchy sequences converge to elements of $\mathcal{M}$. If $\overline{u}$ is such a class then there exists $u \in \mathcal{M}$ such that $\lim_{n \to \infty} u^{(n)} = u$ if $\{u^{(n)}\} \in \overline{u}$. Note that $u, u, \ldots, u, \ldots \in \overline{u}$ (stationary sequence). The map $u \leftrightarrow \overline{u}$ is a 1-1 map of $\mathcal{M}$ onto $\overline{\mathcal{S}}$. It is an isometry since

$$\overline{\rho}(\overline{u}, \overline{v}) = \lim_{n \to \infty} \rho(u^{(n)}, v^{(n)}) = \rho(u, v)$$

for $\overline{u}, \overline{v} \in \overline{\mathcal{S}}$, with $\{u^{(n)}\} = \{u\} \in \overline{u}, \{v^{(n)}\} = \{v\} \in \overline{v}$.   □

5. $\mathcal{M}$ is dense in $\overline{\mathcal{M}}$.

Proof Let $\overline{u} \in \overline{\mathcal{M}}, \{u^{(n)}\} \in \overline{u}$. Consider

$$\overline{s}^{(k)} = \{u^{(k)}, u^{(k)}, \ldots, u^{(k)}, \ldots\} \in \overline{\mathcal{S}} = \mathcal{M}, \ k = 1, 2, \ldots$$

Then $\overline{\rho}(\overline{u}, \overline{s}^{(k)}) = \lim_{n \to \infty} \rho(u^{(n)}, u^{(k)})$. But $\{u^{(n)}\}$ is Cauchy in $\mathcal{M}$. Therefore, given $\epsilon > 0$, if we choose $k > N(\epsilon)$ we have $\overline{\rho}(\overline{u}, \overline{s}^{(k)}) < \epsilon$.   □

6. $\overline{\mathcal{M}}$ is complete.

Proof Let $\{\overline{v}^{(k)}\}$ be a Cauchy sequence in $\overline{\mathcal{M}}$. For each k choose $\overline{s}^{(k)} = \{u^{(k)}, u^{(k)}, \ldots, u^{(k)}, \ldots\} \in \overline{\mathcal{S}} = \mathcal{M}$, (a stationary sequence), such that $\overline{\rho}(\overline{v}^{(k)}, \overline{s}^{(k)}) < 1/k, \ k = 1, 2, \ldots$. Then

$$\rho(u^{(j)}, u^{(k)}) = \overline{\rho}(\overline{s}^{(j)}, \overline{s}^{(k)}) \leq \overline{\rho}(\overline{s}^{(j)}, \overline{v}^{(j)}) + \overline{\rho}(\overline{v}^{(j)}, \overline{v}^{(k)})$$
$$+ \overline{\rho}(\overline{v}^{(k)}, \overline{s}^{(k)}) \to 0$$

as $j, k \to \infty$. Therefore $\overline{u} = \{u^{(k)}\}$ is Cauchy in $\mathcal{M}$. Now

$$\overline{\rho}(\overline{u}, \overline{v}^{(k)}) \le \overline{\rho}(\overline{u}, \overline{s}^{(k)}) + \overline{\rho}(\overline{s}^{(k)}, \overline{v}^{(k)}) \to 0$$

as $k \to \infty$. Therefore $\lim_{k \to \infty} \overline{v}^{(k)} = \overline{u}$.   □

Remark: Step 6 is crucial and the justification of the entire construction.

1.4.1 Completion of a normed linear space

Here $\mathcal{B}$ is a normed linear space with norm $\rho(u, v) = ||u - v||$. We will show how to extend it to a complete normed linear space, called a Banach space. This is now easy, because of our construction of the completion of a metric space.

Definition 1.37 Let $\mathcal{S}$ be a subspace of the normed linear space $\mathcal{B}$. $\mathcal{S}$ is a dense subspace of $\mathcal{B}$ if it is a dense subset of $\mathcal{B}$. $\mathcal{S}$ is a closed subspace of $\mathcal{B}$ if every Cauchy sequence $\{u^{(n)}\}$ in $\mathcal{S}$ converges to an element of $\mathcal{S}$. (Note: If $\mathcal{B}$ is a Banach space then so is $\mathcal{S}$.)

Theorem 1.38 *An incomplete normed linear space $\mathcal{B}$ can be extended to a Banach space $\overline{\mathcal{B}}$ such that $\mathcal{B}$ is a dense subspace of $\overline{\mathcal{B}}$.*

Proof By Theorem 1.35 we can extend the metric space $\mathcal{B}$ to a complete metric space $\overline{\mathcal{B}}$ such that $\mathcal{B}$ is dense in $\overline{\mathcal{B}}$.

1. $\overline{\mathcal{B}}$ is a vector space.
   1. $\overline{u}, \overline{v} \in \overline{\mathcal{B}} \longrightarrow \overline{u} + \overline{v} \in \overline{\mathcal{B}}$.
      If $\{u^{(n)}\} \in \overline{u}, \{v^{(n)}\} \in \overline{v}$, define $\overline{u} + \overline{v} = \overline{u + v}$ as the equivalence class containing $\{u^{(n)} + v^{(n)}\}$. Now $\{u^{(n)} + v^{(n)}\}$ is Cauchy because $||(u^{(n)} + v^{(n)}) - (u^{(m)} - v^{(m)})|| \le ||u^{(n)} - u^{(m)}|| + ||v^{(n)} - v^{(m)}|| \to 0$ as $n, m \to \infty$. It is easy to check that addition is well defined.
   2. $\alpha \in \mathbb{C}, \overline{u} \in \overline{\mathcal{B}} \longrightarrow \alpha \overline{u} \in \overline{\mathcal{B}}$.
      If $\{u^{(n)}\} \in \overline{u}$, define $\alpha \overline{u} \in \overline{\mathcal{B}}$ as the equivalence class containing $\{\alpha u^{(n)}\}$, Cauchy because $||\alpha u^{(n)} - \alpha u^{(m)}|| \le |\alpha| \, ||u^{(n)} - u^{(m)}||$.

2. $\overline{\mathcal{B}}$ is a Banach space.
   Define the norm $||\overline{u}||'$ on $\overline{\mathcal{B}}$ by $||\overline{u}||' = \overline{\rho}(\overline{u}, \overline{\Theta}) = \lim_{n \to \infty} ||u^{(n)}||$ where $\overline{\Theta}$ is the equivalence class containing $\{\Theta, \Theta, \ldots\}$. Positivity is easy. Let $\alpha \in \mathbb{C}$, $\{u_n\} \in \overline{u}$. Then $||\alpha \overline{u}||' = \overline{\rho}(\alpha \overline{u}, \overline{\Theta}) = \lim_{n \to \infty} ||\alpha u^{(n)}||$ $= |\alpha| \, \lim_{n \to \infty} ||u^{(n)}|| = |\alpha| \, \overline{\rho}(\overline{u}, \overline{\Theta}) = |\alpha| \, ||\overline{u}||'$. We have $||\overline{u} + \overline{v}||' = \overline{\rho}(\overline{u} + \overline{v}, \overline{\Theta}) \le \overline{\rho}(\overline{u} + \overline{v}, \overline{v}) + \overline{\rho}(\overline{v}, \overline{\Theta}) = ||\overline{u}||' + ||\overline{v}||'$, because $\overline{\rho}(\overline{u} + \overline{v}, \overline{v}) = \lim_{n \to \infty} ||(u^{(n)} + v^{(n)}) - v^{(n)}|| = \lim_{n \to \infty} ||u^{(n)}|| = ||\overline{u}||'$.   □

1.4.2 Completion of an inner product space

Here $\mathcal{H}$ is an inner product space with inner product (u, v) and norm $\rho(u, v) = ||u - v||$. We will show how to extend it to a complete inner product space, called a *Hilbert Space*.

Theorem 1.39 *Let $\mathcal{H}$ be an inner product space and $\{u^{(n)}\}, \{v^{(n)}\}$ convergent sequences in $\mathcal{H}$ with $\lim_{n \to \infty} u^{(n)} = u$, $\lim_{n \to \infty} v^{(n)} = v$. Then*

$$\lim_{n \to \infty} (u^{(n)}, v^{(n)}) = (u, v).$$

Proof We must first show that $||u^{(n)}||$ is bounded for all n. We have $\{u^{(n)}\}$ converges $\longrightarrow ||u^{(n)}|| \leq ||u^{(n)} - u|| + ||u|| < \epsilon + ||u||$ for $n > N(\epsilon)$. Set $K = \max\{||u^{(1)}||, \ldots, ||u^{(N(\epsilon))}||, \epsilon + ||u||\}$. Then $||u^{(n)}|| \leq K$ for all n. Thus

$$|(u, v) - (u^{(n)}, v^{(n)})| = |(u - u^{(n)}, v) + (u^{(n)}, v - v^{(n)})|$$

$$\leq ||u - u^{(n)}|| \cdot ||v|| + ||u^{(n)}|| \cdot ||v - v^{(n)}|| \to 0$$

as $n \to \infty$. $\qquad\qquad\qquad\qquad\qquad\qquad\qquad\qquad\qquad\qquad\quad\square$

Theorem 1.40 *Let $\mathcal{H}$ be an incomplete inner product space. We can extend $\mathcal{H}$ to a Hilbert space $\overline{\mathcal{H}}$ such that $\mathcal{H}$ is a dense subspace of $\overline{\mathcal{H}}$.*

Proof $\mathcal{H}$ is a normed linear space with norm $||u|| = \sqrt{(u, u)}$. Therefore we can extend $\mathcal{H}$ to a Banach space $\overline{\mathcal{H}}$ such that $\mathcal{H}$ is dense in $\overline{\mathcal{H}}$. We claim that $\overline{\mathcal{H}}$ is a Hilbert space. Let $\overline{u}, \overline{v} \in \overline{\mathcal{H}}$ and let $\{u^{(n)}\}, \{\tilde{u}^{(n)}\} \in \overline{u}$, $\{v^{(n)}\}, \{\tilde{v}^{(n)}\} \in \overline{v}$. We define an inner product on $\overline{\mathcal{H}}$ by $(\overline{u}, \overline{v})' = \lim_{n \to \infty} (u^{(n)}, v^{(n)})$. The limit exists since $|(u^{(n)}, v^{(n)}) - (u^{(m)}, v^{(m)})| = |(u^{(m)}, v^{(n)} - v^{(m)}) + (u^{(n)} - u^{(m)}, v^{(m)}) + (u^{(n)} - u^{(m)}, v^{(n)} - v^{(m)})| \leq ||u^{(m)}|| \cdot ||v^{(n)} - v^{(m)}|| + ||u^{(n)} - u^{(m)}|| \cdot ||v^{(m)}|| + ||u^{(n)} - u^{(m)}|| \cdot ||v^{(n)} - v^{(m)}|| \to 0$ as $n, m \to \infty$. The limit is unique because $|(u^{(n)}, v^{(n)}) - (\tilde{u}^{(n)}, \tilde{v}^{(n)})| \to 0$ as $n, m \to \infty$. We can easily verify that $(\cdot, \cdot)'$ is an inner product on $\overline{\mathcal{H}}$ and $|| \cdot ||' = \sqrt{(\cdot, \cdot)'}$. $\qquad\square$

1.5 Hilbert spaces, L_2 and ℓ_2

A Hilbert space is an inner product space for which every Cauchy sequence in the norm converges to an element of the space.

Example 1.41 ℓ_2

The elements take the form

$$u = (\ldots, u_{-1}, u_0, u_1, \ldots), \quad u_i \in \mathbb{C}$$

such that $\sum_{i=-\infty}^{\infty} |u_i|^2 < \infty$. For

$$v = (\ldots, v_{-1}, v_0, v_1, \ldots) \in \ell_2,$$

we define vector addition and scalar multiplication by

$$u + v = (\ldots, u_{-1} + v_{-1}, u_0 + v_0, u_1 + v_1, \ldots)$$

and

$$\alpha u = (\ldots, \alpha u_{-1}, \alpha u_0, \alpha u_1, \ldots).$$

The zero vector is $\Theta = (\ldots, 0, 0, 0, \ldots)$ and the inner product is defined by $(u, v) = \sum_{i=-\infty}^{\infty} u_i \bar{v}_i$. We have to verify that these definitions make sense. Note that $2|ab| \le |a|^2 + |b|^2$ for any $a, b \in \mathbb{C}$. The inner product is well defined because $|(u, v)| \le \sum_{i=-\infty}^{\infty} |u_i \bar{v}_i| \le \frac{1}{2}(\sum_{i=-\infty}^{\infty} |u_i|^2 + \sum_{i=-\infty}^{\infty} |v_i|^2) < \infty$. Note that $|u_i + v_i|^2 \le |u_i|^2 + 2|u_i| \, |v_i| + |v_i|^2 \le 2(|u_i|^2 + |v_i|^2)$. Thus if $u, v \in \ell_2$ we have $||u+v||^2 \le 2||u||^2 + 2||v||^2 < \infty$, so $u + v \in \ell_2$.

Theorem 1.42 ℓ_2 *is a Hilbert space.*

Proof We have to show that ℓ_2 is complete. Let $\{u^{(n)}\}$ be Cauchy in ℓ_2,

$$u^{(n)} = (\ldots, u_{-1}^{(n)}, u_0^{(n)}, u_1^{(n)}, \ldots).$$

Thus, given any $\epsilon > 0$ there exists an integer $N(\epsilon)$ such that $||u^{(n)} - u^{(m)}|| < \epsilon$ whenever $n, m > N(\epsilon)$. Thus

$$\sum_{i=-\infty}^{\infty} |u_i^{(n)} - u_i^{(m)}|^2 < \epsilon^2. \tag{1.8}$$

Hence, for fixed i we have $|u_i^{(n)} - u_i^{(m)}| < \epsilon$. This means that for each i, $\{u_i^{(n)}\}$ is a Cauchy sequence in $\mathbb{C}$. Since $\mathbb{C}$ is complete, there exists $u_i \in \mathbb{C}$ such that $\lim_{n \to \infty} u_i^{(n)} = u_i$ for all integers i. Now set $u = (\ldots, u_{-1}, u_0, u_1, \ldots)$. We claim that $u \in \ell_2$ and $\lim_{n \to \infty} u^{(n)} = u$. It follows from (1.8) that for any fixed k, $\sum_{i=-k}^{k} |u_i^{(n)} - u_i^{(m)}|^2 < \epsilon^2$ for $n, m > N(\epsilon)$. Now let $m \to \infty$ and get $\sum_{i=-k}^{k} |u_i^{(n)} - u_i|^2 \le \epsilon^2$ for all k and for $n > N(\epsilon)$. Next let $k \to \infty$ and get $\sum_{i=-\infty}^{\infty} |u_i^{(n)} - u_i|^2 \le \epsilon^2$ for $n > N(\epsilon)$. This implies

$$||u^{(n)} - u|| < \epsilon \tag{1.9}$$

for $n > N(\epsilon)$. Thus, $u^{(n)} - u \in \ell_2$ for $n > N(\epsilon)$, so $u = (u - u^{(n)}) + u^{(n)} \in \ell_2$. Finally, (1.9) implies that $\lim_{n \to \infty} u^{(n)} = u$. $\square$

Example 1.43 $L_2[a, b]$

Recall that $C_2(a, b)$ is the set of all complex-valued functions $u(t)$ continuous on the open interval (a, b) of the real line, such that $\int_a^b |u(t)|^2 \, dt < \infty$, (Riemann integral). We define an inner product by

$$(u, v) = \int_a^b u(t)\overline{v}(t) \, dt, \qquad u, v \in C_2^{(n)}(a, b).$$

We verify that this is an inner product space. First, from the inequality $|u(x) + v(x)|^2 \leq 2|u(x)|^2 + 2|v(x)|^2$ we have $||u + v||^2 \leq 2||u||^2 + 2||v||^2$, so if $u, v \in C_2(a, b)$ then $u + v \in C_2(a, b)$. Second, $|u(x)\overline{v}(x)| \leq \frac{1}{2}(|u(x)|^2 + |v(x)|^2)$, so $|(u, v)| \leq \int_a^b |u(t)\overline{v}(t)| \, dt \leq \frac{1}{2}(||u||^2 + ||v||^2) < \infty$ and the inner product is well defined.

Now $C_2(a, b)$ is not complete, but it is dense in a Hilbert space $\overline{C}_2(a, b) = L_2[a, b]$. In most of this book, we will normalize to the case $a = 0, b = 2\pi$. We will show that the functions $e_n(t) = e^{int}/\sqrt{2\pi}$, $n = 0, \pm 1, \pm 2, \ldots$ form a basis for $L_2[0, 2\pi]$. This is a countable (rather than a continuum) basis. Hilbert spaces with countable bases are called separable, and we will be concerned only with separable Hilbert spaces.

1.5.1 The Riemann integral and the Lebesgue integral

Recall that $S(I)$ is the normed linear space of all real or complex-valued step functions on the (bounded or unbounded) interval I on the real line. A function $s(t)$ is a *step function* on I if there are a finite number of non-intersecting bounded subintervals $I_1, \ldots, I_m$ of I and numbers $c_1, \ldots, c_m$ such that $s(t) = c_k$ for $t \in I_k$, $k = 1, \ldots, m$ and $s(t) = 0$ for $t \in I - \cup_{k=1}^m I_k$. The integral of a step function is defined as $\int_I s(t)dt \equiv \sum_{k=1}^m c_k \ell(I_k)$ where $\ell(I_k) = $ length of $I_k = b - a$ if $I_k = [a, b]$ or $[a, b)$, or $(a, b]$ or (a, b). The norm is defined by $||s|| = \int_I |s(t)|dt$. We identify $s_1, s_2 \in S(I)^1$, $s_1 \sim s_2$ if $s_1(t) = s_2(t)$ except at a finite number of points. (This is needed to satisfy property 1 of the norm.) We let $S^1(I)$ be the space of equivalence classes of step functions in $S(I)$. Then $S^1(I)$ is a normed linear space with norm $|| \cdot ||$.

The space $L_1(I)$ of Lebesgue integrable functions on I is the completion of $S^1(I)$ in this norm. $L_1(I)$ is a Banach space. Every element u of $L_1(I)$ is an equivalence class of Cauchy sequences of step functions $\{s_n\}$,

$\int_I |s_j - s_k| dt \to 0$ as $j, k \to \infty$. (Recall $\{s'_n\} \sim \{s_n\}$ if $\int_I |s'_k - s_n| dt \to 0$ as $n \to \infty$).

We will show in §1.5.2 that we can associate equivalence classes of functions $f(t)$ on I with each equivalence class of step functions $\{s_n\}$, [75, 92]. The Lebesgue integral of f is defined by

$$\int_{I \text{ Lebesgue}} f(t)dt = \lim_{n \to \infty} \int_I s_n(t)dt,$$

and its norm by

$$\|f\| = \int_{I \text{ Lebesgue}} |f(t)|dt = \lim_{n \to \infty} \int_I |s_n(t)|dt.$$

How does this definition relate to Riemann integrable functions? To see this we take $I = [a, b]$, a closed bounded interval, and let $f(t)$ be a real bounded function on $[a, b]$. Recall that we have already defined the integral of a step function.

Definition 1.44 f is Riemann integrable on $[a, b]$ if for every $\epsilon > 0$, there exist step functions $r, s \in S[a, b]$ such that $r(t) \leq f(t) \leq s(t)$ for all $t \in [a, b]$, and $0 \leq \int_a^b (s - r)dt < \epsilon$.

Example 1.45 Divide $[a, b]$ by a grid of n points $a = t_0 < t_1 < \cdots < t_n = b$ such that $t_j - t_{j-1} = (b - a)/n$, $j = 1, \ldots, n$. Let $M_j = \sup_{t \in [t_{j-1}, t_j]} f(t)$, $m_j = \inf_{t \in [t_{j-1}, t_j]} f(t)$ and set

$$s_n(t) = \begin{cases} M_j & t \in [t_{j-1}, t_j) \\ 0 & t \notin [a, b) \end{cases},$$

$$r_n(t) = \begin{cases} m_j & t \in [t_{j-1}, t_j) \\ 0 & t \notin [a, b) \end{cases}.$$

Here, $\int_a^b s_n(t)dt$ is an upper Darboux sum and $\int_a^b r_n(t)dt$ is a lower Darboux sum. If f is Riemann integrable then the sequences of step functions $\{r_n\}, \{s_n\}$ satisfy $r_n \leq f \leq s_n$ on $[a, b]$, for $n = 1, 2, \ldots$ and $\int_a^b (s_n - r_n)dt \to 0$ as $n \to \infty$. The Riemann integral is defined by

$$\int_{a \text{ Riemann}}^b f\, dt = \lim_{n \to \infty} \int s_n\, dt = \lim_{n \to \infty} \int r_n\, dt =$$

$$\inf_{\text{upper Darboux sums}} \int s\, dt = \sup_{\text{lower Darboux sums}} \int t\, dt.$$

Note that

$$\sum_{j=1}^{n} M_j(t_j - t_{j-1}) \geq \int_a^b{}_{\text{Riemann}} f\, dt \geq \sum_{j=1}^{n} m_j(t_j - t_{j-1}).$$

Note also that

$$r_j - r_k \leq s_k - r_k, \qquad r_k - r_j \leq s_j - r_j$$

because every "upper" function is $\geq$ every "lower" function. Thus

$$\int |r_j - r_k|dt \leq \int (s_k - r_k)dt + \int (s_j - r_j)dt \to 0$$

as $j, k \to \infty$. It follows that $\{r_n\}$ and similarly $\{s_n\}$ are Cauchy sequences in the norm, equivalent because $\lim_{n\to\infty} \int (s_n - r_n)dt = 0$. The Riemann integral is just the integral of basic calculus that we all know and love.

Theorem 1.46 *If f is Riemann integrable on $I = [a, b]$ then it is also Lebesgue integrable and*

$$\int_I{}_{\text{Riemann}} f(t)dt = \int_I{}_{\text{Lebesgue}} f(t)dt = \lim_{n\to\infty} \int_I s_n(t)dt.$$

The following is a simple example to show that the space of Riemann integrable functions isn't complete. Consider the closed interval $I = [0, 1]$ and let $r_1, r_2, \ldots$ be an enumeration of the rational numbers in $[0, 1]$. Define the sequence of step functions $\{s_n\}$ by

$$s_n(t) = \begin{cases} 1 & \text{if } t = r_1, r_2, \ldots, r_n \\ 0 & \text{otherwise.} \end{cases}$$

Note that

- $s_1(t) \leq s_2(t) \leq \cdots$ for all $t \in [0, 1]$.
- s_n is a step function.
- The pointwise limit

$$f(t) = \lim_{n\to\infty} s_n(t) = \begin{cases} 1 & \text{if } t \text{ is rational} \\ 0 & \text{otherwise.} \end{cases}$$

- $\{s_n\}$ is Cauchy in the norm. Indeed $\int_0^1 |s_j - s_k|dt = 0$ for all $j, k = 1, 2, \ldots$.
- f is Lebesgue integrable with $\int_0^1{}_{\text{Lebesgue}} f(t)dt = \lim_{n\to\infty} \int_0^1 s_n(t)dt = 0$.

- f is not Riemann integrable because *every* upper Darboux sum for f is 1 and every lower Darboux sum is 0. One can't make $1 - 0 < \epsilon$ for $\epsilon < 1$!

Recall that $S(I)$ is the space of all real or complex-valued step functions on the (bounded or unbounded) interval I on the real line with real inner product $(s_1, s_2) = \int_I s_1(t)\bar{s}_2(t)dt$. We identify $s_1, s_2 \in S(I)$, $s_1 \sim s_2$ if $s_1(t) = s_2(t)$ except at a finite number of points. (This is needed to satisfy property 4 of the inner product.) Now we let $S^2(I)$ be the space of equivalence classes of step functions in $S(I)$. Then $S^2(I)$ is an inner product space with norm $||s||^2 = \int_I |s(t)|^2 dt$.

The space $L_2(I)$ of *Lebesgue square-integrable functions* on I is the completion of $S^2(I)$ in this norm. $L_2(I)$ is a Hilbert space. Every element u of $L_2(I)$ is an equivalence class of Cauchy sequences of step functions $\{s_n\}$, $\int_I |s_j - s_k|^2 dt \to 0$ as $j, k \to \infty$. (Recall $\{s'_n\} \sim \{s_n\}$ if $\int_I |s'_k - s_n|^2 dt \to 0$ as $n \to \infty$.)

As shown in §1.5.2 we can associate equivalence classes of functions $f(t)$ on I with each equivalence class of step functions $\{s_n\}$. The Lebesgue integral of $f_1, f_2 \in L_2(I)$ is defined by $(f_1, f_2) = \int_{I\text{Lebesgue}} f_1(t)f_2\, dt = \lim_{n \to \infty} \int_I s_n^{(1)}(t)s_n^{(2)}(t)dt$, [4, 91, 92, 93].

How does this definition relate to Riemann square integrable functions, see Definition 1.44? In a manner similar to our treatment of $L_1(I)$ one can show that if the function f is Riemann square integrable on I, then it is Cauchy square integrable and

$$\int_{I\text{Lebesgue}} |f(t)|^2 dt = \int_{I\text{Riemann}} |f(t)|^2 dt.$$

1.5.2 Digging deeper: associating functions with Cauchy sequences

We turn now from the abstract notion of the elements of the Lebesgue space as equivalence class of Cauchy sequences of step functions to an identification with actual functions. How should we identify a function with an equivalence class of Cauchy sequences of step functions in $S^1(I)$? Let $\{s_j\}$ be a Cauchy sequence in $S^1(I)$.

- Does $f(t) = \lim_{j \to \infty} s_j(t)$ always exist pointwise for all $t \in I$? **Answer**: no.
- Is there a subsequence $\{s_{j'}\}$ of $\{s_j\}$ such that $f(t) = \lim_{j' \to \infty} s_{j'}(t)$ exists? **Answer**: almost.

Example 1.47 Let $I = (0, 1]$ and $f(t) = 0$ for $t \in I$. We will define a particular Cauchy sequence $\{s_j\}$ of step functions that converge to f in the norm. Any positive integer j can be written uniquely as $j = 2^p + q$ for $p = 0, 1, \ldots$ and $q = 0, 1, \ldots, 2^p - 1$. Now set

$$s_{2^p+q}(t) = \begin{cases} 1 & \text{for } \frac{q}{2^p} < t \leq \frac{q+1}{2^p} \\ 0 & \text{otherwise.} \end{cases}$$

Then

$$\int_I |f - s_j| dt = \int_I |s_{2^p+q}| dt = \frac{1}{2^p} \to 0$$

as $p \to \infty$ (or $j \to \infty$) so $s_j \to f$ in the norm as $j \to \infty$. However $\lim_{j \to \infty} s_j(t)$ doesn't exist for any $t \in I$. Indeed for any fixed $t_0 \in I$ and any j there are always integers $j_1, j_2 > j$ such that $s_{j_1}(t_0) = 0$ and $s_{j_2}(t_0) = 1$. Note, however, that we can find pointwise convergent subsequences of $\{s_j\}$. In particular, for the subsequence of values $j = 2^p$ we have $\lim_{p \to \infty} s_{2^p}(t) = f(t) = 0$ for all $t \in I$.

Definition 1.48 A subset N of the real line is a null set (set of measure zero) if for every $\epsilon > 0$ there exists a countable collection of open intervals $\{I_n = (a_n, b_n)\}$ such that $N \subset \cup_{n=1}^\infty I_n$ and $\sum_{n=1}^\infty \ell(I_n) \leq \epsilon$. (Here $\ell(I_n) = b_n - a_n$.)

Examples 1.49 1. Any finite set is null.

2. Any countable set is null.

Proof Let $N = \{r_1, r_2, \ldots, r_n, \ldots\}$. Given $\epsilon > 0$ let $I_n = (r_n - \frac{\epsilon}{2^{n+1}}, r_n + \frac{\epsilon}{2^{n+1}})$, $n = 1, 2, \ldots$ Then $N \subset \cup_{n=1}^\infty I_n$ and $\sum_{n=1}^\infty \ell(I_n) = \sum_{n=1}^\infty \frac{\epsilon}{2^n} = \epsilon$. □

3. The rationals form a null set.

4. The set $I = (a, b)$ with $a < b$ is not null.

5. A countable union of null sets is null.

6. There exist null sets which are not countable. The most famous example is probably the Cantor set.

The Cantor set is defined as

$$C = \{x \in [0, 1] : x = \sum_{n=1}^\infty \frac{c_n}{3^n}, c_n = 0, 2\}. \tag{1.10}$$

Note that any real number $x \in [0, 1]$ can be given a ternary representation $x = \sum_{n=1}^\infty \frac{c_n}{3^n} = .c_1 c_2 c_3 c_4 \cdots$, for $c_n = 0, 1, 2$. For example $1 = \sum_{n=1}^\infty 2/3^n = .2222 \cdots$. (Just as with decimal expansions of real numbers, this representation is not quite unique, e.g., $1/3 = .10000 \cdots =$

.02222···, but we could adopt a convention to make it unique.) The Cantor set consists of those real numbers whose ternary representation doesn't contain any $c_n = 1$. We are more familiar with the binary expansion of any real number $y \in [0, 1]$, $y = \sum_{n=1}^{\infty} \frac{b_n}{2^n} = .b_1b_2b_3b_4 \cdots$, where $b_n = 0, 1$.

- C is uncountable.

  Proof The map $C \Longrightarrow [0, 1]$ defined by

  $$\sum_{n=1}^{\infty} \frac{c_n}{3^n} \Longrightarrow \sum_{n=1}^{\infty} \frac{\frac{1}{2}c_n}{2^n}$$

  is 1-1 and onto. Since the real line contains an uncountable number of points, C is also uncountable. $\qquad\square$

- C is a null set.

  Proof We can see this geometrically, from the triadic representation. The points in the open middle third $(1/3, 2/3)$ of the interval $[0, 1]$ don't lie in the Cantor set, so we can remove this third. Then we can remove the open middle thirds $(1/9, 2/9)$ and $(7/9, 8/9)$ of the remaining intervals, etc. After k steps there remains a set C_k that is the union of 2^k intervals of total length $(2/3)^k$. $C = \cap_{k=1}^{\infty} C_k$, so for each k, C can be covered by 2^k open intervals of length $\leq 2(2/3)^k$. Since this goes to zero as $k \to \infty$ we see that C is a null set. $\qquad\square$

Definition 1.50 If a property holds for all real numbers t except for $t \in N$, a null set, we say that the property holds almost everywhere (a.e.).

The following technical lemmas show us how to associate a function with a Cauchy sequence of step functions.

Lemma 1.51 *Let* $\{s_k\}$ *be a Cauchy sequence in* $S^1(I)$. *Then there exist strictly increasing sequences* $\{n_k\}$ *of positive integers such that*

$$\sum_{k=1}^{\infty} \int_I |s_{n_{k+1}} - s_{n_k}| dt < \infty. \tag{1.11}$$

For every such sequence $\{n_k\}$ *the subsequence* $\{s_{n_k}(t)\}$ *converges pointwise a.e. in* I.

Proof Choose $n_k > n_{k-1}$, and so large that $\int_I |s_m - s_n| dt \leq 1/2^k$ for all $m, n \geq n_k$, $k = 1, 2, \ldots$. Then $\int_I |s_{n_{k+1}} - s_{n_k}| dt \leq 1/2^k$, $k = 1, 2, \ldots$ and $\sum_{k=1}^{\infty} \int_I |s_{n_{k+1}} - s_{n_k}| dt \leq \sum_{k=1}^{\infty} 1/2^k < \infty$. Now assume $\{s_{n_k}\}$ is an arbitrary subsequence of $\{s_n\}$ such that (1.11) converges. Set

$$u_k = |s_{n_1}| + |s_{n_2} - s_{n_1}| + \cdots + |s_{n_k} - s_{n_{k-1}}|, \quad k \geq 2, \qquad u_1 = |s_{n_1}|.$$

Then $0 \leq u_1(t) \leq u_2(t) \leq \cdots$ for all $t \in I$ and u_k is a step function. By (1.11) there exists $M > 0$ such that $\int_I u_k dt \leq M$ for all k. We will show that $\lim_{k \to \infty} u_k(t)$ exists for almost all $t \in I$. Given $\epsilon > 0$, let $\mathcal{R}_k(\epsilon) = \{t \in I : u_k(t) \geq M/\epsilon\}$. Clearly:

1. $\mathcal{R}_k(\epsilon) = \mathcal{R}_k$ is the union of finitely many non-intersecting intervals.
2. Let ϵ_k be the sum of the lengths of the intervals in $\mathcal{R}_k$. Then $\epsilon_k \leq \epsilon$ because $u_k(t) \geq \frac{M}{\epsilon} \chi_{\mathcal{R}_k}(t) \implies \int u_k dt \geq \int \frac{M}{\epsilon} \chi_{\mathcal{R}_k} dt \implies M \geq \frac{M}{\epsilon} \epsilon_k$ where $\chi_S(t)$ is the characteristic function of the set S, i.e.,

$$\chi_S(t) = \begin{cases} 1 & \text{if } t \in S \\ 0 & \text{if } t \notin S. \end{cases}$$

3. $u_k \geq u_{k-1} \implies \mathcal{R}_k \supseteq \mathcal{R}_{k-1}$.
4. Let

$$\mathcal{R} = \mathcal{R}(\epsilon) = \cup_{k \geq 1} \mathcal{R}_k(\epsilon) = \mathcal{R}_1 \cup (\mathcal{R}_2 - \mathcal{R}_1) \cup \cdots \cup (\mathcal{R}_k - \mathcal{R}_{k-1}) \cup \cdots$$

Then $\mathcal{R}_k - \mathcal{R}_{k-1}$ can be represented as the union of finitely many non-intersecting intervals of total length $\epsilon_k - \epsilon_{k-1}$.

5. It follows that $\mathcal{R}(\epsilon)$ is the union of countably many non-intersecting intervals. The sum of the lengths of the intervals in $\mathcal{R}(\epsilon)$ is

$$\epsilon_1 + (\epsilon_2 - \epsilon_1) + \cdots + (\epsilon_k - \epsilon_{k-1}) + \cdots = \lim_{k \to \infty} \epsilon_k \leq \epsilon.$$

Conclusion: $u_k(t) < M/\epsilon,\ k = 1, 2, \ldots, \forall t \in I - \mathcal{R}(\epsilon) \implies \lim_{k \to \infty} u_k(t)$ exists $\forall t \in I - \mathcal{R}(\epsilon)$. The points of divergence are covered by the intervals of $\mathcal{R}(\epsilon)$ (of total length $\leq \epsilon$). But ϵ is arbitrary so the points of divergence form a null set N so $\lim_{k \to \infty} u_k(t)$ exists a.e.

Consider

$$s_{n_1}(t) + (s_{n_2}(t) - s_{n_1}(t)) + \cdots + (s_{n_k}(t) - s_{n_{k-1}}(t)) + \cdots \qquad (1.12)$$

Now

$$|s_{n_1}(t)| + |s_{n_2}(t) - s_{n_1}(t)| + \cdots + |s_{n_k}(t) - s_{n_{k-1}}(t)| + \cdots = \lim_{k \to \infty} u_k(t)$$

exists $\forall t \notin N$. Therefore (1.12) converges $\forall t \notin N \implies \lim_{k \to \infty} s_{n_k}(t)$ exists a.e. $\qquad \square$

Lemma 1.52 *Let $\{s_k\}, \{s'_k\}$ be equivalent Cauchy sequences in $S^1(I)$, possibly the same sequence. Let $\{s_{p_k}\}, \{s'_{q_k}\}$ be subsequences of $\{s_k\}$ and $\{s'_k\}$ that converge pointwise a.e. on I. Then $\lim_{k \to \infty} \left(s_{p_k}(t) - s'_{q_k}(t) \right) = 0$ pointwise a.e. on I.*

Proof Let $v_k = s_{p_k} - s'_{q_k}$, $k = 1, 2, \ldots$. Then $\{v_k\}$ is a Cauchy sequence and $\lim_{k \to \infty} v_k(t)$ exists pointwise $\forall t \notin N_1$ where N_1 is some null set. Also,

$$\int_I |v_k| dt =$$

$$\int_J |s_{p_k} - s'_{q_k}| dt \leq \int_I |s_{p_k} - s_k| dt + \int_I |s_k - s'_k| dt + \int_I |s'_k - s'_{q_k}| dt \to 0$$

as $k \to \infty$. Let $\{k_\ell\}$ be an increasing sequence of positive integers such that $\sum_{\ell=1}^\infty \int_I |v_{k_\ell}| dt < \infty$. Then by lemma 1.51, $\sum_{\ell=1}^\infty |v_{k_\ell}(t)|$ converges $\forall t \notin N_2$ where N_2 is a null set. So $\lim_{\ell=\infty} v_{k_\ell}(t) = 0$ pointwise $\forall t \notin N_2 \implies$ for all $t \notin N_1 \cup N_2$ (a null set) we have $\lim_{k \to \infty} v_k(t) = \lim_{\ell \to \infty} v_{k_\ell}(t) = 0$. $\qquad \square$

We want to associate an equivalence class of functions (equal a.e.) with each equivalence class of Cauchy sequences of step functions. How can we do this uniquely? In particular, let $\{s_k\}$ be a sequence in $S^1(I)$ such that $s_k \to 0$ a.e. as $k \to \infty$. How can we guarantee that $\int_I s_k dt \to 0$ as $k \to \infty$?

Example 1.53 Let $\alpha > 0$, $I = (0, 1)$. Set

$$s_k(t) = \begin{cases} 0 & \text{if } t \geq \frac{1}{k} \\ \alpha k & \text{if } 0 < t < \frac{1}{k}, \ k \text{ odd} \\ k & \text{if } 0 < t < \frac{1}{k}, \ k \text{ even.} \end{cases}$$

Remarks:

- $s_k(t) \to 0$ as $k \to \infty$, for every t.
-
$$\int_I s_k dt = \begin{cases} \alpha, & k \text{ odd} \\ 1, & k \text{ even.} \end{cases}$$

- $\lim_{k \to \infty} \int_I s_k dt = 1$, if $\alpha = 1$. Otherwise the limit doesn't exist.

The next lemma and the basic theorem to follow give conditions that guarantee uniqueness.

Lemma 1.54 *Let $\{s_k\}$, $\{t_k\}$ be Cauchy sequences in $S^1(I)$ that converge pointwise a.e. in I to limit functions that are equal a.e. in I. Then $\{s_k\} \sim \{t_k\}$, i.e., $\int_I |s_k - t_k| dt \to 0$ as $k \to \infty$.*

Proof This is an immediate consequence of the following basic theorem.

Theorem 1.55 *Let $\{s_k\}$ be a sequence in $S^1(I)$ such that $\lim_{k\to\infty} s_k(t) = 0$ a.e. in I. Suppose either*

1. *the s_j are real and*

$$s_1(t) \geq s_2(t) \geq \cdots \geq s_n(t) \geq \cdots \geq 0$$

$\forall t \in I$, or

2. *for every $\epsilon > 0$ there exists an integer $N(\epsilon)$ such that $\int_I |s_j - s_k|\, dt < \epsilon$ whenever, $j, k \geq N(\epsilon)$, i.e., $\{s_k\}$ is Cauchy in the norm.*

Then $\int_I |s_k|\, dt \to 0$ as $k \to \infty$. (Note: $|\int_I s_k dt| \leq \int_J |s_k|\, dt$, so $\int_I s_k dt \to 0$ as $k \to \infty$.)

Proof

1.

$$s_1 \geq s_2 \geq \cdots \geq 0, \qquad s_k(t) \to 0 \text{ a.e. as } k \to \infty.$$

Given $\epsilon > 0$, let $M = \max s_1(t)$, and let $[a, b]$ be the smallest closed interval outside of which $s_1(t) = 0$. We can assume $a \neq b$.

Let N be the null set consisting of the points t where either $\lim_{k\to\infty} s_k(t)$ is not zero or where the limit doesn't exist, plus the points where one or more of the functions $s_1, s_2, \ldots$ is discontinuous. Let $\mathcal{I} = \{I_1, I_2, \ldots\}$ be a countable family of open intervals that cover N and such that $\sum_{k=1}^\infty \ell(I_k) \leq \epsilon$. Choose $t_0 \in (a, b)$ such that $t_0 \notin \cup_{k\geq 1} I_k$. Since $s_k(t_0) \to 0$ as $K \to \infty$ there exists a smallest index $h = h(\epsilon, t_0)$ such that $s_k(t_0) \leq \epsilon\ \forall k \geq h$. Since t_0 is a point of continuity of s_h there exists an open interval in $[a, b]$ that contains t_0 and on which $s_h(t) \leq \epsilon$. Let $I(t_0) = I(t_0, \epsilon)$ be the largest such interval. Then $s_k(t) \leq \epsilon$ on $I(t_0)\ \forall k \geq h(t_0)$. Let $\mathcal{G} = \{I(t) : t \in [a, b] - \cup_n I_n\}$. Let $\mathcal{H} = \mathcal{H}(\epsilon)$ be the family consisting of the intervals of $\mathcal{I}$ and those of $\mathcal{G}$. Now $\mathcal{H}$ forms a covering of $[a, b]$ by open intervals. Therefore, by the Heine–Borel theorem (see [55, 92] if you are not familiar with this theorem) we can find a finite subfamily $\mathcal{H}'$ of $\mathcal{H}$ that covers $[a, b]$.

$$\mathcal{H}' = \{I(t_1), I(t_2), \ldots, I(t_n), I_1, I_2, \ldots, I_m\}.$$

On $I(t_1)$, $1 \leq i \leq n$, we have $s_k(t) \leq \epsilon\ \forall k \geq h(t_i)$. Let $p = \max\{h(t_1), \ldots, h(t_n)\}$. Then $s_k(t) \leq \epsilon$ for $k \geq p$ on every interval

$I(t_i)$, $1 \le i \le n$. On I_k, $1 \le k \le m$, we know only $s_k(t) \le s_1(t) \le M$. Therefore, for $k \ge p$,

$$\int_I s_k dt \le \int_I s_p dt \le \int_I \left[\epsilon \chi_{I(t_1) \cup \cdots \cup I(t_n)} + M \chi_{I_1 \cup \cdots \cup I_m} \right] dt$$

$$\le \epsilon(b-a) + M\epsilon \qquad \text{since} \quad s_p \le \epsilon \chi_{J(t_1) \cup \cdots \cup I(t_n)} + M \chi_{I_1 \cup \cdots \cup I_m}.$$

(Note that the latter are step functions.) But ϵ is arbitrary, so $\lim_{k \to \infty} s_k dt$

2.

$$\lim_{k \to \infty} s_k(t) = 0 \quad \text{a.e.}, \quad \int_I |s_k - s_\ell| dt \to 0 \text{ as } k, \ell \to \infty.$$

Now

$$\left| \int_I |s_k| dt - \int |s_\ell| dt \right| = \left| \int_I (|s_k| - |s_\ell|) dt \right| \le \int_I \big| \, |s_k| - |s_\ell| \, \big| \, dt$$

$$\le \int_I |s_k - s_\ell| dt \to 0$$

as $k, \ell \to \infty$. So $\{ \int_I |s_k| dt \}$ is a Cauchy sequence of real numbers $\Longrightarrow$ $\int_I |s_k| dt \to A$ as $k \to \infty$.

We will show that $A = 0$ by making use of part 1 of the proof. Let $n_1 < n_2 < \cdots < n_k < \cdots$ be a strictly increasing sequence of positive integers. Set $v_1(t) = |s_{n_1}(t)|$, and $v_k = |s_{n_1}| \wedge |s_{n_2}| \wedge \cdots \wedge |s_{n_k}|$, $k = 2, 3, \ldots$, i.e., $v_k(t) = \min\{ |s_{n_1}(t)|, |s_{n_2}(t)|, \ldots, |s_{n_k}(t)| \}$.

Remarks:

- $v_k \in S^1(I)$.
- $v_k = v_{k-1} \wedge |s_{n_k}|$.
- $v_1 \ge v_2 \ge v_3 \ge \cdots \ge 0$.
- $v_k(t) \le |s_{n_k}(t)| \to 0$ a.e. as $k \to \infty$. Therefore, by part 1, $\lim_{k \to \infty} \int_I v_k dt = 0$.
- $-v_k + v_{k-1} = -v_{k-1} \wedge |s_{n_k}| + v_{k-1} \Longrightarrow$

$$v_{k-1}(t) - v_k(t) =$$

$$\begin{cases} 0 & \text{if } |s_{n_k}(t)| \ge v_{k-1}(t) \\ t_{k-1}(t) - |s_{n_k}(t)| \le |s_{n_{k-1}}(t)| - |s_{n_k}(t)| \\ \qquad \le |s_{n_{k-1}}(t) - s_{n_k}(t)| & \text{otherwise.} \end{cases}$$

- $\int_I (v_{k-1} - v_k) dt \le \int_I |s_{n_{k-1}} - s_{n_k}| dt = \delta_k$ (definition).

Therefore,

$$\int_I |s_{n_1}| dt = \int_I v_1 dt = \int_I [v_2 + (v_1 - v_2)] dt \leq \int_I v_2 dt + \delta_2 \quad (1.13)$$

$$= \int_I [v_3 + (v_2 - v_3)] dt + \delta_2 \leq \int_I v_3 dt + \delta_2 + \delta_3$$

$$\leq \cdots \leq \int_I v_p dt + \delta_2 + \delta_3 + \cdots + \delta_p,$$

for $p = 2, 3, \ldots$ Given $\epsilon > 0$ choose n_1 so large that $\int_I |s_{n_1}| dt > A - \epsilon$ and $\int_I |s_{n_1} - s_n| dt < \epsilon/2 \; \forall n > n_1$. For $k \geq 2$ choose $n_k > n_{k-1}$ so large that $\int_I |s_{n_k} - s_n| dt < \epsilon/2^k \; \forall n > n_k$. Then $\delta_k < \epsilon/2^{k-1}$, for $k = 2, 3, \ldots$. Choose p so large that $\int_I v_p dt < \epsilon$. Then (1.13) $\Longrightarrow$

$$A - \epsilon < \int_I |s_{n_1}| dt < \epsilon + \frac{\epsilon}{2} + \frac{\epsilon}{2^2} + \cdots + \frac{\epsilon}{2^{p-1}} < 2\epsilon,$$

$\Longrightarrow 0 \leq A < 3\epsilon$. But ϵ is arbitrary, so $A = 0$. $\qquad\square$

1.5.3 Pointwise convergence and L_1 and L_2 convergence

Now we can clarify what happens when we have a Cauchy sequence of functions in $L_1(I)$ that also converges pointwise a.e.

Theorem 1.56 *Let $\{g_n\}$ be a Cauchy sequence in $L_1(I)$ such that $g_n(t) \to f(t)$ a.e. as $n \to \infty$. Then $f \in L_1(I)$ and $\int_I f \, dt = \lim_{n \to \infty} \int_I g_n dt$. Also, $\lim_{n \to \infty} \int_I |f - g_n| dt = 0$.*

Proof Note that $f(t) = g_1(t) + \sum_{n=1}^{\infty} [g_{n+1}(t) - g_n(t)] = \lim_{n \to \infty} g_n(t)$ a.e. and, by passing to a subsequence if necessary, we can assume that $\int_I |g_1| \, dt + \sum_{n=1}^{\infty} \int_I |g_{n+1} - g_n| \, dt < \infty$. Similarly we can find step functions $s_{n,j}$ such that

1. $g_{n+1}(t) - g_n(t) = s_{n1}(t) + \sum_{j=1}^{\infty} [s_{n,j+1}(t) - s_{n,j}(t)] = \lim_{j \to \infty} s_{n,j+1}(t)$, except on a set N_n of measure zero
2. $\int_I |s_{n,1}| \, dt + \sum_{j=1}^{\infty} \int_I |s_{n,j+1} - s_{n,j}| \, dt < \infty$.

For each n, choose $\epsilon_n > 0$. By passing to a further subsequence of step functions if necessary, we can also assume that

1. $\int_I |s_{n1} - (g_{n+1} - g_n)| \, dt < \epsilon_n$
2. $\sum_{j=1}^{\infty} \int_I |s_{n,j+1} - s_{n,j}| \, dt < \epsilon_n$.

Thus, by the triangle inequality, we have

$$\int_I |s_{n,1}| \, dt + \sum_{j=1}^{\infty} \int_I |s_{n,j+1} - s_{n,j}| \, dt \le \int_I |g_{n+1} - g_n| \, dt$$

$$+ \int_I |(g_{n+1} - g_n) - s_{n,1}| \, dt + \epsilon_n \le \int_I |g_{n+1} - g_n| \, dt + 2\epsilon_n.$$

Now set $\epsilon_n = 1/2^{n+1}$ and note that

$$\sum_{n=1}^{\infty} \int_I |s_{n,1}| \, dt + \sum_{n,j=1}^{\infty} \int_I |s_{n,j+1} - s_{n,j}| dt \le \sum_{n=1}^{\infty} \int_I |g_{n+1} - g_n| dt$$

$$+ \sum_{n=1}^{\infty} \frac{1}{2^n} < \infty.$$

This implies that $\sum_{n,j=1}^{\infty} s_{n,j}(t)$ converges absolutely $\forall t \notin N$, where N is a null set. Therefore, if $t \notin (\cup_n N_n) \cup N$ then $g_1(t) + \sum_{n=1}^{\infty}(g_{n+1}(t) - g_n(t)) = g_1(t) + \sum_{n,j=1}^{\infty}(s_{n,j+1} - s_{n,j})(t) = f(t)$. Since $g_1(t)$ can be expressed as the limit of a Cauchy sequence of step functions, it follows that f is the limit of a (double-indexed) Cauchy sequence of step functions, so

$$f \in L_1(I) \qquad \text{and} \qquad \int_I f \, dt = \lim_{n \to \infty} \int_I g_n dt.$$

$\square$

We can easily modify the preceding results to relate pointwise convergence to convergence in $L_2(I)$. Recall that $S^2(I)$ is the space of all real or complex-valued step functions on I with inner product $(s_1, s_2) = \int_I s_1(t)\bar{s}_2(t)dt$. More precisely, $S^2(I)$ is the space of equivalence classes of step functions in $S(I)$ where two step functions are equivalent provided they differ at most for a finite number of points. Then $S^2(I)$ is an inner product space with norm $||s||^2 = \int_I |s(t)|^2 dt$. The space of $L_2(I)$ is the completion of $S^2(I)$ with respect to this norm. $L_2(I)$ is a Hilbert space. Every element u of $L_2(I)$ is an equivalence class of Cauchy sequences of step functions $\{s_n\}$, $\int_J |s_j - s_k|^2 dt \to 0$ as $j, k \to \infty$. (Recall $\{s'_n\} \sim \{s_n\}$ if $\int_J |s'_k - s_n|^2 dt \to 0$ as $n \to \infty$.) Now by a slight modification of Lemmas 1.51, 1.52 and 1.54 we can identify an equivalence class X of Cauchy sequences of step functions with an equivalence class F of functions $f(t)$ that are equal a.e. Indeed we will show that $f \in F \iff$ there exists a Cauchy sequence $\{s_n\} \in X$ and an increasing sequence of integers $n_1 < n_2 \cdots$ such that $s_{n_k} \to f$ a.e. as $k \to \infty$. This result is an immediate consequence of the following lemma.

Lemma 1.57 *If $\{s_n\}$ is a Cauchy sequence in $S^2(I)$ then $\{|s_n|^2\}$ is a Cauchy sequence in $S^1(I)$.*

Proof Given $\epsilon > 0$ we have

$$|s_n| = |s_m - (s_m - s_n)|$$

$$\leq |s_m| + |s_m - s_n| \leq \begin{cases} (1 + \epsilon)|s_m| & \text{if } |s_m - s_n| \leq \epsilon|s_m| \\ (1 + \frac{1}{\epsilon})|s_m - s_n| & \text{if } |s_m - s_n| \geq \epsilon|s_m|, \end{cases}$$

which implies

$$|s_n|^2 \leq (1 + \epsilon)^2|s_m|^2 + \left(1 + \frac{1}{\epsilon}\right)^2 |s_m - s_n|^2,$$

so

$$\left||s_n|^2 - |s_m|^2\right| \leq [(1 + \epsilon)^2 - 1]|s_m|^2 + \left(1 + \frac{1}{\epsilon}\right)^2 |s_m - s_n|^2.$$

Integrating this inequality over I and using that fact that since $\{s_m\}$ is Cauchy in $S^2(I)$ there exists a finite number M such that $\int_I |s_m|^2 dt \leq M$, we obtain

$$\int_I \left||s_n|^2 - |s_m|^2\right| dt \leq [(1 + \epsilon)^2 - 1]M + \left(1 + \frac{1}{\epsilon}\right)^2 ||s_m - s_n||^2.$$

Now, given $\epsilon' > 0$ we can choose ϵ so small that the first term on the right-hand side of this inequality is $< \epsilon'/2$. Then we can choose n, m so large that the second term on the right-hand side of this inequality is also $< \epsilon'/2$. Thus $\int_J \left||s_n|^2 - |s_m|^2\right| dt \to 0$ as $n, m \to \infty$. $\square$

Though we shall not do this here, the arguments used for the Banach spaces $L_1(I)$, $L_2(I)$ can be generalized to construct the Banach spaces $L_p(I)$ for p any positive real number. Here the norm is $||f||_p$ where $||f||_p^p = \int_I |f(t)|^p \, dt$. The space is the completion of the space of step functions on I with respect to this norm.

1.6 Orthogonal projections, Gram–Schmidt orthogonalization

Orthogonality is the mathematical formalization of the geometrical property of perpendicularity, as adapted to general inner product spaces. In signal processing, bases consisting of mutually orthogonal elements play an essential role in a broad range of theoretical and applied applications such as the design of numerical algorithms. The fact of the matter is that computations become dramatically simpler and less prone to numerical instabilities when performed in orthogonal coordinate systems. For example, the orthogonality of eigenvector and eigenfunction bases for symmetric matrices and self-adjoint boundary value problems is the key to understanding the dynamics of many discrete and continuous mechanical, thermodynamical and electrical systems as well as almost all of Fourier analysis and signal processing. The Gram–Schmidt process will convert an arbitrary basis for an inner product space into an orthogonal basis. In function space, the Gram–Schmidt algorithm is employed to construct orthogonal and biorthogonal polynomials. Orthogonal polynomials play a fundamental role in approximation theory, integrable systems, random matrix theory, computer vision, imaging and harmonic analysis to name just a few areas.

1.6.1 Orthogonality, orthonormal bases

Definition 1.58 Two vectors u, v in an inner product space $\mathcal{H}$ are called orthogonal, $u \perp v$, if $(u, v) = 0$. Similarly, two sets $\mathcal{M}, \mathcal{N} \subset \mathcal{H}$ are orthogonal, $\mathcal{M} \perp \mathcal{N}$, if $(u, v) = 0$ for all $u \in \mathcal{M}$, $v \in \mathcal{N}$.

Definition 1.59 Let $\mathcal{S}$ be a nonempty subset of the inner product space $\mathcal{H}$. We define $\mathcal{S}^\perp = \{u \in \mathcal{H} : u \perp \mathcal{S}\}$.

Lemma 1.60 $\mathcal{S}^\perp$ *is a closed subspace of* $\mathcal{H}$ *in the sense that if* $\{u^{(n)}\}$ *is a sequence in* $\mathcal{S}^\perp$*, and* $u^{(n)} \to u \in \mathcal{H}$ *as* $n \to \infty$*, then* $u \in \mathcal{S}^\perp$.

Proof

1. $\mathcal{S}^\perp$ is a subspace. Let $u, v \in \mathcal{S}^\perp$, $\alpha, \beta \in C$, then $(\alpha u + \beta v, w) = \alpha(u, w) + \beta(v, w) = 0$ for all $w \in \mathcal{S}$, so $\alpha u + \beta v \in \mathcal{S}^\perp$.

2. $\mathcal{S}^\perp$ is closed. Suppose $\{u^{(n)}\} \subset \mathcal{S}^\perp$, $\lim_{n \to \infty} u^{(n)} = u \in \mathcal{H}$. Then

   $(u, v) = (\lim_{n \to \infty} u^{(n)}, v) = \lim_{n \to \infty}(u^{(n)}, v) = 0$ for all $v \in \mathcal{S} \implies u \in \mathcal{S}^\perp$. $\qquad\square$

1.6.2 Orthonormal bases for finite-dimensional inner product spaces

Let $\mathcal{H}$ be an n-dimensional inner product space, (say $\mathcal{H}_n$). A vector $u \in \mathcal{H}$ is a *unit vector* if $||u|| = 1$. The vectors in a finite subset $\{u^{(1)}, \ldots, u^{(k)}\} \subset \mathcal{H}$ are *mutually orthogonal* if $u^{(i)} \perp u^{(j)}$ for $i \neq j$. The finite subset is *orthonormal (ON)* if $u^{(i)} \perp u^{(j)}$ for $i \neq j$, and $||u^i|| = 1$. Orthonormal bases for $\mathcal{H}$ are especially convenient because the expansion coefficients of any vector in terms of the basis can be calculated easily from the inner product.

Theorem 1.61 *Let* $\{u^{(1)}, \ldots, u^{(n)}\}$ *be an ON basis for* $\mathcal{H}$. *If* $u \in \mathcal{H}$ *then*

$$u = \alpha_1 u^{(1)} + \alpha_2 u^{(2)} + \cdots + \alpha_n u^{(n)}$$

where $\alpha_i = (u, u^{(i)})$, $i = 1, \ldots, n$.

Proof

$$(u, u^{(i)}) = (\alpha_1 u^{(1)} + \alpha_2 u^{(2)} + \cdots + \alpha_n u^{(n)}, u^{(i)}) = \sum_j \alpha_j (u^{(j)}, u^{(i)}) = \alpha_i.$$

$\square$

Example 1.62 Consider $\mathcal{H}_3$. The set $e^{(1)} = (1, 0, 0), e^{(2)} = (0, 1, 0), e^{(3)} = (0, 0, 1)$ is an ON basis. The set $u^{(1)} = (1, 0, 0)$, $u^{(2)} = (1, 1, 0)$, $u^{(3)} = (1, 1, 1)$ is a basis, but not ON. The set $v^{(1)} = (1, 0, 0)$, $v^{(2)} = (0, 2, 0)$, $v^{(3)} = (0, 0, 3)$ is an orthogonal basis, but not ON.

Exercise 1.6 Write the vector $w = (1, -2, 3)$ as a linear combination of each of the bases $\{e^{(i)}\}$, $\{u^{(i)}\}$, $\{v^{(i)}\}$ of Example 1.62.

The following are very familiar results from geometry, where the inner product is the dot product, but apply generally:

Corollary 1.63 *For* $u, v \in \mathcal{H}$ *and* $\{u^{(i)}\}$ *an ON basis for* $\mathcal{H}$:

- $(u, v) = (\alpha_1 u^{(1)} + \alpha_2 u^{(2)} + \cdots + \alpha_n u^{(n)}, \beta_1 u^{(1)} + \beta_2 u^{(2)} + \cdots + \beta_n u^{(n)}) = \sum_{i=1}^{n} (u, u^{(i)})(u^{(i)}, v)$.
- $||u||^2 = \sum_{i=1}^{n} |(u, u^{(i)})|^2$. *Parseval's equality.*

Lemma 1.64 *If* $u \perp v$ *then* $||u + v||^2 = ||u||^2 + ||v||^2$. *Pythagorean Theorem.*

Lemma 1.65 *If* u, v *belong to the real inner product space* $\mathcal{H}$ *then* $||u + v||^2 = ||u||^2 + ||v||^2 + 2(u, v)$. *Law of Cosines.*

Does every n-dimensional inner product space have an ON basis? Yes! Recall that $[u^{(1)}, u^{(2)}, \ldots, u^{(m)}]$ is the subspace of $\mathcal{H}$ spanned by all linear combinations of the vectors $u^{(1)}, u^{(2)}, \ldots, u^{(m)}$.

Theorem 1.66 *(Gram–Schmidt) let* $\{u^{(1)}, u^{(2)}, \ldots, u^{(n)}\}$ *be an (ordered) basis for the inner product space* $\mathcal{H}$, *where* $n \geq 1$. *There exists an ON basis* $\{f^{(1)}, f^{(2)}, \ldots, f^{(n)}\}$ *for* $\mathcal{H}$ *such that*

$$[u^{(1)}, u^{(2)}, \ldots, u^{(m)}] = [f^{(1)}, f^{(2)}, \ldots, f^{(m)}]$$

for each $m = 1, 2, \ldots, n$.

Proof Define $f^{(1)}$ by $f^{(1)} = u^{(1)}/||u^{(1)}||$. This implies $||f^{(1)}|| = 1$ and $[u^{(1)}] = [f^{(1)}]$. Now set $g^{(2)} = u^{(2)} - \alpha f^{(1)} \neq \Theta$. We determine the constant α by requiring that $(g^{(2)}, f^{(1)}) = 0$. But $(g^{(2)}, f^{(1)}) = (u^{(2)}, f^{(1)}) - \alpha$ so $\alpha = (u^{(2)}, f^{(1)})$. Now define $f^{(2)}$ by $f^{(2)} = g^{(2)}/||g^{(2)}||$. At this point we have $(f^{(i)}, f^{(j)}) = \delta_{ij}$ for $1 \leq i, j \leq 2$ and $[u^{(1)}, u^{(2)}] = [f^{(1)}, f^{(2)}]$.

We proceed by induction. Assume we have constructed an ON set $\{f^{(1)}, \ldots, f^{(m)}\}$ such that $[f^{(1)}, \cdots, f^{(k)}] = [u^{(1)}, \ldots, u^{(k)}]$ for $k = 1, 2, \ldots, m$. Set $g^{(m+1)} = u^{(m+1)} - \alpha_1 f^{(1)} - \alpha_2 f^{(2)} - \cdots - \alpha_m f^{(m)} \neq \Theta$. Determine the constants α_i by the requirement $(g^{(m+1)}, f^{(i)}) = 0 = (u^{(m+1)}, f^{(i)}) - \alpha_i$, $1 \leq i \leq m$. Set $f^{(m+1)} = g^{(m+1)}/||g^{(m+1)}||$. Then $\{f^{(1)}, \ldots, f^{(m+1)}\}$ is ON. □

Let $\mathcal{W}$ be a subspace of $\mathcal{H}$ and let $\{f^{(1)}, f^{(2)}, \cdots, f^{(m)}\}$ be an ON basis for $\mathcal{W}$. Let $u \in \mathcal{H}$. We say that the vector $u' = \sum_{i=1}^{m}(u, f^{(i)})f^{(i)} \in \mathcal{W}$ is the projection of u on $\mathcal{W}$.

Example 1.67 The vectors

$$u^{(1)} = (1, -1, 0), \quad u^{(2)} = (1, 0, 1),$$

form a basis for a two-dimensional subspace V of $\mathbb{R}^3$. Use the Gram–Schmidt process on these vectors to obtain an orthonormal basis for V. **Solution**: Let $f^{(1)}, f^{(2)}$ be the ON basis vectors. Then $||u^{(1)}|| = 2$, so $f^{(1)} = \frac{1}{\sqrt{2}}(1, -1, 0)$. Now $(u^{(2)}, f^{(1)}) = 1/\sqrt{2}$ so $(g^{(2)}, f^{(1)}) = 0$ where $g^{(2)} = u^{(2)} - (u^{(2)}, f^{(1)})f^{(1)}$. Thus

$$g^{(2)} = \left(\frac{1}{2}, \frac{1}{2}, 1\right), \quad ||g^{(2)}||^2 = \frac{3}{2} \longrightarrow f^{(2)} = \sqrt{\frac{2}{3}}\left(\frac{1}{2}, \frac{1}{2}, 1\right),$$

to within multiplication by ± 1.

Theorem 1.68 *If $u \in \mathcal{H}$ there exist unique vectors $u' \in \mathcal{W}$, $u'' \in \mathcal{W}^{\perp}$ such that $u = u' + u''$.*

Proof

1. Existence: Let $\{f^{(1)}, f^{(2)}, \ldots, f^{(m)}\}$ be an ON basis for $\mathcal{W}$, set $u' = \sum_{i=1}^{m}(u, f^{(i)})f^{(i)} \in \mathcal{W}$ and $u'' = u - u'$. Now $(u'', f^{(i)}) = (u, f^{(i)}) - (u, f^{(i)}) = 0$, $1 \leq i \leq m$, so $(u'', v) = 0$ for all $v \in \mathcal{W}$. Thus $u'' \in \mathcal{W}^{\perp}$.
2. Uniqueness: Suppose $u = u' + u'' = v' + v''$ where $u', v' \in \mathcal{W}$, $u'', v'' \in \mathcal{W}^{\perp}$. Then $u' - v' = v'' - u'' \in \mathcal{W} \cap \mathcal{W}^{\perp} \implies (u' - v', u' - v') = 0 = ||u' - v'||^2 \implies u' = v', u'' = v''$. □

Corollary 1.69 *Bessel's Inequality. Let $\{f^{(1)}, \ldots, f^{(m)}\}$ be an ON set in $\mathcal{H}$. If $u \in \mathcal{H}$ then $||u||^2 \geq \sum_{i=1}^{m}|(u, f^{(i)})|^2$.*

Proof Set $W = [f^{(1)}, \ldots, f^{(m)}]$. Then $u = u' + u''$ where $u' \in W$, $u'' \in W^\perp$ and $u' = \sum_{i=1}^{m} (u, f^{(i)}) f^{(i)}$. Therefore $||u||^2 = (u' + u'', u' + u'') = ||u'||^2 + ||u''||^2 \geq ||u'||^2 = (u', u') = \sum_{i=1}^{m} |(u, f^{(i)})|^2$. $\square$

Note that this inequality holds even if m is infinite.

The projection of $u \in \mathcal{H}$ onto the subspace W has invariant meaning, i.e., it is basis independent. Also, it solves an important minimization problem: u' is the vector in W that is closest to u.

Theorem 1.70 $\min_{v \in W} ||u-v|| = ||u-u'||$ *and the minimum is achieved if and only if* $v = u'$.

Proof Let $v \in W$ and let $\{f^{(1)}, f^{(2)}, \ldots, f^{(m)}\}$ be an ON basis for W. Then $v = \sum_{i=1}^{m} \alpha_i f^{(i)}$ for $\alpha_i = (v, f^{(i)})$ and $||u - v|| = ||u - \sum_{i=1}^{m} \alpha_i f^{(i)}||^2 = (u - \sum_{i=1}^{m} \alpha_i f^{(i)}, u - \sum_{i=1}^{m} \alpha_i f^{(i)}) = ||u||^2 - \sum_{i=1}^{m} \bar{\alpha}_i (u, f^{(i)}) - \sum_{i=1}^{m} \alpha_i (f^{(i)}, u) + \sum_{i=1}^{m} |\alpha_i|^2 = ||u - \sum_{i=1}^{m} (u, f^{(i)}) f^{(i)}||^2 + \sum_{i=1}^{m} |(u, f^{(i)}) - \alpha_i|^2 \geq ||u - u'||^2$. Equality is obtained if and only if $\alpha_i = (u, f^{(i)})$, for $1 \leq i \leq m$. $\square$

To indicate the uniqueness of the decomposition of each vector in $\mathcal{H}$ into the sum of a vector in W and $W^\perp$, we write $\mathcal{H} = W \oplus W^\perp$, a direct sum decomposition.

1.6.3 Orthonormal systems in an infinite-dimensional separable Hilbert space

Let $\mathcal{H}$ be a separable Hilbert space. (We have in mind spaces such as ℓ_2 and $L_2[0, 2\pi]$.)

The idea of an orthogonal projection extends to infinite-dimensional inner product spaces, but there is a problem. If the infinite-dimensional subspace W of $\mathcal{H}$ isn't closed, the concept may not make sense.

For example, let $\mathcal{H} = \ell_2$ and let W be the subspace consisting of elements of the form $(\ldots, \alpha_{-1}, \alpha_0, \alpha_1, \ldots)$ such that $\alpha_i = 0$ for $i = 1, 0, -1, -2, \ldots$ and there are a *finite* number of nonzero components α_i for $i \geq 2$. Choose $u = (\ldots, \beta_{-1}, \beta_0, \beta_1, \ldots)$ such that $\beta_i = 0$ for $i = 0, -1, -2, \cdots$ and $\beta_n = 1/n$ for $n = 1, 2, \ldots$ Then $u \in \ell_2$ but the projection of u on W is undefined. If W is closed, however, i.e., if every Cauchy sequence $\{u^{(n)}\}$ in W converges to an element of W, the problem disappears.

Theorem 1.71 *Let W be a closed subspace of the inner product space $\mathcal{H}$ and let $u \in H$. Set $d = \inf_{v \in W} ||u - v||$. Then there exists a unique $\bar{u} \in W$ such that $||u - \bar{u}|| = d$, ($\bar{u}$ is called the projection of u on W). Furthermore $u - \bar{u} \perp W$ and this characterizes $\bar{u}$.*

Proof Clearly there exists a sequence $\{v^{(n)}\} \in \mathcal{W}$ such that $||u-v^{(n)}|| = d_n$ with $\lim_{n\to\infty} d_n = d$. We will show that $\{v^{(n)}\}$ is Cauchy. To do this it will be convenient to make use of the parallelogram law for inner product spaces:

$$||x + y||^2 + ||x - y||^2 = 2||x||^2 + 2||y||^2,$$

see Exercise 1.23. Set $x = u - v^{(n)}$, $y = u - v^{(m)}$ in the parallelogram law and rewrite it to obtain the identity

$$||v^{(m)} - v^{(n)}||^2 = 2||u - v^{(m)}||^2 + 2||u - v^{(n)}||^2 - 4||u - \frac{v^{(m)} + v^{(n)}}{2}||^2$$
$$\leq 2d_m^2 + 2d_n^2 - 4d^2,$$

since by definition of d we must have $||u - (v^{(m)} + v^{(n)})/2|| \geq d$. Thus in the limit

$$0 \leq ||v^{(m)} - v_{(n)}||^2 \leq 2d_m^2 + 2d_n^2 - 4d^2 \longrightarrow 0$$

as $m, n \longrightarrow +\infty$. This means that $||v^{(m)} - v^{(n)}|| \longrightarrow 0$ as m, n go to ∞, so the sequence $\{v^{(n)}\}$ is Cauchy.

Since $\mathcal{W}$ is closed, there exists $\bar{u} \in \mathcal{W}$ such that $\lim_{n\to\infty} v^{(n)} = \bar{u}$. Also, $||u - \bar{u}|| = ||u - \lim_{n\to\infty} v^{(n)}|| = \lim_{n\to\infty} ||u - v^{(n)}|| = \lim_{n\to\infty} d_n = d$. Furthermore, for any $v \in \mathcal{W}$, $(u - \bar{u}, v) = \lim_{n\to\infty}(u - v^{(n)}, v) = 0 \implies u - \bar{u} \perp \mathcal{W}$.

Conversely, if $u - \bar{u} \perp \mathcal{W}$ and $v \in \mathcal{W}$ then $||u - v||^2 = ||(u - \bar{u}) + (\bar{u} - v)||^2 = ||u - \bar{u}||^2 + ||\bar{u} - v||^2 = d^2 + ||\bar{u} - v||^2$. Therefore $||u - v||^2 \geq d^2$ and $= d^2$ if and only if $\bar{u} = v$. Thus $\bar{u}$ is unique. $\square$

Corollary 1.72 *Let $\mathcal{W}$ be a closed subspace of the Hilbert space $\mathcal{H}$ and let $u \in \mathcal{H}$. Then there exist unique vectors $\bar{u} \in \mathcal{W}$, $\bar{v} \in \mathcal{W}^\perp$, such that $u = \bar{u} + \bar{v}$. We write this direct sum decomposition as $\mathcal{H} = \mathcal{W} \oplus \mathcal{W}^\perp$.*

Corollary 1.73 *A subspace $\mathcal{M} \subseteq \mathcal{H}$ is dense in $\mathcal{H}$ if and only if $u \perp \mathcal{M}$ for $u \in \mathcal{H}$ implies $u = \Theta$.*

Proof $\mathcal{M}$ dense in $\mathcal{H} \implies \overline{\mathcal{M}} = \mathcal{H}$. Suppose $u \perp \mathcal{M}$. Then there exists a sequence $\{u^{(n)}\}$ in $\mathcal{M}$ such that $\lim_{n\to\infty} u^{(n)} = u$ and $(u, u^{(n)}) = 0$ for all n. Thus $(u, u) = \lim_{n\to\infty}(u, u^{(n)}) = 0 \implies u = \Theta$.

Conversely, suppose $u \perp \mathcal{M} \implies u = \Theta$. If $\mathcal{M}$ isn't dense in $\mathcal{H}$ then $\overline{\mathcal{M}} \neq \mathcal{H} \implies$ there is a $u \in \mathcal{H}$ such that $u \notin \overline{\mathcal{M}}$. Therefore there exists a $\bar{u} \in \overline{\mathcal{M}}$ such that $v = u - \bar{u} \neq \Theta$ belongs to $\overline{\mathcal{M}}^\perp \implies v \perp \mathcal{M}$. Impossible! $\square$

Now we are ready to study ON systems on an infinite-dimensional (but separable) Hilbert space $\mathcal{H}$. If $\{v^{(n)}\}$ is a sequence in $\mathcal{H}$, we say that

$\sum_{n=1}^{\infty} v^{(n)} = v \in \mathcal{H}$ if the partial sums $\sum_{n=1}^{k} v^{(n)} = u^{(k)}$ form a Cauchy sequence and $\lim_{k \to \infty} u^{(k)} = v$. This is called *convergence in the mean* or *convergence in the norm*, rather than pointwise convergence of functions. (For Hilbert spaces of functions, such as $L_2[0, 2\pi]$, we need to distinguish this mean convergence from pointwise or uniform convergence.)

The following results are just slight extensions of results that we have proved for ON sets in finite-dimensional inner-product spaces. The sequence $u^{(1)}, u^{(2)}, \ldots \in \mathcal{H}$ is *orthonormal* (ON) if $(u^{(i)}, u^{(j)}) = \delta_{ij}$. (Note that an ON sequence need not be a basis for $\mathcal{H}$.) Given $u \in \mathcal{H}$, the numbers $\alpha_j = (u, u^{(j)})$ are the *Fourier coefficients* of u with respect to this sequence.

Lemma 1.74 $u = \sum_{n=1}^{\infty} \alpha_n u^{(n)} \implies \alpha_n = (u, u^{(n)})$.

Given a fixed ON system $\{u^{(n)}\}$, a positive integer N and $u \in \mathcal{H}$ the projection theorem tells us that we can minimize the "error" $||u - \sum_{n=1}^{N} \alpha_n u^{(n)}||$ of approximating u by choosing $\alpha_n = (u, u^{(n)})$, i.e., as the Fourier coefficients. Moreover,

Corollary 1.75 $\sum_{n=1}^{N} |(u, u^{(n)})|^2 \leq ||u||^2$ *for any N.*

Corollary 1.76 $\sum_{n=1}^{\infty} |(u, u^{(n)})|^2 \leq ||u||^2$, *Bessel's inequality.*

Theorem 1.77 *Given the ON system $\{u^{(n)}\} \in \mathcal{H}$, then $\sum_{n=1}^{\infty} \beta_n u^{(n)}$ converges in the norm if and only if $\sum_{n=1}^{\infty} |\beta_n|^2 < \infty$.*

Proof Let $v^{(k)} = \sum_{n=1}^{k} \beta_n u^{(n)}$. $\sum_{n=1}^{\infty} \beta_n u^{(n)}$ converges if and only if $\{v^{(k)}\}$ is Cauchy in $\mathcal{H}$. For $k \geq \ell$,

$$||v^{(k)} - v^{(\ell)}||^2 = || \sum_{n=\ell+1}^{k} \beta_n u^{(n)} ||^2 = \sum_{n=\ell+1}^{k} |\beta_n|^2. \qquad (1.14)$$

Set $t_k = \sum_{n=1}^{k} |\beta_n|^2$. Then $(1.14) \implies \{v^{(k)}\}$ is Cauchy in $\mathcal{H}$ if and only if $\{t_k\}$ is a Cauchy sequence of real numbers, if and only if $\sum_{n=1}^{\infty} |\beta_n|^2 < \infty$. $\square$

Definition 1.78 A subset $\mathcal{K}$ of $\mathcal{H}$ is *complete* if for every $u \in \mathcal{H}$ and $\epsilon > 0$ there are elements $u^{(1)}, u^{(2)}, \ldots, u^{(N)} \in \mathcal{K}$ and $\alpha_1, \ldots, \alpha_N \in C$ such that $||u - \sum_{n=1}^{N} \alpha_n u^{(n)}|| < \epsilon$, i.e., if the subspace $\hat{\mathcal{K}}$ formed by taking all finite linear combinations of elements of $\mathcal{K}$ is dense in $\mathcal{H}$.

Theorem 1.79 *The following are equivalent for any ON sequence $\{u^{(n)}\}$ in $\mathcal{H}$.*

1. $\{u^{(n)}\}$ *is complete* ($\{u^{(n)}\}$ *is an ON basis for* $\mathcal{H}$).

2. *Every* $u \in \mathcal{H}$ *can be written uniquely in the form* $u = \sum_{n=1}^{\infty} \alpha_n u^{(n)}$, $\alpha_n = (u, u^{(n)})$.

3. *For every* $u \in \mathcal{H}$, $||u||^2 = \sum_{n=1}^{\infty} |(u, u^{(n)})|^2$, *Parseval's equality*.

4. *If* $u \perp \{u^{(n)}\}$ *then* $u = \Theta$.

Proof

1. $1 \Longrightarrow 2$ $\{u^{(n)}\}$ complete $\Longrightarrow$ given $u \in \mathcal{H}$ and $\epsilon > 0$ there is an integer N and constants $\{\alpha_n\}$ such that $||u - \sum_{n=1}^{N} \alpha_n u^{(n)}|| < \epsilon \Longrightarrow ||u - \sum_{n=1}^{k} \alpha_n u^{(n)}|| < \epsilon$ for all $k \geq N$. Clearly $\sum_{n=1}^{\infty} (u, u^{(n)}) u^{(n)} \in \mathcal{H}$ since $\sum_{n=1}^{\infty} |(u, u^{(n)})|^2 \leq ||u||^2 < \infty$. Therefore $u = \sum_{n=1}^{\infty} (u, u^{(n)}) u^{(n)}$. Uniqueness obvious.

2. $2 \Longrightarrow 3$ Suppose $u = \sum_{n=1}^{\infty} \alpha_n u^{(n)}$, $\alpha_n = (u, u^{(n)})$. Then, $||u - \sum_{n=1}^{k} \alpha_n u^{(n)}||^2 = ||u||^2 - \sum_{n=1}^{k} |(u, u^{(n)})|^2 \to 0$ as $k \to \infty$. Hence $||u||^2 = \sum_{n=1}^{\infty} |(u, u^{(n)})|^2$.

3. $3 \Longrightarrow 4$ Suppose $u \perp \{u^{(n)}\}$. Then $||u||^2 = \sum_{n=1}^{\infty} |(u, u^{(n)})|^2 = 0$ so $u = \Theta$.

4. $4 \Longrightarrow 1$ Let $\tilde{\mathcal{M}}$ be the dense subspace of $\mathcal{H}$ formed from all finite linear combinations of $u^{(1)}, u^{(2)}, \ldots$. Then given $v \in \mathcal{H}$ and $\epsilon > 0$ there exists a $\sum_{n=1}^{N} \alpha_n u^{(n)} \in \tilde{\mathcal{M}}$ such that $||v - \sum_{n=1}^{N} \alpha_n u^{(n)}|| < \epsilon$. $\square$

1.7 Linear operators and matrices, LS approximations

We assume that the reader has a basic familiarity with linear algebra and move rapidly to derive the basic results we need for applications. Let V, W be vector spaces over $\mathbb{F}$ (either the real or the complex field).

Definition 1.80 A *linear transformation* (or linear operator) from V to W is a function $\mathbf{T} : V \to W$, defined for all $v \in V$ that satisfies $\mathbf{T}(\alpha u + \beta v) = \alpha \mathbf{T} u + \beta \mathbf{T} v$ for all $u, v \in V$, $\alpha, \beta \in \mathbb{F}$. Here, the set $R(\mathbf{T}) = \{\mathbf{T} u : u \in V\}$ is called the *range* of $\mathbf{T}$. The dimension r of $R(\mathbf{T})$ is called the rank. The set $N(\mathbf{T}) = \{u \in V : \mathbf{T} u = \Theta\}$ is called the *null space* or *kernel* of $\mathbf{T}$. The dimension k of $N(\mathbf{T})$ is called the *nullity*.

Lemma 1.81 $R(\mathbf{T})$ *is a subspace of* W *and* $N(\mathbf{T})$ *is a subspace of* V.

Proof Let $w = \mathbf{T}u, z = \mathbf{T}v \in R(\mathbf{T})$ and let $\alpha, \beta \in \mathbb{F}$. Then $\alpha w + \beta z = \mathbf{T}(\alpha u + \beta v) \in R(\mathbf{T})$. Similarly, if $u, v \in N(\mathbf{T})$ and $\alpha, \beta \in \mathbb{F}$ then $\mathbf{T}(\alpha u + \beta v) = \alpha\Theta + \beta\Theta = \Theta$, so $\alpha u + \beta v \in N(\mathbf{T})$. $\square$

Note that an operator $\mathbf{T} : V \to W$ is 1-1 or injective if and only if its null space is zero dimensional. If $\mathbf{T}$ is 1-1 and if $R(\mathbf{T}) = W$ then $\mathbf{T}$ is invertible and for any $w \in W$ the equation $\mathbf{T}v = w$ always has a unique solution $v \in V$. The linear operator $\mathbf{T}^{-1} : W \to V$ defined by $\mathbf{T}^{-1}w = v$ is called the inverse operator to $\mathbf{T}$.

If V is n-dimensional with basis $v^{(1)}, \ldots, v^{(n)}$ and W is m-dimensional with basis $w^{(1)}, \ldots, w^{(m)}$ then $\mathbf{T}$ is completely determined by its matrix representation $T = (T_{jk})$ with respect to these two bases:

$$\mathbf{T}v^{(k)} = \sum_{j=1}^{m} T_{jk}w^{(j)}, \qquad k = 1, 2, \ldots, n.$$

If $v \in V$ and $v = \sum_{k=1}^{n} \alpha_k v^{(k)}$ then the action $\mathbf{T}v = w$ is given by

$$\mathbf{T}v = \mathbf{T}\left(\sum_{k=1}^{n} \alpha_k v^{(k)}\right) = \sum_{k=1}^{n} \alpha_k \mathbf{T}v^{(k)} = \sum_{j=1}^{m}\sum_{k=1}^{n} (T_{jk}\alpha_k)w^{(j)} = \sum_{j=1}^{m} \beta_j w^{(j)} = w$$

Thus the coefficients β_j of w are given by $\beta_j = \sum_{k=1}^{n} T_{jk}\alpha_k$, $j = 1, \ldots, m$. In matrix notation, one writes this as

$$\begin{pmatrix} T_{11} & \cdots & T_{1n} \\ \vdots & \ddots & \vdots \\ T_{m1} & \cdots & T_{mn} \end{pmatrix} \begin{pmatrix} \alpha_1 \\ \vdots \\ \alpha_n \end{pmatrix} = \begin{pmatrix} \beta_1 \\ \vdots \\ \beta_m \end{pmatrix},$$

or

$$Ta = b.$$

The matrix $T = (T_{jk})$ has m rows and n columns, i.e., it is $m \times n$, whereas the vector $a = (\alpha_k)$ is $n \times 1$ and the vector $b = (\beta_j)$ is $m \times 1$. If V and W are Hilbert spaces with ON bases, we shall sometimes represent operators $\mathbf{T}$ by matrices with an infinite number of rows and columns.

Let V, W, X be vector spaces over F, and $\mathbf{T}, \mathbf{U}$ be linear operators $\mathbf{T} : V \to W$, $\mathbf{U} : W \to X$. The *product* $\mathbf{UT}$ of these two operators is the composition $\mathbf{UT} : V \to X$ defined by $\mathbf{UT}v = \mathbf{U}(\mathbf{T}v)$ for all $v \in V$. The *sum* $\mathbf{T} + \mathbf{T}'$ of two linear operators $\mathbf{T} : V \to W$, $\mathbf{T}' : V \to W$, is defined by $(\mathbf{T} + \mathbf{T}')v = \mathbf{T}v + \mathbf{T}'v$. If $\alpha \in F$ the *scalar multiple* of $\mathbf{T}$ is the linear operator $\alpha\mathbf{T}$ defined by $(\alpha\mathbf{T})v = \alpha(\mathbf{T}v)$.

Suppose V is n-dimensional with basis $v^{(1)}, \ldots, v^{(n)}$, W is m-dimensional with basis $w^{(1)}, \ldots, w^{(m)}$ and X is p-dimensional with basis $x^{(1)}, \ldots, x^{(p)}$.

Then $\mathbf{T}$ has matrix representation $T = (T_{jk})$, $\mathbf{U}$ has matrix representation $U = (U_{\ell j})$,

$$\mathbf{U}w^{(j)} = \sum_{\ell=1}^{p} U_{\ell j}x^{(\ell)}, \quad j = 1, 2, \ldots, m,$$

and $\mathbf{Y} = \mathbf{UT}$ has matrix representation $Y = (Y_{\ell k})$ given by

$$\mathbf{Y}v^{(k)} = \mathbf{UT}v^{(k)} = \sum_{\ell=1}^{p} Y_{\ell k}x^{(\ell)}, \qquad k = 1, 2, \ldots, n,$$

A straightforward computation gives $Y_{\ell k} = \sum_{j=1}^{m} U_{\ell j}T_{jk}$, $\ell = 1, \ldots, p$, $k = 1, \ldots, n$. In matrix notation, one writes this as

$$\begin{pmatrix} U_{11} & \cdots & U_{1m} \\ \vdots & \ddots & \vdots \\ U_{p1} & \cdots & U_{pm} \end{pmatrix} \begin{pmatrix} T_{11} & \cdots & T_{1n} \\ \vdots & \ddots & \vdots \\ T_{m1} & \cdots & T_{mn} \end{pmatrix} = \begin{pmatrix} Y_{11} & \cdots & Y_{1n} \\ \vdots & \ddots & \vdots \\ Y_{p1} & \cdots & Y_{pn} \end{pmatrix},$$

or

$$UT = Y.$$

Here, U is $p \times m$, T is $m \times n$ and Y is $p \times n$. Similarly, if $\mathbf{T} : V \to W$, $\mathbf{T'} : V \to W$, have $m \times n$ matrix representations T, T' respectively, then $\mathbf{T} + \mathbf{T'}$ has $m \times n$ matrix representation $T + T'$, the sum of the two matrices. The scalar multiple of $\alpha\mathbf{T}$ has matrix representation αT where $(\alpha T)_{jk} = \alpha T_{jk}$.

Now let us return to our operator $\mathbf{T} : V \to W$ and suppose that both V and W are complex inner product spaces, with inner products $(\cdot, \cdot)_V, (\cdot, \cdot)_W$, respectively. Then $\mathbf{T}$ induces a linear operator $\mathbf{T}^* : W \to V$ and defined by

$$(\mathbf{T}v, w)_W = (v, \mathbf{T}^*w)_V, \qquad v \in V, w \in W.$$

To show that $\mathbf{T}^*$ exists, we will compute its matrix T^*. Suppose that $v^{(1)}, \ldots, v^{(n)}$ is an ON basis for V and $w^{(1)}, \ldots, w^{(m)}$ is an ON basis for W. Then, for $k = 1, \ldots, n, j = 1, \ldots, m$, we have

$$T_{jk} = (\mathbf{T}v^{(k)}, w^{(j)})_W = (v^{(k)}, \mathbf{T}^*w^{(j)})_V = \bar{T}^*{}_{kj}.$$

Thus the operator $\mathbf{T}^*$ (the *adjoint operator* to $\mathbf{T}$) has the adjoint matrix to T: $T^*_{kj} = \bar{T}_{jk}$. In matrix notation this is written $T^* = \bar{T}^{\mathrm{tr}}$ where the ^{tr} stands for the matrix transpose (interchange of rows and columns). For a real inner product space the complex conjugate is dropped and the adjoint matrix is just the transpose.

There are fundamental relations between the linear transformations $\mathbf{T} : V \to W$ and $\mathbf{T}^* : W \to V$ that apply directly to the solution of systems of $m \times n$ linear equations $Ta = b$. Recall that $N(\mathbf{T})$ and $R(\mathbf{T}^*)$ are subspaces of V with dimensions k, r^*, respectively. Similarly, $N(\mathbf{T}^*)$ and $R(\mathbf{T})$ are subspaces of W with dimensions k^*, r, respectively.

Theorem 1.82 *The Fundamental Theorem of Linear Algebra. Let* $\mathbf{T}$: $V \to W$ *as defined above. Then, [94],*

1. $n = k + r, \quad m = k^* + r^*,$
2. $V = N(\mathbf{T}) \oplus R(\mathbf{T}^*), \quad$ i.e., $\quad R(\mathbf{T}^*) = N(\mathbf{T})^{\perp},$
3. $W = N(\mathbf{T}^*) \oplus R(\mathbf{T}), \quad$ i.e., $\quad N(\mathbf{T}^*) = R(\mathbf{T})^{\perp},$
4. $r = r^*.$

Proof Let $\{v^{(1)}, \ldots, v^{(k)}\}$ be an ON basis for $N(\mathbf{T})$ in V. Using the Gram–Schmidt process we can supplement this set with vectors $v^{(k+1)}, \ldots, v^{(n)}$ in V such that $\{v^{(1)}, \ldots, v^{(n)}\}$ is an ON basis for V. Now set $w^{(j)} = \mathbf{T}v^{(j)}$, $j = k + 1, \ldots, n$. We claim that $\{w^{(k+1)}, \ldots, w^{(n)}\}$ is a basis for $R(\mathbf{T})$ in W. For any constants α_j we have

$$\mathbf{T}(\alpha_1 v^{(1)} + \cdots + \alpha_n v^{(n)}) = \sum_{i=k+1}^{n} \alpha_i \mathbf{T}v^{(i)} = \sum_{i=k+1}^{n} \alpha_i w^{(i)},$$

so the $\{w^{(j)}\}$ form a spanning set for $R(\mathbf{T})$. Further the set $\{w^{(j)}\}$ is linearly independent, since if $\sum_{j=k+1}^{n} \alpha_j w^{(j)} = \Theta$ then

$$\mathbf{T}(\sum_{j=k+1}^{n} \alpha_j v^{(j)}) = \sum_{j=k+1}^{n} \alpha_j \mathbf{T}v^{(j)} = \sum_{j=k+1}^{n} \alpha_j w^{(j)} = \Theta,$$

so $\sum_{j=k+1}^{n} \alpha_j v^{(j)} \in N(\mathbf{T})$, which is possible only if $\alpha_{k+1} = \cdots = \alpha_n = 0$. Thus dim $R(\mathbf{T}) = n - k = r$. The dual argument, using the fact that $\mathbf{T}^{**} = \mathbf{T}$, yields dim $R(\mathbf{T}^*) = m - k^* = r^*$.

For the remaining parts of the proof we first apply the projection theorem to V and the subspace $N(\mathbf{T})$: $V = N(\mathbf{T}) \oplus N(\mathbf{T})^{\perp}$. We will show that $R(\mathbf{T}^*) = N(\mathbf{T})^{\perp}$. Let $\tilde{v} \in R(\mathbf{T}^*) \subseteq V$ and $\hat{v} \in N(\mathbf{T}) \subseteq V$. Then there is a $\tilde{w} \in W$ such that $\tilde{v} = \mathbf{T}^* \tilde{w}$ and we have

$$(\hat{v}, \tilde{v})_V = (\hat{v}, \mathbf{T}^* \tilde{w})_V = (\mathbf{T}\hat{v}, \tilde{w})_W = (\Theta, \tilde{w})_W = 0,$$

so $\tilde{v} \in N(\mathbf{T})^{\perp}$. Thus $R(\mathbf{T}^*) \subseteq N(\mathbf{T})^{\perp}$. This means that $R(\mathbf{T}^*)^{\perp} \supseteq (N(\mathbf{T})^{\perp})^{\perp} = N(\mathbf{T})$. Counting dimensions we have $n - r^* \geq k$. A dual argument applied to the projection $W = N(\mathbf{T}^*) \oplus N(\mathbf{T}^*)^{\perp}$ yields $R(\mathbf{T})^{\perp} \supseteq (N(\mathbf{T}^*)^{\perp})^{\perp} = N(\mathbf{T}^*)$. Counting dimensions we get $m - r \geq k^*$.

Thus we have the inequalities $n \geq k + r^*$, $m \geq k^* + r$, or by adding $m + n \geq (k + k^*) + (r + r^*)$ where strict equality in the last expression holds if and only if there is strict equality in the first two expressions. However, the proof of part 1 shows that indeed $m + n = (k^* + r^*) + (k + r) = (k + k^*) + (r + r^*)$. Since $R(\mathbf{T}^*)^\perp \supseteq N(\mathbf{T})$ and the two spaces have the same dimension, they must be equal. A similar argument gives $R(\mathbf{T})^\perp = N(\mathbf{T}^*)$.

To prove 4) we note from 1) that $n = k + r$ whereas from the proof of 3) $n = k + r^*$. Thus $r = r^*$. □

Note from Theorem 1.82 that the rank of $\mathbf{T}$ is equal to the rank of $\mathbf{T}^*$. This result has an important implication for the matrices T, T^* of these operators with respect to bases in V and W. The rank of $\mathbf{T}$ is just the number of linearly independent column vectors of T, i.e., the dimension of the vector space of n-tuples (the column space) spanned by the column vectors of T. The rank of $\mathbf{T}^*$ is the number of linearly independent column vectors of T^*, which is the same as the number of linearly independent row vectors of T, i.e., the dimension of the vector space of m-tuples (the row space) spanned by the row vectors of T. This is true even for matrices with complex entries because taking the complex conjugate does not alter the linear independence of row or column vectors. Note also that any $m \times n$ matrix T can be realized as the matrix of some operator $\mathbf{T} : V \to W$ with respect to an appropriate basis.

Corollary 1.83 *The dimension of the row space of an $m \times n$ matrix T is equal to the dimension of its column space.*

There are some special operators and matrices that we will meet often in this book. Suppose that $v^{(1)}, \ldots, v^{(m)}$ is a basis for V.

1. The identity operator $\mathbf{I} : V \to V$ is defined by $\mathbf{I}v = v$ for all $v \in V$. The matrix of $\mathbf{I}$ is $I = (\delta_{jh})$ where $\delta_{jj} = 1$ and $\delta_{jh} = 0$ if $j \neq h$, $1 \leq j, h \leq m$. If $\mathbf{T}$ is invertible with inverse $\mathbf{T}^{-1}$ then $\mathbf{T}\mathbf{T}^{-1} = \mathbf{T}^{-1}\mathbf{T} = \mathbf{I}$. In this case the matrix T of $\mathbf{T}$ with respect to the basis has rank m and T^{-1} is the matrix of $\mathbf{T}^{-1}$, where $TT^{-1} = T^{-1}T = I$. We say that T is nonsingular with inverse matrix T^{-1},

2. The *zero operator* $\mathbf{Z} : V \to V$ is defined by $\mathbf{Z}v = \Theta$ for all $v \in V$. The $n \times n$ matrix of $\mathbf{Z}$ has all matrix elements 0.

3. An operator $\mathbf{U} : V \to V$ that preserves the inner product, $(\mathbf{U}v, \mathbf{U}u) = (v, u)$ for all $u, v \in V$ is called *unitary*. The matrix U of a unitary operator is characterized by the matrix equation $UU^* = I$.

4. If V is a real inner product space, the operators $\mathbf{O} : V \to V$ that preserve the inner product, $(\mathbf{O}v, \mathbf{O}u) = (v, u)$ for all $u, v \in V$ are called orthogonal. The matrix O of an orthogonal operator is characterized by the matrix equation $OO^{\text{tr}} = I$.

5. An operator $\mathbf{T} : V \to V$ that is equal to its adjoint, i.e., $\mathbf{T} = \mathbf{T}^*$ is called *self-adjoint* if V is a complex vector space and *symmetric* if V is real. The $m \times m$ matrix T of a self-adjoint operator is characterized by the matrix equation $T^* = T$ (the matrix is self-adjoint) for complex spaces and $T = T^{\text{tr}}$ (the matrix is symmetric) for real spaces.

1.7.1 The singular value decomposition

We recall some important results from linear algebra concerning the diagonalization of self-adjoint operators, e.g. [94]. Suppose that $\mathbf{T} : V \to V$ is an operator on a complex inner product space V. We say that a nonzero $v^\lambda \in V$ is an *eigenvector* of $\mathbf{T}$ with eigenvalue $\lambda \in \mathbb{C}$ if $\mathbf{T}v^\lambda = \lambda v^\lambda$. This definition is also valid for real vector spaces if $\lambda \in \mathbb{R}$.

Now suppose $\mathbf{T}$ is self-adjoint. Then its eigenvalues must be real. Indeed, since $\mathbf{T} = \mathbf{T}^*$ we have

$$\lambda ||v^\lambda||^2 = < \mathbf{T}v^\lambda, v^\lambda > = < v^\lambda, \mathbf{T}^* v^\lambda > = \overline{\lambda} ||v^\lambda||^2,$$

so $\overline{\lambda} = \lambda$. If λ_1, λ_2 are distinct eigenvalues for $\mathbf{T}$, then the corresponding eigenvectors are orthogonal. Indeed

$$\lambda_1 < v^{\lambda_1}, v^{\lambda_2} > = < \mathbf{T}v^{\lambda_1}, v^{\lambda_2} > = < v^{\lambda_1}, \mathbf{T}^* v^{\lambda_2} > = \lambda_2 < v^{\lambda_1}, v^{\lambda_2} >,$$

so $(\lambda_1 - \lambda_2) < v^{\lambda_1}, v^{\lambda_2} > = 0$, which implies $< v^{\lambda_1}, v^{\lambda_2} > = 0$. A fundamental result from linear algebra, the Spectral Theorem for self-adjoint operators, states that if $\mathbf{T}$ is self-adjoint then V has an ON basis $\{v^{\lambda_i}\}$ of eigenvectors of $\mathbf{T}$:

$$\mathbf{T}v^{\lambda_j} = \lambda_j v^{\lambda_j}, \quad j = 1, \ldots, m.$$

The matrix of $\mathbf{T}$ with respect to this basis is diagonal: $D = (\delta_{jk}\lambda_j)$. If V is real and $\mathbf{T}$ is symmetric the same result holds.

Now we give the matrix version of this result for self-adjoint operators. Let T be the self-adjoint matrix of $\mathbf{T}$ with respect to the ON basis $\{f^{(j)}\}$ of V: $T_{jk} = (\mathbf{T}f^{(k)}, f^{(j)})$. We expand this basis in terms of the ON basis of eigenvectors: $f^{(k)} = \sum_{h=1}^m U_{hk} v^{\lambda_h}$ for $j = 1, \ldots, m$. Note that the $m \times m$ matrix U is unitary, so that $U^{-1} = U^*$.

$$T_{jk} = (\mathbf{T}f^{(k)}, f^{(j)}) = \sum_{h,\ell} U_{hk}\overline{U_{\ell j}} < \mathbf{T}v^{\lambda_h}, v^{\lambda_\ell} > = \sum_{h,\ell} U_{hk}\overline{U_{\ell j}}\delta_{h\ell}\lambda_h.$$

Thus $T = U^{-1}DU$ or $D = UTU^{-1}$ where $U^{-1} = U^*$ and D is diagonal. In the real case $U = O$ is real orthogonal and we have $T = O^{-1}DO$ where $O^{-1} = O^{\mathrm{tr}}$.

By definition, an operator $\mathbf{T} : V \to W$ cannot have eigenvectors. Similarly it makes no direct sense to talk about eigenvectors and eigenvalues of nonsquare matrices. However, there is a sense in which $\mathbf{T}$ and its matrix T with respect to a pair of ON bases can be diagonalized. This is the Singular Value Decomposition (SVD) that has proven extremely useful for applications. Let $\mathbf{T} : V \to W$ be a linear operator of rank r from the inner product space V to the inner product space W, with adjoint operator $\mathbf{T}^* : W \to V$.

Theorem 1.84 *(Singular Value Decomposition) There is an ON basis* $\{v^{(1)}, \ldots, v^{(n)}\}$ *for V and an ON basis* $\{w^{(1)}, \ldots, w^{(m)}\}$ *for W such that*

$$\mathbf{T}v^{(\ell)} = \sigma_\ell w^{(\ell)}, \qquad \mathbf{T}^* w^{(\ell)} = \sigma_\ell v^{(\ell)}, \qquad \ell = 1, \ldots, r, \qquad (1.15)$$

$$\mathbf{T}v^{(h)} = \Theta, \ h = r+1, \ldots, n, \qquad \mathbf{T}^* w^{(j)} = \Theta, \ j = r+1, \ldots, m.$$

Here

$$\sigma_1 \geq \sigma_2 \geq \cdots \geq \sigma_r > 0$$

are the singular values of $\mathbf{T}$.

Proof Recall from Theorem 1.82 that $V = R(\mathbf{T}^*) \oplus N(\mathbf{T})$ and $W = R(\mathbf{T}) \oplus N(\mathbf{T}^*)$, where $r = \dim R(\mathbf{T}^*) = \dim R(\mathbf{T})$, $k = \dim N(\mathbf{T})$, $k^* = \dim N(\mathbf{T}^*)$ and $k + r = n$, $k^* + r = m$. Note that the operator $\mathbf{T}_{R(\mathbf{T}^*)} : R(\mathbf{T}^*) \to R(\mathbf{T})$, maps the subspace $R(\mathbf{T}^*)$ of V 1-1 onto the subspace $R(\mathbf{T})$ of W. Similarly, the operator $\mathbf{T}^*_{R(\mathbf{T})} : R(\mathbf{T}) \to R(\mathbf{T}^*)$, maps the subspace $R(\mathbf{T})$ of W 1-1 onto the subspace $R(\mathbf{T}^*)$ of V. Further, the compositions

$$(\mathbf{T}^*\mathbf{T})_{R(\mathbf{T}^*)} : R(\mathbf{T}^*) \to R(\mathbf{T}^*), \quad (\mathbf{T}\mathbf{T}^*)_{R(\mathbf{T})} : R(\mathbf{T}) \to R(\mathbf{T}),$$

are 1-1 maps of $R(\mathbf{T}^*)$ and $R(\mathbf{T})$, respectively, onto themselves. Both of these operators are self-adjoint. Indeed if $u^{(1)}, u^{(2)} \in R(\mathbf{T}^*)$ then

$$< \mathbf{T}^*\mathbf{T}u^{(1)}, u^{(2)} >_V = < \mathbf{T}u^{(1)}, \mathbf{T}u^{(2)} >_V = < u^{(1)}, \mathbf{T}^*\mathbf{T}u^{(2)} >_V,$$

with an analogous demonstration for $\mathbf{T}\mathbf{T}^*$.

It follows from the Spectral Theorem for self-adjoint operators that there is an ON basis $\{v^{(1)}, \ldots, v^{(r)}\}$ for $R(\mathbf{T}^*)$ consisting of eigenvectors of $\mathbf{T}^*\mathbf{T}$:

$$\mathbf{T}^*\mathbf{T}v^{(j)} = \lambda_j v^{(j)}, \quad j = 1, \ldots, r.$$

Moreover, the eigenvalues λ_j are strictly positive. Indeed,

$$0 < ||\mathbf{T}v^{(j)}||_V^2 = < \mathbf{T}v^{(j)}, \mathbf{T}v^{(j)} >_V = < v^{(j)}, \mathbf{T}^*\mathbf{T}v^{(j)} >_V$$

$$= \lambda_j ||v^{(j)}||_V^2 = \lambda_j.$$

Thus we can set $\lambda_j = \sigma_j^2$ where $\sigma_j > 0$. We reorder the labels on our eigenvectors so that $\sigma_1 \geq \sigma_2 \geq \cdots \geq \sigma_r$. Note that

$$(\mathbf{T}\mathbf{T}^*)\mathbf{T}v^{(j)} = \mathbf{T}(\mathbf{T}^*\mathbf{T}v^{(j)}) = \sigma^2 \mathbf{T}v^{(j)},$$

so $\mathbf{T}v^{(j)} \in R(\mathbf{T})$ is an eigenvector of $\mathbf{T}\mathbf{T}^*$ with eigenvalue σ_j^2. Now we renormalize these new eigenvectors by defining $w^{(j)} = \sigma_j^{-1}\mathbf{T}v^{(j)}$. Thus,

$$\mathbf{T}v^{(j)} = \sigma_j w^{(j)}, \quad \mathbf{T}^*w^{(j)} = \sigma_j^{-1}\mathbf{T}^*\mathbf{T}v^{(j)} = \sigma_j v^{(j)},$$

and

$$< w^{(j)}, w^{(\ell)} >_W = \frac{1}{\sigma_j \sigma_\ell} < \mathbf{T}v^{(j)}, \mathbf{T}v^{(\ell)} >_W = \frac{1}{\sigma_j \sigma_\ell} < v^{(j)}, \mathbf{T}^*\mathbf{T}v^{(\ell)} >_W$$

$$= \frac{\sigma_\ell}{\sigma_j} < v^{(j)}, v^{(\ell)} >_W = \delta_{j\ell},$$

so $\{w^{(j)}\}$ is an ON basis for $R(\mathbf{T})$.

Next we choose an ON basis $\{v^{(1)}, \ldots, v^{(n)}\}$ for V such that the first r vectors are the constructed ON eigenvector basis for $R(\mathbf{T}^*)$ and the last k an ON basis for $N(\mathbf{T})$. Similarly, we choose an ON basis $\{w^{(1)}, \ldots, w^{(m)}\}$ for W such that the first r vectors are the constructed ON eigenvector basis for $R(\mathbf{T})$ and the last k^* an ON basis for $N(\mathbf{T}^*)$. These vectors satisfy relations (1.15). $\qquad\qquad\square$

For the matrix version of the theorem we consider the $m \times n$ matrix $T = (T_{ts})$ of $\mathbf{T}$ with respect to the ON bases $\{e^{(s)}, 1 \leq s \leq n\}$ for V and $\{f^{(t)}, 1 \leq t \leq m\}$ for W: $T_{ts} = < \mathbf{T}e^{(s)}, f^{(t)} >_W$. We expand these bases in terms of the ON bases from Theorem 1.84: $e^{(s)} = \sum_{h=1}^n A_{hs}v^{(h)}$ for $s = 1, \ldots, n$ and $f^{(t)} = \sum_{q=1}^m B_{qt}w^{(q)}$ for $t = 1, \ldots, m$. Then

$$T_{ts} = < \mathbf{T}e^{(s)}, f^{(t)} >_W = \sum_{h,q=1}^{n,m} A_{hs}\overline{B_{qt}} < \mathbf{T}v^{(h)}, w^{(q)} >_W$$

$$= \sum_{h,q=1}^r A_{hs}\overline{B_{qt}}\sigma_h\delta_{hq}.$$

Thus, $T = B^*SA$ where A is an $n \times n$ unitary matrix and B is an $m \times m$ unitary matrix. The $m \times n$ matrix S takes the form

$$
S = \begin{pmatrix}
\sigma_1 & 0 & \cdots & \cdots & \cdots & 0 \\
0 & \sigma_2 & \cdots & \cdots & \cdots & 0 \\
\vdots & \vdots & \ddots & & \vdots & \vdots \\
0 & 0 & \cdots & \sigma_r & 0 & \cdots \\
\vdots & \vdots & \vdots & & \vdots & \vdots \\
0 & 0 & \cdots & \cdots & \cdots & 0
\end{pmatrix}
$$

Example 1.85 We consider an example where T is a real 3×4 matrix, so B and A will be orthogonal. Here

$$
T = \begin{pmatrix}
-1.7489 & -.3837 & 0.2643 & 1.5590 \\
-3.9454 & -.8777 & 0.5777 & 3.5049 \\
4.9900 & 1.1081 & -0.7404 & -4.4343
\end{pmatrix}.
$$

It is an easy exercise with the software program MATLAB to obtain the singular value decomposition

$$
T = B^{\mathrm{tr}}SA =
$$

$$
\begin{pmatrix}
-0.2654 & 0.8001 & 0.5379 \\
-0.5979 & -0.5743 & 0.5592 \\
0.7564 & -0.1732 & 0.6308
\end{pmatrix}
\begin{pmatrix}
9 & 0 & 0 & 0 \\
0 & 0.01 & 0 & 0 \\
0 & 0 & 0.003 & 0
\end{pmatrix} \times
$$

$$
\begin{pmatrix}
0.7330 & 0.1627 & -0.1084 & -0.6515 \\
0.2275 & 0.5127 & 0.7883 & 0.2529 \\
0.2063 & 0.5996 & -0.6043 & 0.4824 \\
0.6069 & -0.5925 & 0.0408 & 0.5281
\end{pmatrix}.
$$

Note that in this example $\sigma_1 = 9$ is much larger than the other two singular values. If the matrix elements of T are values obtained from an experiment, we would have reason to suspect that the two smallest singular values are effects due to "noise" or experimental error, and that the basic information in the data is contained in the σ_1. If that is the case we could replace S by a new matrix $\tilde{S}$ of singular values $\tilde{\sigma}_1 = 9$, $\tilde{\sigma}_2 = \tilde{\sigma}_3 = 0$ and recompute the matrix

$$
\tilde{T} = B^{\mathrm{tr}}\tilde{S}A = \begin{pmatrix}
-1.7510 & -0.3888 & 0.2589 & 1.5562 \\
-3.9445 & -0.8757 & 0.5833 & 3.5056 \\
4.9900 & 1.1079 & -0.7379 & -4.4348
\end{pmatrix}.
$$

Here $\tilde{T}$ agrees with T to 2 decimal place accuracy and has a much more transparent structure. Whereas 12 matrix elements are needed to store the content of T, only 8 are needed for $\tilde{T}$: the first row of B, the first row of A and σ_1. Thus we have achieved data compression.

The singular value decomposition displays the underlying structure of a matrix. In many practical applications only the largest singular values are retained and the smallest ones are considered as "noise" and set equal to 0. For large matrices that data compression can be considerable. In the extreme case where only one singular value is retained for an $m \times n$ matrix, only $m + n + 1$ numbers are needed to recompute the altered matrix, rather than mn.

1.7.2 Bounded operators on Hilbert spaces

In this section we present a few concepts and results from functional analysis that are needed for the study of wavelets, [75, 93].

An operator $\mathbf{T} \colon \mathcal{H} \to \mathcal{K}$ from the Hilbert space $\mathcal{H}$ to the Hilbert Space $\mathcal{K}$ is said to be *bounded* if it maps the unit ball $||u||_{\mathcal{H}} \leq 1$ to a bounded set in $\mathcal{K}$. This means that there is a finite positive number N such that

$$||\mathbf{T}u||_{\mathcal{K}} \leq N \quad \text{whenever} \quad ||u||_{\mathcal{H}} \leq 1.$$

The *norm* $||\mathbf{T}||$ of a bounded operator is its least bound:

$$||\mathbf{T}|| = \sup_{||u||_{\mathcal{H}} \leq 1} ||\mathbf{T}u||_{\mathcal{K}} = \sup_{||u||_{\mathcal{H}}=1} ||\mathbf{T}u||_{\mathcal{K}}. \qquad (1.16)$$

Lemma 1.86 *Let* $\mathbf{T} : \mathcal{H} \to \mathcal{K}$ *be a bounded operator.*

1. $||\mathbf{T}u||_{\mathcal{K}} \leq ||\mathbf{T}|| \cdot ||u||_{\mathcal{H}}$ *for all* $u \in \mathcal{H}$.
2. *If* $\mathbf{S} : \mathcal{L} \to \mathcal{H}$ *is a bounded operator from the Hilbert space* $\mathcal{L}$ *to* $\mathcal{H}$, *then* $\mathbf{TS} : \mathcal{L} \to \mathcal{K}$ *is a bounded operator with* $||\mathbf{TS}|| \leq ||\mathbf{T}|| \cdot ||\mathbf{S}||$.[6]

Proof

1. The result is obvious for $u = \Theta$. If u is nonzero, then $v = ||u||_{\mathcal{H}}^{-1}u$ has norm 1. Thus $||\mathbf{T}v||_{\mathcal{K}} \leq ||\mathbf{T}||$. The result follows from multiplying both sides of the inequality by $||u||_{\mathcal{H}}$.
2. From part 1, $||\mathbf{TS}w||_{\mathcal{K}} = ||\mathbf{T}(\mathbf{S}w)||_{\mathcal{K}} \leq ||\mathbf{T}|| \cdot ||\mathbf{S}w||_{\mathcal{H}} \leq ||\mathbf{T}|| \cdot ||\mathbf{S}|| \cdot ||w||_{\mathcal{L}}$. Hence $||\mathbf{TS}|| \leq ||\mathbf{T}|| \cdot ||\mathbf{S}||$.

□

A special bounded operator is the bounded linear functional $\mathbf{f} : \mathcal{H} \to \mathbb{C}$, where $\mathbb{C}$ is the one-dimensional vector space of complex numbers (with the absolute value $|\cdot|$ as the norm). Thus $\mathbf{f}(u)$ is a complex number

---

[6] Equality holds for operators which lie in a sigma algebra.

for each $u \in \mathcal{H}$ and $\mathbf{f}(\alpha u + \beta v) = \alpha \mathbf{f}(u) + \beta \mathbf{f}(v)$ for all scalars α, β and $u, v \in \mathcal{H}$. The *norm* of a bounded linear functional is defined in the usual way:

$$||\mathbf{f}|| = \sup_{||u||_{\mathcal{H}} = 1} |\mathbf{f}(u)|. \tag{1.17}$$

For fixed $v \in \mathcal{H}$ the inner product $\mathbf{f}(u) \equiv (u, v)$ is an important example of a bounded linear functional. The linearity is obvious and the functional is bounded since $|\mathbf{f}(u)| = |(u, v)| \leq ||u|| \cdot ||v||$. Indeed it is easy to show that $||\mathbf{f}|| = ||v||$. A very useful fact is that all bounded linear functionals on Hilbert spaces can be represented as inner products. This important result, the Riesz representation theorem, relies on the fact that a Hilbert space is complete. It is an elegant application of the projection theorem.

Theorem 1.87 *(Riesz representation theorem) Let $\mathbf{f}$ be a bounded linear functional on the Hilbert space $\mathcal{H}$. Then there is a vector $v \in \mathcal{H}$ such that $\mathbf{f}(u) = (u, v)$ for all $u \in \mathcal{H}$.*

Proof

- Let $\mathcal{N} = \{w \in \mathcal{H} : \mathbf{f}(w) = 0\}$ be the null space of $\mathbf{f}$. Then $\mathcal{N}$ is a closed linear subspace of $\mathcal{H}$. Indeed if $w^{(1)}, w^{(2)} \in \mathcal{N}$ and $\alpha, \beta \in \mathbb{C}$ we have $\mathbf{f}(\alpha w^{(1)} + \beta w^{(2)}) = \alpha \mathbf{f}(w^{(1)}) + \beta \mathbf{f}(w^{(2)}) = 0$, so $\alpha w^{(1)} + \beta w^{(2)} \in \mathcal{N}$. If $\{w^{(n)}\}$ is a Cauchy sequence of vectors in $\mathcal{N}$, i.e., $\mathbf{f}(w^{(n)}) = 0$, with $w^{(n)} \to w \in \mathcal{H}$ as $n \to \infty$ then

  $$|\mathbf{f}(w)| = |\mathbf{f}(w) - \mathbf{f}(w^{(n)})| = |\mathbf{f}(w - w^{(n)})| \leq ||\mathbf{f}|| \cdot ||w - w^{(n)}|| \to 0$$

  as $n \to \infty$. Thus $\mathbf{f}(w) = 0$ and $w \in \mathcal{N}$, so $\mathcal{N}$ is closed.

- If $\mathbf{f}$ is the zero functional, then the theorem holds with $v = \Theta$, the zero vector. If $\mathbf{f}$ is not zero, then there is a vector $u^0 \in \mathcal{H}$ such that $\mathbf{f}(u^0) = 1$. By the projection theorem we can decompose u^0 uniquely in the form $u^0 = v^0 + w^0$ where $w^0 \in \mathcal{N}$ and $v^0 \perp \mathcal{N}$. Then $1 = \mathbf{f}(u^0) = \mathbf{f}(v^0) + \mathbf{f}(w^0) = \mathbf{f}(v^0)$.

- Every $u \in \mathcal{H}$ can be expressed uniquely in the form $u = \mathbf{f}(u)v_0 + w$ for $w \in \mathcal{N}$. Indeed $\mathbf{f}(u - \mathbf{f}(u)v^0) = \mathbf{f}(u) - \mathbf{f}(u)\mathbf{f}(v^0) = 0$ so $u - \mathbf{f}(u)v^0 \in \mathcal{N}$.

- Let $v = ||v^0||^{-2}v^0$. Then $v \perp \mathcal{N}$ and

  $$(u, v) = (\mathbf{f}(u)v^0 + w, v) = \mathbf{f}(u)(v^0, v) = \mathbf{f}(u)||v^0||^{-2}(v^0, v^0) = \mathbf{f}(u).$$

  $\square$

We may define adjoints of bounded operators on general Hilbert spaces, in analogy with our construction of adjoints of operators on finite-dimensional inner product spaces. We return to our bounded operator $\mathbf{T} : \mathcal{H} \to \mathcal{K}$. For any $v \in \mathcal{K}$ we define the linear functional $\mathbf{f}_v(u) = (\mathbf{T}u, v)_{\mathcal{K}}$ on $\mathcal{H}$. The functional is bounded because for $||u||_{\mathcal{H}} = 1$ we have

$$|\mathbf{f}_v(u)| = |(\mathbf{T}u, v)_{\mathcal{K}}| \leq ||\mathbf{T}u||_{\mathcal{K}} \cdot ||v||_{\mathcal{K}} \leq ||\mathbf{T}|| \cdot ||v||_{\mathcal{K}}.$$

By Theorem 1.87 there is a unique vector $v^* \in \mathcal{H}$ such that

$$\mathbf{f}_v(u) \equiv (\mathbf{T}u, v)_{\mathcal{K}} = (u, v^*)_{\mathcal{H}},$$

for all $u \in \mathcal{H}$. We write this element as $v^* = \mathbf{T}^*v$. Thus $\mathbf{T}$ induces an operator $\mathbf{T}^* : \mathcal{K} \to \mathcal{H}$ and defined uniquely by

$$(\mathbf{T}u, v)_{\mathcal{K}} = (u, \mathbf{T}^*v)_{\mathcal{H}}, \qquad v \in \mathcal{H}, w \in \mathcal{K}.$$

Note that the above calculations allow us to prove *Hölder–Minkowski's* inequalities. See Exercise 1.19, Exercise 1.21 and Theorem 1.21.

Lemma 1.88 1. $\mathbf{T}^*$ *is a linear operator from* $\mathcal{K}$ *to* $\mathcal{H}$.
2. $\mathbf{T}^*$ *is a bounded operator.*
3. $||\mathbf{T}^*||^2 = ||\mathbf{T}||^2 = ||\mathbf{T}\mathbf{T}^*|| = ||\mathbf{T}^*\mathbf{T}||.$

Proof

1. Let $v \in \mathcal{K}$ and $\alpha \in \mathbb{C}$. Then

$$(u, \mathbf{T}^*\alpha v)_{\mathcal{H}} = (\mathbf{T}u, \alpha v)_{\mathcal{K}} = \overline{\alpha}(\mathbf{T}u, v)_{\mathcal{K}} = \overline{\alpha}(u, \mathbf{T}^*v)_{\mathcal{H}}$$

so $\mathbf{T}^*(\alpha v) = \alpha \mathbf{T}^*v$. Now let $v^1, v^2 \in \mathcal{K}$. Then

$$(u, \mathbf{T}^*[v^1 + v^2])_{\mathcal{H}} = (\mathbf{T}u, [v^1 + v^2])_{\mathcal{K}} =$$

$$(\mathbf{T}u, v^1)_{\mathcal{K}} + (\mathbf{T}u, v^2)_{\mathcal{K}} = (u, \mathbf{T}^*v^1 + \mathbf{T}^*v^2)_{\mathcal{H}}$$

so $\mathbf{T}^*(v^1 + v^2) = \mathbf{T}^*v^1 + \mathbf{T}^*v^2$.
2. Set $u = \mathbf{T}^*v$ in the defining equation $(\mathbf{T}u, v)_{\mathcal{K}} = (u, \mathbf{T}^*v)_{\mathcal{H}}$. Then

$$||\mathbf{T}^*v||_{\mathcal{H}}^2 = (\mathbf{T}^*v, \mathbf{T}^*v)_{\mathcal{H}} =$$

$$(\mathbf{T}\mathbf{T}^*v, v)_{\mathcal{K}} \leq ||\mathbf{T}\mathbf{T}^*v||_{\mathcal{K}}||v||_{\mathcal{K}} \leq ||\mathbf{T}|| \cdot ||\mathbf{T}^*v||_{\mathcal{H}}||v||_{\mathcal{K}}.$$

Canceling the common factor $||\mathbf{T}^*v||_{\mathcal{H}}$ from the far left and far right-hand sides of these inequalities, we obtain

$$||\mathbf{T}^*v||_{\mathcal{H}} \leq ||\mathbf{T}|| \cdot ||v||_{\mathcal{K}},$$

so $\mathbf{T}^*$ is bounded.

3. From the last inequality of the proof of 2 we have $||\mathbf{T}^*|| \leq ||\mathbf{T}||$. However, if we set $v = \mathbf{T}u$ in the defining equation $(\mathbf{T}u, v)_{\mathcal{K}} = (u, \mathbf{T}^*v)_{\mathcal{H}}$, then we obtain an analogous inequality

$$||\mathbf{T}u||_{\mathcal{K}} \leq ||\mathbf{T}^*|| \cdot ||u||_{\mathcal{H}}.$$

This implies $||\mathbf{T}|| \leq ||\mathbf{T}^*||$. Thus $||\mathbf{T}|| = ||\mathbf{T}^*||$. From the proof of part 2 we have

$$||\mathbf{T}^*v||_{\mathcal{H}}^2 = (\mathbf{T}\mathbf{T}^*v, v)_{\mathcal{K}}. \tag{1.18}$$

Applying the Cauchy–Schwarz inequality to the right-hand side of this identity we have

$$||\mathbf{T}^*v||_{\mathcal{H}}^2 \leq ||\mathbf{T}\mathbf{T}^*v||_{\mathcal{K}} ||v||_{\mathcal{K}} \leq ||\mathbf{T}\mathbf{T}^*|| \cdot ||v||_{\mathcal{K}}^2,$$

so $||\mathbf{T}^*||^2 \leq ||\mathbf{T}\mathbf{T}^*||$. But from Lemma 1.86 we have $||\mathbf{T}\mathbf{T}^*|| \leq ||\mathbf{T}|| \cdot ||\mathbf{T}^*||$, so

$$||\mathbf{T}^*||^2 \leq ||\mathbf{T}\mathbf{T}^*|| \leq ||\mathbf{T}|| \cdot ||\mathbf{T}^*|| = ||\mathbf{T}^*||^2.$$

An analogous proof, switching the roles of u and v, yields

$$||\mathbf{T}||^2 \leq ||\mathbf{T}^*\mathbf{T}|| \leq ||\mathbf{T}|| \cdot ||\mathbf{T}^*|| = ||\mathbf{T}||^2.$$

$\square$

1.7.3 Least squares approximations

Most physical systems seek stability. For example in a mechanical system, equilibrium configurations minimize the total potential energy of the system; in electrical circuits, the current adjusts itself to minimize power. In optics and relativity, light rays follow the paths of minimal distance, the geodesics on curved space-time. In random matrix theory, approximation theory and harmonic analysis, suitably scaled zeros of extremal polynomials, eigenvalues of large random ensembles, interpolation points and points of good quadrature often converge weakly to minimizers of minimal energy problems. Not surprisingly, in data analysis, a common way of fitting a function to a prescribed set of data points is to first minimize the least squares error which serves to quantify the overall deviation between the data and sampled function values. Thus we are given a system of equations to solve of the form $Ax = b$ or

$$\begin{pmatrix} A_{11} & \cdots & A_{1n} \\ \vdots & \ddots & \vdots \\ A_{m1} & \cdots & A_{mn} \end{pmatrix} \begin{pmatrix} x_1 \\ \vdots \\ x_n \end{pmatrix} = \begin{pmatrix} b_1 \\ \vdots \\ b_m \end{pmatrix}. \tag{1.19}$$

Here $b = \{b_1, \ldots, b_m\}$ are m measured quantities, the $m \times n$ matrix $A = (A_{jk})$ is known, and we have to compute the n quantities $x = \{x_1, \ldots, x_n\}$. Since b is measured experimentally, there may be errors in these quantities. This will induce errors in the calculated vector x. Indeed for some measured values of b there may be no solution x.

Example 1.89 Consider the 3×2 system

$$\begin{pmatrix} 3 & 1 \\ 1 & 1 \\ 1 & 2 \end{pmatrix} \begin{pmatrix} x_1 \\ x_2 \end{pmatrix} = \begin{pmatrix} 0 \\ 2 \\ b_3 \end{pmatrix}.$$

If $b_3 = 5$ then this system has the unique solution $x_1 = -1, x_2 = 3$. However, if $b_3 = 5 + \epsilon$ for ϵ small but nonzero, then there is no solution!

We want to guarantee an (approximate) solution of (1.19) for all vectors b and matrices A. Let's embed our problem into the inner product spaces V and W above. That is A is the matrix of the operator $\mathbf{A} : V \to W$, b is the component vector of a given $w \in W$ (with respect to the $\{w^{(j)}\}$ basis) and x is the component vector of $v \in V$ (with respect to the $\{v^{(k)}\}$ basis), which is to be computed. Now the original equation $Ax = b$ becomes $\mathbf{A}v = w$.

Let us try to find an approximate solution v of the equation $\mathbf{A}v = w$ such that the norm of the error $||w - \mathbf{A}v||_W$ is minimized. If the original problem has an exact solution then the error will be zero; otherwise we will find a solution v^0 with minimum (least squares) error. The square of the error will be

$$\epsilon^2 = \min_{v \in V} ||w - \mathbf{A}v||_W^2 = ||w - \mathbf{A}v^0||_W^2.$$

This may not determine v^0 uniquely, but it will uniquely determine $\mathbf{A}v^0$.

We can easily solve this problem via the projection theorem. Recall that the range of $\mathbf{A}$, $R(\mathbf{A}) = \{\mathbf{A}u : u \in V\}$ is a subspace of W. We need to find the point on $R(\mathbf{A})$ that is closest in norm to w. By the projection theorem, that point is just the projection of w on $R(\mathbf{A})$, i.e., the point $\mathbf{A}v^0 \in R(\mathbf{A})$ such that $w - \mathbf{A}v^0 \perp R(\mathbf{A})$. This means that

$$(w - \mathbf{A}v^0, \mathbf{A}v)_W = 0$$

for all $v \in V$. Now, using the adjoint operator, we have

$$(w - \mathbf{A}v^0, \mathbf{A}v)_W = (\mathbf{A}^*[w - \mathbf{A}v^0], v)_V = (\mathbf{A}^*w - \mathbf{A}^*\mathbf{A}v^0, v)_V = 0$$

for all $v \in V$. This is possible if and only if

$$\mathbf{A}^* \mathbf{A} v^0 = \mathbf{A}^* w.$$

In matrix notation, our equation for the least squares solution x^0 is

$$A^* A x^0 = A^* b. \tag{1.20}$$

The original system was rectangular; it involved m equations for n unknowns. Furthermore, in general it had no solution. Here however, the $n \times n$ matrix $A^* A$ is square and there are n equations for the n unknowns $x^0 = \{x_1, \ldots, x_n\}$. If the matrix A is real, then Equations (1.20) become $A^{\mathrm{tr}} A x^0 = A^{\mathrm{tr}} b$. This problem always has a solution x^0 and $A x^0$ is unique. Moreover, if A is 1-1 (so necessarily $m \geq n$) then the $n \times n$ matrix $A^* A$ is invertible and the unique solution is

$$x^0 = (A^* A)^{-1} A^* b. \tag{1.21}$$

Example 1.90 Consider Example 1.89 with $b_3 = 6$, a 3×2 system. Then the equation $A^* A x = A^* b$ becomes

$$\begin{pmatrix} 11 & 6 \\ 6 & 6 \end{pmatrix} \begin{pmatrix} x_1 \\ x_2 \end{pmatrix} = \begin{pmatrix} 8 \\ 14 \end{pmatrix}.$$

Inverting the 2×2 matrix $A^* A$ we obtain the least squares solution

$$\begin{pmatrix} x_1 \\ x_2 \end{pmatrix} = \begin{pmatrix} 1/5 & -1/5 \\ -1/5 & 11/30 \end{pmatrix} \begin{pmatrix} 8 \\ 14 \end{pmatrix},$$

so $x_1 = -6/5$, $x_2 = 53/15$.

We could use another norm to find the best approximation of (1.19), e.g, we could find the x such that $\|Ax - b\|_p$ is a minimum for some ℓ_p-norm. (It is easy to see that a minimum must exist for any $p \geq 1$.) However, only for the case $p = 2$ is there an inner product which permits the explicit analytic solution (1.21). In all other cases numerical methods must be used. On the other hand, some of these norms possess special advantages for certain kinds of signal processing problems. In particular we shall find the ℓ_1 norm to be especially useful in dealing with sparse signals, discrete signals that contain only a few nonzero terms. Moreover, minimization of $\|Ax - b\|_1$ is an example of a linear programming problem and there are very fast numerical algorithms (based on the simplex method) to find the solutions.

Example 1.91 We consider the same problem as in Example 1.90, except that here we minimize the ℓ_1 norm

$$f(x) = ||Ax - b||_1 = |3x_1 + x_2| + |x_1 + x_2 - 2| + |x_1 + 2x_2 - 6| = |y_1| + |y_2| + |y_3|.$$

Keys to the direct solution of this problem are the lines $3x_1 + x_2 = 0$, $x_1 + x_2 - 2 = 0$ and $x_1 + 2x_2 - 6 = 0$ in the $x_1 - x_2$ plane. They divide the plane into up to eight regions based on the eight possibilities $(y_1, y_2, y_3 \geq 0)$, $(y_1, y_2 \geq 0, y_3 \leq 0), \ldots, (y_1, y_2, y_3 \leq 0)$. The regions overlap only on the boundary lines. In each region $f(x)$ is a linear function. Thus in the first region we have

$$3x_1 + x_2 \geq 0, \quad x_1 + x_2 \geq 2, \quad x_1 + 2x_2 \geq 6, \quad f(x) = 5x_1 + 4x_2 - 6.$$

This region is actually bounded below by the lines $y_1 = 0$ and $y_3 = 0$ because if $y_1 \geq 0$ and $y_3 \geq 0$ we can verify that $y_2 \geq 0$. The lines $f(x) = c$ for any c are not parallel to $y_1 = 0$ or $y_3 = 0$ so the minimum of $f(x)$ in this region must occur at the point of intersection of the two lines $y_1 = y_3 = 0$. This intersection point occurs at $(x_1, x_2) = (-6/5, 18/5)$. Here $f(-6/5, 18/5) = |y_2| = 2/5$. Similarly, in the other regions the minimum of f never occurs at an interior point but always on the boundary at a vertex where two lines intersect. We have only to check the three intersection points. If $y_1 = y_2 = 0$ then $(x_1, x_2) = (-1, 3)$ and $f = |y_3| = 1$. If $y_2 = y_3 = 0$ then $(x_1, x_2) = (-2, 4)$ and $f = |y_1| = 2$. Thus the minimum value of f occurs at $(x_1, x_2) = (-6/5, 18/5)$. Note that this answer differs from the ℓ_2 solution. In this case two of the three equations $y_1 = y_2 = y_3 = 0$ are satisfied.

In the general case where we minimize the expression

$$f(x) = ||Ax - b||_1 = |y_1| + |y_2| + |y_3| + \cdots + |y_m|$$

with $y_j = \sum_{k=1}^{n} A_{jk} x_k$, $j = 1, \ldots, n$ and the rank of A is r, we can always achieve the minimum of f at an x such that at least r of the y_j vanish. For the least squares solution it may be that none of the y_j vanish.

In our previous discussion of the $m \times n$ system $Ax = b$ we have mostly considered cases where the system was *overdetermined*, i.e., where the number of conditions m was greater than the number of unknowns n. As we have seen, these equations are typically inconsistent. They have no exact solution, but we can obtain approximate solutions via minimization with respect to an appropriate norm. If the number of conditions m is less than the number of unknowns n, then the system is called *underdetermined.* In this case the null space of A has strictly positive dimension, so if there is a solution at all then there are infinitely many solutions. In signal processing this situation occurs when we sample an n-component signal m times where m is much less than n. Then typically there are many signals x that could yield the same sample $b = Ax$. If one solution is x^o then the general solution is $x = x^0 + v$ where $Av = \Theta$, i.e., v belongs to the null space $N(A)$ of A. How should we choose an "optimal" reconstruction of x?

There is no one best answer to the problem we have posed for underdetermined systems. The approach depends on the nature of the problem. One way to proceed mathematically would be to choose the solution with least ℓ_p norm. Thus the solution would be $x = x^0 + v^0$ where $v^0 \in N(A)$ and

$$||x^0 + v^0||_p = \min_{v \in N(A)} ||x^0 + v||_p.$$

This method can be justified in many applications, particularly for $p = 2$ and $p = 1$. Later we shall show that the ℓ_1 minimization is exactly the right approach in compressive sampling. A very simple example will help demonstrate the difference between ℓ_1 and ℓ_2 minimization for underdetermined systems of equations. Consider the case $m = 1$, $n = 3$ with single equation

$$3x_1 + 2x_2 + x_3 = 6.$$

Via Gaussian elimination we can see that the general solution of this equation is

$$x_1 = 2 - \frac{2}{3}c_1 - \frac{1}{3}c_2, \quad x_2 = c_1, \quad x_3 = c_2,$$

where c_1, c_2 are arbitrary. Geometrically, the solution is a plane in x_1, x_2, x_3-space that cuts the x_j-axes at $x_1 = 2$, $x_2 = 3$ and $x_3 = 6$. In the ℓ_1 norm the unit ball is a cube with vertices $(\pm 1, 0, 0), (0, \pm 1, 0)$ and $(0, 0, \pm 1)$. To minimize the ℓ_1 norm we must find the point on the plane such that $||x||_1 = |2 - \frac{2}{3}c_1 - \frac{1}{3}c_2| + |c_1| + |c_2|$ is a minmum. By dilating the unit ball we see that the minimum value occurs at the vertex $(2, 0, 0)$. On the other hand, in the ℓ_2 norm the unit ball is a sphere centered at the origin with radius 1. To minimize the ℓ_2 norm we must find the point on the plane such that $||x||_2^2 = (2 - \frac{2}{3}c_1 - \frac{1}{3}c_2)^2 + (c_1)^2 + (c_2)^2$ is a minmum. Here the solution is the point $(9/7, 6/7, 3/7)$. Thus the

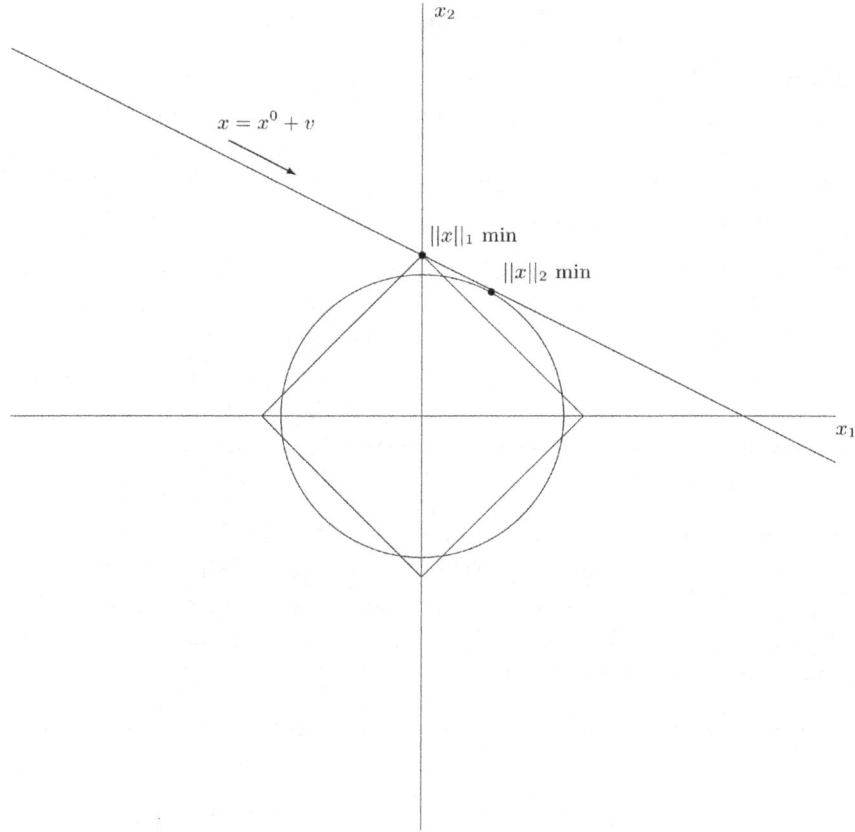

Figure 1.2 ℓ_1 and ℓ_2 minimization for an underdetermined system.

ℓ_1 solution has only one nonzero term, whereas the ℓ_2 solution has all terms nonzero.

Figure 1.2 will help to fix the ideas. The ℓ_2 minimum occurs at the point on the set $x^0 + v$ whose Euclidean distance to the origin is a minimum, whereas the ℓ_1 minimum occurs at points where some of the components of x are zero. In general, if $Ax = b$ is an underdetermined $m \times n$ system of rank $r \leq m$ with at least one solution, then there is always an ℓ_1 minimal solution with at least $n - r$ zero components, whereas an ℓ_2 minimal solution may have no zero components. This property of the ℓ_1 norm plays an important role in compressive sampling.

1.8 Additional exercises

Exercise 1.7 (a) Let V be a vector space. Convince yourself that V and $\{\Theta\}$ are subspaces of V and every subspace W of V is a vector space over the same field as V.

(b) Show that there are only two nontrivial subspaces of $\mathbb{R}^3$. (1) A plane passing through the origin and (2) A line passing through the origin.

Exercise 1.8 (a) Prove or give a counterexample to the following statement: If $v^{(1)}, \ldots, v^{(k)}$ are elements of a vector space V and do not span V, then $v^{(1)}, \ldots, v^{(k)}$ are linearly independent.

(b) Prove that if $v^{(1)}, \ldots, v^{(m)}$ are linearly independent, then any subset $v^{(1)}, \ldots, v^{(k)}$ with $k < m$ is also linearly independent.

(c) Does the same hold true for linearly dependent vectors? If not, give a counterexample.

Exercise 1.9 Show that the basic monomials $\{1, x, x^2, x^3, \ldots, x^n\}$ are linearly independent in the space of polynomials in x. Hint: Use the Fundamental Theorem of Algebra which states that a non-zero polynomial of degree $n \geq 1$ has at most n distinct real roots (and exactly n complex roots, not necessarily distinct).

Exercise 1.10 The Wronskian of a pair of differentiable, real-valued functions f and g is the scalar function

$$W[f(x), g(x)] = \det \begin{pmatrix} f(x) & g(x) \\ f'(x) & g'(x) \end{pmatrix}$$

$$= f(x)g'(x) - f'(x)g(x).$$

(a) Prove that if f, g are linearly dependent, then $W[f(x), g(x)] \equiv 0$.

(b) Prove that if $W[f(x), g(x)] \not\equiv 0$, then f, g are linearly independent.

(c) Let $f(x) = x^3$ and $g(x) = |x|^3$. Prove that f and g are twice continuously differentiable on $\mathbb{R}$, are linearly independent but $W[f(x), g(x)] \equiv 0$. Thus (b) is sufficient but not necessary. Indeed, one can show the following: $W[f(x), g(x)] \equiv 0$ iff f, g satisfy a second-order linear ordinary differential equation.

Exercise 1.11 Prove the following theorem. Suppose that V is an $n \geq 1$ dimensional vector space. Then the following hold.

(a) Every set of more than n elements of V is linearly dependent.

(b) No set of less than n elements spans V.

(c) A set of n elements forms a basis iff it spans V.

(d) A set of n elements forms a basis iff it is linearly independent.

Exercise 1.12 The Legendre polynomials $P_n(x)$, $n = 0, 1, \ldots$ are the ON set of polynomials on the real space $L_2[-1, 1]$ with inner product

$$(f, g) = \int_{-1}^{1} f(x)g(x)dx,$$

obtained by applying the Gram–Schmidt process to the monomials $1, x, x^2,$ $x^3, \ldots$ and defined uniquely by the requirement that the coefficient of t^n in $P_n(t)$ is positive. (In fact they form an ON *basis* for $L_2[-1, 1]$.) Compute the first four of these polynomials.

Exercise 1.13 Using the facts from the preceding exercise, show that the Legendre polynomials must satisfy a three-term recurrence relation

$$xP_n(x) = a_n P_{n+1}(x) + b_n P_n(x) + c_n P_{n-1}(x), \qquad n = 0, 1, 2, \ldots$$

where we take $P_{-1}(x) \equiv 0$. (Note: If you had a general sequence of polynomials $\{p_n(x)\}$, where the highest-order term on $p_n(x)$ was a nonzero multiple of t^n, then the best you could say was that

$$xp_n(x) = \sum_{j=0}^{n+1} \alpha_j p_j(x).$$

What is special about orthogonal polynomials that leads to three-term relations?) What can you say about a_n, b_n, c_n without doing any detailed computations?

Exercise 1.14 Let $L_2[\mathbb{R}, \omega(x)]$ be the space of square integrable functions on the real line, with respect to the weight function $\omega(x) = e^{-x^2}$. The inner product on this space is thus

$$(f, g) = \int_{-\infty}^{\infty} f(x)\overline{g(x)}\omega(x)dx.$$

The Hermite polynomials $H_n(x)$, $n = 0, 1, \ldots$ are the ON set of polynomials on $L_2[\mathbb{R}, \omega(x)]$, obtained by applying the Gram–Schmidt process to the monomials $1, x, x^2, x^3, \ldots$ and defined uniquely by the requirement that the coefficient of x^n in $H_n(x)$ is positive. (In fact they form an ON *basis* for $L_2[\mathbb{R}, \omega(x)]$.) Compute the first four of these polynomials. NOTE: In the study of Fourier transforms we will show that $\int_{-\infty}^{\infty} e^{-isx}e^{-x^2}dt = \sqrt{\pi}e^{-s^2/4}$. You can use this result, if you wish, to simplify the calculations.

Exercise 1.15 Note that in the last problem, $H_0(x), H_2(x)$ contained only even powers of x and $H_1(x), H_3(x)$ contained only odd powers. Can you find a simple proof, using only the uniqueness of the Gram–Schmidt process, of the fact that $H_n(-x) = (-1)^n H_n(x)$ for all n?

Exercise 1.16 Use least squares to fit a straight line of the form $y = bx + c$ to the data

$$
\begin{aligned}
x &= 0 \quad 1 \quad 3 \quad 4 \\
y &= 0 \quad 8 \quad 8 \quad 20
\end{aligned}
$$

in order to estimate the value of y when $x = 2.0$. Hint: Write the problem in the form

$$
\begin{pmatrix} 0 \\ 8 \\ 8 \\ 20 \end{pmatrix} = \begin{pmatrix} 0 & 1 \\ 1 & 1 \\ 3 & 1 \\ 4 & 1 \end{pmatrix} \begin{pmatrix} b \\ c \end{pmatrix}.
$$

Exercise 1.17 Repeat the previous problem to find the best least squares fit of the data to a parabola of the form $y = ax^2 + bx + c$. Again, estimate the value of y when $x = 2.0$.

Exercise 1.18 Project the function $f(t) = t$ onto the subspace of $L_2[0,1]$ spanned by the functions $\phi(t), \psi(t), \psi(2t), \psi(2t-1)$, where

$$
\phi(t) = \begin{cases} 1, & \text{for } 0 \le t \le 1 \\ 0, & \text{otherwise} \end{cases}
\qquad
\psi(t) = \begin{cases} 1, & \text{for } 0 \le t \le \frac{1}{2} \\ -1, & \text{for } \frac{1}{2} \le t < 1 \\ 0, & \text{otherwise.} \end{cases}
$$

(This is related to the Haar wavelet expansion for f.)

Exercise 1.19 A vector space pair $V := (V, ||\cdot||)$ is called a quasi-normed space if for every $x, y \in V$ and $\alpha \in \mathbb{F}$, there exists a map $||\cdot|| : V \to [0, \infty)$ such that

(i) $||x|| > 0$ if $x \ne 0$ and $||0|| = 0$. (positivity)

(ii) $||\alpha x|| = |\alpha| ||x||$. (homogeneity)

(iii) $||x + y|| \le C(||x|| + ||y||)$ for some $C > 0$ independent of x and y.

If $C = 1$ in (iii), then V is just a normed space since (iii) is just the triangle inequality. If $C = 1$, $||x|| = 0$ but $x \ne 0$, then V is called a semi-normed space.

(a) Let $n \geq 1$ and for $x = (x_1, x_2, \ldots, x_n) \in \mathbb{R}^n$ and $y = (y_1, y_2, \ldots, y_n) \in \mathbb{R}^n$ let $(x, y) := \sum_{j=1}^{n} x_j y_j$. Recall and prove that $(\cdot, \cdot)$ defines the Euclidean inner product on the finite-dimensional vector space $\mathbb{R}^n$ with Euclidean norm $(x_1^2 + \cdots + x_n^2)^{1/2}$. Note that the Euclidean inner product is exactly the same as the dot product we know from the study of vectors. In this sense, inner product is a natural generalization of dot product.

(b) Use the triangle inequality to show that any metric is a continuous mapping. From this deduce that any norm and inner product are continuous mappings.

(c) Let $0 < p < \infty$, $[a, b] \subset \mathbb{R}$ and consider the infinite-dimensional vector space $L_p[a, b]$ of p integrable functions $f : [a, b] \to \mathbb{R}$, i.e., for which $\int_a^b |f(x)|^p dx < \infty$ where we identify functions equal if they are equal almost everywhere. Use the Cauchy–Schwarz inequality to show that $(f, g) = \int_a^b f(x)g(x)dx$ is finite and also that $(\cdot, \cdot)$, defines an inner product on $L_2[a, b]$. Hint: you need to show that

$$\int_a^b |f(x)g(x)| dx \leq \left(\int_a^b |f(x)|^2 dx \right)^{1/2} \left(\int_a^b |g(x)|^2 dx \right)^{1/2}.$$

Let $1/p + 1/q = 1$, $p > 1$. The Hölder–Minkowski inequality

$$\int_a^b |f(x)g(x)| dx \leq \left(\int_a^b |f(x)|^p dx \right)^{1/p} \left(\int_a^b |g(x)|^q dx \right)^{1/q}$$

for $f \in L_p[a, b]$, $g \in L_q[a, b]$, generalizes the former inequality. Prove this first for step functions using the Hölder inequality, Theorem 1.21, and then in general by approximating $L_p[a, b]$ and $L_q[a, b]$ functions by step functions. Note that without the almost everywhere identification above, $L_p[a, b]$ is a semi-normed space.

(d) Show that $L_p[a, b]$ is a complete metric space for all $0 < p < \infty$ by defining $d(f, g) := \left(\int_a^b |f(x) - g(x)|^p dx \right)^{1/p}$.

(e) Show that $L_p[a, b]$ is a complete normed space (Banach space) for all $1 \leq p < \infty$ and a quasi-normed space for $0 < p < 1$ by defining $\|f\| := \left(\int_a^b |f(x)|^p dx \right)^{1/p}$.

(f) Show that $L_p[a, b]$ is a complete inner product space (Hilbert space) iff $p = 2$.

(g) $L_\infty[a, b]$ is defined as the space of essentially bounded functions $f : [a, b] \to \mathbb{R}$ for which $\|f\|_\infty [a, b] := \sup_{x \in [a,b]} |f(x)| < \infty$. $C[a, b]$ is the space of continuous functions $f : [a, b] \to \mathbb{R}$ with the same norm. Show that both these spaces are Banach spaces.

Exercise 1.20 Prove that

$$(f, g) := \int_a^b [f(x)g(x) + f'(x)g'(x)]dx$$

defines an inner product on $C^1[a, b]$, the space of real-valued continuously differentiable functions on the interval $[a, b]$. The induced norm is called the *Sobolev* H^1 norm.

Exercise 1.21 Let $1 \leq p < \infty$ and consider the infinite-dimensional space $l_p(\mathbb{R})$ of all real sequences $\{x_i\}_{i=1}^\infty$ so that $(\sum_{i=1}^\infty |x_i|^p)^{1/p} < \infty$. Also define

$$l_\infty(\mathbb{R}) := \sup \{|x_1|, |x_2|, \dots\}.$$

Then show the following.

(a) $(x_i, y_i) = \sum_{i=1}^\infty x_i y_i$ defines an inner product on $l_p(\mathbb{R})$ iff $p = 2$.
(b) $l_p(\mathbb{R})$ is a complete normed space for all $1 \leq p \leq \infty$.
(c) $l_p(\mathbb{R})$ is a quasi-normed, complete space for $0 < p < 1$.
(d) Recall the Hölder–Minkowski inequality for $l_p(\mathbb{R})$, $1 < p < \infty$. See Theorem 1.21. In particular, prove that

$$(x_1 + x_2 + \cdots + x_n)^2 \leq n(x_1^2 + x_2^2 + \cdots + x_n^2)$$

for all real numbers $x_1, \dots, x_n$. When does equality hold in the above?

Exercise 1.22 Let V be an inner product space with induced norm $|| \cdot ||$. If $x \perp y$ show that

$$||x|| + ||y|| \leq \sqrt{2}||x + y||, \ x, y \in V.$$

Exercise 1.23 Let V be an inner product space with induced norm $|| \cdot ||$. Prove the parallelogram law:

$$||x + y||^2 + ||x - y||^2 = 2|x||^2 + 2||y||^2, \ x, y \in V.$$

The parallelogram law does not hold for normed spaces in general. Indeed, inner product spaces are completely characterized as normed spaces satisfying the parallelogram law.

Exercise 1.24 Show that for $0 < p < \infty$, $f, g \in L_p[a, b]$ and $x_i, y_i \in l_p(\mathbb{F})$, the Minkowski and triangle inequalities yield the following:

- For all $f, g \in L_p[a, b]$, there exists $C > 0$ independent of f and g such that

$$(\int_a^b |f(x) + g(x)|^p dx)^{1/p} \leq C[(\int_a^b |f(x)|^p dx)^{1/p} + (\int_a^b |g(x)|^p dx)^{1/p}].$$

- For all $x_i, y_i \in l_p(\mathbb{F})$, there exists $C' > 0$ independent of x_i and y_i such that

$$(\sum_{i=1}^\infty |x_i + y_i|^p)^{1/p} \leq C'[(\sum_{i=1}^\infty |x_i|^p)^{1/p} + (\sum_{i=1}^\infty |y_i|^p)^{1/p}].$$

Exercise 1.25 An $n \times n$ matrix K is called *positive definite* if it is symmetric, $K^T = K$ and K satisfies the positivity condition $x^T K x > 0$ for all $x \neq 0 \in \mathbb{R}^n$. K is is called *positive semi definite* if $x^T K x \geq 0$ for all $x \in \mathbb{R}^n$. Show the following:

- Every inner product on $\mathbb{R}^n$ is given by

$$(x, y) = x^T K y, \ x, y \in \mathbb{R}^n$$

where K is symmetric and a positive definite matrix.
- If K is positive definite, then K is nonsingular.
- Every diagonal matrix is positive definite (semi positive definite) iff all its diagonal entries are positive (nonnegative). What is the associated inner product?

Exercise 1.26 • Let V be an inner product space and let $v_1, v_2, \ldots, v_n \in V$. The associated *Gram matrix*

$$K := \begin{pmatrix} (v_1, v_1) & (v_1, v_2) & \ldots & (v_1, v_n) \\ (v_2, v_1) & (v_2, v_2) & \ldots & (v_2, v_n) \\ \cdot & \cdot & \cdot & \cdot \\ \cdot & \cdot & \cdot & \cdot \\ \cdot & \cdot & \cdot & \cdot \\ (v_n, v_1) & (v_n, v_2) & \ldots & (v_n, v_n) \end{pmatrix}$$

is the $n \times n$ matrix whose entries are the inner products between selective vector space elements. Show that K is positive semi definite and positive definite iff all $v_1, \ldots, v_n$ are linearly independent.

- Given an $m \times n$ matrix V, $n \leq m$, show that the following are equivalent:

  - The $m \times n$ Gram matrix $A^T A$ is positive definite.
  - A has linearly independent columns.
  - A has rank n.
  - $N(A) = \{0\}$.

Exercise 1.27 Given $n \geq 1$, a Hilbert matrix is of the form

$$K(n) := \begin{pmatrix} 1 & 1/2 & \cdots & 1/n \\ 1/2 & 1/3 & \cdots & 1/(n+1) \\ \cdot & \cdot & \cdot & \cdot \\ \cdot & \cdot & \cdot & \cdot \\ \cdot & \cdot & \cdot & \cdot \\ 1/n & 1/(n+1) & \cdots & 1/(2n-1) \end{pmatrix}.$$

- Show that in $L_2[0,1]$, the Gram matrix from the monomials $1, x, x^2$ is the Hilbert matrix $K(3)$.

- More generally, show that the Gram matrix corresponding to the monomials $1, x, \ldots, x^n$ has entries $\frac{1}{i+j-1}$, $i,j = 1, \ldots, n+1$ and is $K(n+1)$.

- Deduce that $K(n)$ is positive definite and nonsingular.

Exercise 1.28 • Use Exercise 1.25 to prove: If K is a symmetric, positive definite $n \times n$, $n \geq 1$ matrix, $f \in \mathbb{R}^n$ and $c \in \mathbb{R}$, then the quadratic function

$$x^T K x - 2x^T f + c$$

has a unique minimizer $c - f^T K^{-1} f = c - f^T x^* = c - (x^*)^T K x^*$ which is a solution to the linear system $Kx = f$.

- Prove the following: Let $v_1, \ldots, v_n$ form a basis for a subspace $V \subset \mathbb{R}^m$, $n, m \geq 1$. Then given $b \in \mathbb{R}^m$, the closest point $v^* \in V$ is the solution $x^* = K^{-1}f$ of $Kx = f$ where K is the Gram matrix, whose (i,j)th entry is (v_i, v_j) and $f \in \mathbb{R}^n$ is the vector whose ith entry is the inner product (v_i, b). Show also that the distance

$$||v^* - b|| = \sqrt{||b||^2 - f^T x^*}.$$

Exercise 1.29 Suppose $w : \mathbb{R} \to (0, \infty)$ is continuous and for each fixed $j \geq 0$, the *moments*, $\mu_j := \int_{\mathbb{R}} x^j w(x)dx$ are finite. Use Gram–Schmidt to construct for $j \geq 0$, a sequence of orthogonal polynomials P_j of degree at most j such that $\int_{\mathbb{R}} P_i(x)P_j(x) = \delta_{ij}$ where δ_{ij} is 1 when $i = j$ and 0 otherwise. Recall the P_j are orthogonal if we only require

$\int_{\mathbb{R}} P_i(x)P_j(x) = 0$, $i \neq j$. Now consider the *Hankel* matrix

$$K := \begin{pmatrix} \mu_0 & \mu_1 & \cdots \\ \mu_1 & \mu_2 & \cdots \\ \mu_2 & \mu_3 & \cdots \\ \cdot & \cdot & \cdots \\ \cdot & \cdot & \cdots \end{pmatrix}.$$

Prove that K is positive-semi definite and hence that P_j are unique. The sequence $\{P_j\}$ defines the unique orthonormal polynomials with respect to the weight w.

Exercise 1.30 • For $n \geq 1$, let $t_1, t_2, \ldots, t_{n+1}$ be $n + 1$ distinct points in $\mathbb{R}$. Define the $1 \leq k \leq n + 1$th *Lagrange interpolating polynomial* by

$$L_k(t) := \frac{\prod_{j=1}^{n+1}(t - t_j)}{(t_k - t_j)}, \ t \in \mathbb{R}.$$

Verify that for each fixed k, L_k is of exact degree n and $L_k(t_i) = \delta_{ik}$.

• Prove the following: Given a collection of points (say data points), $y_1, \ldots, y_{n+1}$ in $\mathbb{R}$, the polynomial $p(t) := \sum_{j=1}^{n+1} y_j L_j(t)$, $t \in \mathbb{R}$ is of degree at most n and is the unique polynomial of degree at most n interpolating the pairs (t_j, y_j), $j = 1, \ldots, n + 1$, i.e. $p(t_j) = y_j$ for every $j \geq 0$.

• Now let y be an arbitrary polynomial of full degree n on $\mathbb{R}$ with coefficients c_j, $0 \leq j \leq n$. Show that the total least squares error between $m \geq 1$ data points $t_1, \ldots, t_m$ and the sample values $y(t_i)$, $1 \leq i \leq m$ is $||y - Ax||^2$ where $x = [c_1, \ldots, c_m]^T$, $y = [y_1, \ldots, y_m]^T$ and

$$A := \begin{pmatrix} 1 & t_1 & t_1^2 & \cdots & t_1^n \\ 1 & t_2 & t_2^2 & \cdots & t_2^n \\ \cdot & \cdot & \cdot & \cdot & \cdot \\ \cdot & \cdot & \cdot & \cdot & \cdot \\ \cdot & \cdot & \cdot & \cdot & \cdot \\ 1 & t_m & t_m^2 & \cdots & t_m^n \end{pmatrix}$$

which is called the Vandermonde matrix.[7]

• Show that

$$det A = \prod_{1 \leq i < j \leq n+1} (t_i - t_j)$$

---

[7] The Vandermonde matrix is named after the eighteenth-century French mathematician, scientist and musicologist Alexander-Theophile Vandermonde.

when A is of size $n + 1 \times n + 1$.

_____

For additional background material concerning the linear algebra and functional analysis topics in this chapter, see [55, 76, 89, 102].

2
Analytic tools

In this chapter we develop simple analytic tools that provide essential examples, approximation methods and convergence procedures for the study of linear transform analysis and its applications, particularly to signal processing. They are not directly part of the theoretical development of transform analysis, but rather related results and techniques needed at critical points in the development. Thus, particular improper integrals are needed in Fourier analysis and infinite products are employed in the cascade algorithm of wavelet analysis. Depending on the interests of the reader, parts of this chapter can be omitted at a first reading.

2.1 Improper integrals

In order to proceed, we need the notion of an improper Riemann integral. Henceforth and throughout, by a Riemann integral on a bounded closed interval $[a, b]$, we mean the usual definite integral, $\int_a^b f(x)dx$ with upper limits and lower limits on the real line. This motivates the following:

Definition 2.1 $R[a, b]$ is the class of all functions $f : [a, b] \to \mathbb{R}$ which are bounded and Riemann integrable on a finite interval $[a, b]$.

In this chapter, we shall define integrals of functions on infinite intervals and also integrals of unbounded functions on finite intervals.

2.1.1 Integrals on $[a, \infty)$

Definition 2.2 Let $f : [a, \infty) \to \mathbb{R}$. Assume that for every $s > a$, $f \in R[a, s]$. Then we define

$$\int_a^\infty f(x)dx = \lim_{s \to \infty} \int_a^s f(x)dx$$

if the limit exists and is finite. (Note that for every fixed s, $\int_a^s f(x)dx$ is well defined since f is bounded and Riemann integrable on $[a, s]$.) In this case, we say that $\int_a^\infty f(x)dx$ is *convergent*. Otherwise, we say that the integral is *divergent*. We call such an integral an *improper integral of the first kind*.

Remarks (a) If $\int_a^\infty f(x)dx = \pm\infty$, we say that the integral diverges to $\pm\infty$.

(b) Suppose that $b > a$. Then for $s > b$,

$$\int_a^s f(x)dx = \int_a^b f(x)dx + \int_b^s f(x)dx$$

so, letting $s \to \infty$, we see that $\int_a^\infty f(x)dx$ is convergent iff $\int_b^\infty f(x)dx$ is convergent.

(c) When f is nonnegative, then we can write $\int_a^\infty f(x)dx < \infty$ to indicate that the integral converges.

(d) We note that everything we have said above, and in what follows, applies to functions $f : [a, b] \to \mathbb{C}$ if we agree to write $f = \text{Re}f + i\text{Im}f$ where both $\text{Re}f$ and $\text{Im}f$ are real valued with domain $[a, b]$. In what follows, we choose to work with real valued functions, keeping this fact in mind.

Example 2.3 For what values of $a \in \mathbb{R}$ is $\int_1^\infty \frac{dx}{x^a}$ convergent?

Solution Let $f(x) := 1/x^a$, $x \in (1, \infty)$. Then it is clear that $f \in R[1, s]$ for every $s > 1$. Now:

$$\int_1^s f(x)dx = \begin{cases} \log s, & a = 1 \\ \frac{s^{1-a}}{1-a} + \frac{1}{a-1}, & a \neq 1 \end{cases}.$$

Hence

$$\lim_{s \to \infty} \int_1^s f(x)dx = \begin{cases} \infty, & a \leq 1 \\ \frac{1}{a-1}, & a > 1 \end{cases}$$

and the integral converges iff $a > 1$.

Note the equivalent well-known fact (1) from series of positive numbers: $\sum_{n=1}^{\infty} 1/n^a$ converges iff $a > 1$. This follows from the integral test for series:

Let J be a positive integer and let $f : [J, \infty) \to [0, \infty)$ be decreasing. Then $\sum_{n=J}^{\infty}$ converges iff $\int_{J}^{\infty}$ converges. The fact of the matter is that f decreases on $[1, \infty)$. Thus we deduce that there exists a positive constant C depending only on a such that

$$\frac{1}{C} \int_{1}^{\infty} f(x)dx \leq \sum_{n=1}^{\infty} \frac{1}{n^a} \leq C \int_{1}^{\infty} f(x)dx$$

from which (1) follows.

Exercise 2.1 Use the integral test and Example (2.3) to show that $\sum_{n=2}^{\infty} \frac{1}{n(\log n)^a}$ converges iff $a > 1$.

We may also define integrals over $(-\infty, b]$ and $(-\infty, \infty)$. We have:

Definition 2.4 Let $f : (-\infty, b] \to \mathbb{R}$. Suppose that for every $s < b$, $f \in R[s, b]$. Then we define

$$\int_{-\infty}^{b} f(x)dx = \lim_{s \to -\infty} \int_{s}^{b} f(x)dx$$

if the limit exists and is finite. In this case, we say that the integral is convergent; otherwise it is divergent.

Definition 2.5 Let $f : (-\infty, \infty) \to \mathbb{R}$. Assume that for every $-\infty < a < b < \infty$, $f \in R[a, b]$. Then we define

$$\int_{-\infty}^{\infty} f(x)dx = \int_{-\infty}^{0} f(x)dx + \int_{0}^{\infty} f(x)dx$$

provided that both the integrals on the right-hand side are finite. We say that the integral is convergent; otherwise it is divergent.

Exercise 2.2 Are $\int_{-\infty}^{\infty} e^{-|x|}dx$ and $\int_{-\infty}^{\infty} xdx$ convergent? If so, evaluate them.

Solution $e^{-|x|}$ is continuous and bounded on $\mathbb{R}$. So $e^{-|x|}$ is in $R[a, b]$ for every $-\infty < a < b < \infty$. Next, note that

$$\int_{0}^{\infty} e^{|x|}dx = \lim_{s \to \infty} (-e^{-s} + 1) = 1$$

and similarly,

$$\int_{-\infty}^{0} e^{|x|}dx = \lim_{s \to -\infty} (1 - e^s) = 1.$$

Hence $\int_{-\infty}^{\infty} e^{|x|}dx = 2$. On the other hand, we have

$$\lim_{s \to \infty} \int_{0}^{s} xdx = \infty$$

and so $\int_{-\infty}^{\infty} xdx$ diverges. Note also that for every finite s, we have $\int_{-s}^{s} xdx = 0$. (We note that this latter observation, allows us to define principal-valued, infinite integrals, see Exercise 2.16.)

2.1.2 Absolute convergence and conditional convergence

Just as one treats absolutely convergent series and conditionally convergent series, one can do the same for integrals.

Definition 2.6 Let $f : [a, \infty) \to \mathbb{R}$. If $f \in R[a, s]$ for every $s > a$ and $\int_{a}^{\infty} |f(x)|dx$ converges, then we say that $\int_{a}^{\infty} f(x)dx$ is *absolutely convergent*. If $\int_{a}^{\infty} f(x)dx$ converges but $\int_{a}^{\infty} |f(x)|dx$ diverges, then we say that $\int_{a}^{\infty} f(x)dx$ is *conditionally convergent*.

We have the following lemma which we state (without proof [92]).

Lemma 2.7 *Cauchy criterion for limits at infinity Let $F : (a, \infty) \to \mathbb{R}$ be a function. The following are equivalent:*

(a) $\lim_{x \to \infty} F(x)$ *exists and is finite.*
(b) $\forall \varepsilon > 0$, *there exists $C = C(\varepsilon) \geq a$ such that for $s, t \geq C$, we have* $|F(s) - F(t)| < \varepsilon$.

Exercise 2.3 In Chapter 1, we defined completeness of a metric space. Formulate Lemma 2.7 into a statement about completeness of the real numbers.

As a direct consequence of Lemma 2.7, we obtain:

Theorem 2.8 *Let $f : [a, \infty) \to \mathbb{R}$ and $f \in R[a, s]$, $\forall s > a$. Then $\int_{a}^{\infty} f(x)dx$ is convergent iff $\forall \varepsilon > 0$, there exists $C = C(\varepsilon)$ such that for $s, t \geq C$, we have*

$$\left| \int_{a}^{s} f(x)dx - \int_{a}^{t} f(x)dx \right| < \varepsilon.$$

We know that if $\int_a^\infty f(x)dx$ converges absolutely, then it converges. Another criterion, besides the Cauchy criterion above, for testing convergence of integrals is given by the following:

Theorem 2.9 *Let $f, g : [a, \infty) \to \mathbb{R}$ and suppose that $f, g \in R[a, s], \forall s > a$. Assume that g is nonnegative and $\int_a^\infty g(x)dx$ converges and there exist nonnegative constants C, C' such that for $x \geq C'$, $|f(x)| \leq Cg(x)$. Then $\int_a^\infty f(x)dx$ converges.*

Exercise 2.4 Prove Theorem 2.9 using the fact that a monotone (i.e. all terms have the same sign) bounded sequence converges. What is this limit?

Examples 2.10 Show that $\int_1^\infty \frac{\sin(3x)\log x}{x^2}dx$ is absolutely convergent (hence convergent), and $\int_1^\infty \frac{\sin(x)}{x}dx$ is convergent.

Indeed, given any fixed $1 > \delta > 0$, one has for x sufficiently large that $\left|\frac{\sin(3x)\log x}{x^2}\right| \leq \frac{1}{2x^{2-\delta}}$. Thus the first integral converges absolutely by Example 2.3. For the second integral, integrate by parts. Indeed, let $s > 1$, and write

$$\int_1^s \frac{\sin(x)}{x}dx = \cos(1) - \frac{\cos(s)}{s} - \int_1^s \frac{\cos(x)}{x^2}dx.$$

Using the method of the first integral in this example, we see that $\int_1^s \frac{\cos(x)}{x^2}dx$ is absolutely convergent as $s \to \infty$. Hence $\int_1^\infty \frac{\sin(x)}{x}dx$ converges. But this integral is not absolutely convergent as we will demonstrate in Section 2.3.

2.1.3 Improper integrals on finite intervals

This section deals with integrals of functions that are unbounded on a finite interval. We have in analogy to Definition 2.2, the following:

Definition 2.11 Let $f : [a, b] \to \mathbb{R}$ and $f \in R[s, b]$ for every $s \in (a, b)$ but f not bounded at a. Thus $f \notin R[a, b]$. We define

$$\int_a^b f(x)dx = \lim_{s \to a^+} \int_s^a f(x)dx$$

if the limit exists and is finite. In this case, we say that the integral is convergent; otherwise it is divergent. (Such integrals are called *improper integrals of the second kind*.)

We have the following:

Example 2.12 For what values of $a > 0$, does $\int_0^1 \frac{dx}{x^a}$ converge?

It is straightforward to show that for any $s \in (0, 1)$, one has

$$\lim_{s \to 0^+} \int_s^1 x^a dx = \left\{ \begin{array}{ll} \frac{1}{1-a}, & a < 1 \\ \infty, & a \geq 1 \end{array} \right.$$

which implies that the integral in question converges only for $a < 1$. (Clearly this integral is trivially finite for $a \leq 0$.)

Much as in Definition 2.11, we may define improper integrals of functions $f : [a, b] \to \mathbb{R}$ that satisfy $f \in R[a, s]$ for every $s \in (a, b)$ but $f \notin R[a, b]$ and also for functions unbounded at both endpoints. We have:

Definition 2.13 Let $f : [a, b] \to \mathbb{R}$ and assume that $f \in R[c, d]$ for all $a < c < d < b$ but $f \notin R[a, d]$ and $f \notin R[c, b]$. Then we choose $\alpha \in (a, b)$ and define

$$\int_a^b f(x)dx = \int_a^\alpha f(x)dx + \int_\alpha^b f(x)dx$$

if both improper integrals on the right-hand side converge. In this case, as before, we say that the integral is convergent; otherwise it is divergent.

Exercise 2.5 Show, using Definition 2.13 that $\int_{-1}^1 \frac{dx}{\sqrt{1-x^2}} = \pi$.

Remark Just as for improper integrals of the first kind, we may define absolute and conditional convergence for improper integrals of the second kind.

Exercise 2.6 (a) Show that $\int_0^1 \frac{\sin(\ln x)}{\sqrt{x}} dx$ is absolutely convergent.

(b) Show that $\int_0^{\pi/2} 1/\sqrt{t} \cos(t) dt$ is divergent. (Hint: Split the integral into two integrals from $[0, \pi/4]$ and from $[\pi/4, \pi/2]$. For the first integral use that $\cos(t)$ is decreasing and for the second integral expand $\cos(t)$ about $\pi/2$.)

(c) Show that $\int_{-1}^1 \frac{1}{x} dx$ is divergent.

2.2 The gamma functions and beta functions

Sometimes, we have integrals that are improper of both first (infinite interval) and second (unbounded function) kind. Important examples are given by the gamma functions and beta functions.

Definition 2.14 The *gamma* function Γ is defined by way of

$$\Gamma(x) := \int_0^\infty e^{-t}t^{x-1}dt, \ x > 0.$$

The later integral is often called Euler's integral for Γ. It is easy to see that Γ is a convergent improper integral for $x > 0$.

Next we have:

Definition 2.15 For $p, q > 0$, the Beta function is:

$$B(p,q) := \int_0^1 t^{p-1}(1-t)^{q-1}dt.$$

Then using Example 2.12, it is straightforward to see that $B(p,q)$ is a proper Riemann integral. Moreover, if $0 < p < 1$ and/or $0 < q < 1$, then $B(p,q)$ is a convergent improper Riemann integral. We have

Lemma 2.16 *For $p, q > 0$,*

$$B(p,q) := 2\int_0^{\pi/2} (\sin\psi)^{2p-1}(\cos\psi)^{2q-1}d\psi.$$

Proof We pick $0 < c < d < 1$ and make a substitution $t = \sin 2(\psi)$ in the integral. This gives,

$$\int_c^d t^{p-1}(1-t)^{q-1}dt = 2\int_{\arcsin(\sqrt{c})}^{\arcsin(\sqrt{d})} (\sin(\psi)^{2p-1}(\cos(\psi))^{2q-1}d\psi.$$

Now let $c \to 0^+$ and $d \to 1_-$. $\qquad\qquad\qquad\qquad\qquad \square$

The following classical result relates the functions Γ and B.

Theorem 2.17 *For $p, q > 0$, we have $B(p,q) = \frac{\Gamma(p)\Gamma(q)}{\Gamma(p+q)}$.*

Proof Exercise. (Hint: Write out the right-hand side, make a substitution $t = x^2$ in each integral and evaluate the resulting double integral over a $1/4$ circle using polar coordinates.)

2.3 The sinc function and its improper relatives

In this section we compute some improper Riemann integrals that are important both in Fourier analysis and in applications to such fields as wavelets and splines. First of all we will need the value of the improper Riemann integral $\int_0^\infty \frac{\sin x}{x}dx$. The function sinc x is one of the most important that occurs in this book, both in the theory and in the applications.

Definition 2.18 sinc $x = \begin{cases} \frac{\sin \pi x}{\pi x} & \text{for } x \neq 0 \\ 1 & \text{for } x = 0. \end{cases}$

The sinc function is one of the few we will study that is *not* Lebesgue integrable. Indeed we will show that the $L_1[0, \infty]$ norm of the sinc function is infinite. (The $L_2[0, \infty]$ norm is finite.) However, the sinc function is improper Riemann integrable because it is related to a (conditionally) convergent alternating series. Computing the integral is easy if you know about contour integration. If not, here is a direct verification (with some tricks).

Lemma 2.19 *The sinc function doesn't belong to $L_1[0, \infty]$. However the improper Riemann integral $I = \lim_{N \to \infty} \int_0^{N\pi} \frac{\sin x}{x} dx$ does converge and*

$$I = \int_0^\infty \frac{\sin x}{x} dx = \pi \int_0^\infty \text{sinc } x \, dx = \frac{\pi}{2}. \tag{2.1}$$

Proof Set $A_k = \int_{k\pi}^{(k+1)\pi} |\frac{\sin x}{x}| dx$. Note that

$$\frac{2}{k+1} = \int_{k\pi}^{(k+1)\pi} \frac{|\sin x|}{(k+1)\pi} dx < \int_{k\pi}^{(k+1)\pi} |\frac{\sin x}{x}| dx$$

$$< \int_{k\pi}^{(k+1)\pi} \frac{|\sin x|}{k\pi} dx| = \frac{2}{k},$$

so $\frac{2}{(k+1)\pi} < A_k < \frac{2}{(k)\pi}$ for $k > 0$. Thus $A_k \to 0$ as $k \to \infty$ and $A_{k+1} < A_k$, so $I = \sum_{k=0}^\infty (-1)^k A_k$ converges. (Indeed, it is easy to see that the even sums $E_h = \sum_{k=0}^{2h} (-1)^k A_k$ form a decreasing sequence $E_0 > E_1 > \cdots > E_h > \cdots > I$ of upper bounds for I and the odd sums $O_h = \sum_{k=0}^{2h+1} (-1)^k A_k$ form an increasing sequence $O_0 < O_1 < \cdots < O_h < \cdots < I$ of lower bounds for I. Moreover $E_h - O_h = A_{2h+1} \to 0$ as $h \to \infty$.) However,

$$\int_\pi^\infty |\frac{\sin x}{x}| dx = \sum_{k=1}^\infty A_k > \frac{2}{\pi}(1 + \frac{1}{2} + \frac{1}{3} + \cdots + \frac{1}{k} + \cdots)$$

diverges, since the harmonic series diverges.

Now consider the function

$$G(t) = \int_0^\infty e^{-tx} \frac{\sin x}{x} dx, \tag{2.2}$$

defined for $\operatorname{Re} t \geq 0$. Note that

$$G(t) = \sum_{k=0}^{\infty} (-1)^k \int_{k\pi}^{(k+1)\pi} (-1)^k e^{-tx} \frac{\sin x}{x} dx$$

converges uniformly on $[0, \infty]$ which implies that $G(t)$ is continuous for $t \geq 0$ and infinitely differentiable for $t > 0$. Also $G(0) = \lim_{t \to 0, t > 0} G(t) = G(+0)$.

Now, $G'(t) = -\int_0^\infty e^{-tx} \sin x \, dx$. Integrating by parts we have

$$\int_0^\infty e^{-tx} \sin x \, dx = \frac{-e^{-tx} \sin x}{t} \Big|_0^\infty + \int_0^\infty e^{-tx} \cos x \, dx = \frac{-e^{-tx} \cos x}{t^2} \Big|_0^\infty$$
$$- \int_0^\infty \frac{e^{-tx} \sin x}{t^2} dx.$$

Hence, $\int_0^\infty e^{-tx} \sin x \, dx = \frac{1}{1+t^2}$ and $G'(t) = \frac{-1}{1+t^2}$. Integrating this equation we have $G(t) = -\arctan t + c$ where c is the integration constant. However, from the integral expression for G we have $\lim_{t \to +\infty} G(t) = 0$, so $c = \pi/2$. Therefore $G(0) = \pi/2$. $\qquad \Box$

Remark: We can use this construction to compute some other important integrals. Consider the integral $\int_0^\infty \frac{\sin x}{x} e^{-i\lambda x} dx$ for λ a real number. (If integrals of complex functions are new to you, see Exercise 2.13.) Taking real and imaginary parts of this expression and using the trigonometric identities

$$\sin \alpha \sin \beta = \frac{1}{2} [\cos(\alpha - \beta) - \cos(\alpha + \beta)],$$
$$\sin \alpha \cos \beta = \frac{1}{2} [\sin(\alpha - \beta) + \sin(\alpha + \beta)],$$

we can mimic the construction of the lemma to show that the improper Riemann integral converges for $|\lambda| \neq 1$. Then

$$\int_0^\infty \frac{\sin x}{x} e^{i\lambda x} dx = \lim_{\epsilon \to 0, \epsilon > 0} G(i\lambda + \epsilon)$$

where, as we have shown, $G(t) = -\arctan t + \frac{\pi}{2} = \frac{i}{2} \ln \frac{1+it}{1-it} + \frac{\pi}{2}$. Using the property that $\ln(e^{i\theta}) = i\theta$ for $-\pi \leq \theta < \pi$ and taking the limit carefully, we find

$$\int_0^\infty \frac{\sin x}{x} e^{-i\lambda x} dx = \begin{cases} \frac{\pi}{2} + \frac{i}{2} \ln \left| \frac{1-\lambda}{1+\lambda} \right| & \text{for } |\lambda| < 1 \\ \frac{i}{2} \ln \left| \frac{1-\lambda}{1+\lambda} \right| & \text{for } |\lambda| > 1. \end{cases}$$

Noting that $\int_{-\infty}^\infty \frac{\sin x}{x} e^{-i\lambda x} dx = 2 \int_0^\infty \frac{\sin x}{x} \cos(\lambda x) dx$ we find

$$\int_{-\infty}^\infty \frac{\sin x}{x} e^{-i\lambda x} dx = \begin{cases} \pi & \text{for } |\lambda| < 1 \\ \frac{\pi}{2} & \text{for } |\lambda| = 1 \\ 0 & \text{for } |\lambda| > 1. \end{cases} \tag{2.3}$$

It is easy to compute the L_2 norm of sinc x:

Lemma 2.20

$$\int_0^\infty \frac{\sin^2 x}{x^2}dx = \pi \int_0^\infty \text{sinc}^2 x \, dx = \frac{\pi}{2}. \qquad (2.4)$$

Proof Integrate by parts.

$$\int_0^\infty \frac{\sin^2 x}{x^2}dx = -\frac{\sin^2 x}{x}\Big|_0^\infty + \int_0^\infty \frac{2\sin x \cos x}{x}dx = \int_0^\infty \frac{\sin 2x}{x}dx$$

$$= \int_0^\infty \frac{\sin y}{y}dy = \frac{\pi}{2}.$$

$\square$

Here is a more complicated proof, using the same technique as for Lemma 13. Set $G(t) = \int_0^\infty e^{-tx}(\frac{\sin x}{x})^2 dx$, defined for $t \geq 0$. Now, $G''(t) = \int_0^\infty e^{-tx}\sin^2 x \, dx$ for $t > 0$. Integrating by parts we have

$$\int_0^\infty e^{-tx}\sin^2 x \, dx = 2\int_0^\infty \frac{e^{-tx}}{t^2}dx - 4\int_0^\infty \frac{e^{-tx}\sin x}{t^2}dx.$$

Hence, $G''(t) = \frac{2}{4t+t^3}$. Integrating this equation twice we have

$$G(t) = \frac{1}{2}t\ln t - \frac{1}{4}t\ln(4+t^2) - \arctan\frac{t}{2} + bt + c$$

where b, c are the integration constants. However, from the integral expression for G we have $\lim_{t\to+\infty} G(t) = \lim_{t\to+\infty} G'(t) = 0$, so $b = 0, c = \pi/2$. Therefore $G(0) = \pi/2$. $\square$

Again we can use this construction to compute some other important integrals. Consider the integral $\int_0^\infty (\frac{\sin x}{x})^2 e^{-i\lambda x}dx$ for λ a real number. Then

$$\int_0^\infty (\frac{\sin x}{x})^2 e^{i\lambda x}dx = \lim_{\epsilon\to 0, \epsilon>0} G(i\lambda + \epsilon)$$

where,

$$G(t) = \frac{1}{2}t\ln t - \frac{1}{4}t\ln(4+t^2) - \arctan\frac{t}{2} + \frac{\pi}{2}.$$

Using the property that $\ln(e^{i\theta}) = i\theta$ for $-\pi \leq \theta < \pi$ and taking the limit carefully, we find

$$\int_0^\infty (\frac{\sin x}{x})^2 e^{-i\lambda x}dx =$$

$$\begin{cases} \frac{i\lambda}{2}(\ln|\lambda| \pm i\frac{\pi}{2}) - \frac{i\lambda}{4}\ln|4 - \lambda^2| + \frac{i}{2}\ln|\frac{2-\lambda}{2+\lambda}| + \frac{\pi}{2} & \text{for } \pm\lambda \leq 2 \\ \frac{i\lambda}{2}(\ln|\lambda| \pm i\frac{\pi}{2}) - \frac{i\lambda}{4}(\ln|4 - \lambda^2| \pm i\pi) + \frac{i}{2}\ln|\frac{2-\lambda}{2+\lambda}| & \text{for } \pm\lambda > 2. \end{cases}$$

Noting that $\int_{-\infty}^{\infty}(\frac{\sin x}{x})^2 e^{-i\lambda x}dx = 2\int_0^\infty (\frac{\sin x}{x})^2 \cos(\lambda x)dx$ we find

$$\int_{-\infty}^{\infty}(\frac{\sin x}{x})^2 e^{-i\lambda x}dx = \begin{cases} \pi(1 - \frac{|\lambda|}{2}) & \text{for}|\lambda| < 2 \\ 0 & \text{for}|\lambda| \geq 2. \end{cases} \tag{2.5}$$

2.4 Infinite products

The next sections deal with the infinite products used in various areas of computer vision, signal processing, approximation theory and wavelets. We have already developed the notion of an improper integral rigorously which has, as one of its motivations, the idea of adding infinitely often. In what follows, we will develop the notion of multiplying and dividing infinitely often as well.

2.4.1 Basics

Consider some $r \in (0, \infty)$ and $\Pi_{k=1}^n r = r^n$. We know that

$$r^n \to \begin{cases} 0, & r < 1 \\ \infty, & r > 1 \end{cases} \quad \text{as } n \to \infty.$$

This suggests that if we wish to multiply more and more numbers, their product can only remain bounded and tend to a finite number, if they approach 1. This motivates:

Definition 2.21 Let $\{a_k\}_{k=1}^\infty \in \mathbb{R}$. For $n \geq 1$, we define the nth partial product

$$P_n = \Pi_{k=1}^n(1 + a_k) = (1 + a_1)\cdots(1 + a_n).$$

We assume $a_k \neq -1, \forall k$.

(a) If

$$P = \lim_{n\to\infty} P_n$$

exists, and is nonzero, then we say that the infinite product $\Pi_{k=1}^\infty(1 + a_k)$ converges and has value P.

(b) Otherwise, we say that the product diverges. In the special case, when $\lim_{n\to\infty} P_n = 0$, then we say the product diverges to zero. We remark that dropping/adding finitely many nonzero terms does not change convergence.

Example 2.22 Is $\Pi_{k=2}^{\infty}(1 - 1/k^2)$ convergent? If so, evaluate it.

Solution Note that we start for $k = 2$, since $1 - 1/k^2 = 0$ for $k = 1$. We form the partial product P_n:

$$\Pi_{k=2}^{n}(1 - \frac{1}{k^2}) = \Pi_{k=2}^{n}\left(\frac{(k-1)(k+1)}{k^2}\right) = \frac{\Pi_{k=2}^{n}(k-1)\Pi_{k=2}^{n}(k+1)}{\left(\Pi_{k=2}^{n}k\right)^2}$$

$$= \frac{[1 \cdot 2 \cdot 3 \cdots (n-2)(n-1)][3 \cdot 4 \cdot 5 \cdots n(n+1)]}{[2 \cdot 3 \cdots (n-1)n]2[3 \cdot 4 \cdot 5 \cdots (n-1)n]} = \frac{n+1}{2n}.$$

Thus

$$\Pi_{k=2}^{\infty}(1 - 1/k^2) = \lim_{n \to \infty} \Pi_{k=2}^{n}(1 - 1/k^2) = 1/2$$

converges.

Example 2.23 Is $\Pi_{k=2}^{\infty}(1 - 1/k)$ convergent? If so, evaluate it.

Solution Form

$$P_n = \Pi_{k=2}^{\infty}(1 - 1/k) = \Pi_{k=2}^{n}\left(\frac{k-1}{k}\right) = \frac{\Pi_{k=2}^{n}(k-1)}{\Pi_{k=2}^{n}k}$$

$$= \frac{1 \cdot 2 \cdot 3 \cdots (n-2)(n-1)}{2 \cdot 3 \cdots (n-1)n} = 1/n$$

So $\Pi_{k=2}^{\infty}(1 - 1/k)$ diverges to zero.

Example 2.24 Let

$$a_k = \left\{ \begin{array}{rl} 1, & k \text{ even} \\ -1/2, & k \text{ odd}. \end{array} \right.$$

Does $\Pi_{k=1}^{\infty}(1 + a_k)$ converge?

Solution By induction on n, it is easy to see that

$$P_n = \left\{ \begin{array}{rl} 1/2, & n \text{ odd} \\ 1, & n \text{ even} \end{array} \right.$$

and so $\lim_{n \to \infty} P_n$ doesn't exist.

This example, helps to motivate

Theorem 2.25 A necessary condition for convergence *If* $P = \Pi_{k=1}^{\infty}(1 + a_k)$ *converges, then* $\lim_{k \to \infty} a_k = 0$.

Proof Our hypothesis ensures that $P \neq 0$ and $P_n = \Pi_{k=1}^{n}(1 + a_k)$ has $\lim_{n \to \infty} P_n = P$. Then

$$1 = \frac{\lim_{n \to \infty} P_n}{\lim_{n \to \infty} P_{n-1}} = \lim_{n \to \infty} \frac{P_n}{P_{n-1}} = \lim_{n \to \infty} (1 + a_n).$$

$\square$

The condition above is necessary but not sufficient, since with $a_n = -1/n$, the product does not converge.

Definition 2.26 We say that $\Pi_{k=1}^{\infty}(1 + a_k)$ converges absolutely iff $\Pi_{k=1}^{\infty}(1 + |a_k|)$ converges.

This motivates:

Theorem 2.27 *Let $a_k \neq -1$, $k \geq 1$. The following are equivalent:*

(a) $\Pi_{k=1}^{\infty}(1 + a_k)$ *converges absolutely.*
(b) $\sum_{k=1}^{\infty} a_k$ *converges absolutely.*

Proof Set $p_k = |a_k|$ and

$$\mathcal{Q}_n = \sum_{k=1}^{n} p_k, \qquad P_n = \Pi_{k=1}^{n}(1 + p_k).$$

Now

$$1 + \mathcal{Q}_n \leq P_n = \Pi_{k=1}^{n}(1 + p_k) \leq e^{\sum_{k=1}^{n} p_k} = e^{\mathcal{Q}_n},$$

where the last inequality is a consequence of $1 + p_k \leq e^{p_k}$. (The left-hand side is just the first two terms in the power series of e^{p_k}, and the power series contains only nonnegative terms.) Thus the infinite product converges if and only if the power series converges. □

Corollary 2.28 *Let $a_k \neq -1$, $k \geq 1$. Then if $\Pi_{k=1}^{\infty}(1 + a_k)$ converges absolutely, it converges.*

Proof We have that $\lim_{k \to \infty} |a_k| = 0$. So there exists $L > 0$ such that

$$|a_k| \leq 1/2, \ k \geq L.$$

But by the inequality in the previous proof, we have

$$|\log(1 + a_k)| \leq 3/2|a_k|, \ k \geq L.$$

Then much as before, we have that

$$\sum_{k=L}^{\infty} |\log(1 + a_k)| \leq 3/2 \sum_{k=L}^{\infty} |a_k| < \infty$$

so that $\sum_{k=L}^{\infty} \ln(1 + a_k)$ converges to say finite S. Then clearly, as $\ln(x)$ is continuous away from 0, we deduce taking limits that

$$\lim_{n \to \infty} \Pi_{k=L}^{n}(1 + a_k) = \exp(S) \neq 0$$

which of course implies that $\Pi_{k=1}^{\infty}(1 + a_k)$ converges. □

Example 2.29 For what value of $p > 0$ does $\Pi_{k=2}^{\infty}(1 + a_k)$ converge with $a_k = (-1)^k/k^p$, $k \geq 2$? Note that

$$\sum_{k=2}^{\infty} |a_k| = \sum_{k=2}^{\infty} \frac{1}{k^p} < \infty$$

if $p > 1$. If $p = 1$, then one easily observes that if

$$P_n = \Pi_{k=1}^{n}(1 + a_k), \; n \geq 1,$$

then

$$P_n = \begin{cases} 1, & n\,\text{even} \\ 1 - 1/(2n+1), & n\,\text{odd}. \end{cases}$$

So the product converges also for $p = 1$ but not absolutely. In fact, one can show that the product converges for all $p > 0$.

2.5 Additional exercises

Exercise 2.7 Test the following integrals for convergence/absolute convergence/conditional convergence:

(a) $\int_{10}^{\infty} \frac{1}{x(\ln x)(\ln \ln x)^a} dx, a > 0$.

(b) $\int_{1}^{\infty} \frac{\cos x}{x} dx$.

(c) $\int_{-\infty}^{\infty} P(x)e^{-|x|} dx$, P a polynomial.

(d) $\int_{-\infty}^{\infty} P(x)\ln(1 + x^2)e^{-x^2} dx$, P a polynomial.

(e) $\int_{1}^{\infty} \frac{\cos x \sin x}{x} dx$.

(f) $\int_{-\infty}^{\infty} \text{sign}(x) dx$.

(g) $\int_{1}^{\infty} \frac{\cos 2x}{x} dx$.

(h) $\int_{9}^{\infty} \frac{\ln x}{x(\ln \ln x)^{100}} dx$.

Exercise 2.8 Let $f : [a, \infty) \to \mathbb{R}$ with $f \in R[a, s]$, $\forall s > a$ and suppose that $\int_{a}^{\infty} f(x) dx$ converges. Show that $\forall \; \varepsilon > 0$, there exists $B_\varepsilon > 0$ such that

$$s \geq B_\varepsilon \text{ implies } \left| \int_{s}^{\infty} f(x) dx \right| < \varepsilon.$$

Exercise 2.9 Show that if $f, g : [a, \infty) \to \mathbb{R}$ and $\int_{a}^{\infty} f(x) dx$ and $\int_{a}^{\infty} g(x) dx$ converge, then $\forall \alpha, \beta \in \mathbb{R}$, $\int_{a}^{\infty} [\alpha f(x) + \beta g(x)] dx$ converges to $\alpha \int_{a}^{\infty} f(x) dx + \beta \int_{a}^{\infty} g(x) dx$.

Exercise 2.10 The substitution rule for improper integrals.
Suppose that $g : [a, \infty) \to \mathbb{R}$ is monotone, increasing and continuously
differentiable and $\lim_{t \to \infty} g(t) = \infty$. Suppose that $f : [g(a), \infty) \to \mathbb{R}$
and $f \in R[a, s], \forall s > g(a)$. Show that $\int_a^\infty f(g(t))g'(t)dt$ converges iff
$\int_{g(a)}^\infty f(x)dx$ converges and if either converges, both are equal.

Exercise 2.11 For what values of a do the series below converge?

$$\sum_{n=1}^\infty \frac{1}{n^a}, \quad \sum_{n=2}^\infty \frac{1}{n(\ln n)^a}, \quad \sum_{n=3}^\infty \frac{1}{n(\ln n)(\ln \ln n)^a}.$$

Exercise 2.12 Use integration by parts for the integral $\int_1^s \frac{\sin(x)}{x}dx$ to
establish the convergence of the improper integral $\int_1^\infty \frac{\sin(x)}{x}dx$.

Exercise 2.13 If $f = f_1 + if_2$ is a complex-valued function on $[a, \infty)$,
we define

$$\int_a^\infty f(x)dx = \int_a^\infty f_1(x)dx + i \int_a^\infty f_2(x)dx$$

whenever both integrals on the right-hand side are defined and finite.
Show that

$$\int_1^\infty \frac{e^{ix}}{x}dx$$

converges.

Exercise 2.14 Suppose that $f : [-a, a] \to \mathbb{R}$ and $f \in R[-s, s] \ \forall \ 0 <
s < a$, but $f \notin R[-a, a]$. Suppose furthermore that f is even or odd.
Show that $\int_{-a}^a f(x)dx$ converges iff $\int_0^a f(x)dx$ converges and

$$\int_{-a}^a f(x)dx = \begin{cases} 2 \int_0^a f(x)dx, & \text{if } f \text{ even} \\ 0, & \text{if } f \text{ odd}. \end{cases}$$

Exercise 2.15 Test the following integrals for convergence/absolute
convergence/conditional convergence. Also say whether or not they are
improper.

(a) $\int_{-1}^1 |x|^{-a}dx, a > 0$.

(b) $\int_0^1 \frac{\sin(1/x)}{x}dx$.

(c) $\int_0^1 (t(1-t))^{-a}dt, a > 0$.

(d) $\int_0^1 \frac{\sin x}{x}dx$.

(e) $\int_1^\infty e^{-x}(x-1)^{-2/3}dx$.

(f) $\int_{-\infty}^\infty e^{-x^2}|x|^{-1}dx$.

(g) The Beta function, Definition 33:

$$B(p,q) = \int_0^1 t^{p-1}(1-t)^{q-1}dt.$$

Exercise 2.16 Cauchy Principal Value. Sometimes, even when $\int_a^b f(x)dx$ diverges for a function f that is unbounded at some $c \in (a,b)$, we may still define what is called the *Cauchy Principal Value Integral.* Suppose that $f \in R[a, c-\varepsilon]$ and $f \in R[c+\varepsilon, b]$ for all small enough $\varepsilon > 0$ but f is unbounded at c. Define

$$\text{P.V.} \int_a^b f(x)dx := \lim_{\varepsilon \to 0^+} \left[\int_a^{c-\varepsilon} f(x)dx + \int_{c+\varepsilon}^b f(x)dx \right]$$

if the limit exists. (P.V. stands for Principal Value).

(a) Show that $\int_{-1}^1 \frac{1}{x}dx$ diverges but P.V. $\int_{-1}^1 \frac{1}{x}dx = 0$.

(b) Show that P.V. $\int_a^b \frac{1}{x-c}dx = \ln\left(\frac{b-c}{c-a}\right)$ if $c \in (a,b)$.

(c) Show that if f is odd and $f \in R[\varepsilon, 1]$, $\forall\, 0 < \varepsilon < 1$, then P.V. $\int_{-1}^1 f(x)dx = 0$.

Exercise 2.17 (a) Use integration by parts to show that

$$\Gamma(x+1) = x\Gamma(x), \ x > 0.$$

This is called the **difference equation** for $\Gamma(x)$.

(b) Show that $\Gamma(1) = 1$ and deduce that for $n \geq 0$, $\Gamma(n+1) = n!$. Thus Γ generalizes the factorial function.

(c) Show that for $n \geq 1$,

$$\Gamma(n+1/2) = (n-1/2)(n-3/2)\cdots(3/2)(1/2)\pi^{1/2}.$$

Exercise 2.18 (a) Show that

$$B(p,q) = \int_0^\infty y^{p-1}(1+y)^{-p-q}dy.$$

(b) Show that

$$B(p,q) = (1/2)^{p+q-1}\int_{-1}^1 (1+x)^{p-1}(1-x)^{q-1}dx.$$

Also $B(p,q) = B(q,p)$.

(c) Show that

$$B(p,q) = B(p+1,q) + B(p,q+1).$$

(d) Write out $B(m,n)$ for m and n positive integers.

Exercise 2.19 Large O and little o notation Let f and $h : [a, \infty) \to \mathbb{R}$, $g : [a, \infty) \to (0, \infty)$. We write $f(x) = O(g(x))$, $x \to \infty$ iff there exists $C > 0$ independent of x such that

$$|f(x)| \leq Cg(x), \; \forall \text{ large enough } x.$$

We write $f(x) = o(g(x))$, $x \to \infty$ iff $\lim_{x \to \infty} f(x)/g(x) = 0$. Quite often, we take $g \equiv 1$. In this case, $f(x) = O(1)$, $x \to \infty$ means $|f(x)| \leq C$ for some C independent of x, i.e. f is uniformly bounded above in absolute value for large enough x and $f(x) = o(1)$, $x \to \infty$ means $\lim_{x \to \infty} f(x) = 0$.

(a) Show that if $f(x) = O(1)$ and $h(x) = o(1)$ as $x \to \infty$, then $f(x)h(x) = o(1)$, $x \to \infty$ and $f(x) + h(x) = O(1)$, $x \to \infty$. Can we say anything about the ratio $h(x)/f(x)$ in general for large enough x?
(b) Suppose that $f, g \in R[a, s]$, $\forall s > a$ and $\int_a^\infty g(x)dx$ converges. Show that if $f(x) = O(g(x))$, $x \to \infty$, then $\int_a^\infty f(x)dx$ converges absolutely.
(c) Show that if also h is nonnegative and

$$f(x) = O(g(x)), \; x \to \infty, \; g(x) = O(h(x)), \; x \to \infty$$

then this relation is transitive, that is

$$f(x) = O(h(x)), \; x \to \infty.$$

Exercise 2.20 Show that

$$\Pi_{n=2}^\infty \left(1 - \frac{2}{n(n+1)} \right) = 1/3.$$

Exercise 2.21 Show that

$$\Pi_{n=1}^\infty \left(1 + \frac{6}{(n+1)(2n+9)} \right) = 21/8.$$

Exercise 2.22 Show that

$$\Pi_{n=0}^\infty (1 + x^{2^n}) = \frac{1}{1-x}, \; |x| < 1.$$

Hint: Multiply the partial products by $1 - x$.

Exercise 2.23 Let $a, b \in \mathbb{C}$ and suppose that they are not negative integers. Show that

$$\Pi_{n=1}^\infty \left(1 + \frac{a - b - 1}{(n + a)(n + b)} \right) = \frac{a}{b+1}.$$

Exercise 2.24 Is $\Pi_{n=0}^\infty (1 + n^{-1})$ convergent?

Exercise 2.25 Prove the inequality $0 \leq e^u - 1 \leq 3u$ for $u \in (0,1)$ and deduce that $\Pi_{n=1}^{\infty} \left(1 + \frac{1}{n}[e^{1/n} - 1]\right)$ converges.

Exercise 2.26 The **Euler–Mascheroni Constant**. Let

$$H_n = \sum_{k=1}^{n} \frac{1}{k}, \ n \geq 1.$$

Show that $1/k \geq 1/x$, $x \in [k, k+1]$ and hence that

$$\frac{1}{k} \geq \int_k^{k+1} \frac{1}{x} dx.$$

Deduce that

$$H_n \geq \int_1^{n+1} \frac{1}{x} dx = \ln(n+1)$$

and that

$$[H_n - \ln(n+1)] - [H_n - \ln n] = \frac{1}{n+1} + \ln\left(1 - \frac{1}{n+1}\right)$$

$$= -\sum_{k=2}^{\infty} \left(\frac{1}{n+1}\right)^k /k < 0.$$

Deduce that $H_n - \ln n$ decreases as n increases and thus has a non-negative limit γ, called the *Euler–Mascheroni constant* and with value approximately 0.5772... not known to be either rational or irrational. Finally show that

$$\Pi_{k=1}^n (1+1/k)e^{-1/k} = \Pi_{k=1}^n (1+1/k)\Pi_{k=1}^n e^{-1/k} = \frac{n+1}{n} \exp(-(H_n - \ln n)).$$

Deduce that

$$\Pi_{k=1}^{\infty} (1 + 1/k)e^{-1/k} = e^{-\gamma}.$$

Many functions have nice infinite product expansions which can be derived from first principles. The following examples, utilizing contour integration in the complex plane, illustrate some of these nice expansions.

Exercise 2.27 The aim of this exercise is to establish Euler's reflection formula:

$$\Gamma(x)\Gamma(1-x) = \frac{\pi}{\sin(\pi x)}, \ \mathrm{Re}(x) > 0.$$

Step 1: Set $y = 1 - x$, $0 < x < 1$ in the integral

$$\int_0^\infty \frac{s^{x-1}}{(1+s)^{x+y}} ds = \frac{\Gamma(x)\Gamma(y)}{\Gamma(x+y)}$$

to obtain

$$\Gamma(x)\Gamma(1-x) = \int_0^\infty \frac{t^{x-1}}{1+t} dt.$$

Step 2: Let now C consist of two circles about the origin of radii R and ε respectively which are joined along the negative real axis from R to $-\varepsilon$. Show that

$$\int_C \frac{z^{x-1}}{1-z} dz = -2i\pi$$

where z^{x-1} takes its principal value.

Now let us write,

$$-2i\pi = \int_{-\pi}^\pi \frac{iR^x \exp(ix\theta)}{1 - R\exp(i\theta)} d\theta + \int_R^\varepsilon \frac{t^{x-1} \exp(ix\pi)}{1+t} dt$$
$$+ \int_{-\pi}^\pi \frac{i\varepsilon^x \exp(ix\theta)}{1 - \varepsilon\exp(i\theta)} d\theta + \int_\varepsilon^R \frac{t^{x-1} \exp(-ix\pi)}{1+t} dt.$$

Let $R \to \infty$ and $\varepsilon \to 0^+$ to deduce the result for $0 < x < 1$ and then for Re $x > 0$ by continuation.

Exercise 2.28 Using the previous exercise, derive the following expansions for those arguments for which the RHS is meaningful.

(1)

$$\sin(x) = x\Pi_{j=1}^\infty \left(1 - (x/j\pi)^2\right).$$

(2)

$$\cos(x) = \Pi_{j=1}^\infty \left(1 - \left(\frac{x}{(j-1/2)\pi}\right)^2\right).$$

(3) Euler's product formula for the gamma function
$$x\Gamma(x) = \Pi_{n=1}^\infty \left[(1+1/n)^x (1+x/n)^{-1}\right].$$

(4) Weierstrass's product formula for the gamma function
$$(\Gamma(x))^{-1} = xe^{x\gamma}\Pi_{k=1}^\infty \left[(1+x/k)e^{-x/k}\right].$$

Here, γ is the *Euler–Mascheroni constant*.

3
Fourier series

Jean Baptiste Joseph Fourier (1768–1830) was a French mathematician, physicist and engineer, and the founder of Fourier analysis. In 1822 he made the claim, seemingly preposterous at the time, that any function of t, continuous or discontinuous, could be represented as a linear combination of functions $\sin nt$. This was a dramatic distinction from Taylor series. While not strictly true in fact, this claim was true in spirit and it led to the modern theory of Fourier analysis, with wide applications to science and engineering.

3.1 Definitions, real Fourier series and complex Fourier series

We have observed that the functions $e_n(t) = e^{int}/\sqrt{2\pi}$, $n = 0, \pm 1, \pm 2, \ldots$ form an ON set in the Hilbert space $L_2[0, 2\pi]$ of square-integrable functions on the interval $[0, 2\pi]$. In fact we shall show that these functions form an ON basis. Here the inner product is

$$(u, v) = \int_0^{2\pi} u(t)\overline{v}(t)\ dt, \qquad u, v \in L_2[0, 2\pi].$$

We will study this ON set and the completeness and convergence of expansions in the basis, both pointwise and in the norm. Before we get started, it is convenient to assume that $L_2[0, 2\pi]$ consists of square-integrable functions on the unit circle, rather than on an interval of the real line. Thus we will replace every function $f(t)$ on the interval $[0, 2\pi]$ by a function $f^*(t)$ such that $f^*(0) = f^*(2\pi)$ and $f^*(t) = f(t)$ for $0 \leq t < 2\pi$. Then we will extend f^* to all $-\infty < t < \infty$ by requiring periodicity: $f^*(t + 2\pi) = f^*(t)$. This will not affect the values of any integrals over the interval $[0, 2\pi]$, though it may change the value of f at one point. Thus, from now on our functions will be assumed 2π-periodic. One reason for this assumption is:

Lemma 3.1 *Suppose f is $2\pi - periodic$ and integrable. Then for any real number a*

$$\int_a^{2\pi+a} f(t)dt = \int_0^{2\pi} f(t)dt.$$

Proof Each side of the identity is just the integral of f over one period. For an analytic proof, note that

$$\int_0^{2\pi} f(t)dt = \int_0^a f(t)dt + \int_a^{2\pi} f(t)dt = \int_0^a f(t+2\pi)dt + \int_a^{2\pi} f(t)dt$$
$$= \int_{2\pi}^{2\pi+a} f(t)dt + \int_a^{2\pi} f(t)dt = \int_a^{2\pi+a} f(t)dt.$$

□

Thus we can transfer all our integrals to any interval of length 2π without altering the results.

Exercise 3.1 Let $g(a) = \int_a^{2\pi+a} f(t)dt$ and give a new proof of of Lemma 3.1 based on a computation of the derivative $g'(a)$.

For students who don't have a background in complex variable theory we will define the complex exponential in terms of real sines and cosines, and derive some of its basic properties directly. Let $z = x + iy$ be a complex number, where x and y are real. (Here and in all that follows, $i = \sqrt{-1}$.) Then $\bar{z} = x - iy$.

Definition 3.2 $e^z = \exp(x)(\cos y + i \sin y)$.

Lemma 3.3 *Properties of the complex exponential: $e^{z_1} e^{z_2} = e^{z_1+z_2}$, $|e^z| = \exp(x)$, $\overline{e^z} = e^{\bar{z}} = \exp(x)(\cos y - i \sin y)$.*

Exercise 3.2 Verify Lemma 3.3. You will need the addition formulas for sines and cosines.

Simple consequences for the basis functions $e_n(t) = e^{int}/\sqrt{2\pi}$, $n = 0, \pm 1, \pm 2, \ldots$ where t is real, are given by

Lemma 3.4 *Properties of e^{int}: $e^{in(t+2\pi)} = e^{int}$, $|e^{int}| = 1$, $\overline{e^{int}} = e^{-int}$, $e^{imt}e^{int} = e^{i(m+n)t}$, $e^0 = 1$,*

$$\frac{d}{dt}e^{int} = ine^{int}, \quad \cos nt = \frac{e^{int} + e^{-int}}{2}, \quad \sin nt = \frac{e^{int} - e^{-int}}{2i}.$$

Lemma 3.5 $(e_n, e_m) = \delta_{nm}$.

Proof If $n \neq m$ then

$$(e_n, e_m) = \frac{1}{2\pi} \int_0^{2\pi} e^{i(n-m)t} dt = \frac{1}{2\pi} \frac{e^{i(n-m)t}}{i(n-m)} \Big|_0^{2\pi} = 0.$$

If $n = m$ then $(e_n, e_m) = \frac{1}{2\pi} \int_0^{2\pi} 1 \, dt = 1$. $\qquad\qquad\qquad\square$

Since $\{e_n\}$ is an ON set, we can project any $f \in L_2[0, 2\pi]$ on the closed subspace generated by this set to get the Fourier expansion

$$f(t) \sim \sum_{n=-\infty}^{\infty} (f, e_n) e_n(t),$$

or

$$f(t) \sim \sum_{n=-\infty}^{\infty} c_n e^{int}, \qquad c_n = \frac{1}{2\pi} \int_0^{2\pi} f(t) e^{-int} dt. \qquad (3.1)$$

This is the *complex version* of Fourier series. (For now the $\sim$ just denotes that the right-hand side is the Fourier series of the left-hand side. In what sense the Fourier series represents the function is a matter to be resolved.) From our study of Hilbert spaces we already know that Bessel's inequality holds: $(f, f) \geq \sum_{n=-\infty}^{\infty} |(f, e_n)|^2$ or

$$\frac{1}{2\pi} \int_0^{2\pi} |f(t)|^2 dt \geq \sum_{n=-\infty}^{\infty} |c_n|^2. \qquad (3.2)$$

An immediate consequence is the Riemann–Lebesgue Lemma.

Lemma 3.6 *(Riemann–Lebesgue, weak form)* $\lim_{|n| \to \infty} \int_0^{2\pi} f(t) e^{-int} dt = 0.$

Thus, as $|n|$ gets large the Fourier coefficients go to 0.

If f is a real-valued function then $\bar{c}_n = c_{-n}$ for all n. If we set

$$c_n = \frac{a_n - ib_n}{2}, \qquad n = 0, 1, 2, \ldots$$

$$c_{-n} = \frac{a_n + ib_n}{2}, \qquad n = 1, 2, \ldots$$

and rearrange terms, we get the *real version* of Fourier series:

$$f(t) \sim \frac{a_0}{2} + \sum_{n=1}^{\infty} (a_n \cos nt + b_n \sin nt), \qquad (3.3)$$

$$a_n = \frac{1}{\pi} \int_0^{2\pi} f(t) \cos nt \, dt \qquad b_n = \frac{1}{\pi} \int_0^{2\pi} f(t) \sin nt \, dt$$

with Bessel inequality

$$\frac{1}{\pi}\int_0^{2\pi}|f(t)|^2dt \geq \frac{|a_0|^2}{2} + \sum_{n=1}^{\infty}(|a_n|^2 + |b_n|^2),$$

and Riemann–Lebesgue Lemma

$$\lim_{n\to\infty}\int_0^{2\pi}f(t)\cos(nt)dt = \lim_{n\to\infty}\int_0^{2\pi}f(t)\sin(nt)dt = 0.$$

Remark: The set $\{\frac{1}{\sqrt{2\pi}}, \frac{1}{\sqrt{\pi}}\cos nt, \frac{1}{\sqrt{\pi}}\sin nt\}$ for $n = 1, 2, \ldots$ is also ON in $L_2[0, 2\pi]$, as is easy to check, so (3.3) is the correct Fourier expansion in this basis for complex functions $f(t)$, as well as real functions.

Later we will prove the following fundamental result called Parseval's equality:

Theorem 3.7 *Let $f \in L_2[0, 2\pi]$. Then*

$$(f, f) = \sum_{n=-\infty}^{\infty}|(f, e_n)|^2.$$

In terms of the complex and real versions of Fourier series this reads

$$\frac{1}{2\pi}\int_0^{2\pi}|f(t)|^2dt = \sum_{n=-\infty}^{\infty}|c_n|^2 \tag{3.4}$$

or

$$\frac{1}{\pi}\int_0^{2\pi}|f(t)|^2dt = \frac{|a_0|^2}{2} + \sum_{n=1}^{\infty}(|a_n|^2 + |b_n|^2).$$

Let $f \in L_2[0, 2\pi]$ and remember that we are assuming that all such functions satisfy $f(t + 2\pi) = f(t)$. We say that f is *piecewise continuous* on $[0, 2\pi]$ if it is continuous except for a finite number of discontinuities. Furthermore, at each t the limits $f(t + 0) = \lim_{h\to 0, h>0} f(t + h)$ and $f(t - 0) = \lim_{h\to 0, h>0} f(t - h)$ exist. NOTE: At a point t of continuity of f we have $f(t + 0) = f(t - 0)$, whereas at a point of discontinuity $f(t + 0) \neq f(t - 0)$ and $f(t + 0) - f(t - 0)$ is the magnitude of the jump discontinuity.

Theorem 3.8 *Suppose*

- *$f(t)$ is periodic with period 2π.*
- *$f(t)$ is piecewise continuous on $[0, 2\pi]$.*
- *$f'(t)$ is piecewise continuous on $[0, 2\pi]$.*

Then the Fourier series of $f(t)$ converges to $\frac{f(t+0)+f(t-0)}{2}$ at each point t.

We will first illustrate this theorem and then prove it.

3.2 Examples

We will use the real version of Fourier series for these examples. The transformation to the complex version is elementary.

1. Let

$$f(t) = \begin{cases} 0, & t = 0 \\ \frac{\pi - t}{2}, & 0 < t < 2\pi \\ 0, & t = 2\pi \end{cases}$$

and $f(t + 2\pi) = f(t)$. We have $a_0 = \frac{1}{\pi} \int_0^{2\pi} \frac{\pi-t}{2}\, dt = 0$ and for $n \geq 1$,

$$a_n = \frac{1}{\pi} \int_0^{2\pi} \frac{\pi - t}{2} \cos nt\, dt = \frac{\frac{\pi-t}{2} \sin nt}{n\pi} \Big|_0^{2\pi} + \frac{1}{2\pi n} \int_0^{2\pi} \sin nt\, dt = 0,$$

$$b_n = \frac{1}{\pi} \int_0^{2\pi} \frac{\pi - t}{2} \sin nt\, dt = -\frac{\frac{\pi-t}{2} \cos nt}{n\pi} \Big|_0^{2\pi} - \frac{1}{2\pi n} \int_0^{2\pi} \cos nt\, dt = \frac{1}{n}.$$

Therefore,

$$\frac{\pi - t}{2} = \sum_{n=1}^{\infty} \frac{\sin nt}{n}, \qquad 0 < t < 2\pi.$$

By setting $t = \pi/2$ in this expansion we get an alternating series for $\pi/4$:

$$\frac{\pi}{4} = 1 - \frac{1}{3} + \frac{1}{5} - \frac{1}{7} + \frac{1}{9} - \cdots$$

Parseval's identity gives

$$\frac{\pi^2}{6} = \sum_{n=1}^{\infty} \frac{1}{n^2}.$$

2. Let

$$f(t) = \begin{cases} \frac{1}{2}, & t = 0 \\ 1, & 0 < t < \pi \\ \frac{1}{2}, & t = \pi \\ 0 & \pi < t < 2\pi \end{cases}$$

and $f(t + 2\pi) = f(t)$ (a step function). We have $a_0 = \frac{1}{\pi} \int_0^\pi dt = 1$, and for $n \geq 1$,

$$a_n = \frac{1}{\pi} \int_0^\pi \cos nt \, dt = \frac{\sin nt}{n\pi} \Big|_0^\pi = 0,$$

$$b_n = \frac{1}{\pi} \int_0^\pi \sin nt \, dt = -\frac{\cos nt}{n\pi} \Big|_0^\pi = \frac{(-1)^{n+1} + 1}{n\pi} = \begin{cases} \frac{2}{\pi n}, & n \text{ odd} \\ 0, & n \text{ even}. \end{cases}$$

Therefore,

$$f(t) = \frac{1}{2} + \frac{2}{\pi} \sum_{j=1}^\infty \frac{\sin(2j-1)t}{2j-1}.$$

For $0 < t < \pi$ this gives

$$\frac{\pi}{4} = \sin t + \frac{\sin 3t}{3} + \frac{\sin 5t}{5} + \cdots,$$

and for $\pi < t < 2\pi$ it gives

$$-\frac{\pi}{4} = \sin t + \frac{\sin 3t}{3} + \frac{\sin 5t}{5} + \cdots$$

Parseval's equality becomes

$$\frac{\pi^2}{8} = \sum_{j=1}^\infty \frac{1}{(2j-1)^2}.$$

3.3 Intervals of varying length, odd and even functions

Although it is convenient to base Fourier series on an interval of length 2π there is no necessity to do so. Suppose we wish to look at functions $f(x)$ in $L_2[\alpha, \beta]$. We simply make the change of variables

$$t = \frac{2\pi(x - \alpha)}{\beta - \alpha} = 2\pi u$$

in our previous formulas. Every function $f(x) \in L_2[\alpha, \beta]$ is uniquely associated with a function $\hat{f}(t) \in L_2[0, 2\pi]$ by the formula $f(x) = \hat{f}(\frac{2\pi(x-\alpha)}{\beta-\alpha})$. The set $\{\frac{1}{\sqrt{\beta-\alpha}}, \sqrt{\frac{2}{\beta-\alpha}} \cos \frac{2\pi n(x-\alpha)}{\beta-\alpha}, \sqrt{\frac{2}{\beta-\alpha}} \sin \frac{2\pi n(x-\alpha)}{\beta-\alpha}\}$ for $n = 1, 2, \ldots$ is an ON basis for $L_2[\alpha, \beta]$, The real Fourier expansion is, for $u = (x - \alpha)/(\beta - \alpha)$,

$$f(x) \sim \frac{a_0}{2} + \sum_{n=1}^{\infty}(a_n \cos 2\pi nu + b_n \sin 2\pi nu), \tag{3.5}$$

$$a_n = \frac{2}{\beta - \alpha} \int_{\alpha}^{\beta} f(x) \cos 2\pi nu \, dx, \quad b_n = \frac{2}{\beta - \alpha} \int_{\alpha}^{\beta} f(x) \sin 2\pi nu \, dx$$

with Parseval's equality

$$\frac{2}{\beta - \alpha} \int_{\alpha}^{\beta} |f(x)|^2 dx = \frac{|a_0|^2}{2} + \sum_{n=1}^{\infty}(|a_n|^2 + |b_n|^2).$$

For our next variant of Fourier series it is convenient to consider the interval $[-\pi, \pi]$ and the Hilbert space $L_2[-\pi, \pi]$. This makes no difference in the formulas for the coefficients, since all elements of the space are 2π-periodic. Now suppose $f(t)$ is defined and square integrable on the interval $[0, \pi]$. We define $F(t) \in L_2[-\pi, \pi]$ by

$$F(t) = \begin{cases} f(t) & \text{on } [0, \pi] \\ f(-t) & \text{for } -\pi < t < 0. \end{cases}$$

The function F has been constructed so that it is *even*, i.e., $F(-t) = F(t)$. For an even function the coefficients $b_n = \frac{1}{\pi}\int_{-\pi}^{\pi} F(t) \sin nt \, dt = 0$ so

$$F(t) \sim \frac{a_0}{2} + \sum_{n=1}^{\infty} a_n \cos nt$$

on $[-\pi, \pi]$ or

$$f(t) \sim \frac{a_0}{2} + \sum_{n=1}^{\infty} a_n \cos nt, \quad \text{for } 0 \le t \le \pi \tag{3.6}$$

$$a_n = \frac{1}{\pi} \int_{-\pi}^{\pi} F(t) \cos nt \, dt = \frac{2}{\pi} \int_{0}^{\pi} f(t) \cos nt \, dt.$$

Here, (3.6) is called the *Fourier cosine series* of f.

We can also extend the function $f(t)$ from the interval $[0, \pi]$ to an odd function on the interval $[-\pi, \pi]$. We define $G(t) \in L_2[-\pi, \pi]$ by

$$G(t) = \begin{cases} f(t) & \text{on } (0, \pi] \\ 0 & \text{for } t = 0 \\ -f(-t) & \text{for } -\pi < t < 0. \end{cases}$$

The function G has been constructed so that it is *odd*, i.e., $G(-t) = -G(t)$. For an odd function the coefficients $a_n = \frac{1}{\pi} \int_{-\pi}^{\pi} G(t) \cos nt \, dt = 0$ so

$$G(t) \sim \sum_{n=1}^{\infty} b_n \sin nt$$

on $[-\pi, \pi]$ or

$$f(t) \sim \sum_{n=1}^{\infty} b_n \sin nt, \quad \text{for } 0 < t \le \pi, \tag{3.7}$$

$$b_n = \frac{1}{\pi} \int_{-\pi}^{\pi} G(t) \sin nt \, dt = \frac{2}{\pi} \int_{0}^{\pi} f(t) \sin nt \, dt.$$

Here, (3.7) is called the *Fourier sine series* of f.

Example 3.9 Let $f(t) = t$, $0 \le t \le \pi$.

1. Fourier Sine series.

$$b_n = \frac{2}{\pi} \int_{0}^{\pi} t \sin nt \, dt = \frac{-2t \cos nt}{n\pi} \Big|_{0}^{\pi} + \frac{2}{n\pi} \int_{0}^{\pi} \cos nt \, dt = \frac{2(-1)^{n+1}}{n}.$$

Therefore,

$$t = \sum_{n=1}^{\infty} \frac{2(-1)^{n+1}}{n} \sin nt, \quad 0 < t < \pi.$$

2. Fourier Cosine series.

$$a_n = \frac{2}{\pi} \int_{0}^{\pi} t \cos nt \, dt = \frac{2t \sin nt}{n\pi} \Big|_{0}^{\pi} - \frac{2}{n\pi} \int_{0}^{\pi} \sin nt \, dt = \frac{2[(-1)^n - 1]}{n^2 \pi},$$

for $n \ge 1$ and $a_0 = \frac{2}{\pi} \int_{0}^{\pi} t \, dt = \pi$, so

$$t = \frac{\pi}{2} - \frac{4}{\pi} \sum_{j=1}^{\infty} \frac{\cos(2j-1)t}{(2j-1)^2}, \quad 0 < t < \pi.$$

3.4 Convergence results

In this section we will prove Theorem 3.8 on pointwise convergence. Let f be a complex-valued function such that

- $f(t)$ is periodic with period 2π.
- $f(t)$ is piecewise continuous on $[0, 2\pi]$.
- $f'(t)$ is piecewise continuous on $[0, 2\pi]$.

Expanding f in a Fourier series (real form) we have

$$f(t) \sim \frac{a_0}{2} + \sum_{n=1}^{\infty}(a_n \cos nt + b_n \sin nt) = S(t), \tag{3.8}$$

$$a_n = \frac{1}{\pi}\int_0^{2\pi} f(x)\cos nx \, dx, \qquad b_n = \frac{1}{\pi}\int_0^{2\pi} f(x)\sin nx \, dx.$$

For a fixed t we want to understand the conditions under which the Fourier series converges to a number $S(t)$, and the relationship between this number and f. To be more precise, let

$$S_k(t) = \frac{a_0}{2} + \sum_{n=1}^{k}(a_n \cos nt + b_n \sin nt)$$

be the kth partial sum of the Fourier series. This is a finite sum, a *trigonometric polynomial*, so it is well defined for all $t \in \mathbb{R}$. Now we have

$$S(t) = \lim_{k \to \infty} S_k(t),$$

if the limit exists. To better understand the properties of $S_k(t)$ in the limit, we will recast this finite sum as a single integral. Substituting the expressions for the Fourier coefficients a_n, b_n into the finite sum we find

$$S_k(t) = \frac{1}{\pi}\int_0^{2\pi}\left[\frac{1}{2} + \sum_{n=1}^{k}(\cos nx \cos nt + \sin nx \sin nt)\right]f(x)dx$$

$$= \frac{1}{\pi}\int_0^{2\pi}\left[\frac{1}{2} + \sum_{n=1}^{k}\cos[n(t-x)]\right]f(x)dx$$

$$= \frac{1}{\pi}\int_0^{2\pi} D_k(t-x)f(x)dx. \tag{3.9}$$

We can find a simpler form for the Dirichlet kernel

$$D_k(t) = \frac{1}{2} + \sum_{n=1}^{k}\cos nt = -\frac{1}{2} + \sum_{m=0}^{k}\cos mt = -\frac{1}{2} + \frac{1}{2}\sum_{m=0}^{k}\left[e^{itm} + e^{-itm}\right].$$

Recalling the geometric series

$$\sum_{m=0}^{k}(e^{it})^m = \frac{(e^{it})^{k+1} - 1}{e^{it} - 1}$$

we find, for $e^{it} \neq 1$, $D_k(t) =$

$$-\frac{1}{2} + \frac{1}{2}\left[\frac{e^{it(k+1)} - 1}{e^{it} - 1} + \frac{e^{-it(k+1)} - 1}{e^{-it} - 1}\right] = \frac{e^{itk} + e^{-itk} - e^{it(k+1)} - e^{-it(k+1)}}{2(e^{it} - 1)(e^{-it} - 1)}$$

$$= \frac{-\cos t(k+1) + \cos tk}{2\sin^2 \frac{t}{2}} = \frac{\sin(k+\frac{1}{2})t}{2\sin \frac{t}{2}}, \tag{3.10}$$

where we have made use of standard addition and double angle formulas for the trigonometric functions. For $e^{it} = 1$ we have $\cos mt = 1$ so it is clear that $D_k(0) = k + 1/2$.

Exercise 3.3 Verify the details of the derivation of the expression (3.10) for $D_k(t)$.

Note that D_k has the properties:

- $D_k(t) = D_k(t + 2\pi)$
- $D_k(-t) = D_k(t)$
- $D_k(t)$ is defined and differentiable for all t and $D_k(0) = k + \frac{1}{2}$.

From these properties it follows that the integrand of (3.9) is a 2π-periodic function of x, so that we can take the integral over any full 2π-period:

$$S_k(t) = \frac{1}{\pi} \int_{a-\pi}^{a+\pi} D_k(t-x)f(x)dx,$$

for any real number a. Let us set $a = t$ and fix a δ such that $0 < \delta < \pi$. (Think of δ as a very small positive number.) We break up the integral as follows:

$$S_k(t) = \frac{1}{\pi} \left(\int_{t-\pi}^{t-\delta} + \int_{t+\delta}^{t+\pi} \right) D_k(t-x)f(x)dx + \frac{1}{\pi} \int_{t-\delta}^{t+\delta} D_k(t-x)f(x)dx.$$

For fixed t we can write $D_k(t-x)$ in the form

$$D_k(t-x) = \frac{f_1(x,t)\cos k(t-x) + f_2(x,t)\sin k(t-x)}{\sin[\frac{1}{2}(t-x)]}$$

$$= g_1(x,t)\cos k(t-x) + g_2(x,t)\sin k(t-x).$$

In the interval $[t-\pi, t-\delta]$ the functions g_1, g_2 are bounded. Thus the functions

$$G_\ell(x,t) = \begin{cases} g_\ell(x,t) & \text{for } x \in [t-\pi, t-\delta] \\ 0 & \text{elsewhere} \end{cases}, \qquad \ell = 1,2$$

are elements of $L_2[-\pi, \pi]$ (and its 2π-periodic extension). Hence, by the Riemann–Lebesgue Lemma, applied to the ON basis determined by the orthogonal functions $\cos k(t-x), \sin k(t-x)$ for fixed t, the first integral goes to 0 as $k \to \infty$. A similar argument shows that the integral over the interval $[t + \delta, t + \pi]$ goes to 0 as $k \to \infty$. (This argument doesn't hold for the interval $[t - \delta, t + \delta]$ because the term $\sin[\frac{1}{2}(t - x)]$ vanishes in the interval, so that the G_ℓ are not square integrable.) Thus,

$$\lim_{k \to \infty} S_k(t) = \lim_{k \to \infty} \frac{1}{\pi} \int_{t-\delta}^{t+\delta} D_k(t - x) f(x) dx, \qquad (3.11)$$

where,

$$D_k(t) = \frac{\sin(k + \frac{1}{2})t}{2 \sin \frac{t}{2}}.$$

Theorem 3.10 *The sum $S(t)$ of the Fourier series of f at t is completely determined by the behavior of f in an arbitrarily small interval $(t - \delta, t + \delta)$ about t.*

This is a remarkable fact! It is called the Localization Theorem. Although the Fourier coefficients contain information about all of the values of f over the interval $[0, 2\pi)$, only the local behavior of f affects the convergence at a specific point t.

Now we are ready to prove a basic pointwise convergence result. It makes use of another simple property of the kernel function $D_k(t)$:

Lemma 3.11

$$\int_0^{2\pi} D_k(x) dx = \pi.$$

Proof

$$\int_0^{2\pi} D_k(x) dx = \int_0^{2\pi} (\frac{1}{2} + \sum_{n=1}^{k} \cos nx) dx = \pi,$$

since each of the last k functions integrates to 0. □

Theorem 3.12 *Suppose*

- $f(t)$ *is periodic with period* 2π.
- $f(t)$ *is piecewise continuous on* $[0, 2\pi]$.
- $f'(t)$ *is piecewise continuous on* $[0, 2\pi]$.

Then the Fourier series of $f(t)$ converges to $\frac{f(t+0)+f(t-0)}{2}$ at each point t.

Proof It will be convenient to modify f, if necessary, so that

$$f(t) = \frac{f(t + 0) + f(t - 0)}{2}$$

at each point t. This condition affects the definition of f only at a finite number of points of discontinuity. It doesn't change any integrals and the values of the Fourier coefficients. Then, using Lemma 3.11 we can write

$$
\begin{aligned}
S_k(t) - f(t) &= \frac{1}{\pi} \int_0^{2\pi} D_k(t-x)[f(x) - f(t)]dx \\
&= \frac{1}{\pi} \int_0^{\pi} D_k(x)[f(t+x) + f(t-x) - 2f(t)]dx \\
&= \frac{1}{\pi} \int_0^{\pi} \frac{[f(t+x) + f(t-x) - 2f(t)]}{2 \sin \frac{x}{2}} \sin(k + \frac{1}{2})x \, dx \\
&= \frac{1}{\pi} \int_0^{\pi} [H_1(t,x) \sin kx + H_2(t,x) \cos kx]dx.
\end{aligned}
$$

From the assumptions, H_1, H_2 are square integrable in x. Indeed, we can use L'Hospital's rule and the assumptions that f and f' are piecewise continuous to show that the limit

$$
\lim_{x \to 0+} \frac{[f(t+x) + f(t-x) - 2f(t)]}{2 \sin \frac{x}{2}}
$$

exists. Thus H_1, H_2 are bounded for $x \to 0+$. Then, by the Riemann–Lebesgue Lemma, the last expression goes to 0 as $k \to \infty$:

$$
\lim_{k \to \infty} [S_k(t) - f(t)] = 0.
$$

$\square$

Exercise 3.4 Suppose f is piecewise continuous and 2π-periodic. For any point t define the *right-hand derivative* $f'_R(t)$ and the *left-hand derivative* $f'_L(t)$ of f by

$$
f'_R(t) = \lim_{u \to t+} \frac{f(u) - f(t+0)}{u - t}, \quad f'_L(t) = \lim_{u \to t-} \frac{f(u) - f(t-0)}{u - t},
$$

respectively. Show that in the proof of Theorem 3.12 we can drop the requirement for f' to be piecewise continuous and the conclusion of the theorem will still hold at any point t such that both $f'_R(t)$ and $f'_L(t)$ exist.

Exercise 3.5 Show that if f and f' are piecewise continuous then for any point t we have $f'(t+0) = f'_R(t)$ and $f'(t-0) = f'_L(t)$. Hint: By the mean value theorem of calculus, for $u > t$ and u sufficiently close to t there is a point c such that $t < c < u$ and

$$\frac{f(u) - f(t+0)}{u - t} = f'(c).$$

Exercise 3.6 Let

$$f(t) = \begin{cases} t^2 \sin(\frac{1}{t}) & \text{for } t \neq 0 \\ 0 & \text{for } t = 0. \end{cases}$$

Show that f is continuous for all t and that $f'_R(0) = f'_L(0) = 0$. Show that $f'(t+0)$ and $f'(t-0)$ do not exist for $t = 0$. Hence, argue that f' is not a piecewise continuous function. This shows that the result of Exercise 3.4 is a true strengthening of Theorem 3.12.

Exercise 3.7 Let

$$f(t) = \begin{cases} \frac{2 \sin \frac{t}{2}}{t} & \text{if } 0 < |t| \leq \pi, \\ 1 & \text{if } t = 0. \end{cases}$$

Extend f to be a 2π-periodic function on the entire real line. Verify that f satisfies the hypotheses of Theorem 3.12 and is continuous at $t = 0$. Apply the Localization Theorem (3.11) to f at $t = 0$ to give a new evaluation of the improper integral $\int_0^\infty \frac{\sin x}{x} dx = \pi/2$.

3.4.1 Uniform pointwise convergence

We have shown that for functions f with the properties:

- $f(t)$ is periodic with period 2π,
- $f(t)$ is piecewise continuous on $[0, 2\pi]$,
- $f'(t)$ is piecewise continuous on $[0, 2\pi]$,

then at each point t the partial sums of the Fourier series of f,

$$f(t) \sim \frac{a_0}{2} + \sum_{n=1}^\infty (a_n \cos nt + b_n \sin nt) = S(t), \qquad a_n = \frac{1}{\pi} \int_0^{2\pi} f(t) \cos nt$$

$$b_n = \frac{1}{\pi} \int_0^{2\pi} f(t) \sin nt \, dt,$$

converge to $\frac{f(t+0)+f(t-0)}{2}$:

$$S_k(t) = \frac{a_0}{2} + \sum_{n=1}^k (a_n \cos nt + b_n \sin nt), \qquad \lim_{k \to \infty} S_k(t) = \frac{f(t+0) + f(t-0}{2}$$

(If we require that f satisfies $f(t) = \frac{f(t+0)+f(t-0)}{2}$ at each point then the series will converge to f everywhere. In this section we will make this requirement.) Now we want to examine the rate of convergence.

We know that for every $\epsilon > 0$ we can find an integer $N(\epsilon, t)$ such that $|S_k(t) - f(t)| < \epsilon$ for every $k > N(\epsilon, t)$. Then the finite sum trigonometric polynomial $S_k(t)$ will approximate $f(t)$ with an error $< \epsilon$. However, in general N depends on the point t; we have to recompute it for each t. What we would prefer is *uniform convergence*. The Fourier series of f will converge to f *uniformly* if for every $\epsilon > 0$ we can find an integer $N(\epsilon)$ such that $|S_k(t) - f(t)| < \epsilon$ for every $k > N(\epsilon)$ and *for all* t. Then the finite sum trigonometric polynomial $S_k(t)$ will approximate $f(t)$ everywhere with an error $< \epsilon$.

We cannot achieve uniform convergence for all functions f in the class above. The partial sums are continuous functions of t. Recall from calculus that if a sequence of continuous functions converges uniformly, the limit function is also continuous. Thus for any function f with discontinuities, we cannot have uniform convergence of the Fourier series.

If f is continuous, however, then we do have uniform convergence.

Theorem 3.13 *Assume f has the properties:*

- $f(t)$ *is periodic with period 2π,*
- $f(t)$ *is continuous on $[0, 2\pi]$,*
- $f'(t)$ *is piecewise continuous on $[0, 2\pi]$.*

Then the Fourier series of f converges uniformly.

Proof Consider the Fourier series of both f and f':

$$f(t) \sim \frac{a_0}{2} + \sum_{n=1}^{\infty}(a_n \cos nt + b_n \sin nt), \qquad a_n = \frac{1}{\pi}\int_0^{2\pi} f(t) \cos nt \, dt$$

$$b_n = \frac{1}{\pi}\int_0^{2\pi} f(t) \sin nt \, dt,$$

$$f'(t) \sim \frac{A_0}{2} + \sum_{n=1}^{\infty}(A_n \cos nt + B_n \sin nt), \qquad A_n = \frac{1}{\pi}\int_0^{2\pi} f'(t) \cos nt \, dt$$

$$B_n = \frac{1}{\pi}\int_0^{2\pi} f'(t) \sin nt \, dt.$$

Now

$$A_n = \frac{1}{\pi}\int_0^{2\pi} f'(t) \cos nt \, dt = \frac{1}{\pi}f(t) \cos nt \Big|_0^{2\pi} + \frac{n}{\pi}\int_0^{2\pi} f(t) \sin nt \, dt$$

$$= \begin{cases} nb_n, & n \geq 1 \\ 0, & n = 0. \end{cases}$$

(We have used the fact that $f(0) = f(2\pi)$.) Similarly, $B_n =$

$$\frac{1}{\pi} \int_0^{2\pi} f'(t) \sin nt \, dt = \frac{1}{\pi} f(t) \sin nt \Big|_0^{2\pi} - \frac{n}{\pi} \int_0^{2\pi} f(t) \cos nt \, dt = -na_n.$$

Using Bessel's inequality for f' we have

$$\sum_{n=1}^{\infty} (|A_n|^2 + |B_n|^2) \leq \frac{1}{\pi} \int_0^{2\pi} |f'(t)|^2 dt < \infty,$$

hence

$$\sum_{n=1}^{\infty} n^2 (|a_n|^2 + |b_n|^2) \leq \infty.$$

Now

$$\frac{\sum_{n=1}^m |a_n|}{\sum_{n=1}^m |b_n|} \leq \sum_{n=1}^m \sqrt{|a_n|^2 + |b_n|^2} = \sum_{n=1}^m \frac{1}{n} \sqrt{|A_n|^2 + |B_n|^2}$$

$$\leq (\sum_{n=1}^m \frac{1}{n^2})(\sum_{n=1}^m (|A_n|^2 + |B_n|^2))$$

which converges as $m \to \infty$. (We have used the Schwarz inequality for the last step.) Hence $\sum_{n=1}^{\infty} |a_n| < \infty$, $\sum_{n=1}^{\infty} |b_n| < \infty$. Now

$$\left| \frac{a_0}{2} + \sum_{n=1}^{\infty} (a_n \cos nt + b_n \sin nt) \right| \leq \left| \frac{a_0}{2} \right| + \sum_{n=1}^{\infty} (|a_n \cos nt| + |b_n \sin nt|)$$

$$\leq \left| \frac{a_0}{2} \right| + \sum_{n=1}^{\infty} (|a_n| + |b_n|) < \infty,$$

so the series converges uniformly and absolutely. □

Corollary 3.14 *For f satisfying the hypotheses of the preceding theorem*

$$\frac{|a_0|^2}{2} + \sum_{n=1}^{\infty} (|a_n|^2 + |b_n|^2) = \frac{1}{\pi} \int_0^{2\pi} |f(t)|^2 dt.$$

Proof This is Parseval's Theorem. The Fourier series of f converges uniformly, so for any $\epsilon > 0$ there is an integer $N(\epsilon)$ such that $|S_k(t) - f(t)| < \epsilon$ for every $k > N(\epsilon)$ and *for all t*. Thus

$$\int_0^{2\pi} |S_k(t) - f(t)|^2 \, dt = ||S_k - f||^2$$

$$= ||f||^2 - \pi \left(\frac{|a_0|^2}{2} + \sum_{n=1}^k (|a_n|^2 + |b_n|^2) \right) < 2\pi \epsilon^2$$

for $k > N(\epsilon)$. □

Remark: Parseval's Theorem actually holds for any $f \in L_2[0, 2\pi]$, as we shall show later.

Remark: As the proof of the preceding theorem illustrates, differentiability of a function improves convergence of its Fourier series. The more derivatives the faster the convergence. There are famous examples to show that continuity alone is *not* sufficient for convergence, [91, 92].

3.5 More on pointwise convergence, Gibbs phenomena

We return to our first Example of Fourier series:

$$h(t) = \begin{cases} 0, & t = 0 \\ \frac{\pi - t}{2}, & 0 < t < 2\pi \\ 0, & t = 2\pi \end{cases}$$

and $h(t + 2\pi) = h(t)$. In this case, $a_n = 0$ for all n and $b_n = 1/n$. Therefore,

$$\frac{\pi - t}{2} = \sum_{n=1}^{\infty} \frac{\sin nt}{n}, \qquad 0 < t < 2\pi.$$

The function h has a discontinuity at $t = 0$ so the convergence of this series can't be uniform. Let's examine this case carefully. What happens to the partial sums near the discontinuity?

Here, $S_k(t) = \sum_{n=1}^{k} \frac{\sin nt}{n}$ so

$$S'_k(t) = \sum_{n=1}^{k} \cos nt = D_k(t) - \frac{1}{2} = \frac{\sin(k + \frac{1}{2})t}{2 \sin \frac{t}{2}} - \frac{1}{2} = \frac{\sin \frac{kt}{2} \cos \frac{(k+1)t}{2}}{\sin \frac{t}{2}}.$$

Thus, since $S_k(0) = 0$ we have

$$S_k(t) = \int_0^t S'_k(x) dx = \int_0^t \left(\frac{\sin \frac{kx}{2} \cos \frac{(k+1)x}{2}}{2 \sin \frac{x}{2}} \right) dx.$$

Note that $S'_k(0) > 0$ so that S_k starts out at 0 for $t = 0$ and then increases. Looking at the derivative of S_k we see that the first maximum is at the critical point $t_k = \frac{\pi}{k+1}$ (the first zero of $\cos \frac{(k+1)x}{2}$ as x increases from 0). Here, $h(t_k) = \frac{\pi - t_k}{2}$. The error is

$$S_k(t_k) - h(t_k) = \int_0^{t_k} \frac{\sin(k + \frac{1}{2})x}{2 \sin \frac{x}{2}} dx - \frac{\pi}{2}$$

$$= \int_0^{t_k} \frac{\sin(k+\frac{1}{2})x}{x} dx + \int_0^{t_k} \left(\frac{1}{2\sin \frac{x}{2}} - \frac{1}{x} \right) \sin(k+\frac{1}{2})x \, dx - \frac{\pi}{2}$$

$$= I(t_k) + J(t_k) - \frac{\pi}{2}$$

where $I(t_k) =$

$$\int_0^{t_k} \frac{\sin(k+\frac{1}{2})x}{x} dx = \int_0^{(k+\frac{1}{2})t_k} \frac{\sin u}{u} du \rightarrow \int_0^{\pi} \frac{\sin u}{u} du \approx 1.851397052$$

(according to MAPLE). Also

$$J(t_k) = \int_0^{t_k} \left(\frac{1}{2\sin \frac{x}{2}} - \frac{1}{x} \right) \sin(k+\frac{1}{2})x \, dx.$$

Note that the integrand is bounded near $x = 0$ (indeed it goes to 0 as $x \rightarrow 0$) and the interval of integration goes to 0 as $k \rightarrow \infty$. Hence we have $J(t_k) \rightarrow 0$ as $k \rightarrow \infty$. We conclude that

$$\lim_{k \rightarrow \infty} (S_k(t_k) - h(t_k)) \approx 1.851397052 - \frac{\pi}{2} \approx 0.280600725$$

To sum up, $\lim_{k \rightarrow \infty} S_k(t_k) \approx 1.851397052$ whereas $\lim_{k \rightarrow \infty} h(t_k) = \frac{\pi}{2} \approx 1.570796327$. The partial sum is overshooting the correct value by about 17.86359697%! This is called the *Gibbs Phenomenon*.

To understand it we need to look more carefully at the convergence properties of the partial sums $S_k(t) = \sum_{n=1}^{k} \frac{\sin nt}{n}$ for all t.

First some preparatory calculations. Consider the geometric series $E_k(t) = \sum_{n=1}^{k} e^{int} = \frac{e^{it}(1-e^{ikt})}{1-e^{it}}$.

Lemma 3.15 *For $0 < t < 2\pi$,*

$$|E_k(t)| \leq \frac{2}{|1 - e^{it}|} = \frac{1}{\sin \frac{t}{2}}.$$

Note that $S_k(t)$ is the imaginary part of the complex series $\sum_{n=1}^{k} \frac{e^{int}}{n}$.

Lemma 3.16 *Let $0 < \alpha < \beta < 2\pi$. The series $\sum_{n=1}^{\infty} \frac{e^{int}}{n}$ converges uniformly for all t in the interval $[\alpha, \beta]$.*

Proof (tricky)

$$\sum_{n=j}^{k} \frac{e^{int}}{n} = \sum_{n=j}^{k} \frac{E_n(t) - E_{n-1}(t)}{n} = \sum_{n=j}^{k} \frac{E_n(t)}{n} - \sum_{n=j}^{k} \frac{E_n(t)}{n+1} - \frac{E_{j-1}(t)}{j} + \frac{E_k(t)}{k+1}$$

so

$$\left|\sum_{n=j}^{k} \frac{e^{int}}{n}\right| \leq \frac{1}{\sin \frac{t}{2}} \left(\sum_{n=j}^{k} (\frac{1}{n} - \frac{1}{n+1}) + \frac{1}{j} + \frac{1}{k+1}\right) = \frac{2}{j \sin \frac{t}{2}}.$$

This implies by the Cauchy Criterion that $\sum_{n=j}^{k} \frac{e^{int}}{n}$ converges uniformly on $[\alpha, \beta]$. □

This shows that the Fourier series for $h(t)$ converges uniformly on any closed interval that doesn't contain the discontinuities at $t = 2\pi\ell$, $\ell = 0, \pm 1, \pm 2, \ldots$ Next we will show that the partial sums $S_k(t)$ are bounded for *all* t and all k. Thus, even though there is an overshoot near the discontinuities, the overshoot is strictly bounded.

From Lemma 3.16 on uniform convergence above we already know that the partial sums are bounded on any closed interval not containing a discontinuity. Also, $S_k(0) = 0$ and $S_k(-t) = -S_k(t)$, so it suffices to consider the interval $0 < t < \pi/2$.

We will use the facts that $2/\pi \leq \sin t/t \leq 1$ for $0 < t \leq \pi/2$. The right-hand inequality is a basic calculus fact and the left-hand one is obtained by solving the calculus problem of minimizing $\sin t/t$ over the interval $0 < t < \pi/2$. Note that

$$\left|\sum_{n=1}^{k} \frac{\sin nt}{n}\right| \leq \left|\sum_{1 \leq n < 1/t} \frac{t \sin nt}{nt}\right| + \left|\sum_{1/t \leq n \leq k} \frac{\sin nt}{n}\right|.$$

Using the calculus inequalities and the lemma, we have

$$\left|\sum_{n=1}^{k} \frac{\sin nt}{n}\right| \leq \sum_{1 \leq n < 1/t} t + \frac{2}{\frac{1}{t}\sin \frac{t}{2}} \leq t \cdot \frac{1}{t} + \frac{2}{\frac{1}{t}\frac{t}{2}\frac{2}{\pi}} = 1 + 2\pi.$$

Thus the partial sums are uniformly bounded for all t and all k.

We conclude that the Fourier series for $h(t)$ converges uniformly to $h(t)$ in any closed interval not including a discontinuity. Furthermore the partial sums of the Fourier series are uniformly bounded. At each discontinuity $t_N = 2\pi N$ of h the partial sums S_k overshoot $h(t_N + 0)$ by about 17.9% (approaching from the right) as $k \to \infty$ and undershoot $h(t_N - 0)$ by the same amount. This is illustrated in Figures 3.1 and 3.2.

All of the work that we have put into this single example will pay off, because the facts that have emerged are of broad validity. Indeed we can consider any function f satisfying our usual conditions as the sum of a continuous function for which the convergence is uniform everywhere and a finite number of translated and scaled copies of $h(t)$.

Fourier series

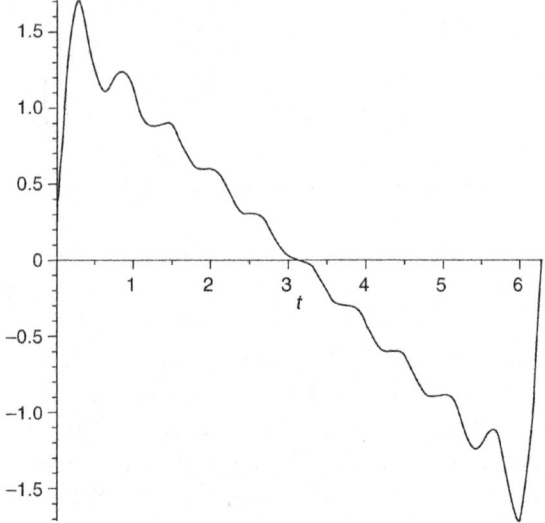

Figure 3.1 Plot of the sum of the first 10 nonzero terms in the Fourier series for $h(t)$ on the interval $0 \leq t \leq 2\pi$. Note the overshoot and undershoot around the discontinuities.

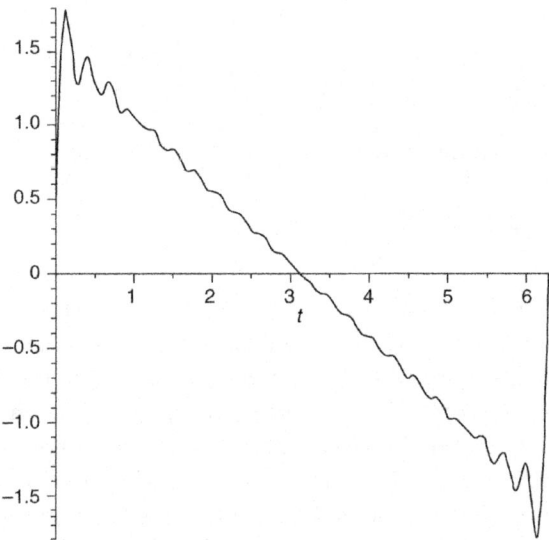

Figure 3.2 Plot of the sum of the first 22 nonzero terms in the Fourier series for $h(t)$ on the interval $0 \leq t \leq 2\pi$. Note that as the number of terms in the partial sum increases, the width of the overshoot/undershoot region narrows, but that the magnitude of the error stays the same.

Theorem 3.17 *Let f be a complex-valued function such that*

- $f(t)$ *is periodic with period* 2π.
- $f(t)$ *is piecewise continuous on* $[0, 2\pi]$.
- $f'(t)$ *is piecewise continuous on* $[0, 2\pi]$.
- $f(t) = \frac{f(t+0)+f(t-0)}{2}$ *at each point* t.

Then

$$f(t) = \frac{a_0}{2} + \sum_{n=1}^{\infty} (a_n \cos nt + b_n \sin nt)$$

pointwise. The convergence of the series is uniform on every closed interval in which f is continuous.

Proof Let $x_1, x_2, \ldots, x_\ell$ be the points of discontinuity of f in $[0, 2\pi)$. Set $s(x_j) = f(x_j + 0) - f(x_j - 0)$. Then the function

$$g(t) = f(t) - \sum_{j=1}^{\ell} \frac{s(x_j)}{\pi} h(t - x_j)$$

is everywhere continuous and also satisfies all of the hypotheses of the theorem. Indeed, at the discontinuity x_j of f we have $g(x_j - 0) =$

$$f(x_j - 0) - \frac{2s(x_j)}{\pi} h(-0) = f(x_j - 0) - \frac{f(x_j + 0) - f(x_j - 0)}{\pi} \left(\frac{-\pi}{2}\right)$$

$$= \frac{f(x_j - 0) + f(x_j + 0)}{2} = f(x_j).$$

Similarly, $g(x_j + 0) = f(x_j)$. Therefore $g(t)$ can be expanded in a Fourier series that converges absolutely and uniformly. However, $\sum_{j=1}^{\ell} \frac{s(x_j)}{\pi} h(t - x_j)$ can be expanded in a Fourier series that converges pointwise and uniformly in every closed interval that doesn't include a discontinuity. But

$$f(t) = g(t) + \sum_{j=1}^{\ell} \frac{s(x_j)}{\pi} h(t - x_j),$$

and the conclusion follows. $\square$

Corollary 3.18 *Parseval's Theorem. For f satisfying the hypotheses of the preceding theorem*

$$\frac{|a_0|^2}{2} + \sum_{n=1}^{\infty} (|a_n|^2 + |b_n|^2) = \frac{1}{\pi} \int_0^{2\pi} |f(t)|^2 dt.$$

Proof As in the proof of the theorem, let $x_1, x_2, \ldots, x_\ell$ be the points of discontinuity of f in $[0, 2\pi)$ and set $s(x_j) = f(x_j+0) - f(x_j-0)$. Choose $a \geq 0$ such that the discontinuities are contained in the interior of $I = (-a, 2\pi - a)$. From our earlier results we know that the partial sums of the Fourier series of h are uniformly bounded with bound $M > 0$. Choose $P = \sup_{t \in [0, 2\pi]} |f(t)|$. Then $|S_k(t) - f(t)|^2 \leq (M + P)^2$ for all t and all k. Given $\epsilon > 0$ choose non-overlapping open intervals $I_1, I_2, \ldots, I_\ell \subset I$ such that $x_j \in I_j$ and $(\sum_{j=1}^\ell |I_j|)(M+P)^2 < \epsilon/2$. Here, $|I_j|$ is the length of the interval I_j. Now the Fourier series of f converges uniformly on the closed set $A = [-a, 2\pi - a] - I_1 \cup I_2 \cup \cdots \cup I_\ell$. Choose an integer $N(\epsilon)$ such that $|S_k(t) - f(t)|^2 < \epsilon/4\pi$ for all $t \in A, k \geq N(\epsilon)$. Then

$$\int_0^{2\pi} |S_k(t) - f(t)|^2 dt = \int_{-a}^{2\pi - a} |S_k(t) - f(t)|^2 dt =$$

$$\int_A |S_k(t) - f(t)|^2 dt + \int_{I_1 \cup I_2 \cup \cdots \cup I_\ell} |S_k(t) - f(t)|^2 dt$$

$$< 2\pi \frac{\epsilon}{4\pi} + (\sum_{j=1}^\ell |I_j|)(M + P)^2 < \frac{\epsilon}{2} + \frac{\epsilon}{2} = \epsilon.$$

Thus $\lim_{k \to \infty} \|S_k - f\| = 0$ and the partial sums converge to f in the mean.

Furthermore,

$$\epsilon > \int_0^{2\pi} |S_k(t) - f(t)|^2 \, dt = \|S_k - f\|^2 = \|f\|^2 - \pi(\frac{|a_0|^2}{2} + \sum_{n=1}^k (|a_n|^2 + |b_n|^2))$$

for $k > N(\epsilon)$.   □

Exercise 3.8 Expand

$$f(x) = \begin{cases} 0 & -\pi < x \leq -\frac{\pi}{2} \\ 1 & -\frac{\pi}{2} < x \leq \frac{\pi}{2} \\ 0 & \frac{\pi}{2} < x \leq \pi \end{cases}$$

in a Fourier series on the interval $-\pi \leq x \leq \pi$. Plot both f and the partial sums S_k for $k = 5, 10, 20, 40$. Observe how the partial sums approximate f. What accounts for the slow rate of convergence?

3.6 Further properties of Fourier series

The convergence theorem and the version of the Parseval identity proved in the previous section apply to step functions on $[0, 2\pi]$. However, we already know that the space of step functions on $[0, 2\pi]$ is dense in $L_2[0, 2\pi]$. Since every step function is the limit in the norm of the partial sums of its Fourier series, this means that the space of all finite linear combinations of the functions $\{e^{int}\}$ is dense in $L_2[0, 2\pi]$. Hence $\{e^{int}/\sqrt{2\pi}\}$ is an ON basis for $L_2[0, 2\pi]$ and we have

Theorem 3.19 *Parseval's Equality (strong form) [Plancherel Theorem]. If $f \in L_2[0, 2\pi]$ then*

$$\frac{|a_0|^2}{2} + \sum_{n=1}^{\infty}(|a_n|^2 + |b_n|^2) = \frac{1}{\pi}\int_0^{2\pi}|f(t)|^2 dt,$$

where a_n, b_n are the Fourier coefficients of f.

Integration of a Fourier series term-by-term yields a series with improved convergence.

Theorem 3.20 *Let f be a complex-valued function such that*

- $f(t)$ *is periodic with period 2π.*
- $f(t)$ *is piecewise continuous on $[0, 2\pi]$.*
- $f'(t)$ *is piecewise continuous on $[0, 2\pi]$.*

Let

$$f(t) \sim \frac{a_0}{2} + \sum_{n=1}^{\infty}(a_n \cos nt + b_n \sin nt)$$

be the Fourier series of f. Then

$$\int_0^t f(x)dx = \frac{a_0}{2}t + \sum_{n=1}^{\infty}\frac{1}{n}[a_n \sin nt - b_n(\cos nt - 1)]$$

where the convergence is uniform on the interval $[0, 2\pi]$.

Proof Let $F(t) = \int_0^t f(x)dx - \frac{a_0}{2}t$. Then

- $F(2\pi) = \int_0^{2\pi} f(x)dx - \frac{a_0}{2}(2\pi) = 0 = F(0)$.
- $F(t)$ is continuous on $[0, 2\pi]$.
- $F'(t) = f(t) - \frac{a_0}{2}$ is piecewise continuous on $[0, 2\pi]$.

Thus the Fourier series of F converges to F uniformly and absolutely on $[0, 2\pi]$:

$$F(t) = \frac{A_0}{2} + \sum_{n=1}^{\infty}(A_n \cos nt + B_n \sin nt).$$

Now

$$A_n = \frac{1}{\pi}\int_0^{2\pi} F(t) \cos nt\ dt = \left.\frac{F(t)\sin nt}{n\pi}\right|_0^{2\pi} - \frac{1}{n\pi}\int_0^{2\pi}(f(t) - \frac{a_0}{2})\sin nt\ dt$$

$$= -\frac{b_n}{n}, \qquad n \neq 0,$$

and

$$B_n = \frac{1}{\pi}\int_0^{2\pi} F(t)\sin nt\ dt = -\left.\frac{F(t)\cos nt}{n\pi}\right|_0^{2\pi} + \frac{1}{n\pi}\int_0^{2\pi}(f(t) - \frac{a_0}{2})\cos nt\ dt$$

$$= \frac{a_n}{n}.$$

Therefore,

$$F(t) = \frac{A_0}{2} + \sum_{n=1}^{\infty}(-\frac{b_n}{n}\cos nt + \frac{a_n}{n}\sin nt),$$

$$F(2\pi) = 0 = \frac{A_0}{2} - \sum_{n=1}^{\infty}\frac{b_n}{n}.$$

Solving for A_0 we find

$$F(t) = \int_0^t f(x)dx - \frac{a_0}{2}t = \sum_{n=1}^{\infty}\frac{1}{n}[a_n \sin nt - b_n(\cos nt - 1)].$$

$\square$

Example 3.21 Let

$$f(t) = \begin{cases} \frac{\pi - t}{2} & 0 < t < 2\pi \\ 0 & t = 0, 2\pi. \end{cases}$$

Then

$$\frac{\pi - t}{2} \sim \sum_{n=1}^{\infty}\frac{\sin nt}{n}.$$

Integrating term-by term we find

$$\frac{2\pi t - t^2}{4} = -\sum_{n=1}^{\infty}\frac{1}{n^2}(\cos nt - 1), \quad 0 \le t \le 2\pi.$$

Differentiation of Fourier series, however, makes them less smooth and may not be allowed. For example, differentiating the Fourier series

$$\frac{\pi - t}{2} \sim \sum_{n=1}^{\infty} \frac{\sin nt}{n},$$

formally term-by term we get

$$-\frac{1}{2} \sim \sum_{n=1}^{\infty} \cos nt,$$

which doesn't converge on $[0, 2\pi]$. In fact it can't possibly be a Fourier series for an element of $L_2[0, 2\pi]$. (Why?)

If f is sufficiently smooth and periodic it *is* OK to differentiate term-by-term to get a new Fourier series.

Theorem 3.22 *Let f be a complex-valued function such that*

- *$f(t)$ is periodic with period 2π.*
- *$f(t)$ is continuous on $[0, 2\pi]$.*
- *$f'(t)$ is piecewise continuous on $[0, 2\pi]$.*
- *$f''(t)$ is piecewise continuous on $[0, 2\pi]$.*

Let

$$f(t) = \frac{a_0}{2} + \sum_{n=1}^{\infty} (a_n \cos nt + b_n \sin nt)$$

be the Fourier series of f. Then at each point $t \in [0, 2\pi]$ where $f''(t)$ exists we have

$$f'(t) = \sum_{n=1}^{\infty} n \left(-a_n \sin nt + b_n \cos nt \right).$$

Proof By the Fourier convergence theorem the Fourier series of f' converges to $\frac{f'(t_0+0)+f'(t_0-0)}{2}$ at each point t_0. If $f''(t_0)$ exists at the point then the Fourier series converges to $f'(t_0)$, where

$$f'(t) \sim \frac{A_0}{2} + \sum_{n=1}^{\infty} (A_n \cos nt + B_n \sin nt).$$

Now

$$A_n = \frac{1}{\pi} \int_0^{2\pi} f'(t) \cos nt \, dt = \left. \frac{f(t) \cos nt}{\pi} \right|_0^{2\pi} + \frac{n}{\pi} \int_0^{2\pi} f(t) \sin nt \, dt$$

$$= nb_n,$$

$A_0 = \frac{1}{\pi} \int_0^{2\pi} f'(t)dt = \frac{1}{\pi}(f(2\pi) - f(0)) = 0$ (where, if necessary, we adjust the interval of length 2π so that f' is continuous at the endpoints) and

$$B_n = \frac{1}{\pi} \int_0^{2\pi} f'(t) \sin nt \; dt = \left. \frac{f(t) \sin nt}{\pi} \right|_0^{2\pi} - \frac{n}{\pi} \int_0^{2\pi} f(t) \cos nt \; dt$$

$$= -n a_n.$$

Therefore,

$$f'(t) \sim \sum_{n=1}^{\infty} n(b_n \cos nt - a_n \sin nt).$$

$\square$

Note the importance of the requirement in the theorem that f is continuous everywhere and periodic, so that the boundary terms vanish in the integration by parts formulas for A_n and B_n. Thus it is OK to differentiate the Fourier series

$$f(t) = \frac{2\pi t - t^2}{4} - \frac{\pi^2}{6} = -\sum_{n=1}^{\infty} \frac{1}{n^2} \cos nt, \quad 0 \le t \le 2\pi$$

term-by-term, where $f(0) = f(2\pi)$, to get

$$f'(t) = \frac{\pi - t}{2} \sim \sum_{n=1}^{\infty} \frac{\sin nt}{n}.$$

However, even though $f'(t)$ is infinitely differentiable for $0 < t < 2\pi$ we have $f'(0) \ne f'(2\pi)$, so we cannot differentiate the series again.

Exercise 3.9 Consider Example 3.9: $f(t) = t$, $0 \le t \le \pi$. Is it permissible to differentiate the Fourier sine series for this function term-by-term? If so, say what the differentiated series converges to at each point. Is it permissible to differentiate the Fourier cosine series for the function $f(t)$ term-by-term? If so, say what the differentiated series converges to at each point.

3.7 Digging deeper: arithmetic summability and Fejér's theorem

We know that the kth partial sum of the Fourier series of a square integrable function f:

$$S_k(t) = \frac{a_0}{2} + \sum_{n=1}^{k} (a_n \cos nt + b_n \sin nt)$$

is the trigonometric polynomial of order k that best approximates f in the Hilbert space sense. However, the limit of the partial sums

$$S(t) = \lim_{k \to \infty} S_k(t),$$

doesn't necessarily converge pointwise. We have proved pointwise convergence for piecewise smooth functions, but if, say, all we know is that f is continuous then pointwise convergence is much harder to establish. Indeed there are examples of continuous functions whose Fourier series diverge at uncountably many points ([103], Chapter 8). Furthermore we have seen that at points of discontinuity the Gibbs phenomenon occurs and the partial sums overshoot the function values. In this section we will look at another way to recapture $f(t)$ from its Fourier coefficients, by Cesàro sums (arithmetic means). This method is surprisingly simple, gives uniform convergence for continuous functions $f(t)$ and avoids most of the Gibbs phenomenon difficulties.

The basic idea is to use the arithmetic means of the partial sums to approximate f. Recall that the kth partial sum of $f(t)$ is defined by

$$S_k(t) = \frac{1}{2\pi} \int_0^{2\pi} f(x)dx \quad +$$

$$\frac{1}{\pi} \sum_{n=1}^{k} \left(\int_0^{2\pi} f(x) \cos nx \, dx \cos nt + \int_0^{2\pi} f(x) \sin nx \, dx \sin nt \right),$$

so

$$S_k(t) = \frac{1}{\pi} \int_0^{2\pi} \left[\frac{1}{2} + \sum_{n=1}^{k} (\cos nx \cos nt + \sin nx \sin nt) \right] f(x)dx$$

$$= \frac{1}{\pi} \int_0^{2\pi} \left[\frac{1}{2} + \sum_{n=1}^{k} \cos[n(t - x)] \right] f(x)dx$$

$$= \frac{1}{\pi} \int_0^{2\pi} D_k(t - x)f(x)dx.$$

where the kernel $D_k(t) = \frac{1}{2} + \sum_{n=1}^{k} \cos nt = -\frac{1}{2} + \sum_{m=0}^{k} \cos mt$. Further,

$$D_k(t) = \frac{\cos kt - \cos(k+1)t}{4 \sin^2 \frac{t}{2}} = \frac{\sin(k + \frac{1}{2})t}{2 \sin \frac{t}{2}}.$$

Rather than use the partial sums $S_k(t)$ to approximate $f(t)$ we use the arithmetic means $\sigma_k(t)$ of these partial sums:

$$\sigma_k(t) = \frac{S_0(t) + S_1(t) + \cdots + S_{k-1}(t)}{k}, \qquad k = 1, 2, \ldots \qquad (3.12)$$

Then we have

$$\sigma_k(t) = \frac{1}{k\pi} \sum_{j=0}^{k-1} \int_0^{2\pi} D_j(t-x)f(x)dx = \int_0^{2\pi} \left[\frac{1}{k\pi} \sum_{j=0}^{k-1} D_j(t-x) \right] f(x)dx$$

$$= \frac{1}{\pi} \int_0^{2\pi} F_k(t-x)f(x)dx \qquad (3.13)$$

where

$$F_k(t) = \frac{1}{k} \sum_{j=0}^{k-1} D_j(t) = \frac{1}{k} \sum_{j=0}^{k-1} \frac{\sin(j+\frac{1}{2})t}{2\sin\frac{t}{2}}.$$

Lemma 3.23

$$F_k(t) = \frac{1}{k} \left(\frac{\sin kt/2}{\sin t/2} \right)^2.$$

Proof Using the geometric series, we have

$$\sum_{j=0}^{k-1} e^{i(j+\frac{1}{2})t} = e^{i\frac{t}{2}} \frac{e^{ikt}-1}{e^{it}-1} = e^{i\frac{kt}{2}} \frac{\sin\frac{kt}{2}}{\sin\frac{t}{2}}.$$

Taking the imaginary part of this identity we find

$$F_k(t) = \frac{1}{k\sin\frac{t}{2}} \sum_{j=0}^{k-1} \sin(j+\frac{1}{2})t = \frac{1}{k} \left(\frac{\sin kt/2}{\sin t/2} \right)^2.$$

$\square$

Note that F has the properties:

- $F_k(t) = F_k(t + 2\pi)$
- $F_k(-t) = F_k(t)$
- $F_k(t)$ is defined and differentiable for all t and $F_k(0) = k$
- $F_k(t) \geq 0$.

From these properties it follows that the integrand of (3.13) is a 2π-periodic function of x, so that we can take the integral over any full 2π-period. Finally, we can change variables and divide up the integral, in analogy with our study of the Dirichlet kernel $D_k(t)$, and obtain the following simple expression for the arithmetic means:

Lemma 3.24

$$\sigma_k(t) = \frac{2}{k\pi} \int_0^{\pi/2} \frac{f(t+2x) + f(t-2x)}{2} \left(\frac{\sin kx}{\sin x} \right)^2 dx.$$

Exercise 3.10 Derive Lemma 3.24 from expression (3.13) and Lemma 3.23.

Lemma 3.25

$$\frac{2}{k\pi} \int_0^{\pi/2} \left(\frac{\sin kx}{\sin x} \right)^2 dx = 1.$$

Proof Let $f(t) \equiv 1$ for all t. Then $\sigma_k(t) \equiv 1$ for all k and t. Substituting into the expression from Lemma 3.24 we obtain the result. $\square$

Theorem 3.26 *(Fejér) Suppose $f(t) \in L_1[0, 2\pi]$, periodic with period 2π and let*

$$\sigma(t) = \lim_{x \to 0_+} \frac{f(t+x) + f(t-x)}{2} = \frac{f(t+0) + f(t-0)}{2}$$

whenever the limit exists. For any t such that $\sigma(t)$ is defined we have

$$\lim_{k \to \infty} \sigma_k(t) = \sigma(t) = \frac{f(t+0) + f(t-0)}{2}.$$

Proof From Lemmas 3.24 and 3.25 we have

$$\sigma_k(t) - \sigma(t) = \frac{2}{k\pi} \int_0^{\pi/2} \left[\frac{f(t+2x) + f(t-2x)}{2} - \sigma(t) \right] \left(\frac{\sin kx}{\sin x} \right)^2 dx.$$

For any t for which $\sigma(t)$ is defined, let $G_t(x) = \frac{f(t+2x) + f(t-2x)}{2} - \sigma(t)$. Then $G_t(x) \to 0$ as $t \to 0$ through positive values. Thus, given $\epsilon > 0$ there is a $\delta < \pi/2$ such that $|G_t(x)| < \epsilon/2$ whenever $0 < x \leq \delta$. We have

$$\sigma_k(t) - \sigma(t) = \frac{2}{k\pi} \int_0^{\delta} G_t(x) \left(\frac{\sin kx}{\sin x} \right)^2 dx + \frac{2}{k\pi} \int_\delta^{\pi/2} G_t(x) \left(\frac{\sin kx}{\sin x} \right)^2 dx.$$

Now

$$\left| \frac{2}{k\pi} \int_0^{\delta} G_t(x) \left(\frac{\sin kx}{\sin x} \right)^2 dx \right| \leq \frac{\epsilon}{k\pi} \int_0^{\pi/2} \left(\frac{\sin kx}{\sin x} \right)^2 dx = \frac{\epsilon}{2}$$

and

$$\left| \frac{2}{k\pi} \int_\delta^{\pi/2} G_t(x) \left(\frac{\sin kx}{\sin x} \right)^2 dx \right| \leq \frac{2}{k\pi \sin^2 \delta} \int_\delta^{\pi/2} |G_t(x)| dx \leq \frac{2I}{k\pi \sin^2 \delta},$$

where $I = \int_0^{\pi/2} |G_t(x)| dx$. This last integral exists because F is in L_1. Now choose K so large that $2I/(K\pi \sin^2 \delta) < \epsilon/2$. Then if $k \geq K$ we have $|\sigma_k(t) - \sigma(t)| \leq$

$$\left| \frac{2}{k\pi} \int_0^{\delta} G_t(x) \left(\frac{\sin kx}{\sin x} \right)^2 dx \right| + \left| \frac{2}{k\pi} \int_\delta^{\pi/2} G_t(x) \left(\frac{\sin kx}{\sin x} \right)^2 dx \right| < \epsilon.$$

    $\square$

Corollary 3.27 *Suppose $f(t)$ satisfies the hypotheses of the theorem and also is continuous on the closed interval $[a, b]$. Then the sequence of arithmetic means $\sigma_k(t)$ converges uniformly to $f(t)$ on $[a, b]$.*

Proof If f is continuous on the closed bounded interval $[a, b]$ then it is uniformly continuous on that interval and the function G_t is bounded on $[a, b]$ with upper bound M, independent of t. Furthermore one can determine the δ in the preceding theorem so that $|G_t(x)| < \epsilon/2$ whenever $0 < x \leq \delta$ and uniformly for all $t \in [a, b]$. Thus we can conclude that $\sigma_k \to \sigma$, uniformly on $[a, b]$. Since f is continuous on $[a, b]$ we have $\sigma(t) = f(t)$ for all $t \in [a, b]$.   □

Corollary 3.28 *(Weierstrass approximation theorem) Suppose $f(t)$ is real and continuous on the closed interval $[a, b]$. Then for any $\epsilon > 0$ there exists a polynomial $p(t)$ such that*

$$|f(t) - p(t)| < \epsilon$$

for every $t \in [a, b]$.

Sketch of proof Using the methods of Section 3.3 we can find a linear transformation to map $[a, b]$ 1-1 on a closed subinterval $[a', b']$ of $[0, 2\pi]$, such that $0 < a' < b' < 2\pi$. This transformation will take polynomials in t to polynomials. Thus, without loss of generality, we can assume $0 < a < b < 2\pi$. Let $g(t) = f(t)$ for $a \leq t \leq b$ and define $g(t)$ outside that interval so that it is continuous at $t = a, b$ and is periodic with period 2π. Then from the first corollary to Fejér's theorem, given an $\epsilon > 0$ there is an integer N and arithmetic sum

$$\sigma(t) = \frac{A_0}{2} + \sum_{j=1}^{N}(A_j \cos jt + B_j \sin jt)$$

such that $|f(t) - \sigma(t)| = |g(t) - \sigma(t)| < \epsilon/2$ for $a \leq t \leq b$. Now $\sigma(t)$ is a trigonometric polynomial and it determines a power series expansion in t about the origin that converges uniformly on every finite interval. The partial sums of this power series determine a series of polynomials $\{p_n(t)\}$ of order n such that $p_n \to \sigma$ uniformly on $[a, b]$, Thus there is an M such that $|\sigma(t) - p_M(t)| < \epsilon/2$ for all $t \in [a, b]$. Hence

$$|f(t) - p_M(t)| \leq |f(t) - \sigma(t)| + |\sigma(t) - p_M(t)| < \epsilon$$

for all $t \in [a, b]$.   □

This important result implies not only that a continuous function on a bounded interval can be approximated uniformly by a polynomial function but also (since the convergence is uniform) that continuous functions on bounded domains can be approximated with arbitrary accuracy in the L_2 norm on that domain. Indeed the space of polynomials is dense in that Hilbert space.

Another important offshoot of approximation by arithmetic sums is that the Gibbs phenomenon doesn't occur. This follows easily from the next result.

Lemma 3.29 *Suppose the 2π-periodic function $f(t) \in L_2[-\pi, \pi]$ is bounded, with $M = \sup_{t \in [-\pi, \pi]} |f(t)|$. Then $|\sigma_n(t)| \leq M$ for all t.*

Proof From Lemma 3.13 and Lemma 3.24 we have $|\sigma_k(t)| \leq$

$$\frac{1}{2k\pi} \int_0^{2\pi} |f(t+x)| \left(\frac{\sin kx/2}{\sin x/2}\right)^2 dx \leq \frac{M}{2k\pi} \int_0^{2\pi} \left(\frac{\sin kx/2}{\sin x/2}\right)^2 dx = M.$$

$\square$

Now consider the example which has been our prototype for the Gibbs phenomenon:

$$h(t) = \begin{cases} 0, & t = 0 \\ \frac{\pi - t}{2}, & 0 < t < 2\pi \\ 0, & t = 2\pi \end{cases}$$

and $h(t + 2\pi) = h(t)$. Here the ordinary Fourier series gives

$$\frac{\pi - t}{2} = \sum_{n=1}^{\infty} \frac{\sin nt}{n}, \qquad 0 < t < 2\pi$$

and this series exhibits the Gibbs phenomenon near the simple discontinuities at integer multiples of 2π. Furthermore the supremum of $|h(t)|$ is $\pi/2$ and it approaches the values $\pm \pi/2$ near the discontinuities. However, the lemma shows that $|\sigma(t)| \leq \pi/2$ for all n and t. Thus the arithmetic sums never overshoot or undershoot as t approaches the discontinuities, so there is no Gibbs phenomenon in the arithmetic series for this example. This is illustrated in Figure 3.3, which should be compared with Figure 3.1.

In fact, the example is universal; there is no Gibbs phenomenon for arithmetic sums. To see this, we can mimic the proof of Theorem 3.17. This then shows that the arithmetic sums for all piecewise smooth functions converge uniformly except in arbitrarily small neighborhoods of the discontinuities of these functions. In the neighborhood of each

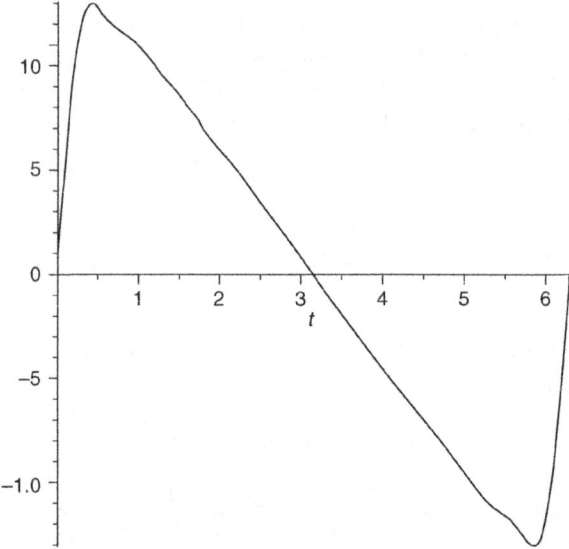

Figure 3.3 Plot of the average of the first 10 partial sums for the Fourier series of $h(t)$ on the interval $0 \le t \le 2\pi$. Note that the overshoot disappears.

discontinuity the arithmetic sums behave exactly as does the series for $h(t)$. Thus there is no overshooting or undershooting.

Exercise 3.11 Expand

$$f(x) = \begin{cases} 0 & -\pi < x \le -\frac{\pi}{2} \\ 1 & -\frac{\pi}{2} < x \le \frac{\pi}{2} \\ 0 & \frac{\pi}{2} < x \le \pi \end{cases}$$

in a Fourier series on the interval $-\pi \le x \le \pi$. Plot both f and the arithmetic sums σ_k for $k = 5, 10, 20, 40$. Observe how the arithmetic sums approximate f. Compare with Exercise 3.8.

Remark The pointwise convergence criteria for the arithmetic means are much more general (and the proofs of the theorems are simpler) than for the case of ordinary Fourier series. Further, they provide a means of getting around the most serious problems caused by the Gibbs phenomenon. The technical reason for this is that the kernel function $F_k(t)$ is nonnegative. Why don't we drop ordinary Fourier series and just use the arithmetic means? There are a number of reasons, one being that the arithmetic means $\sigma_k(t)$ are not the best L_2 approximations for order k, whereas the $S_k(t)$ *are* the best L_2 approximations. There is no

Parseval theorem for arithmetic means. Further, once the approximation $S_k(t)$ is computed for ordinary Fourier series, in order to get the next level of approximation one needs only to compute two more constants:

$$S_{k+1}(t) = S_k(t) + a_{k+1}\cos(k+1)t + b_{k+1}\sin(k+1)t.$$

However, for the arithmetic means, in order to update $\sigma_k(t)$ to $\sigma_{k+1}(t)$ one must recompute *ALL* of the expansion coefficients. This is a serious practical difficulty.

Fourier Series in $L_p(0 < p \le \infty)$ There is an old mathematical saying that L_1, L_2 and L_∞ were invented by God and man invented all else. Among the many manifestations of this for Fourier series and more generally orthogonal expansions on the real line, is the best approximation property in L_2 which ensures polynomials are dense in the space of continuous functions (the generalized Weierstrass approximation problem). But since it is part of the mathematician's spirit to take important tools out of their natural domain, so it is not surprising that much effort has been devoted to studying mean convergence of orthogonal expansions on the line. Many fundamental advances have ensued. For example, the boundedness of the Hilbert transform in L_p for $1 < p < \infty$ was established in order to prove that Fourier series converge in L_p. The theory of so-called A_p weights started with Mukenhoupt and Pollard's efforts to prove convergence in L_p of Hermite expansions and general expansions in orthogonal polynomials with respect to exponential weights. We refer the reader to the references [61, 80] and the references cited therein for more on this specialized but interesting topic.

3.8 Additional exercises

Exercise 3.12 Let $f(x) = x$, $x \in [-\pi, \pi]$.

(a) Show that $f(x)$ has the Fourier series

$$\sum_{n=1}^{\infty} \frac{2}{n}(-1)^{n-1}\sin(nx).$$

(b) Let $\alpha > 0$. Show that $f(x) = \exp(\alpha x)$, $x \in [-\pi, \pi]$ has the Fourier series

$$\left(\frac{e^{\alpha\pi} - e^{-\alpha\pi}}{\pi}\right)\left(\frac{1}{2\alpha} + \sum_{k=1}^{\infty}(-1)^k\frac{\alpha\cos(kx) - k\sin(kx)}{\alpha^2 + k^2}\right).$$

(c) Let α be any real number other than an integer. Let $f(x) = \cos(\alpha x)$, $x \in [-\pi, \pi]$. Show that f has a Fourier series

$$\frac{\sin(\alpha\pi)}{\alpha\pi} + \frac{1}{\pi}\sum_{n=1}^{\infty}\left[\frac{\sin(\alpha+n)\pi}{\alpha+n} + \frac{\sin(\alpha-n)\pi}{\alpha-n}\right]\cos(nx).$$

(d) Find the Fourier series of $f(x) = -\alpha\sin(\alpha x)$, $x \in [-\pi, \pi]$. Do you notice any relationship to that in (c)?

(e) Find the Fourier series of $f(x) = |x|$, $x \in [-\pi, \pi]$.

(f) Find the Fourier series of

$$f(x) = \text{sign}(x) = \begin{cases} -1, & x < 0 \\ 0, & x = 0 \\ 1, & x > 0. \end{cases}$$

Do you notice any relationship to that in (e)?

Exercise 3.13 Sum the series $\sum_{n=0}^{\infty} 2^{-n}\cos nt$ and $\sum_{n=1}^{\infty} 2^{-n}\sin nt$. Hint: Take the real and imaginary parts of $\sum_{n=0}^{\infty} 2^{-n}e^{int}$. You should be able to sum this last series directly.

Exercise 3.14 Find the $L_2[-\pi, \pi]$ projection of the function $f_1(x) = x^2$ onto the $(2n+1)$-dimensional subspace spanned by the ON set

$$\{\frac{1}{\sqrt{2\pi}}, \frac{\cos kx}{\sqrt{\pi}}, \frac{\sin kx}{\sqrt{\pi}} : k = 1, \ldots, n\}$$

for $n = 1$. Repeat for $n = 2, 3$. Plot these projections along with f_1. (You can use MATLAB, a computer algebra system, a calculator, etc.) Repeat the whole exercise for $f_2(x) = x^3$. Do you see any marked differences between the graphs in the two cases?

Exercise 3.15 By substituting special values of x in convergent Fourier series, we can often deduce interesting series expansions for various numbers, or just find the sum of important series.

(a) Use the Fourier series for $f(x) = x^2$ to show that

$$\sum_{n=1}^{\infty}\frac{(-1)^{n-1}}{n^2} = \frac{\pi^2}{12}.$$

(b) Prove that the Fourier series for $f(x) = x$ converges in $(-\pi, \pi)$ and hence that

$$\sum_{l=0}^{\infty}\frac{(-1)^l}{2l+1} = \frac{\pi}{4}.$$

(c) Show that the Fourier series of $f(x) = \exp(\alpha x)$, $x \in [-\pi, \pi)$ converges in $[-\pi, \pi)$ to $\exp(\alpha x)$ and at $x = \pi$ to $\frac{\exp(\alpha\pi)+\exp(-\alpha\pi)}{2}$. Hence show that

$$\frac{\alpha\pi}{\tanh(\alpha\pi)} = 1 + \sum_{k=1}^{\infty} \frac{2\alpha^2}{k^2 + \alpha^2}.$$

(d) Show that

$$\frac{\pi}{4} = \frac{1}{2} + \sum_{n=1}^{\infty} \frac{(-1)^{n-1}}{4n^2 - 1}.$$

(e) Plot both $f(s) = x^2$ and the partial sums

$$S_k(x) = \frac{a_0}{2} + \sum_{n=0}^{k} (a_n \cos nx + b_n \sin nx)$$

for $k = 1, 2, 5, 7$. Observe how the partial sums approximate f.

Exercise 3.16 Let $f_1(x) = x$ and $f_2(x) = \pi^2 - 3x^2$, $-\pi \le t \le \pi$. Find the Fourier series of f_1 and f_2 and use them alone to sum the following series:

1. $\sum_{n=1}^{\infty} \frac{(-1)^n}{2n+1}$
2. $\sum_{n=1}^{\infty} \frac{1}{n^2}$
3. $\sum_{n=1}^{\infty} \frac{(-1)^n}{n^2}$
4. $\sum_{n=1}^{\infty} \frac{1}{n^4}$

Exercise 3.17 By applying Parseval's identity to suitable Fourier series:

(a) Show that

$$\sum_{n=1}^{\infty} \frac{1}{n^4} = \frac{\pi^4}{90}.$$

(b) Evaluate

$$\sum_{n=1}^{\infty} \frac{1}{\alpha^2 + n^2}, \quad \alpha > 0.$$

(c) Show that

$$\sum_{l=0}^{\infty} \frac{1}{(2l+1)^4} = \frac{\pi^4}{96}.$$

Exercise 3.18 Is

$$\sum_{n=1}^{\infty} \frac{\cos(nx)}{n^{1/2}}$$

the Fourier series of some square integrable function f on $[-\pi, \pi]$? The same question for the series

$$\sum_{n=1}^{\infty} \frac{\cos(nx)}{\log(n+2)}.$$

4

The Fourier transform

4.1 Fourier transforms as integrals

There are several ways to define the Fourier transform (FT) of a function. In this section, we define it using an integral representation and state some basic uniqueness and inversion properties, without proof. Thereafter, we will define it as a suitable limit of Fourier series, and will prove the results stated here.

Definition 4.1 Let $f : \mathbb{R} \to \mathbb{C}$. The Fourier transform of $f \in L_1(\mathbb{R})$, denoted by $\mathcal{F}[f](\cdot)$, is given by the integral:

$$\mathcal{F}[f](\lambda) := \frac{1}{\sqrt{2\pi}} \int_{-\infty}^{\infty} f(t) \exp(-i\lambda t) dt$$

for $\lambda \in \mathbb{R}$ for which the integral exists.

We have the **Dirichlet condition** for inversion of Fourier integrals.

Theorem 4.2 *Let $f : \mathbb{R} \to \mathbb{R}$. Suppose that (1) $\int_{-\infty}^{\infty} |f| \, dt$ converges and (2) in any finite interval, f, f' are piecewise continuous with at most finitely many maxima/minima/discontinuities. Let $F = \mathcal{F}[f]$. Then if f is continuous at $t \in \mathbb{R}$, we have*

$$f(t) = \frac{1}{\sqrt{2\pi}} \int_{-\infty}^{\infty} F(\lambda) \exp(it\lambda) d\lambda.$$

Moreover, if f is discontinuous at $t \in \mathbb{R}$ and $f(t+0)$ and $f(t-0)$ denote the right and left limits of f at t, then

$$\frac{1}{2}[f(t+0) + f(t-0)] = \frac{1}{\sqrt{2\pi}} \int_{-\infty}^{\infty} F(\lambda) \exp(it\lambda) d\lambda.$$

From the above, we deduce a uniqueness result:

Theorem 4.3 *Let $f, g : \mathbb{R} \to \mathbb{R}$ be continuous, f', g' piecewise continuous. If*

$$\mathcal{F}[f](\lambda) = \mathcal{F}[g](\lambda), \ \forall \lambda$$

then

$$f(t) = g(t), \ \forall t.$$

Proof We have from inversion, easily that

$$f(t) = \frac{1}{\sqrt{2\pi}} \int_{-\infty}^{\infty} \mathcal{F}[f](\lambda) \exp(it\lambda) d\lambda$$

$$= \frac{1}{\sqrt{2\pi}} \int_{-\infty}^{\infty} \mathcal{F}[g](\lambda) \exp(it\lambda) d\lambda$$

$$= g(t).$$

$\square$

Example 4.4 Find the Fourier transform of $f(t) = \exp(-|t|)$ and, using inversion, deduce that $\int_0^{\infty} \frac{dx}{1+x^2} = \frac{\pi}{2}$ and $\int_0^{\infty} \frac{\lambda \sin(\lambda t)}{1+\lambda^2} d\lambda = \frac{\pi \exp(-t)}{2}$, $t > 0$.

Solution We write

$$F(\lambda) = \frac{1}{\sqrt{2\pi}} \int_{-\infty}^{\infty} f(t) \exp(-i\lambda t) dt$$

$$= \frac{1}{\sqrt{2\pi}} \left[\int_{-\infty}^{0} \exp(t(1 - i\lambda)) dt + \int_0^{\infty} \exp(-t(1 + i\lambda)) \right]$$

$$= \sqrt{\frac{2}{\pi}} \frac{1}{1 + \lambda^2}.$$

Now by the inversion formula,

$$\exp(-|t|) = \frac{1}{\sqrt{2\pi}} \int_{-\infty}^{\infty} F(\lambda) \exp(i\lambda t) d\lambda$$

$$= \frac{1}{\pi} \left[\int_0^{\infty} \frac{\exp(i\lambda t) + \exp(-i\lambda t)}{1 + \lambda^2} dt \right]$$

$$= \frac{2}{\pi} \int_0^{\infty} \frac{\cos(\lambda t)}{1 + \lambda^2} d\lambda.$$

This formula holds at $t = 0$, so substituting $t = 0$ into the above gives the first required identity. Differentiating with respect to t as we may for $t > 0$, gives the second required identity. $\square$.

We will show that we can reinterpret Definition 4.1 to obtain the Fourier transform of any complex-valued $f \in L_2(\mathbb{R})$, and that the Fourier transform is unitary on this space:

Theorem 4.5 *If $f, g \in L_2(\mathbb{R})$ then $\mathcal{F}[f], \mathcal{F}[g] \in L_2(\mathbb{R})$ and*

$$\int_{-\infty}^{\infty} f(t)\overline{g}(t)\ dt = \int_{-\infty}^{\infty} \mathcal{F}[f](\lambda)\overline{\mathcal{F}[g](\lambda)}\ d\lambda.$$

This is a result of fundamental importance for applications in signal processing.

4.2 The transform as a limit of Fourier series

We start by constructing the Fourier series (complex form) for functions on an interval $[-\pi L, \pi L]$. The ON basis functions are

$$e_n(t) = \frac{1}{\sqrt{2\pi L}} e^{\frac{int}{L}}, \qquad n = 0, \pm 1, \ldots,$$

and a sufficiently smooth function f of period $2\pi L$ can be expanded as

$$f(t) = \sum_{n=-\infty}^{\infty} \left(\frac{1}{2\pi L} \int_{-\pi L}^{\pi L} f(x) e^{-\frac{inx}{L}} dx \right) e^{\frac{int}{L}}.$$

For purposes of motivation let us abandon periodicity and think of the functions f as differentiable everywhere, vanishing at $t = \pm \pi L$ and identically zero outside $[-\pi L, \pi L]$. We rewrite this as

$$f(t) = \sum_{n=-\infty}^{\infty} e^{\frac{int}{L}} \frac{1}{2\pi L} \hat{f}(\frac{n}{L})$$

which looks like a Riemann sum approximation to the integral

$$f(t) = \frac{1}{2\pi} \int_{-\infty}^{\infty} \hat{f}(\lambda) e^{i\lambda t} d\lambda \tag{4.1}$$

to which it would converge as $L \to \infty$. (Indeed, we are partitioning the λ interval $[-L, L]$ into $2L$ subintervals, each with partition width $1/L$.) Here,

$$\hat{f}(\lambda) = \int_{-\infty}^{\infty} f(t) e^{-i\lambda t} dt. \tag{4.2}$$

Similarly the Parseval formula for f on $[-\pi L, \pi L]$,

$$\int_{-\pi L}^{\pi L} |f(t)|^2 dt = \sum_{n=-\infty}^{\infty} \frac{1}{2\pi L} |\hat{f}(\frac{n}{L})|^2$$

goes in the limit as $L \to \infty$ to the *Plancherel identity*

$$2\pi \int_{-\infty}^{\infty} |f(t)|^2 dt = \int_{-\infty}^{\infty} |\hat{f}(\lambda)|^2 d\lambda. \tag{4.3}$$

Expression (4.2) is called the *Fourier integral* or *Fourier transform* of f. Expression (4.1) is called the *inverse Fourier integral* for f. The Plancherel identity suggests that the Fourier transform is a 1-1 norm-preserving map of the Hilbert space $L_2[\mathbb{R}]$ onto itself (or to another copy of itself). We shall show that this is the case. Furthermore we shall show that the pointwise convergence properties of the inverse Fourier transform are somewhat similar to those of the Fourier series. Although we could make a rigorous justification of the steps in the Riemann sum approximation above, we will follow a different course and treat the convergence in the mean and pointwise convergence issues separately.

A second notation that we shall use is

$$\mathcal{F}[f](\lambda) = \frac{1}{\sqrt{2\pi}} \int_{-\infty}^{\infty} f(t) e^{-i\lambda t} dt = \frac{1}{\sqrt{2\pi}} \hat{f}(\lambda) \tag{4.4}$$

$$\mathcal{F}^*[g](t) = \frac{1}{\sqrt{2\pi}} \int_{-\infty}^{\infty} g(\lambda) e^{i\lambda t} d\lambda \tag{4.5}$$

Note that, formally, $\mathcal{F}^*[\hat{f}](t) = \sqrt{2\pi} f(t)$. The first notation is used more often in the engineering literature. The second notation makes clear that $\mathcal{F}$ and $\mathcal{F}^*$ are linear operators mapping $L_2(\mathbb{R})$ onto itself in one view, and $\mathcal{F}$ mapping the *signal space* onto the *frequency space* with $\mathcal{F}^*$ mapping the frequency space onto the signal space in the other view. In this notation the Plancherel theorem takes the more symmetric form

$$\int_{-\infty}^{\infty} |f(t)|^2 dt = \int_{-\infty}^{\infty} |\mathcal{F}[f](\lambda)|^2 d\lambda.$$

Examples 4.5

1. The box function (or rectangular wave)

$$\Pi(t) = \begin{cases} 1 & \text{if } -\pi < t < \pi \\ \frac{1}{2} & \text{if } t = \pm\pi \\ 0 & \text{otherwise.} \end{cases} \tag{4.6}$$

Then, since $\Pi(t)$ is an even function and $e^{-i\lambda t} = \cos(\lambda t) + i\sin(\lambda t)$, we have

$$\hat{\Pi}(\lambda) = \sqrt{2\pi}\mathcal{F}[\Pi](\lambda) = \int_{-\infty}^{\infty} \Pi(t) e^{-i\lambda t} dt = \int_{-\infty}^{\infty} \Pi(t) \cos(\lambda t) dt$$

$$= \int_{-\pi}^{\pi} \cos(\lambda t)dt = \frac{2\sin(\pi\lambda)}{\lambda} = 2\pi \text{ sinc } \lambda.$$

Thus sinc λ is the Fourier transform of the box function. The inverse Fourier transform is

$$\int_{-\infty}^{\infty} \text{sinc}(\lambda)e^{i\lambda t}d\lambda = \Pi(t), \qquad (4.7)$$

as follows from (2.3). Furthermore, we have

$$\int_{-\infty}^{\infty} |\Pi(t)|^2 dt = 2\pi$$

and

$$\int_{-\infty}^{\infty} |\text{ sinc }(\lambda)|^2 d\lambda = 1$$

from (2.4), so the Plancherel equality is verified in this case. Note that the inverse Fourier transform converged to the midpoint of the discontinuity, just as for Fourier series.

2. A triangular wave.

$$f(t) = \begin{cases} 1+t & \text{if } -1 \le t \le 0 \\ 1-t & \text{if } 0 \le t \le 1 \\ 0 & \text{otherwise.} \end{cases} \qquad (4.8)$$

Then, since f is an even function, we have

$$\hat{f}(\lambda) = \sqrt{2\pi}\mathcal{F}[f](\lambda) = \int_{-\infty}^{\infty} f(t)e^{-i\lambda t}dt = 2\int_0^1 (1-t)\cos(\lambda t)dt$$

$$= \frac{2 - 2\cos\lambda}{\lambda^2}.$$

3. A truncated cosine wave.

$$f(t) = \begin{cases} \cos t & \text{if } -\pi < t < \pi \\ -\frac{1}{2} & \text{if } t = \pm\pi \\ 0 & \text{otherwise.} \end{cases}$$

Then, since the cosine is an even function, we have

$$\hat{f}(\lambda) = \sqrt{2\pi}\mathcal{F}[f](\lambda) = \int_{-\infty}^{\infty} f(t)e^{-i\lambda t}dt = \int_{-\pi}^{\pi} \cos(t)\cos(\lambda t)dt$$

$$= \frac{2\lambda\sin(\lambda\pi)}{1 - \lambda^2}.$$

4. A truncated sine wave.

$$f(t) = \begin{cases} \sin t & \text{if } -\pi \le t \le \pi \\ 0 & \text{otherwise.} \end{cases}$$

Since the sine is an odd function, we have

$$\hat{f}(\lambda) = \sqrt{2\pi}\mathcal{F}[f](\lambda) = \int_{-\infty}^{\infty} f(t)e^{-i\lambda t}dt = -i\int_{-\pi}^{\pi} \sin(t)\sin(\lambda t)dt$$

$$= \frac{-2i\sin(\lambda\pi)}{1-\lambda^2}.$$

5. The Gaussian distribution.

$$f(t) = e^{-t^2/2}, \quad -\infty < t < \infty.$$

This is an even function and it can be shown that

$$\hat{f}(\lambda) = \sqrt{2\pi}e^{-\lambda^2/2},$$

see Exercise 4.27.

NOTE: The Fourier transforms of the discontinuous functions above decay as $1/\lambda$ for $|\lambda| \to \infty$ whereas the Fourier transforms of the continuous functions decay at least as fast as $1/\lambda^2$.

4.2.1 Properties of the Fourier transform

Recall that

$$\mathcal{F}[f](\lambda) = \frac{1}{\sqrt{2\pi}} \int_{-\infty}^{\infty} f(t)e^{-i\lambda t}dt = \frac{1}{\sqrt{2\pi}}\hat{f}(\lambda)$$

$$\mathcal{F}^*[g](t) = \frac{1}{\sqrt{2\pi}} \int_{-\infty}^{\infty} g(\lambda)e^{i\lambda t}d\lambda$$

We list some properties of the Fourier transform that will enable us to build a repertoire of transforms from a few basic examples. Suppose that f, g belong to $L_1(\mathbb{R})$, i.e., $\int_{-\infty}^{\infty} |f(t)|dt < \infty$ with a similar statement for g. We can state the following rules:

- $\mathcal{F}$ and $\mathcal{F}^*$ are linear operators. For $a, b \in \mathbb{C}$ we have

$$\mathcal{F}[af + bg] = a\mathcal{F}[f] + b\mathcal{F}[g], \quad \mathcal{F}^*[af + bg] = a\mathcal{F}^*[f] + b\mathcal{F}^*[g].$$

- Suppose $t^n f(t) \in L_1(\mathbb{R})$ for some positive integer n. Then

$$\mathcal{F}[t^n f(t)](\lambda) = i^n \frac{d^n}{d\lambda^n} \mathcal{F}[f](\lambda).$$

Similarly, suppose $\lambda^n f(\lambda) \in L_1(\mathbb{R})$ for some positive integer n. Then

$$\mathcal{F}^*[\lambda^n f(\lambda)](t) = i^n \frac{d^n}{dt^n} \mathcal{F}^*[f](t).$$

- Suppose the nth derivative $f^{(n)}(t) \in L_1(\mathbb{R})$ and piecewise continuous for some positive integer n, and f and the lower derivatives are all continuous in $(-\infty, \infty)$. Then

$$\mathcal{F}[f^{(n)}](\lambda) = (i\lambda)^n \mathcal{F}[f](\lambda).$$

Similarly, suppose the nth derivative $f^{(n)}(\lambda) \in L_1(\mathbb{R})$ and piecewise continuous for some positive integer n, and f and the lower derivatives are all continuous in $(-\infty, \infty)$. Then

$$\mathcal{F}^*[f^{(n)}](t) = (-it)^n \mathcal{F}^*[f](t).$$

- The Fourier transform of a translated and scaled function is given by

$$\mathcal{F}[f(bt - a)](\lambda) = \frac{1}{b} e^{-i\lambda a/b} \mathcal{F}[f](\frac{\lambda}{b}).$$

Exercise 4.1 Verify the rules for Fourier transforms listed above.

Examples 4.6 • We want to compute the Fourier transform of the rectangular box function with support on $[c, d]$:

$$R(t) = \begin{cases} 1 & \text{if } c < t < d \\ \frac{1}{2} & \text{if } t = c, d \\ 0 & \text{otherwise.} \end{cases}$$

Recall that the box function

$$\Pi(t) = \begin{cases} 1 & \text{if } -\pi < t < \pi \\ \frac{1}{2} & \text{if } t = \pm\pi \\ 0 & \text{otherwise} \end{cases}$$

has the Fourier transform $\hat{\Pi}(\lambda) = 2\pi \operatorname{sinc} \lambda$. But we can obtain R from Π by first translating $t \to s = t - \frac{(c+d)}{2}$ and then rescaling $s \to \frac{2\pi}{d-c} s$:

$$R(t) = \Pi(\frac{2\pi}{d-c} t - \pi \frac{c+d}{d-c}).$$

$$\hat{R}(\lambda) = \frac{4\pi^2}{d-c} e^{i\pi\lambda(c+d)/(d-c)} \operatorname{sinc}(\frac{2\pi\lambda}{d-c}). \tag{4.9}$$

Furthermore, from (2.3) we can check that the inverse Fourier transform of $\hat{R}$ is R, i.e., $\mathcal{F}^*(\mathcal{F})R(t) = R(t)$.

- Consider the truncated sine wave

$$f(t) = \begin{cases} \sin t & \text{if } -\pi \le t \le \pi \\ 0 & \text{otherwise} \end{cases}$$

with

$$\hat{f}(\lambda) = \frac{-2i\sin(\lambda\pi)}{1 - \lambda^2}.$$

Note that the derivative f' of $f(t)$ is just $g(t)$ (except at 2 points) where $g(t)$ is the truncated cosine wave

$$g(t) = \begin{cases} \cos t & \text{if } -\pi < t < \pi \\ -\frac{1}{2} & \text{if } t = \pm\pi \\ 0 & \text{otherwise.} \end{cases}$$

We have computed

$$\hat{g}(\lambda) = \frac{2\lambda\sin(\lambda\pi)}{1 - \lambda^2},$$

so $\hat{g}(\lambda) = (i\lambda)\hat{f}(\lambda)$, as predicted. Reversing this example above we can differentiate the truncated cosine wave to get the truncated sine wave. The prediction for the Fourier transform doesn't work! Why not?

- If $f(t) = e^{-t^2/2}$, then for $g(t) = e^{-(bt-a)^2/2}$ we have $\hat{g}(\lambda) = \frac{1}{b}e^{-i\lambda a/b}e^{-\lambda^2/2b^2}$.

4.2.2 Fourier transform of a convolution

The following property of the Fourier transform is of particular importance in signal processing. Suppose f, g belong to $L_1(\mathbb{R})$.

Definition 4.7 The *convolution* of f and g is the function $f * g$ defined by

$$(f * g)(t) = \int_{-\infty}^{\infty} f(t - x)g(x)dx.$$

Note also that $(f * g)(t) = \int_{-\infty}^{\infty} f(x)g(t - x)dx$, as can be shown by a change of variable.

Lemma 4.8 $f * g \in L_1(\mathbb{R})$ *and*

$$\int_{-\infty}^{\infty} |f * g(t)|dt = \int_{-\infty}^{\infty} |f(x)|dx \int_{-\infty}^{\infty} |g(t)|dt.$$

Sketch of proof

$$\int_{-\infty}^{\infty} |f * g(t)| dt = \int_{-\infty}^{\infty} \left(\int_{-\infty}^{\infty} |f(x)g(t-x)| dx \right) dt =$$

$$\int_{-\infty}^{\infty} \left(\int_{-\infty}^{\infty} |g(t-x)| dt \right) |f(x)| dx = \int_{-\infty}^{\infty} |g(t)| dt \int_{-\infty}^{\infty} |f(x)| dx.$$

$\square$

Theorem 4.9 *Let $h = f * g$. Then*

$$\hat{h}(\lambda) = \hat{f}(\lambda)\hat{g}(\lambda).$$

Sketch of proof

$$\hat{h}(\lambda) = \int_{-\infty}^{\infty} f * g(t) e^{-i\lambda t} dt = \int_{-\infty}^{\infty} \left(\int_{-\infty}^{\infty} f(x)g(t-x) dx \right) e^{-i\lambda t} dt =$$

$$\int_{-\infty}^{\infty} f(x) e^{-i\lambda x} \left(\int_{-\infty}^{\infty} g(t-x) e^{-i\lambda(t-x)} dt \right) dx = \int_{-\infty}^{\infty} f(x) e^{-i\lambda x} dx \ \hat{g}(\lambda)$$

$$= \hat{f}(\lambda)\hat{g}(\lambda).$$

$\square$

Exercise 4.2 Let $f, g, h : \mathbb{R} \to \mathbb{R}$. Let $a, b \in \mathbb{R}$. Show that:

(i) Convolution is linear:

$$(af + bg) * h = a(f * h) + b(g * h).$$

(ii) Convolution is commutative:

$$f * g = g * f.$$

(iii) Convolution is associative:

$$(f * g) * h = f * (g * h).$$

4.3 L_2 convergence of the Fourier transform

In this book our primary interest is in Fourier transforms of functions in the Hilbert space $L_2(\mathbb{R})$. However, the formal definition of the Fourier integral transform,

$$\mathcal{F}[f](\lambda) = \frac{1}{\sqrt{2\pi}} \int_{-\infty}^{\infty} f(t) e^{-i\lambda t} dt \qquad (4.10)$$

doesn't make sense for a general $f \in L_2(\mathbb{R})$. If $f \in L_1(\mathbb{R})$ then f is absolutely integrable and the integral (4.10) converges. However, there are square integrable functions that are not integrable. (Example: $f(t) = \frac{1}{1+|t|}$.) How do we define the transform for such functions?

We will proceed by defining $\mathcal{F}$ on a dense subspace of $f \in L_2(\mathbb{R})$ where the integral makes sense and then take Cauchy sequences of functions in the subspace to define $\mathcal{F}$ on the closure. Since $\mathcal{F}$ preserves inner product, as we shall show, this simple procedure will be effective.

First some comments on integrals of L_2 functions. If $f, g \in L_2(\mathbb{R})$ then the integral $(f, g) = \int_{-\infty}^{\infty} f(t)\overline{g}(t)dt$ necessarily exists, whereas the integral (4.10) may not, because the exponential $e^{-i\lambda t}$ is not an element of L_2. However, the integral of $f \in L_2$ over any finite interval, say $[-N, N]$ does exist. Indeed for N a positive integer, let $\chi_{[-N,N]}$ be the characteristic function for that interval:

$$\chi_{[-N,N]}(t) = \begin{cases} 1 & \text{if } -N \leq t \leq N \\ 0 & \text{otherwise.} \end{cases} \tag{4.11}$$

Then $\chi_{[-N,N]} \in L_2(\mathbb{R})$ so $\int_{-N}^{N} f(t)dt$ exists because, using the Cauchy–Schwarz inequality,

$$\int_{-N}^{N} |f(t)|dt = |(|f|, \chi_{[-N,N]})| \leq \|f\|_{L_2}\|\chi_{[-N,N]}\|_{L_2} = \|f\|_{L_2}\sqrt{2N} < \infty.$$

Now the space of step functions is dense in $L_2(\mathbb{R})$, so we can find a convergent sequence of step functions $\{s_n\}$ such that $\lim_{n\to\infty} \|f - s_n\|_{L_2} = 0$. Note that the sequence of functions $\{f_N = f\chi_{[-N,N]}\}$ converges to f pointwise as $N \to \infty$ and each $f_N \in L_2 \cap L_1$.

Lemma 4.10 $\{f_N\}$ *is a Cauchy sequence in the norm of* $L_2(\mathbb{R})$ *and* $\lim_{n\to\infty} \|f - f_n\|_{L_2} = 0$.

Proof Given $\epsilon > 0$ there is step function s_M such that $\|f - s_M\| < \epsilon/2$. Choose N so large that the support of s_M is contained in $[-N, N]$, i.e., $s_M(t)\chi_{[-N,N]}(t) = s_M(t)$ for all t. Then $\|s_M - f_N\|^2 = \int_{-N}^{N} |s_M(t) - f(t)|^2 dt \leq \int_{-\infty}^{\infty} |s_M(t) - f(t)|^2 dt = \|s_M - f\|^2$, so

$$\|f - f_N\| - \|(f - s_M) + (s_M - f_N)\| \leq \|f - s_M\| + \|s_M - f_N\|$$
$$\leq 2\|f - s_M\| < \epsilon.$$

$\square$

Here we will study the linear mapping $\mathcal{F} : L_2(\mathbb{R}) \to \hat{L}^2[-\infty, \infty]$ from the signal space to the frequency space. We will show that the mapping is *unitary*, i.e., it preserves the inner product and is 1-1 and onto. Moreover, the map $\mathcal{F}^* : \hat{L}^2[-\infty, \infty] \to L_2(\mathbb{R})$ is also a unitary mapping and is the inverse of $\mathcal{F}$:

$$\mathcal{F}^*\mathcal{F} = I_{L_2}, \qquad \mathcal{F}\mathcal{F}^* = I_{\hat{L}_2}$$

where $I_{L_2}, I_{\hat{L}_2}$ are the identity operators on L_2 and $\hat{L}_2$, respectively. We know that the space of step functions is dense in L_2. Hence to show that $\mathcal{F}$ preserves inner product, it is enough to verify this fact for step functions and then go to the limit. Once we have done this, we can define $\mathcal{F}f$ for any $f \in L_2[-\infty, \infty]$. Indeed, if $\{s_n\}$ is a Cauchy sequence of step functions such that $\lim_{n\to\infty} ||f - s_n||_{L_2} = 0$, then $\{\mathcal{F}s_n\}$ is also a Cauchy sequence (in fact, $||s_n - s_m|| = ||\mathcal{F}s_n - \mathcal{F}s_m||$) so we can define $\mathcal{F}f$ by $\mathcal{F}f = \lim_{n\to\infty} \mathcal{F}s_n$. The standard methods of Section 1.3 show that $\mathcal{F}f$ is uniquely defined by this construction. Now the truncated functions f_N have Fourier transforms given by the convergent integrals

$$\mathcal{F}[f_N](\lambda) = \frac{1}{\sqrt{2\pi}} \int_{-N}^{N} f(t) e^{-i\lambda t} dt$$

and $\lim_{N\to\infty} ||f - f_N||_{L_2} = 0$. Since $\mathcal{F}$ preserves inner product we have $||\mathcal{F}f - \mathcal{F}f_N||_{L_2} = ||\mathcal{F}(f - f_N)||_{L_2} = ||f - f_N||_{L_2}$, so $\lim_{N\to\infty} ||\mathcal{F}f - \mathcal{F}f_N||_{L_2} = 0$. We write

$$\mathcal{F}[f](\lambda) = \text{l.i.m.}_{N\to\infty} \mathcal{F}[f_N](\lambda) = \text{l.i.m.}_{N\to\infty} \frac{1}{\sqrt{2\pi}} \int_{-N}^{N} f(t) e^{-i\lambda t} dt$$

where 'l.i.m.' indicates that the convergence is in the mean (Hilbert space) sense, rather than pointwise.

We have already shown that the Fourier transform of the rectangular box function with support on $[c, d]$:

$$R_{c,d}(t) = \begin{cases} 1 & \text{if } c < t < d \\ \frac{1}{2} & \text{if } t = c, d \\ 0 & \text{otherwise} \end{cases}$$

is

$$\mathcal{F}[R_{c,d}](\lambda) = \frac{4\pi^2}{\sqrt{2\pi}(d-c)} e^{i\pi\lambda(c+d)/(d-c)} \text{sinc}(\frac{2\pi\lambda}{d-c})$$

and that $\mathcal{F}^*(\mathcal{F})R_{c,d}(t) = R_{c,d}(t)$. (Since here we are concerned only with convergence in the mean the value of a step function at a particular point is immaterial. Hence for this discussion we can ignore such niceties as the values of step functions at the points of their jump discontinuities.)

Lemma 4.11

$$(R_{a,b}, R_{c,d})_{L_2} = (\mathcal{F}R_{a,b}, \mathcal{F}R_{c,d})_{\hat{L}^2}$$

for all real numbers $a \leq b$ and $c \leq d$.

Proof

$$(\mathcal{F}R_{a,b}, \mathcal{F}R_{c,d})_{\hat{L}^2} = \int_{-\infty}^{\infty} \mathcal{F}[R_{a,b}](\lambda)\overline{\mathcal{F}}[R_{c,d}](\lambda)d\lambda$$

$$= \lim_{N \to \infty} \int_{-N}^{N} \left(\mathcal{F}[R_{a,b}](\lambda) \int_{c}^{d} \frac{e^{i\lambda t}}{\sqrt{2\pi}}dt \right) d\lambda$$

$$= \lim_{N \to \infty} \int_{c}^{d} \left(\int_{-N}^{N} \mathcal{F}[R_{a,b}](\lambda) \frac{e^{i\lambda t}}{\sqrt{2\pi}}d\lambda \right) dt.$$

Now the inside integral is converging to $R_{a,b}$ as $N \to \infty$ in both the pointwise and L_2 sense, as we have shown. Thus

$$(\mathcal{F}R_{a,b}, \mathcal{F}R_{c,d})_{\hat{L}^2} = \int_{c}^{d} R_{a,b}dt = (R_{a,b}, R_{c,d})_{L_2}.$$

$\square$

Since any step functions u, v are finite linear combinations of indicator functions R_{a_j, b_j} with complex coefficients, $u = \sum_j \alpha_j R_{a_j, b_j}$, $v = \sum_k \beta_k R_{c_k, d_k}$ we have

$$(\mathcal{F}u, \mathcal{F}v)_{\hat{L}^2} = \sum_{j,k} \alpha_j \overline{\beta}_k (\mathcal{F}R_{a_j,b_j}, \mathcal{F}R_{c_k,d_k})_{\hat{L}^2}$$

$$= \sum_{j,k} \alpha_j \overline{\beta}_k (R_{a_j,b_j}, R_{c_k,d_k})_{L_2} = (u, v)_{L_2}.$$

Thus $\mathcal{F}$ preserves inner product on step functions, and by taking Cauchy sequences of step functions, we have

Theorem 4.12 *(Plancherel Formula) Let $f, g \in L_2(\mathbb{R})$. Then*

$$(f, g)_{L_2} = (\mathcal{F}f, \mathcal{F}g)_{\hat{L}_2}, \qquad \|f\|_{L_2}^2 = \|\mathcal{F}f\|_{\hat{L}_2}^2.$$

In the engineering notation this reads

$$2\pi \int_{-\infty}^{\infty} f(t)\overline{g}(t)dt = \int_{-\infty}^{\infty} \hat{f}(\lambda)\overline{\hat{g}}(\lambda)d\lambda.$$

Theorem 4.13 *The map $\mathcal{F}^* : \hat{L}_2(\mathbb{R}) \to L_2(\mathbb{R})$ has the following properties:*

1. *It preserves inner product, i.e.,*

$$(\mathcal{F}^*\hat{f}, \mathcal{F}^*\hat{g})_{L_2} = (\hat{f}, \hat{g})_{\hat{L}_2}$$

for all $\hat{f}, \hat{g} \in \hat{L}_2(\mathbb{R})$.
2. *$\mathcal{F}^*$ is the adjoint operator to $\mathcal{F} : L_2(\mathbb{R}) \to \hat{L}_2(\mathbb{R})$, i.e.,*

$$(\mathcal{F}f, \hat{g})_{\hat{L}_2} = (f, \mathcal{F}^*\hat{g})_{L_2},$$

for all $f \in L_2(\mathbb{R})$, $\hat{g} \in \hat{L}_2(\mathbb{R})$.

Proof

1. This follows immediately from the facts that $\mathcal{F}$ preserves inner product and $\overline{\mathcal{F}[\overline{f}]}(\lambda) = \mathcal{F}^*[f](\lambda)$.
2.

$$(\mathcal{F}R_{a,b}, R_{c,d})_{\hat{L}_2} = (R_{a,b}, \mathcal{F}^*R_{c,d})_{L_2}$$

as can be seen by an interchange in the order of integration. Then using the linearity of $\mathcal{F}$ and $\mathcal{F}^*$ we see that

$$(\mathcal{F}u, v)_{\hat{L}_2} = (u, \mathcal{F}^*v)_{L_2},$$

for all step functions u, v. Since the space of step functions is dense in $\hat{L}_2(\mathbb{R})$ and in $L_2(\mathbb{R})$

$\square$

Theorem 4.14 1. *The Fourier transform $\mathcal{F} : L_2(\mathbb{R}) \to \hat{L}_2(\mathbb{R})$ is a unitary transformation, i.e., it preserves the inner product and is 1-1 and onto.*
2. *The adjoint map $\mathcal{F}^* : \hat{L}_2(\mathbb{R}) \to L_2(\mathbb{R})$ is also a unitary mapping.*
3. *$\mathcal{F}^*$ is the inverse operator to $\mathcal{F}$:*

$$\mathcal{F}^*\mathcal{F} = I_{L_2}, \qquad \mathcal{F}\mathcal{F}^* = I_{\hat{L}_2}$$

where $I_{L_2}, I_{\hat{L}_2}$ are the identity operators on L_2 and $\hat{L}_2$, respectively.

Proof

1. The only thing left to prove is that for every $\hat{g} \in \hat{L}_2(\mathbb{R})$ there is a $f \in L_2(\mathbb{R})$ such that $\mathcal{F}f = \hat{g}$, i.e., $\mathcal{R} \equiv \{\mathcal{F}f : f \in L_2(\mathbb{R})\} = \hat{L}_2(\mathbb{R})$. Suppose this isn't true. Then there exists a nonzero $\hat{h} \in \hat{L}_2(\mathbb{R})$ such that $\hat{h} \perp \mathcal{R}$, i.e., $(\mathcal{F}f, \hat{h})_{\hat{L}_2} = 0$ for all $f \in L_2(\mathbb{R})$. But this means that $(f, \mathcal{F}^*\hat{h})_{L_2} = 0$ for all $f \in L_2(\mathbb{R})$, so $\mathcal{F}^*\hat{h} = \Theta$. But then $\|\mathcal{F}^*\hat{h}\|_{L_2} = \|\hat{h}\|_{\hat{L}_2} = 0$ so $\hat{h} = \Theta$, a contradiction.

2. Same proof as for 1.
3. We have shown that $\mathcal{F}\mathcal{F}^*R_{a,b} = \mathcal{F}^*\mathcal{F}R_{a,b} = R_{a,b}$ for all indicator functions $R_{a,b}$. By linearity we have $\mathcal{F}\mathcal{F}^*s = \mathcal{F}^*\mathcal{F}s = s$ for all step functions s. This implies that

$$(\mathcal{F}^*\mathcal{F}f, g)_{L_2} = (f, g)_{L_2}$$

for all $f, g \in L_2(\mathbb{R})$. Thus

$$([\mathcal{F}^*\mathcal{F} - I_{L_2}]f, g)_{L_2} = 0$$

for all $f, g \in L_2(\mathbb{R})$, so $\mathcal{F}^*\mathcal{F} = I_{L_2}$. An analogous argument gives $\mathcal{F}\mathcal{F}^* = I_{\hat{L}_2}$.

$\square$

4.4 The Riemann–Lebesgue lemma and pointwise convergence

Lemma 4.15 *(Riemann–Lebesgue) Suppose f is absolutely Riemann integrable in $(-\infty, \infty)$ (so that $f \in L_1(\mathbb{R})$), and is bounded in any finite subinterval $[a, b]$, and let α, β be real. Then*

$$\lim_{\alpha \to +\infty} \int_{-\infty}^{\infty} f(t) \sin(\alpha t + \beta) dt = 0.$$

Proof Without loss of generality, we can assume that f is real, because we can break up the complex integral into its real and imaginary parts.

1. The statement is true if $f = R_{a,b}$ is an indicator function, for

$$\int_{-\infty}^{\infty} R_{a,b}(t) \sin(\alpha t + \beta) dt = \int_a^b \sin(\alpha t + \beta) dt = \frac{-1}{\alpha} \cos(\alpha t + \beta)|_a^b \to 0$$

as $\alpha \to +\infty$.
2. The statement is true if f is a step function, since a step function is a finite linear combination of indicator functions.
3. The statement is true if f is bounded and Riemann integrable on the finite interval $[a, b]$ and vanishes outside the interval. Indeed given any $\epsilon > 0$ there exist two step functions $\bar{s}$ (Darboux upper sum) and $\underline{s}$ (Darboux lower sum) with support in $[a, b]$ such that $\bar{s}(t) \geq f(t) \geq \underline{s}(t)$ for all $t \in [a, b]$ and $\int_a^b |\bar{s} - \underline{s}| < \epsilon/2$. Then

$$\int_a^b f(t) \sin(\alpha t + \beta) dt =$$

$$\int_a^b [f(t) - \underline{s}(t)] \sin(\alpha t + \beta) dt + \int_a^b \underline{s}(t) \sin(\alpha t + \beta) dt.$$

Now

$$\left| \int_a^b [f(t) - \underline{s}(t)] \sin(\alpha t + \beta) dt \right| \leq \int_a^b |f(t) - \underline{s}(t)| dt \leq \int_a^b |\overline{s} - \underline{s}| < \frac{\epsilon}{2}$$

and (since $\underline{s}$ is a step function), by choosing α sufficiently large we can ensure

$$\left| \int_a^b \underline{s}(t) \sin(\alpha t + \beta) dt \right| < \frac{\epsilon}{2}.$$

Hence

$$\left| \int_a^b f(t) \sin(\alpha t + \beta) dt \right| < \epsilon$$

for α sufficiently large.

4. The statement of the lemma is true in general. Indeed

$$\left| \int_{-\infty}^\infty f(t) \sin(\alpha t + \beta) dt \right| \leq \left| \int_{-\infty}^a f(t) \sin(\alpha t + \beta) dt \right|$$

$$+ \left| \int_a^b f(t) \sin(\alpha t + \beta) dt \right| + \left| \int_b^\infty f(t) \sin(\alpha t + \beta) dt \right|.$$

Given $\epsilon > 0$ we can choose a and b such that the first and third integrals are each $< \epsilon/3$, and we can choose α so large that the second integral is $< \epsilon/3$. Hence the limit exists and is 0.

$\square$

The sinc function has a delta-function property:

Lemma 4.16 *Let $c > 0$, and $F(x)$ a function on $[0, c]$. Suppose*

- *$F(x)$ is piecewise continuous on $[0, c]$*
- *$F'(x)$ is piecewise continuous on $[0, c]$*
- *$F'(+0)$ exists.*

Then

$$\lim_{L \to \infty} \int_0^\delta \frac{\sin Lx}{x} F(x) dx = \frac{\pi}{2} F(+0).$$

Proof We write

$$\int_0^c \frac{\sin kx}{x} F(x) dx = F(+0) \int_0^c \frac{\sin Lx}{x} dx + \int_0^c \frac{F(x) - F(+0)}{x} \sin Lx \, dx.$$

Set $G(x) = \frac{F(x) - F(+0)}{x}$ for $x \in [0, \delta]$ and $G(x) = 0$ elsewhere. Since $F'(+0)$ exists it follows that $G \in L_2$. Hence, by the Riemann–Lebesgue

Lemma, the second integral goes to 0 in the limit as $L \to \infty$. Thus

$$\lim_{L \to \infty} \int_0^c \frac{\sin Lx}{x} F(x) dx = F(+0) \lim_{L \to \infty} \int_0^c \frac{\sin Lx}{x} dx$$

$$= F(+0) \lim_{L \to \infty} \int_0^{Lc} \frac{\sin u}{u} du = \frac{\pi}{2} F(+0).$$

For the last equality we have used our evaluation (2.1) for the integral of the sinc function.   □

We have the Fourier convergence theorem.

Theorem 4.17 *Let f be a complex-valued function such that*

1. $f(t)$ *is absolutely Riemann integrable on* $(-\infty, \infty)$.
2. $f(t)$ *is piecewise continuous on* $(-\infty, \infty)$, *with only a finite number of discontinuities in any bounded interval.*
3. $f'(t)$ *is piecewise continuous on* $(-\infty, \infty)$, *with only a finite number of discontinuities in any bounded interval.*
4. $f(t) = \frac{f(t+0)+f(t-0)}{2}$ *at each point t.*

Let

$$\hat{f}(\lambda) = \int_{-\infty}^{\infty} f(t) e^{-i\lambda t} dt$$

be the Fourier transform of f. Then

$$f(t) = \frac{1}{2\pi} \int_{-\infty}^{\infty} \hat{f}(\lambda) e^{i\lambda t} d\lambda \tag{4.12}$$

for every $t \in (-\infty, \infty)$.

Proof Fix t and for real $L > 0$ set

$$f_L(t) = \int_{-L}^{L} \hat{f}(\lambda) e^{i\lambda t} d\lambda = \frac{1}{2\pi} \int_{-L}^{L} \left[\int_{-\infty}^{\infty} f(x) e^{-i\lambda x} dx \right] e^{i\lambda t} d\lambda$$

$$= \frac{1}{2\pi} \int_{-\infty}^{\infty} f(x) \left[\int_{-L}^{L} e^{i\lambda(t-x)} d\lambda \right] dx = \int_{-\infty}^{\infty} f(x) \Delta_L(t-x) dx,$$

where

$$\Delta_L(x) = \frac{1}{2\pi} \int_{-L}^{L} e^{i\lambda x} d\lambda = \begin{cases} \frac{L}{\pi} & \text{if } x = 0 \\ \frac{\sin Lx}{\pi x} & \text{otherwise.} \end{cases}$$

Here we have interchanged the order of integration, which we can since the integral is absolutely convergent. Indeed

$$\int_{-L}^{L} \left| \int_{-\infty}^{\infty} f(x) e^{-i\lambda x} e^{i\lambda t} \, dx \right| d\lambda \leq \int_{-L}^{L} \int_{-\infty}^{\infty} |f(x)| \, dx \, d\lambda < \infty.$$

We have for any $c > 1$,

$$f_L(t) - f(t) = \int_0^{\infty} \Delta_L(x)[f(t+x) + f(t-x)] \, dx - f(t)$$

$$= \int_0^c \left\{ \frac{f(t+x) + f(t-x)}{\pi x} \right\} \sin Lx \, dx - f(t)$$

$$+ \int_c^{\infty} [f(t+x) + f(t-x)] \frac{\sin Lx}{\pi x} \, dx.$$

Now choose $\epsilon > 0$. Since

$$\left| \int_c^{\infty} [f(t+x) + f(t-x)] \frac{\sin Lx}{\pi x} \, dx \right| \leq \frac{1}{\pi} \int_c^{\infty} |f(t+x) + f(t-x)| \, dx$$

and f is absolutely integrable, by choosing c sufficiently large we can make

$$\left| \int_c^{\infty} [f(t+x) + f(t-x)] \frac{\sin Lx}{\pi x} \, dx \right| < \frac{\epsilon}{2}.$$

On the other hand, by applying Lemma 4.16 to the expression in curly brackets we see that for this c and sufficiently large L we can achieve

$$\left| \int_0^c \left\{ \frac{f(t+x)}{\pi x} \right\} \sin Lx \, dx - \frac{f(t+0)}{2} + \int_0^c \left\{ \frac{f(t-x)}{\pi x} \right\} \sin Lx \, dx - \frac{f(t-0)}{2} \right|$$

$< \epsilon/2$. Thus for any $\epsilon > 0$ we can assure $|f_L(t) - f(t)| < \epsilon$ by choosing L sufficiently large. Hence $\lim_{L \to \infty} f_L(t) = f(t)$. $\square$

NOTE: Condition 4 is just for convenience; redefining f at the discrete points where there is a jump discontinuity doesn't change the value of any of the integrals. The inverse Fourier transform converges to the midpoint of a jump discontinuity, just as does the Fourier series.

Exercise 4.3 Let

$$\Pi(t) = \begin{cases} 1 & \text{for } -\frac{1}{2} < t < \frac{1}{2} \\ 0 & \text{otherwise} \end{cases},$$

the box function on the real line. We will often have the occasion to express the Fourier transform of $f(at + b)$ in terms of the Fourier transform $\hat{f}(\lambda)$ of $f(t)$, where a, b are real parameters. This exercise will give you practice in correct application of the transform.

1. Sketch the graphs of $\Pi(t)$, $\Pi(t-3)$ and $\Pi(2t-3) = \Pi(2(t-3/2))$.
2. Sketch the graphs of $\Pi(t)$, $\Pi(2t)$ and $\Pi(2(t-3))$. *Note:* In the first part a right 3-translate is followed by a 2-dilate; but in the second part a 2-dilate is followed by a right 3-translate. The results are not the same.
3. Find the Fourier transforms of $g_1(t) = \Pi(2t-3)$ and $g_2(t) = \Pi(2(t-3))$ from parts 1 and 2.
4. Set $g(t) = \Pi(2t)$ and check your answers to part 3 by applying the translation rule to

$$g_1(t) = g(t - \frac{3}{2}), \qquad g_2(t) = g(t-3), \qquad \text{noting } g_2(t) = g_1(t - \frac{3}{2}).$$

Exercise 4.4 Assuming that the improper integral $\int_0^\infty (\sin x / x)dx = I$ exists, establish its value (2.1) by first using the Riemann–Lebesgue lemma for Fourier series to show that

$$I = \lim_{k \to \infty} \int_0^{(k+1/2)\pi} \frac{\sin x}{x} dx = \lim_{k \to \infty} \int_0^\pi D_k(u)du$$

where $D_k(u)$ is the Dirichlet kernel function. Then use Lemma 3.11.

Exercise 4.5 Define the *right-hand derivative* $f_R'(t)$ and the *left-hand derivative* $f_L'(t)$ of f by

$$f_R'(t) = \lim_{u \to t+} \frac{f(u) - f(t+0)}{u-t}, \quad f_L'(t) = \lim_{u \to t-} \frac{f(u) - f(t-0)}{u-t},$$

respectively, as in Exercise 3.4. Show that in the proof of Theorem 4.17 we can drop the requirements 3 and 4, and the right-hand side of (4.12) will converge to $\frac{f(t+0)+f(t-0)}{2}$ at any point t such that both $f_R'(t)$ and $f_L'(t)$ exist.

Exercise 4.6 Let $a > 0$. Use the Fourier transforms of $\text{sinc}(x)$ and $\text{sinc}^2(x)$, together with the basic tools of Fourier transform theory, such as Parseval's equation, substitution, ... to show the following. (Use only rules from Fourier transform theory. You shouldn't do any detailed computation such as integration by parts.)

- $\int_{-\infty}^\infty \left(\frac{\sin ax}{x}\right)^3 dx = \frac{3a^2\pi}{4}$
- $\int_{-\infty}^\infty \left(\frac{\sin ax}{x}\right)^4 dx = \frac{2a^3\pi}{3}$

Exercise 4.7 Show that the n-translates of sinc are orthonormal:

$$\int_{-\infty}^\infty \text{sinc}(x-n) \cdot \text{sinc}(x-m) \, dx = \begin{cases} 1 & \text{for } n = m \\ 0 & \text{otherwise,} \end{cases} \quad n, m = 0, \pm, 1, \ldots$$

Exercise 4.8 Let

$$f(x) = \begin{cases} 1 & -2 \le t \le -1 \\ 1 & 1 \le t \le 2 \\ 0 & \text{otherwise.} \end{cases}$$

- Compute the Fourier transform $\hat{f}(\lambda)$ and sketch the graphs of f and $\hat{f}$.
- Compute and sketch the graph of the function with Fourier transform $\hat{f}(-\lambda)$.
- Compute and sketch the graph of the function with Fourier transform $\hat{f}'(\lambda)$.
- Compute and sketch the graph of the function with Fourier transform $\hat{f} * \hat{f}(\lambda)$.
- Compute and sketch the graph of the function with Fourier transform $\hat{f}(\frac{\lambda}{2})$.

Exercise 4.9 Deduce what you can about the Fourier transform $\hat{f}(\lambda)$ if you know that $f(t)$ satisfies the dilation equation

$$f(t) = f(2t) + f(2t - 1).$$

Exercise 4.10 Just as Fourier series have a complex version and a real version, so does the Fourier transform. Under the same assumptions as Theorem 4.17 set

$$\hat{C}(\alpha) = \frac{1}{2}[\hat{f}(\alpha) + \hat{f}(-\alpha)], \quad \hat{S}(\alpha) = \frac{1}{2i}[-\hat{f}(\alpha) + \hat{f}(-\alpha)], \quad \alpha \ge 0,$$

and derive the expansion

$$f(t) = \frac{1}{\pi} \int_0^\infty \left(\hat{C}(\alpha) \cos \alpha t + \hat{S}(\alpha) \sin \alpha t \right) d\alpha, \qquad (4.13)$$

$$\hat{C}(\alpha) = \int_{-\infty}^\infty f(s) \cos \alpha s \, ds, \quad \hat{S}(\alpha) = \int_{-\infty}^\infty f(s) \sin \alpha s \, ds.$$

Show that the transform can be written in a more compact form as

$$f(t) = \frac{1}{\pi} \int_0^\infty d\alpha \int_{-\infty}^\infty f(s) \cos \alpha(s - t) \, ds.$$

Exercise 4.11 There are also Fourier integral analogs of the Fourier cosine series and the Fourier sine series. Let $f(t)$ be defined for all $t \ge 0$ and extend it to an even function on the real line, defined by

$$F(t) = \begin{cases} f(t) & \text{if } t \ge 0, \\ f(-t) & \text{if } t < 0. \end{cases}$$

By applying the results of Exercise 4.10 show that, formally,

$$f(t) = \frac{2}{\pi} \int_0^\infty \cos\alpha t\, d\alpha \int_0^\infty f(s)\cos\alpha s\, ds, \quad t \geq 0. \qquad (4.14)$$

Find conditions on $f(t)$ such that this pointwise expansion is rigorously correct.

Exercise 4.12 Let $f(t)$ be defined for all $t > 0$ and extend it to an odd function on the real line, defined by

$$G(t) = \begin{cases} f(t) & \text{if } t > 0, \\ -f(-t) & \text{if } t < 0. \end{cases}$$

By applying the results of Exercise 4.10 show that, formally,

$$f(t) = \frac{2}{\pi} \int_0^\infty \sin\alpha t\, d\alpha \int_0^\infty f(s)\sin\alpha s\, ds, \quad t > 0. \qquad (4.15)$$

Find conditions on $f(t)$ such that this pointwise expansion is rigorously correct.

Exercise 4.13 Find the Fourier Cosine and Sine transforms of the following functions:

$$f(t) := \begin{cases} 1, & t \in [0, a] \\ 0, & t > a \end{cases}$$

$$f(t) := \begin{cases} \cos(at), & t \in [0, a] \\ 0, & t > a. \end{cases}$$

4.5 Relations between Fourier series and integrals: sampling

For the purposes of Fourier analysis we have been considering signals $f(t)$ as arbitrary $L_2(\mathbb{R})$ functions. In the practice of signal processing, however, one can treat only a finite amount of data. Typically the signal is digitally sampled at regular or irregular discrete time intervals. Then the processed sample alone is used to reconstruct the signal. If the sample isn't altered, then the signal should be recovered exactly. How is this possible? How can one reconstruct a function $f(t)$ exactly from discrete samples? Of course, this is not possible for arbitrary functions $f(t)$. The task isn't hopeless, however, because the signals employed in signal processing, such as voice or images, are not arbitrary. The human voice for example is easily distinguished from static noise or random noise.

One distinguishing characteristic is that the frequencies of sound in the human voice are mostly in a narrow frequency band. In fact, any signal that we can acquire and process with real hardware must be restricted to some finite frequency band. In this section we will explore Shannon–Whittaker–Kotelnikov sampling, one way that the special class of signals restricted in frequency can be sampled and then reproduced exactly. This method is of immense practical importance as it is employed routinely in telephone, radio and TV transmissions, radar, etc. (In later chapters we will study other special structural properties of signal classes, such as sparsity, that can be used to facilitate their processing and efficient reconstruction.) We say that a function $f \in L_2(\mathbb{R})$ is *frequency bandlimited* if there exists a constant $B > 0$ such that $\hat{f}(\lambda) = 0$ for $|\lambda| > B$.

Theorem 4.18 *(Sampling theorem) Suppose f is a function such that*

1. *f satisfies the hypotheses of the Fourier convergence theorem 4.17.*
2. *$\hat{f}$ is continuous and has a piecewise continuous first derivative on its domain.*
3. *There is a $B > 0$ such that $\hat{f}(\lambda) = 0$ for $|\lambda| > B$.*

Then

$$f(t) = \sum_{-\infty}^{\infty} f(\frac{j\pi}{B}) \operatorname{sinc}(\frac{Bt}{\pi} - j), \tag{4.16}$$

where the series converges uniformly on $(-\infty, \infty)$.

(NOTE: The theorem states that for a frequency bandlimited function, to determine the value of the function at all points, it is sufficient to sample the function at the Nyquist rate or sampling rate, i.e., at intervals of π/B. The method of proof is obvious: compute the Fourier series expansion of $\hat{f}(\lambda)$ on the interval $[-B, B]$.)

Proof We have

$$\hat{f}(\lambda) = \sum_{k=-\infty}^{\infty} c_k e^{\frac{i\pi k\lambda}{B}}, \qquad c_k = \frac{1}{2B} \int_{-B}^{B} \hat{f}(\lambda) e^{-\frac{i\pi k\lambda}{B}} d\lambda,$$

where the convergence is uniform on $[-B, B]$. This expansion holds only on the interval: $\hat{f}(\lambda)$ vanishes outside the interval.

Taking the inverse Fourier transform we have

$$f(t) = \frac{1}{2\pi} \int_{-\infty}^{\infty} \hat{f}(\lambda)e^{i\lambda t}d\lambda = \frac{1}{2\pi} \int_{-B}^{B} \hat{f}(\lambda)e^{i\lambda t}d\lambda$$

$$= \frac{1}{2\pi} \int_{-B}^{B} \sum_{k=-\infty}^{\infty} c_k e^{\frac{i(\pi k + tB)\lambda}{B}}d\lambda$$

$$= \frac{1}{2\pi} \sum_{k=-\infty}^{\infty} c_k \int_{-B}^{B} e^{\frac{i(\pi k + tB)\lambda}{B}}d\lambda = \sum_{k=-\infty}^{\infty} c_k \frac{B\sin(Bt + k\pi)}{\pi(Bt + k\pi)}.$$

Now

$$c_k = \frac{1}{2B} \int_{-B}^{B} \hat{f}(\lambda)e^{-\frac{i\pi k\lambda}{B}}d\lambda = \frac{1}{2B} \int_{-\infty}^{\infty} \hat{f}(\lambda)e^{-\frac{i\pi k\lambda}{B}}d\lambda = \frac{\pi}{B}f(-\frac{\pi k}{B}).$$

Hence, setting $k = -j$,

$$f(t) = \sum_{j=-\infty}^{\infty} f(\frac{j\pi}{B})\frac{\sin(Bt - j\pi)}{Bt - j\pi}.$$

$\square$

Exercise 4.14 Use the fact that $f(t) = \text{sinc}(t + a)$ is frequency bandlimited with $B = \pi$ to show that

$$\text{sinc}(t + a) = \sum_{j=-\infty}^{\infty} \text{sinc}(j + a)\text{sinc}(t - j)$$

for all a, t. Derive the identity

$$1 = \sum_{j=-\infty}^{\infty} \text{sinc}^2(t - j).$$

Exercise 4.15 Suppose $f(t)$ satisfies the conditions of Theorem 4.18. Derive the Parseval formula

$$\int_{-\infty}^{\infty} |f(t)|^2 dt = \frac{1}{2\pi} \int_{-\infty}^{\infty} |\hat{f}(\lambda)|^2 d\lambda = \frac{\pi}{B} \sum_{k=-\infty}^{\infty} |f(\frac{\pi k}{B})|^2.$$

Exercise 4.16 Let

$$f(t) = \begin{cases} (\frac{\sin t}{t})^2 & \text{for } t \neq 0 \\ 1 & \text{for } t = 0. \end{cases}$$

Then from Equation (2.5) we have

$$\hat{f}(\lambda) = \begin{cases} \pi(1 - \frac{|\lambda|}{2}) & \text{for } |\lambda| < 2 \\ 0 & \text{for } |\lambda| \geq 2 \end{cases}$$

1. Choose $B = 1$ and use the sampling theorem to write $f(t)$ as a series.
2. Graph the sum of the first 20 terms in the series and compare with the graph of $f(t)$.
3. Repeat the last two items for $B = 2$ and $B = 3$.

There is a trade-off in the choice of B. Choosing it as small as possible reduces the sampling rate, hence the amount of data to be processed or stored. However, if we increase the sampling rate, i.e. *oversample*, the series converges more rapidly. Moreover, sampling at the smallest possible rate leads to numerical instabilities in the reconstruction of the signal. This difficulty is related to the fact that the reconstruction is an expansion in sinc $(Bt/\pi - j) = \sin(Bt - j\pi)/(Bt - j\pi)$. The sinc function is frequency bandlimited, but its Fourier transform is discontinuous, see (2.3), (4.7). This means that the sinc function decays slowly in time, like $1/(Bt - j\pi)$. Summing over j yields the divergent, harmonic series:$\sum_{j=-\infty}^{\infty} |\text{sinc } (Bt/\pi - j)|$. Thus a small error ϵ for each sample can lead to arbitrarily large reconstruction error. Suppose we could replace sinc (t) in the expansion by a frequency bandlimited function $g(t)$ such that $\hat{g}(\lambda)$ was infinitely differentiable. Since all derivatives $\hat{g}^{(n)}(\lambda)$ have compact support it follows from Section 4.2.1 that $t^n g(t)$ is square integrable for all positive integers n. Thus $g(t)$ decays faster than $|t|^{-n}$ as $|t| \to \infty$. This fast decay would prevent the numerical instability.

Exercise 4.17 If the continuous-time band limited signal is $x(t) = \cos t$, what is the period T that gives sampling exactly at the Nyquist (minimal B) rate? What samples $x(nT)$ do you get at this rate? What samples do you get from $x(t) = \sin t$?

In order to employ such functions $g(t)$ in place of the sinc function it will be necessary to oversample. Oversampling will provide us with redundant information but also flexibility in the choice of expansion function, and improved convergence properties. We will now take samples $f(j\pi/aB)$ where $a > 1$. (A typical choice is $a = 2$.) Recall that the support of $\hat{f}$ is contained in the interval $[-B, B] \subset [-aB, aB]$. We choose $g(t)$ such that (1) $\hat{g}(\lambda)$ is arbitrarily differentiable, (2) its support is contained in the interval $[-aB, aB]$ and (3) $\hat{g}(\lambda) = 1$ for $\lambda \in [-B, B]$. Note that there are many possible functions g that could satisfy these requirements, see for example Exercise 4.18. Now we repeat the major steps of the proof of the sampling theorem, but for the interval $[-aB, aB]$. Thus

$$\hat{f}(\lambda) = \sum_{k=-\infty}^{\infty} c_k e^{\frac{i\pi k\lambda}{aB}}, \qquad c_k = \frac{1}{2aB} \int_{-aB}^{aB} \hat{f}(\lambda) e^{-\frac{i\pi k\lambda}{aB}} \, d\lambda.$$

At this point we insert $\hat{g}$ by noting that $\hat{f}(\lambda) = \hat{f}(\lambda)\hat{g}(\lambda)$, since $\hat{g}(\lambda) = 1$ on the support of $\hat{f}$. Thus,

$$\hat{f}(\lambda) = \sum_{k=-\infty}^{\infty} c_k \hat{g}(\lambda) e^{\frac{i\pi k\lambda}{aB}}, \tag{4.17}$$

where from one of the properties listed in Section 4.2.1, $\hat{g}(\lambda)e^{\frac{i\pi k\lambda}{aB}}$, is the Fourier transform of $g(t + \pi k/aB)$. Taking the inverse Fourier transform of both sides (OK since the series on the right converges uniformly) we obtain

$$f(t) = \sum_{j=-\infty}^{\infty} f(\frac{j\pi}{aB})g(t - \pi j/aB). \tag{4.18}$$

Since $|g(t)t^n| \to 0$ as $|t| \to \infty$ for any positive integer n this series converges very rapidly and is not subject to instabilities.

Exercise 4.18 Show that the function

$$h(\lambda) = \left\{ \begin{array}{cc} \exp(\frac{1}{1-\lambda^2}) & \text{if } -1 < \lambda < 1 \\ 0 & \text{if } |\lambda| \geq 1, \end{array} \right.$$

is infinitely differentiable with compact support. In particular compute the derivatives $\frac{d^n}{d\lambda^n}h(\pm 1)$ for all n.

Exercise 4.19 Construct a function $\hat{g}(\lambda)$ which (1) is arbitrarily differentiable, (2) has support contained in the interval $[-4, 4]$ and (3) $\hat{g}(\lambda) = 1$ for $\lambda \in [-1, 1]$. Hint: Consider the convolution $\frac{1}{2c}R_{[-2,2]} * h_2(\lambda)$ where $R_{[-2,2]}$ is the rectangular box function on the interval $[-2, 2]$, $h_2(\lambda) = h(\lambda/2)$, h is defined in Exercise 4.18 and $c = \int_{-\infty}^{\infty} h(\lambda)d\lambda$.

Remark: We should note that, theoretically, it isn't possible to restrict a finite length signal in the time domain $f(t)$ to a finite frequency interval. Since the support of f is bounded, the Fourier transform integral $\hat{f}(\lambda) = \int_{-\infty}^{\infty} f(t)e^{-i\lambda t}dt$ converges for all complex λ and defines $\hat{f}(\lambda)$ as an analytic function for all points in the complex plane. A well-known property of functions analytic in the entire complex plane is that if they vanish along a segment of a curve, say an interval on the real axis, then they must vanish everywhere, [92]. Thus a finite length time signal cannot be frequency bounded unless it is identically 0. For practical transmission of finite signals the frequency bandlimited condition must be relaxed. One way to accomplish this is to replace the bandlimited condition in Theorem 4.18 by the weaker requirement that $|\hat{f}(\lambda)| \leq M/(1 + \lambda^2)$ for some positive constant M. Thus we require that $|\hat{f}(\lambda)|$ decays rapidly for large $|\lambda|$. Now we divide the Fourier transform up into non-overlapping strips, each of which is bandlimited:

$$\hat{f}_n(\lambda) = \begin{cases} \hat{f}(\lambda) & \text{if } B(2n-1) \le \lambda \le B(2n+1), \\ 0 & \text{otherwise.} \end{cases}$$

We define

$$f_n(t) = \frac{1}{2\pi} \int_{-\infty}^{\infty} \hat{f}_n(\lambda) e^{i\lambda t} d\lambda.$$

Then

$$\hat{f}(\lambda) = \sum_{n=-\infty}^{\infty} \hat{f}_n(\lambda) \longrightarrow f(t) = \sum_{-\infty}^{\infty} f_n(t),$$

and the last series converges absolutely and uniformly, since $\sum_{-\infty}^{\infty} |f_n(t)| \le \int_{-\infty}^{\infty} |\hat{f}(\lambda)| \, d\lambda < \infty$. Computing the Fourier series expansion for each strip, just as in the proof of Theorem 4.18, we find

$$f_n(t) = e^{2iBtn} \sum_j f_n(\frac{j\pi}{B}) \operatorname{sinc}(\frac{Bt}{\pi} - j),$$

where this series also converges absolutely and uniformly, due to our assumptions on $\hat{f}$. Now, $|f(t) - \sum_{j=-\infty}^{\infty} f(\frac{j\pi}{B}) \operatorname{sinc}(\frac{Bt}{\pi} - j)| =$

$$|f(t) - \sum_{j=-\infty}^{\infty} \sum_{n=-\infty}^{\infty} f(\frac{j\pi}{B}) \operatorname{sinc}(\frac{Bt}{\pi} - j)| =$$

$$|f(t) - \sum_{n=-\infty}^{\infty} \sum_{j=-\infty}^{\infty} f(\frac{j\pi}{B}) \operatorname{sinc}(\frac{Bt}{\pi} - j)| = |\sum_{n=-\infty}^{\infty} f_n(t) - \sum_{n=-\infty}^{\infty} e^{-2iBtn} f_n(t)|$$

$$= |\sum_{n=-\infty}^{\infty} (1 - e^{-2iBtn}) f_n(t)| \le 2 \sum_{n \ne 0} |f_n(t)| \le 2 \int_{|\lambda| \ge B} |\hat{f}(\lambda)| d\lambda.$$

Up to this point we have not chosen B. From the last inequality we see that for any given $\epsilon > 0$, if we choose B large enough such that $\int_{|\lambda| \ge B} |\hat{f}(\lambda)| d\lambda < \epsilon/2$, then we can assure that

$$|f(t) - \sum_{j=-\infty}^{\infty} f(\frac{j\pi}{B}) \operatorname{sinc}(\frac{Bt}{\pi} - j)| < \epsilon$$

for all t. This computation is valid, assuming that the interchange in order of summation between n and j is valid. From advanced calculus, the interchange is justified if the iterated summation converges absolutely:

$$\sum_{n=-\infty}^{\infty} \sum_{j=-\infty}^{\infty} |f_n(\frac{j\pi}{B})| \, |\text{sinc}(\frac{Bt}{\pi} - j)| < \infty.$$

To see this we use the Cauchy–Schwarz inequality, and results of Exercises 4.14 and 4.15 to obtain

$$\sum_{n=-\infty}^{\infty} \sum_{j=-\infty}^{\infty} |f_n(\frac{j\pi}{B})| \, |\text{sinc}(\frac{Bt}{\pi} - j)| \le \sum_{n=-\infty}^{\infty} \sqrt{\sum_{j=-\infty}^{\infty} |f_n(\frac{j\pi}{B})|^2}$$

$$= \sum_{n=-\infty}^{\infty} \sqrt{\frac{B}{2} \int_{-\infty}^{\infty} |\hat{f}_n(\lambda)|^2 d\lambda}.$$

Now

$$\int_{-\infty}^{\infty} |\hat{f}_n(\lambda)|^2 d\lambda \le \begin{cases} 2BM^2/[1 + B^2(2n-1)^2]^2 & \text{for } n > 0, \\ 2BM^2/[1 + B^2(2n+1)^2]^2 & \text{for } n < 0, \\ 2BM^2 & \text{for } n = 0. \end{cases}$$

Thus the iterated sum converges. We conclude that if the signal $f(t)$ is nearly frequency band limited, in the sense given above, then for any $\epsilon > 0$ we can find B such that the expansion (4.16) holds for f and all t, with error $< \epsilon$.

4.6 Fourier series and Fourier integrals: periodization

Another way to compare the Fourier transform with Fourier series is to periodize a function. The *periodization* of a function $f(t)$ on the real line is the function

$$P[f](t) = \sum_{m=-\infty}^{\infty} f(t + 2\pi m). \tag{4.19}$$

It is easy to see that $P[f]$ is 2π-periodic: $P[f](t) = P[f](t + 2\pi)$, assuming that the series converges. However, this series will not converge in general, so we need to restrict ourselves to functions that decay sufficiently rapidly at infinity. We could consider functions with compact support, say infinitely differentiable. Another useful but larger space of functions is the Schwartz class. We say that $f \in L_2(\mathbb{R})$ belongs to the *Schwartz class* if f is infinitely differentiable everywhere, and there exist constants $C_{n,q}$ (depending on f) such that $|t^n \frac{d^q}{dt^q} f| \leq C_{n,q}$ on R for each $n, q = 0, 1, 2, \ldots$ Then the projection operator P maps an f in the Schwartz class to a continuous function in $L_2[0, 2\pi]$ with period 2π. (However, periodization can be applied to a much larger class of functions, e.g. functions on $L_2(\mathbb{R})$ that decay as c/t^2 as $|t| \to \infty$.) Assume that f is chosen appropriately so that its periodization is a continuous function. Thus we can expand $P[f](t)$ in a Fourier series to obtain

$$P[f](t) = \sum_{n=-\infty}^{\infty} c_n e^{int}$$

where

$$c_n = \frac{1}{2\pi} \int_0^{2\pi} P[f](t) e^{-int} dt = \frac{1}{2\pi} \int_{-\infty}^{\infty} f(t) e^{-int} dt = \frac{1}{2\pi} \hat{f}(n)$$

and $\hat{f}(\lambda)$ is the Fourier transform of $f(t)$. Then,

$$\sum_{n=-\infty}^{\infty} f(t + 2\pi n) = \frac{1}{2\pi} \sum_{n=-\infty}^{\infty} \hat{f}(n) e^{int}, \qquad (4.20)$$

and we see that $P[f](t)$ tells us the value of $\hat{f}$ at the integer points $\lambda = n$, but not in general at the non-integer points. (For $t = 0$, Equation (4.20) is known as the *Poisson summation formula*. If we think of f as a signal, we see that **periodization** (4.19) of f results in a loss of information. However, if f vanishes outside of $[0, 2\pi)$ then $P[f](t) \equiv f(t)$ for $0 \leq t < 2\pi$ and

$$f(t) = \sum_n \hat{f}(n) e^{int}, \quad 0 \leq t < 2\pi$$

without error.)

Exercise 4.20 Let $f(t) = \frac{a}{t^2 + a^2}$ for $a > 0$.

- Show that $\hat{f}(t) = \pi e^{-a|\lambda|}$. Hint: It is easier to work backwards.
- Use the Poisson summation formula to derive the identity

$$\sum_{n=-\infty}^{\infty} \frac{1}{n^2 + a^2} = \frac{\pi}{a} \frac{1 + e^{-2\pi a}}{1 - e^{-2\pi a}}.$$

What happens as $a \to 0+$? Can you obtain the value of $\sum_{n=1}^{\infty} \frac{1}{n^2}$ from this?

4.7 The Fourier integral and the uncertainty principle

The uncertainty principle gives a limit to the degree that a function $f(t)$ can be simultaneously localized in time as well as in frequency. To be precise, we introduce some basic ideas from probability theory.

A continuous probability distribution for the random variable t on the real line $\mathbb{R}$ is a continuous function $\rho(t)$ on $\mathbb{R}$ such that $0 \le \rho(t) \le 1$ and $\int_{-\infty}^{\infty} \rho(t)\, dt = 1$. We also require that $\int_{-\infty}^{\infty} p(t)\rho(t)\, dt$ converges for any polynomial $p(t)$. Here $\int_{t_1}^{t_2} \rho(t)\, dt$ is interpreted as the probability that a sample t taken from $\mathbb{R}$ falls in the interval $t_1 \le t \le t_2$. The expectation (or mean) μ of the distribution is $\mu = E_\rho(t) \equiv \int_{-\infty}^{\infty} t\rho(t)\, dt$ and the standard deviation $\sigma \ge 0$ is defined by

$$\sigma^2 = \int_{-\infty}^{\infty} (t-\mu)^2 \rho(t)\, dt = E_\rho\left((t-\mu)^2\right).$$

Here σ is a measure of the concentration of the distribution about its mean. The most famous continuous distribution is the normal (or Gaussian) distribution function

$$\rho_0(t) = \frac{1}{\sigma\sqrt{2\pi}} e^{-(t-\mu)^2/2\sigma^2} \tag{4.21}$$

where μ is a real parameter and $\sigma > 0$. This is just the bell curve, centered about $t = \mu$. In this case $E_{\rho_0}(t) = \mu$ and $\sigma^2 = E_{\rho_0}((t-\mu)^2)$. The usual notation for the normal distribution with mean μ and standard deviation σ is $N(\mu, \sigma)$.

Every nonzero continuous $f \in L_2(\mathbb{R})$ defines a probability distribution function $\rho(t) = \frac{|f(t)|^2}{||f||^2}$, i.e., $\rho(t) \ge 0$ and $\int_{-\infty}^{\infty} \rho(t)dt = 1$.

Definition 4.19 The *mean* of the distribution defined by f is

$$\mu = E_\rho(t) = \int_{-\infty}^{\infty} t\frac{|f(t)|^2}{||f||^2} dt.$$

The *dispersion* of f about $t_0 \in \mathbb{R}$ is

$$D_{t_0} f = E_\rho((t-t_0)^2) = \int_{-\infty}^{\infty} (t-t_0)^2 \frac{|f(t)|^2}{||f||^2} dt.$$

$D_\mu f = \sigma^2$ is called the variance of f.

The dispersion of f about t_0 is a measure of the extent to which the graph of f is concentrated at t_0. If $f=\delta(x-a)$, the "Dirac delta function," the dispersion is zero. The constant $f(t) \equiv 1$ has infinite dispersion. (However there are no such L_2 functions.) Similarly we can define the dispersion of the Fourier transform of f about some point $\lambda_0 \in \mathbb{R}$:

$$D_{\lambda_0}\hat{f} = \int_{-\infty}^{\infty} (\lambda - \lambda_0)^2 \frac{|\hat{f}(\lambda)|^2}{||\hat{f}||^2} d\lambda.$$

It makes no difference which definition of the Fourier transform that we use, $\hat{f}$ or $\mathcal{F}f$, because the normalization gives the same probability measure.

Example 4.20 Let $f_s(t) = (\frac{2s}{\pi})^{1/4}e^{-st^2}$ for $s > 0$, the Gaussian distribution. From the fact that $\int_{-\infty}^{\infty} e^{-t^2} dt = \sqrt{\pi}$ we see that $||f_s|| = 1$. The Fourier transform of f_s is $\hat{f}_s(\lambda) = (\frac{s}{2\pi})^{1/4}e^{\frac{-\lambda^2}{4s}}$. By plotting some graphs one can see informally that as s increases the graph of f_s concentrates more and more about $t = 0$, i.e., the dispersion $D_0 f_s$ decreases. However, the dispersion of $\hat{f}_s$ increases as s increases. We can't make both values, simultaneously, as small as we would like. Indeed, a straightforward computation gives

$$D_0 f_s = \frac{1}{4s}, \qquad D_0 \hat{f}_s = s,$$

so the product of the variances of f_s and $\hat{f}_s$ is always $1/4$, no matter how we choose s.

We now introduce Heisenberg's inequality and the Uncertainty theorem.

Theorem 4.21 *If $f(t) \neq 0$ and $tf(t)$ belong to $L_2(\mathbb{R})$ then for any $t_0, \lambda_0 \in \mathbb{R}$, we have $(D_{t_0}f)(D_{\lambda_0}\hat{f}) \geq 1/4$.*

Sketch of proof We will give the proof under the added assumptions that $f'(t)$ exists everywhere and also belongs to $L_2(\mathbb{R})$. (In particular this implies that $f(t) \to 0$ as $t \to \pm\infty$.) The main ideas occur there.

Let $g(t) = f(t + t_0)e^{-i\lambda_0 t}$. Then from the rules for Fourier transforms it is straightforward to show that

$$||g|| = ||f||, \quad ||\hat{g}|| = ||\hat{f}||, \quad D_0 g = D_{t_0}f, \quad D_0 \hat{g} = D_{\lambda_0}\hat{f}, \qquad (4.22)$$

so to verify the theorem it is enough to consider the case $t_0 = 0$, $\lambda_0 = 0$.

We make use of the canonical commutation relation of quantum mechanics, the fact that the operations of multiplying a function $f(t)$ by t,

$(Tf(t) = tf(t))$ and of differentiating a function $(Df(t) = f'(t))$ don't commute: $DT - TD = I$. Thus

$$\frac{d}{dt}[tf(t)] - t\left[\frac{d}{dt}f(t)\right] = f(t).$$

Taking the inner product of the left-hand side of this identity with f we obtain

$$\left(\frac{d}{dt}[tf(t)], f(t)\right) - \left(t[\frac{d}{dt}f(t)], f(t)\right) = (f, f) = ||f||^2.$$

Integrating by parts in the first integral (inner product), we can rewrite the identity as

$$-\left([tf(t)], [\frac{d}{dt}f(t)]\right) - \left([\frac{d}{dt}f(t)], [tf(t)]\right) = ||f||^2.$$

The Schwarz inequality and the triangle inequality now yield

$$2||tf(t)|| \cdot ||\frac{d}{dt}f(t)|| \geq ||f||^2. \tag{4.23}$$

From the list of properties of the Fourier transform in Section 4.2.1 and the Plancherel formula, we see that $||\frac{d}{dt}f(t)|| = \frac{1}{\sqrt{2\pi}}||\lambda\hat{f}(\lambda)||$ and $||f|| = \frac{1}{\sqrt{2\pi}}||\hat{f}||$. Then, squaring, we have

$$(D_0 f)(D_0 \hat{f}) \geq \frac{1}{4}.$$

$\square$

NOTE: We see that the Schwarz inequality becomes an equality if and only if $2stf(t) + \frac{d}{dt}f(t) = 0$ for some constant s. Solving this differential equation we find $f(t) = c_0 e^{-st^2}$ where c_0 is the integration constant, and we must have $s > 0$ in order for f to be square integrable. Thus the Heisenberg inequality becomes an equality only for Gaussian distributions.

Exercise 4.21 Verify Equations (4.22).

Exercise 4.22 Let

$$f(t) = \begin{cases} 0 & \text{if } t < 0, \\ \sqrt{2}e^{-t} & \text{if } t \geq 0. \end{cases}$$

Compute $(D_{t_0} f)(D_{\lambda_0} \hat{f})$ for any $t_0, \lambda_0 \in \mathbb{R}$ and compare with Theorem 4.21.

These considerations suggest that for proper understanding of signal analysis we should be looking in the two-dimensional time–frequency space (phase space), rather than the time domain or the frequency domains alone. In subsequent chapters we will study tools such as windowed Fourier transforms and wavelet transforms that probe the full phase space. The Heisenberg inequality also suggests that probabilistic methods have an important role to play in signal analysis, and we shall make this clearer in later sections.

We now look at probabilistic tools.

4.8 Digging deeper

The notion of a probability distribution can easily be extended to n dimensions. A continuous probability distribution (multivariate distribution) for the vector random variables $t = (t_1, t_2, \ldots, t_n) \in \mathbb{R}^n$ is a continuous function $\rho(t) = \rho(t_1, \ldots, t_n)$ on $\mathbb{R}^n$ such that $0 \leq \rho(t) \leq 1$ and $\int_{\mathbb{R}^n} \rho(t)\, dt_1 \cdots dt_n = 1$. We also require that $\int_{-\infty}^{\infty} p(t)\rho(t)\, dt$ converges for any polynomial $p(t_1, \ldots, t_n)$. If S is an open subset of $\mathbb{R}^n$ and $\chi_S(t)$ is the characteristic function of S, i.e.,

$$\chi_S(t) = \begin{cases} 1 & \text{if } t \in S \\ 0 & \text{if } t \notin S, \end{cases}$$

then $\int_S \rho(t)\, dt_1 \cdots dt_n$ is interpreted as the probability that a sample t taken from $\mathbb{R}^n$ lies in the set S. The expectation (or mean) μ_i of random variable t_i is

$$\mu_i = E_\rho(t_i) \equiv \int_{\mathbb{R}^n} t_i \rho(t)\, dt_1 \cdots dt_n, \quad i = 1, 2, \cdots, n$$

and the standard deviation $\sigma_i \geq 0$ is defined by

$$\sigma_i^2 = \int_{\mathbb{R}^n} (t_i - \mu_i)^2 \rho(t)\, dt_1 \cdots dt_n = E_\rho\left((t_i - \mu_i)^2\right).$$

The **covariance matrix** of the distribution is the $n \times n$ symmetric matrix

$$C(i, j) = E\left((t_i - \mu_i)(t_j - \mu_j)\right), \quad 1 \leq i, j \leq n.$$

Note that $\sigma_i^2 = C(i, i)$. In general, the *expectation* of any function $f(t_1, \ldots, t_n)$ of the random variables is defined as $E\left(f(t_1, \ldots, t_n)\right)$.

Exercise 4.23 Show that the eigenvalues of a covariance matrix are nonnegative.

One of the most important multivariate distributions is the multivariate normal distribution

$$\rho(t) = \frac{1}{\sqrt{(2\pi)^n \det(C)}} \exp\left[-\frac{1}{2}(t-\mu)C^{-1}(t-\mu)^{\mathrm{tr}}\right]. \qquad (4.24)$$

Here μ is the column vector with components $\mu_1, \cdots, \mu_n$ where $\mu_i = E(t_i)$ and C is an $n \times n$ nonsingular real symmetric matrix.

Exercise 4.24 Show that C is in fact the covariance matrix of the distribution (4.24). This takes some work and involves making an orthogonal change of coordinates where the new coordinate vectors are the orthonormal eigenvectors of C.

Important special types of multivariate distributions are those in which the random variables are independently distributed, i.e., there are one-variable probability distributions $\rho_1(t_1), \ldots, \rho_n(t_n)$ such that $\rho(t) = \Pi_{i=1}^n \rho_i(t_1)$. If, further, $\rho_1(\tau) = \cdots = \rho_n(\tau)$ for all τ we say that the random variables are independently and identically distributed (iid).

Exercise 4.25 If the random variables are independently distributed, show that

$$C(t_i, t_j) = \sigma_i^2 \delta_{ij}.$$

Example 4.22 If C is a diagonal matrix in the multivariate normal distribution (4.24) then the random variables $t_1, \ldots, t_n$ are independently distributed . If $C = \sigma^2 I$ where I is the $n \times n$ identity matrix and $\mu_i = \mu$ for all i then the distribution is iid, where each random variable is distributed according to the normal distribution $N(\mu, \sigma^2)$, i.e., the Gaussian distribution (4.21) with mean μ and variance σ^2.

An obvious way to construct an iid multivariate distribution is to take a random sample $T_1, \ldots, T_n$ of values of a single random variable t with probability distribution $\rho(t)$. Then the multivariate distribution function for the vector random variable $T = (T_1, \ldots T_n)$ is the function $\rho(T) = \Pi_{i=1}^n \rho(T_i)$. It follows that for any integers $k_1, \ldots, k_n$ we have

$$E(T_1^{k_1} \cdots T_n^{k_n}) = \Pi_{i=1}^n E(T_i^{k_i}).$$

Note that if t has mean μ and standard deviation σ^2 then $E(T_i) = \mu$, $E((T_i - \mu)^2) = \sigma^2$ and $E(T_i^2) = \sigma^2 + \mu^2$.

Exercise 4.26 Show that $E(T_i T_j) = \mu^2 + \sigma^2 \delta_{ij}$.

Now we define the *sample mean* of the random sample as

$$\bar{T} = \frac{1}{n} \sum_{i=1}^{n} T_i.$$

The sample mean is itself a random variable with expectation

$$E(\bar{T}) = E\left(\frac{1}{n}\sum_i T_i\right) = \frac{1}{n}\sum_i E(T_i) = \frac{1}{n}\sum_i \mu = \mu \tag{4.25}$$

and variance

$$E\left((\bar{T} - \mu)^2\right) = \frac{1}{n^2} \sum_{i,j=1}^{n} E(T_i T_j) - \frac{2\mu}{n} \sum_{i=1}^{n} E(T_i) + E(\mu^2) \tag{4.26}$$

$$= \frac{1}{n^2}\left(\sum_{i,j}(\mu^2 + \delta_{ij}\sigma^2)\right) - \frac{2n\mu^2}{n} + \mu^2 = \mu^2 + \frac{n}{n^2}\sigma^2 - \mu^2 = \frac{\sigma^2}{n}.$$

Thus the sample mean has the same mean as a random variable as does t, but its variance is less than the variance of t by the factor $1/n$. This suggests that the distribution of the sample mean is increasingly "peaked" around the mean as n grows. Thus if the original mean is unknown, we can obtain better and better estimates of it by taking many samples. This idea lies behind the Law of Large Numbers that we will prove later.

A possible objection to the argument presented in the previous paragraph is that we already need to know the mean of the distribution to compute the variance of the sample mean. This leads us to the definition of the *sample variance*:

$$S^2 = \frac{1}{n-1} \sum_{i=1}^{n} (T_i - \bar{T})^2, \tag{4.27}$$

where $\bar{T}$ is the sample mean. Here S is the *sample standard deviation*. (The sample mean and sample standard deviation are typically reported as outcomes for an exam in a high enrollment undergraduate math course.) We will explain the factor $n - 1$, rather than n.

Theorem 4.23 *The expectation of the sample variance* S^2 *is* $E(S^2) = \sigma^2$.

Proof $E(S^2) =$

$$\frac{1}{n-1}E(\sum_i (T_i - \bar{T})^2) = \frac{1}{n-1}\left[\sum_i E(T_i^2) - 2E(\bar{T}\sum_i T_i) + nE(\bar{T}^2)\right]$$

$$= \frac{1}{n-1} \left[n(\mu^2 + \sigma^2) - \frac{2}{n} \sum_{i,j} E(T_i T_j) + \frac{1}{n} \sum_{i,j} E(T_i T_j) \right]$$

$$= \frac{1}{n-1} [(n - 2n + n)\mu^2 + (n - 2 + 1)\sigma^2] = \sigma^2. \qquad \square$$

If we had used n as the denominator of the sample variance then the expectation would not have been σ^2.

We continue to explore the extent to which the probability distribution function of a random variable peaks around its mean. For this we make use of a form of the Markov inequality, a simple but powerful result that applies to a wide variety of probability distributions.

Theorem 4.24 *Let x be a nonnegative random variable with continuous probability distribution function $p(x)$ (so that $p(x) = 0$ for $x < 0$) and let d be a positive constant. Then*

$$\Pr(x \geq d) \leq \frac{1}{d} E(x). \qquad (4.28)$$

Proof Here the symbol $\Pr(x \geq d)$ should be interpreted as the probability that a random selection of the variable x has a value $\geq d$. The claim is that this probability is $\leq \frac{1}{d} E(x)$. Now

$$E(x) = \int_0^\infty x p(x) \, dx = \int_0^d x p(x) \, dx + \int_d^\infty x p(x) \, dx$$

$$\geq \int_0^d x p(x) \, dx + d \int_d^\infty p(x) \, dx \geq d \int_d^\infty p(x) \, dx = d \Pr(X \geq d).$$

$$\square$$

We now have Chebyshev's inequality.

Corollary 4.25 *Let t be a random variable with expected value μ and finite variance σ^2. Then for any real number $\alpha > 0$,*

$$\Pr(|t - \mu| \geq \alpha\sigma) \leq \frac{1}{\alpha^2}.$$

Proof We apply the Markov inequality to the random variable $x = (t - \mu)^2$ with $d = (\sigma\alpha)^2$. Thus

$$\Pr\left((t - \mu)^2 \geq (\sigma\alpha)^2\right) \leq \frac{1}{(\sigma\alpha)^2} E\left((t - \mu)^2\right) = \frac{1}{(\alpha)^2}.$$

This is equivalent to the statement of the Chebyshev inequality. $\square$

Example 4.26 Setting $\alpha = 1/\sqrt{2}$ we see that at least half of the sample values will lie in the interval $(\mu - \sigma/\sqrt{2}, \mu + \sigma/\sqrt{2})$.

As a corollary, we have the Law of Large Numbers.

Corollary 4.27 *Let $\rho(t)$ be a probability distribution with mean μ and standard deviation σ'. Take a sequence of independent random samples from this population: $T_1, \ldots, T_n, \ldots$ and let $\bar{T}^{(n)} = \frac{1}{n} \sum_{i=1}^n T_i$ be the sample mean of the first n samples. Then for any $\epsilon > 0$ we have $\lim_{n \to \infty} \Pr(|\bar{T}^{(n)} - \mu| > \epsilon) = 0$.*

Proof From Equations (4.25) and (4.26) we have

$$E(\bar{T}^{(n)}) = \mu, \quad E\left((\bar{T}^{(n)} - \mu)^2\right) = \sigma'^2/n.$$

Applying the Chebyshev inequality to the random variable $\bar{T}^{(n)}$ with $\sigma = \sigma'/\sqrt{n}$, $\alpha = \sqrt{n}\epsilon/\sigma'$ we obtain

$$\Pr\left(\left|\bar{T}^{(n)} - \mu\right| \geq \epsilon\right) \leq \frac{\sigma'^2}{n\epsilon^2}.$$

Thus the probability that the sample mean differs by more than ϵ from the distribution mean gets smaller as n grows and approaches 0. In particular $\lim_{n \to \infty} \Pr(|\bar{T}^{(n)} - \mu| > \epsilon) = 0$. $\square$

This form of the Law of Large Numbers tells us that for any fixed $\epsilon > 0$ and sufficiently large sample size, we can show that the sample average will differ by less than ϵ from the mean with *high probability*, but not with certainty. It shows us that the sample distribution is more and more sharply peaked around the mean as the sample size grows. With modern computers that can easily generate large random samples and compute sample means, this insight forms the basis for many practical applications, as we shall see.

4.9 Additional exercises

Exercise 4.27 Let $f(x) = \exp(-sx^2)$ for fixed $s > 0$ and all x. Verify that $\hat{f}(\lambda) = \sqrt{\pi/s} \exp(-\lambda^2/4s)$. Hint: By differentiating under the integral sign and integration by parts, establish that

$$\frac{d\hat{f}(\lambda)}{d\lambda} = -\frac{\lambda}{2s}\hat{f}(\lambda), \quad \text{so } \hat{f}(\lambda) = C\exp(-\lambda^2/4s)$$

for some constant C. To compute C, note that

$$\left[\int_{-\infty}^{\infty} e^{-sx^2} dx \right]^2 = \int_{-\infty}^{\infty} \int_{-\infty}^{\infty} e^{-s(x^2+y^2)} dx \, dy = \int_{0}^{2\pi} d\theta \int_{0}^{\infty} e^{-sr^2} r \, dr.$$

Exercise 4.28 Let $g : \mathbb{R} \to \mathbb{R}$. Find a function H such that for all x,

$$\frac{1}{\sqrt{2\pi}} \int_{-\infty}^{x} g(t) dt = (H * g)(x).$$

(H is called the Heaviside function.)

Exercise 4.29 Let $f, g : \mathbb{R} \to \mathbb{R}$. Let f' exist. Assuming the convergence of the relevant integrals below, show that

$$(f * g)'(x) = f'(x) * g(x).$$

Exercise 4.30 For $a \in \mathbb{R}$, let

$$f_a(t) = \begin{cases} 0, & t < a \\ 1, & t \geq a. \end{cases}$$

Compute $f_a * f_b$ for $a, b \in \mathbb{R}$. Deduce that

$$(f_a * f_{-a})(x) = \frac{x f_0(x)}{\sqrt{2\pi}}.$$

Does $f_a * (1 - f_b)$ exist? For $a \in \mathbb{R}$, let

$$g_a(t) := \begin{cases} 0, & t < 0 \\ \exp(-at), & t \geq 0. \end{cases}$$

Compute $g_a * g_b$.

Exercise 4.31 Fourier transforms are useful in "deconvolution" or solving "convolution integral equations." Suppose that we are given functions g, h and are given that

$$f * g = h.$$

Our task is to find f in terms of g, h.

(i) Show that

$$\mathcal{F}[f] = \mathcal{F}[h] / \mathcal{F}[g]$$

and hence, if we can find a function k such that

$$\mathcal{F}[h] / \mathcal{F}[g] = \mathcal{F}[k]$$

then $f = k$.

(ii) As an example, suppose that

$$f * \exp(-t^2/2) = (1/2)t \exp(-t^2/4).$$

Find f.

Exercise 4.32 (i) The *Laplace transform* of a function $f : [0, \infty) \to \mathbb{R}$ is defined as

$$\mathcal{L}[f](p) = \int_0^\infty f(t) \exp(-pt)dt$$

whenever the right-hand side makes sense. Show formally, that if we set

$$g(x) := \begin{cases} f(x), & x \geq 0 \\ 0, & x < 0 \end{cases}$$

then

$$\mathcal{L}[f](p) := \sqrt{2\pi}\mathcal{F}[g](-ip).$$

(ii) Let $h : \mathbb{R} \to \mathbb{R}$ and define:

$$h_+(x) := \begin{cases} h(x), & x \geq 0 \\ 0, & x < 0 \end{cases}$$

and

$$h_-(x) := \begin{cases} h(-x), & x > 0 \\ 0, & x \leq 0. \end{cases}$$

Show that $h(x) = h_+(x) + h_-(x)$ and express $\mathcal{F}[h]$ in terms of $\mathcal{L}[h_+]$ and $\mathcal{L}[h_-]$.

For additional background information about Fourier series and the Fourier transform see [55, 90, 96].

5

Compressive sampling

5.1 Introduction

Although it is convenient for conceptual and theoretical purposes to think of signals as general functions of time, in practice they are usually acquired, processed, stored and transmitted as discrete and finite time samples. We need to study this sampling process carefully to determine to what extent a sampling or discretization allows us to reconstruct the original information in the signal. Furthermore, real signals such as speech or images are not arbitrary functions. Depending on the type of signal, they have special structure. No one would confuse the output of a random number generator with human speech. It is also important to understand the extent to which we can compress the basic information in the signal to minimize storage space and maximize transmission time.

Shannon sampling is one approach to these issues. In that approach we model real signals as functions $f(t)$ in $L_2(\mathbb{R})$ that are bandlimited. Thus if the frequency support of $\hat{f}(\omega)$ is contained in the interval $[-\Omega, \Omega]$ and we sample the signal at discrete time intervals with equal spacing less than $1/2\pi\Omega$, i.e., faster than the Nyquist rate, we can reconstruct the original signal exactly from the discrete samples. This method will work provided hardware exists to sample the signal at the required rate. Increasingly this is a problem because modern technologies can generate signals of higher bandwidth than existing hardware can sample.

There are other models for signals that exploit different properties of real signals and can be used as alternatives to Shannon sampling. In this chapter we introduce an alternative model that is based on the sparsity of many real signals. Intuitively we think of a signal as sparse if its expression with respect to some chosen basis has coefficients that are mostly zero (or very small). The content of the signal is in the location and values of the spikes, the nonzero terms. For example the return from a radar signal at an airport is typically null, except for a few spikes locating the positions and velocities of nearby aircraft. A time trace of the sound from a musical instrument might not be sparse, whereas the Fourier transform of the same signal would be sparse. Compressive sampling is an approach to the modeling, processing and storage of sparse signals.

We start with a simple model of a signal as an n-tuple $x \in \mathbb{R}^n$ or $\mathbb{C}^n$. Think of n as large. We are especially interested in the case where x is k-sparse. That is, at most k of the components $x_1, \ldots, x_n$ of x are nonzero. We take k to be small with respect to n. In order to obtain information about the signal x we sample it. Here, a sample is defined by a linear functional f_r. That is, the sample $f_r(x) = r \cdot x$ where $r = (r_1, \ldots, r_n)$ is a given sample vector and $r \cdot x$ is the dot product of r and x. Once r is chosen, the dot product is easy to implement in hardware. In the case where $r_i = \delta_{ij}$ for fixed j and $i = 1, \ldots, n$ the sample would yield the value of the component x_j. Now suppose we take m different samples y_ℓ of x, i.e., we have m distinct sample vectors $r^{(\ell)}$, $\ell = 1, \ldots, m$. We can describe the sampling process by the equation $y = \Phi x$ where the $m \times n$ sampling matrix Φ is defined by

$$
(\Phi_{\ell,j}) = \left(r_j^{(\ell)} \right) = \begin{pmatrix} r^{(1)} \\ \vdots \\ r^{(m)} \end{pmatrix} = (c^{(1)}, \ldots, c^{(n)}) \tag{5.1}
$$

and y is an m-tuple. The n-vectors $r^{(\ell)}$ are the row vectors of Φ whereas the m-vectors $c^{(j)}$ are the column vectors of Φ. Note that $r_j^{(\ell)} = c_\ell^{(j)}$. For compressive sampling m is less than n, so that the system is underdetermined. The problem is to design the sampling matrix Φ so that we can determine the signal x uniquely from m samples. Obviously, this is impossible for arbitrary signals x. It *is* possible if we know in advance that x has some special form. For compressive sampling at its simplest, the only requirement on the signal x is that it is k-sparse, i.e., that at most $k < n$ of the components x_j are nonzero. The only assumption is k-sparsity, not the possible locations of the nonzero components.

A simple example, the bit counter, will illustrate the utility of compressive sampling. Suppose we have a sequence of long n-component signals $x^{(h)}$ that are 1-sparse. Thus, each signal consists of a single spike at some location $j(h)$ with all other components 0. To determine the signal uniquely we need only find the location and the value of the spike.

How can we do this from a minimum number of samples? This problem
has an elegant solution, particularly simple in the case $n = 2^\ell$ for ℓ a
positive integer. For each 1-sparse signal x we number the components
x_j from $j = 0$ to $j = 2^\ell - 1$. Similarly we will number the columns of
Φ, starting from 0. Recall that each j can be written uniquely in binary
numbers as $[j_{\ell-1}, j_{\ell-2} \ldots, j_1, j_0]$ where $j = j_0 2^0 + j_1 2^1 + \cdots + j_{\ell-1} 2^{\ell-1}$
where each $j_s = 0, 1$. We design the $m \times n$ sampling matrix Φ where
$m = \ell + 1$ by first filling the top (0th) row with ones: $\Phi_{0,j} = 1$. For
the remaining $m - 1$ terms in the 0th column of Φ we enter the binary
number for 0, for the remaining $m - 1$ terms in the 1st column we enter the binary number for 1, ..., and for the remaining $m - 1$ terms
in the $(n - 1)$st column we enter the binary number of $n - 1$. Thus
the sampling matrix is given by $\Phi_{ij} = j_{\ell-i}$ for $i = 1, \cdots, \ell$. Computing $y = \Phi x$ we have $y_0 = x_{j(h)}$ the magnitude of the spike, whereas if
$y_0 \neq 0$, $[y_m/y_0, \cdots, y_1/y_0]$ is the location of the spike, written in binary.
For example, suppose $\ell = 3$, $m = 4$, $n = 8$ and the signal x has the
spike $x_6 = 6.1$. Then $y = \Phi x$ becomes

$$
\begin{pmatrix} 6.1 \\ 0 \\ 6.1 \\ 6.1 \end{pmatrix} = \begin{pmatrix} 1 & 1 & 1 & 1 & 1 & 1 & 1 & 1 \\ 0 & 1 & 0 & 1 & 0 & 1 & 0 & 1 \\ 0 & 0 & 1 & 1 & 0 & 0 & 1 & 1 \\ 0 & 0 & 0 & 0 & 1 & 1 & 1 & 1 \end{pmatrix} \begin{pmatrix} 0 \\ 0 \\ 0 \\ 0 \\ 0 \\ 0 \\ 6.1 \\ 0 \end{pmatrix}, \tag{5.2}
$$

and from y we see that the value of the spike is $y_0 = 6.1$ and the location
is $[1, 1, 0] = 6$. We see from this construction that for 1-sparse signals of
large length n we can identify the signal uniquely with only $\log_2 n + 1$
samples; it isn't necessary to sample all n components individually. For
storage or transmission of the information, the compression is impressive.
Rather than store n numbers we need only store $m \approx \log_2 n$ numbers.

Exercise 5.1 Why is the row of 1's needed in the sample matrix for
example (5.2)?.

Exercise 5.2 Using the method of expression (5.2) design a sample
matrix to detect a 1-sparse signal of length $n = 16$. Apply your matrix
to a signal with a spike $x_9 = -2$.

In this chapter we will study the problem, for given k, n with $k < n$,
of how to design $m \times n$ sampling matrices (or encoders) to determine

k-sparse signals x uniquely such that the number of samples m is as small as possible. Later we shall treat the more important case where noise and measuring errors are allowed and we want to find a k-sparse approximation $\tilde{x}$ of x with approximation error and m as small as possible.

This chapter is the most technical of the book and many readers may decide to skim the results and then proceed to the last section where compressive sampling is implemented. However, the basic ideas are relatively simple and we will outline these before entering into details. A fundamental observation is that if $y = \Phi x^{(0)}$ so that y is the sample produced by the signal $x^{(0)}$ then y is also produced by all signals $x = x^{(0)} + u$ where u is a solution of the homogeneous equation $\Phi u = \Theta$, and only by such signals. Now if $x^{(1)}, x^{(2)}$ are k-sparse signals with the same sample $\Phi x^{(1)} = \Phi x^{(2)}$ then $\Phi(x^{(2)} - x^{(1)}) = \Theta$, so $\Phi u = \Theta$ has a solution $u = x^{(2)} - x^{(1)}$ that is at most $2k$-sparse. If we can design a sampling matrix Φ such that its homogeneous equation has no nonzero solutions u that are $2k$-sparse, then for a given sample y there is at most one k-sparse signal $x^{(0)}$ such that $y = \Phi x$. How do we recover $x^{(0)}$? A solution advanced here is to design the sample matrix such that among all signals x with sample y the k-sparse solution $x^{(0)}$ has minimal ℓ_1 norm (but not in general minimal ℓ_2, least squares norm). For such a matrix, if the homogeneous equation has no $2k$-sparse solutions then we can recover a k-sparse signal $x^{(0)}$ from its sample y by solving the minimization problem $x^{(0)} = \text{Argmin}_{\Phi x=y}||x||_1$. For real signals and sample matrices this is a straightforward linear programming problem that can be solved in polynomial time by standard algorithms, freely available even for spreadsheets. This works, basically, because the ℓ_1 minimum always occurs for a vector with a maximal number of zeros.

In summary, we need to address two problems. First, for given k, m, n how do we characterize sample matrices whose homogeneous equation has no $2k$-sparse solutions? For such sample matrices we can, in principle, identify a k-sparse signal, but determining the signal may be impractical. Our second problem is in determining those matrices for which it is easy to recover the k-sparse signal in practice. Here we shall employ ℓ_1 minimization.

Given $k \ll n$ how do we construct an $m \times n$ sample matrix Φ with m as small as possible? The surprising answer is that the bit counter example is somewhat misleading. Matrices with homogeneous equations admitting $2k$-sparse solutions are rather special. An arbitrarily chosen matrix is less likely to have this failing. We can use a random number generator, or coin flipping, to choose the matrix elements of Φ. Using random matrix theory we can show that the smallest m that works is about $m = Ck \log_2(n/k) \sim Ck \log_2(n)$ where C is a constant, close to 1 in practice (though still much larger in theory). We can construct the matrices via random number generators and guarantee that they will "probably" work, e.g., guarantee no more than one failure in 10^9 sample matrices. An implementation of the entire process is given in the last section.

5.2 Algebraic theory of compressive sampling

The sampling problem is closely associated with the structure of the null space $N(\Phi)$ of the $m \times n$ sampling matrix/encoding matrix. Clearly, the rank of Φ is $\leq m$, and $m < n$ since our system is underdetermined. There will be many signals x that will yield the same sample y. Recall that for Φ a complex $m \times n$ matrix and complex signals x we have $N(\Phi) = \{h \in \mathbb{C}^n : \Phi h = \Theta\}$ and $\dim N(\Phi) \geq n - m$. The hyperplane $F(y)$ of signals giving the sample y takes the form

$$F(y) = \{x \in \mathbb{C}^n : \Phi x = y\} = x^{(0)} + N(\Phi)$$

for any $x^{(0)} \in F(y)$. Note that every signal x lies on one and only one of these hyperplanes. Once we have sampled or encoded the signal we will need to try to reconstruct the signal from the sample. We define a decoder as a mapping $\Delta : \mathbb{C}^m \to \mathbb{C}^n$, which could be nonlinear. Given a signal x and a decoder Δ we think of $\hat{x} = \Delta(\Phi(x)) = \Delta(y)$ as our approximation to x, based on the sample. Ideally we would like an encoding-decoding pair Φ, Δ such that $\hat{x} = x$. However, this is not possible for an underdetermined system, unless we restrict the set of signals. Clearly, ensuring that at most one allowable signal x gives a particular sample y (so that we can recover x from y via a suitable decoder) is equivalent to requiring that each hyperplane $F(y)$ contains at most one allowable signal. Initially we restrict to signals that are sparse with respect to the standard basis $e^{(j)}$ for $\mathbb{C}^n$, without any other requirements. More precisely, given a positive integer $k < n$ we will restrict x to the subset Σ_k of k-sparse signals. Here,

$$\Sigma_k = \{x \in \mathbb{C}^n : \#\mathrm{supp}\,(x) \leq k\}, \text{ where supp } (x) = \{i : x_i \neq 0\} \quad (5.3)$$

and #supp is the cardinality of the support of x. In other words, Σ_k is the subset of $\mathbb{C}^n$ consisting of all signals with at most k nonzero components. Note that Σ_k is not a subspace of $\mathbb{C}^n$ because the sum of two k-sparse signals may not be k-sparse. Indeed, if $x^{(1)}, x^{(2)} \in \Sigma_k$, the best that we can guarantee is that $x^{(1)} + x^{(2)} \in \Sigma_{2k}$. Though we have expressed our problem in terms of complex matrices, complex signals and complex null spaces, we could give virtually identical definitions for real matrices, real signals and real null spaces, and the algebraic theory to follow will be virtually the same for both.

Exercise 5.3 Show that even though Σ_k is not a vector space, it can be expressed as the set-theoretical union of subspaces $X_{T_k} = \{x : \text{supp}(x) \subset T_k\}$ where T_k runs over all k-element subsets of the integers $\{1, 2, \ldots, n\}$, i.e.,

$$\Sigma_k = \cup_{T_k} X_{T_k}.$$

If we choose Σ_k for some fixed $k < n$ as our set of allowable signals, in order to recover each k-sparse signal x from its sample y, we need to find the possible $m \times n$ sample matrices Φ such that $y = \Phi x$ uniquely determines x for all $x \in \Sigma_k$. The basic idea behind the algebraic theory of compressive sampling is very simple. If $x^{(1)}, x^{(2)}$ are two k-sparse signals with the same sample y then the signal $x^{(1)} - x^{(2)}$ is an at most $2k$-sparse signal in the null space $N(\Phi)$. Thus to guarantee that each sample corresponds to at most one k-sparse signal, we need only require that the null space of Φ contains no $2k$-sparse signals.

To characterize the sampling matrices we first recall the representation (5.1) of Φ in terms of its column vectors and introduce notation to describe submatrices of Φ. Let T be a set of $\#(T) = \ell$ column indices $1 \leq i_1 < i_2 < \cdots < i_\ell \leq n$. Then $\Phi_T = \left(c^{(i_1)}, \ldots, c^{(i_\ell)}\right)$ is the $m \times \#(T)$ submatrix of Φ formed from the columns indexed by T. Thus, $\Phi_T^* \Phi_T$ is the $\#(T) \times \#(T)$ matrix with (s, t) matrix element given by the dot product $\overline{c^{(i_s)}} \cdot c^{(i_t)}$, $1 \leq s, t \leq \#(T)$.

Theorem 5.1 *Let Φ be a $m \times n$ matrix and k a positive integer $\leq n$. The following are equivalent*

(1) *For all $y \in \mathbb{C}^m$, each $F(y)$ contains at most one element of Σ_k.*

(2) $\Sigma_{2k} \cap N(\Phi) = \{\Theta\}$.

(3) *For any set of column indices T with $\#(T) = 2k$, the matrix Φ_T has rank $2k$.*

(4) *For any set of column indices T with $\#(T) = 2k$, the $2k \times 2k$ matrix $\Phi_T^* \Phi_T$ has all eigenvalues $\lambda_j > 0$ and is invertible. (Here Φ^* is the $n \times m$ adjoint matrix: $\Phi_{ij}^* = \overline{\Phi}_{ji}$.)*

Proof

(1) → (2). Suppose $x \in \Sigma_{2k}$ and $\Phi x = \Theta$. Since x has at most $2k$ nonzero components, we can always find vectors $x^{(1)}, x^{(2)}$ each with at most k nonzero components, and such that $x = x^{(1)} - x^{(2)}$. Now set $y = \Phi x^{(1)}$. Since $\Phi x = \Phi x^{(1)} - \Phi x^{(2)} = \Theta$ we have that both $x^{(1)}, x^{(2)} \in F(y)$. By (1), $x^{(1)} = x^{(2)}$, so $x = \Theta$.

(2) → (3). Let T be a set of indices $1 \le i_1 < i_2 < \cdots < i_{2k} \le n$ with $\#(T) = 2k$ and let $x \in \mathbb{R}^n$ such that $\operatorname{supp}(x) \subseteq T$, so $x \in \Sigma_{2k}$. If $\sum_{\ell=1}^{2k} x_{i_\ell} c^{(i_\ell)} = \Phi x = \Theta$ then $x \in N(\Phi)$, so $x = \Theta$ by (2). Thus the $2k$ column vectors indexed by T must be linearly independent.

(3) → (4). Let T and x be chosen as in the preceding proof and consider the quadratic form $< x, \Phi^* \Phi x > =$

$$\sum_{s,t=1}^{2k} x_{i_s} (c^{(i_s)} \cdot \overline{c^{(i_t)}}) x_{i_t} = (\sum_{s=1}^{2k} x_{i_s} c^{(i_s)}) \cdot \overline{(\sum_{t=1}^{2k} x_{i_t} c^{(i_t)})}$$

$$= \| \sum_{t=1}^{2k} x_{i_t} c^{(i_t)} \|_2^2.$$

By (3), the $2k$ column vectors $c^{(i_t)}$ are linearly independent, so the quadratic form $< x, \Phi^* \Phi x >$ is > 0 for all nonzero x with $\operatorname{supp}(x) \subseteq T$. Note that the matrix $\Phi_T^* \Phi_T$ is self-adjoint. It is a well-known fact from linear algebra that any $N \times N$ self-adjoint matrix A has N real eigenvalues and that these eigenvalues are all positive if and only if the hermitian form $\sum_{i,j=1}^{N} y_i A_{ij} \overline{y}_j > 0$ for all complex nonzero vectors y. Furthermore a self-adjoint matrix with all positive eigenvalues is invertible. This establishes (4).

(4) → (1). Suppose $x^{(1)}, x^{(2)} \in F(y) \cap \Sigma_k$ for some $y \in \mathbb{C}^m$. Setting $x = x^{(1)} - x^{(2)}$ we see that $\Phi x = \Phi x^{(1)} - \Phi x^{(2)} = y - y = \Theta$ and $x \in \Sigma_{2k}$. Let T be the index set $\{i_t\}$ for the support of x. Then the fact that x is in the null space of Φ can be expressed in terms of the column vectors of Φ as $\sum_{t=1}^{2k} x_{i_t} c^{(i_t)} = \Phi x = \Theta$. Thus

$$< x, \Phi^* \Phi x > = \sum_{s,t=1}^{2k} x_{i_s} (c^{(i_s)} \cdot \overline{c^{(i_t)}}) x_{i_t} = \| \sum_{t=1}^{2k} x_{i_t} c^{(i_t)} \|_2^2 = 0.$$

By (4), the eigenvalues of $\Phi_T^* \Phi_T$ are all positive so the only way for this last sum to vanish is if $x = \Theta$. This implies $x^{(1)} = x^{(2)}$.

$\square$

Example 5.2 Consider the $2k \times n$ Vandermonde matrix

$$\Phi = \begin{pmatrix} 1 & 1 & \cdots & 1 \\ a_1 & a_2 & \cdots & a_n \\ a_1^2 & a_2^2 & \cdots & a_n^2 \\ \vdots & \vdots & \vdots & \vdots \\ a_1^{2k-1} & a_2^{2k-1} & \cdots & a_n^{2k-1} \end{pmatrix},$$

where $a_1 < a_2 < \cdots < a_n$. It is an exercise in linear algebra to show that the determinant of a square Vandermonde matrix is equal to $\pm \Pi_{i>j}(a_i - a_j) \neq 0$, i.e., a square Vandermonde matrix is invertible. Thus any $2k \times 2k$ submatrix of Φ has rank $2k$ and Φ satisfies the conditions of Theorem 5.1 with $m = 2k$.

Exercise 5.4 Verify that the matrix $\Phi_T^* \Phi_T$ is self-adjoint.

If the $m \times n$ sample matrix Φ satisfies the requirements of the theorem for some k with $2k \leq m < n$, then we can show the existence of a encoder/decoder system that reproduces any k-sparse signal x. Necessary and sufficient conditions for unique reproduction are that every subset of $2k$ column vectors must be linearly independent. If this is true we have $m \geq 2k$. Now let T be any $2k$-index set that contains the support of x. Then the encoder gives the sample $y = \Phi x = \Phi_T x$, so $\Phi_T^* y = (\Phi_T^* \Phi_T) x$. Since the $2k \times 2k$ matrix $\Phi_T^* \Phi_T$ is invertible, we can recover x from y via the decoding operation

$$x = \Delta(y) = (\Phi_T^* \Phi_T)^{-1} \Phi_T^* y.$$

However, this construction requires us to find some T containing the support of x, a highly nontrivial problem for large n! Thus this appears not to be a very efficient means of implementing compressive sensing. Next we shall pursue a more practical way of finding a k-sparse solution by using the ℓ_1 norm to find x from the minimization problem:

$$x = \Delta(y) = \text{Argmin}_{z \in F(y)} ||z||_1,$$

where by Argmin we mean the vector z that achieves the minimum value of $||z||_1$.

5.3 Analytic theory of compressive sampling

The algebraic solution to the compressed sampling problem presented
in the preceding section, by itself, seems of only modest practical importance. Firstly, there is a significant numerical problem of actually
computing x for large values of n. Also, the solution assumes that the
signals are precisely k-sparse, that the inverse of the matrix $(\Phi_T^* \Phi_T)^{-1}$
can be computed with no error, and we have to find T! Real signals typically have a few spikes with all other components small but not zero.
The numerical computation of the matrix inverse may be unstable for
large n and k. (This is the case with the Vandermonde matrix.) The signals may be partially corrupted by noise. We need to develop analytic
estimates that enable us to determine how well the encoding/decoding
procedure approximates the initial signal. We also need to be concerned
with the design of the decoder, so that it can compute the approximation
efficiently.

5.3.1 Recovering a sparse solution of $y = \Phi x$ via ℓ_1 minimization

Theorem 5.1 gave necessary and sufficient conditions that there was at
most one signal $\hat{x} \in \Sigma_k \cap F(y)$ that produced a given sample y. (Recall
that if $x = \tilde{x}$ is one solution of the equation $y = \Phi x$ then all solutions
are of the form $x \in F(y) = \{\tilde{x} + h : h \in N(\Phi)\}$ where $N(\Phi)$ is the
null space of the $m \times n$ sampling matrix Φ.) However, the theorem did
not lead directly to an efficient method for explicit computation of the
k-sparse solution from the sample y. Here we will study the feasibility
of finding $\hat{x}$ by solving the ℓ_1 minimization problem

$$\hat{x} = \text{Argmin}_{x \in F(y)} ||x||_1. \qquad (5.4)$$

Already, we have seen examples of the special ability of the ℓ_1 norm to
produce solutions x of $y = \Phi x$ with maximal sparsity. Now we will find
necessary and sufficient conditions that this ℓ_1 minimization will lead to
a unique k-sparse solution.

Suppose that the equation $y = \Phi x$ has a k-sparse solution $\hat{x}$ and that
this solution also satisfies (5.4), i.e., it has minimal ℓ_1 norm. This means
that

$$||\hat{x} + h||_1 \geq ||\hat{x}||_1$$

for all $h \in N(\Phi)$. If $T \subset \{1, 2, \ldots, n\}$ is the index set for the support of $\hat{x}$ and T^c is the set of remaining indices, then $\#(T) = k$, $\#(T^c) = n-k$ and the minimization condition can be written in terms of absolute values as

$$\sum_{i \in T} |\hat{x}_i + h_i| + \sum_{i \in T^c} |h_i| \geq \sum_{i \in T} |\hat{x}_i|.$$

Let $S \subset \{1, 2, \ldots, n\}$ and z an n-tuple,. We define z_S as the n-tuple such that $(z_S)_i = z_i$ for $i \in S$ and $(z_S)_i = 0$ for $i \in S^c$. Thus z_S agrees with z for all indices $i \in S$, but the remaining elements are zero. Similarly we define z_{S^c} as the n-tuple such that $(z_{S^c})_i = z_i$ for $i \in S^c$, and $(z_{S^c})_i = 0$ for $i \in S$. Note that $z = z_S + z_{S^c}$.

Definition 5.3 The $m \times n$ complex matrix Φ satisfies the null space property of order k provided for every index set S with $\#(S) = k$ we have

$$||h_S||_1 < ||h_{S^c}||_1, \quad \forall\, h \in N(\Phi),\ h \neq \Theta. \tag{5.5}$$

Theorem 5.4 *Every k-sparse vector $\hat{x}$ is the unique solution of the ℓ_1 minimization problem (5.4) if and only if Φ satisfies the null space property of order k.*

Proof First, suppose every k-sparse $\hat{x}$ is the unique solution of the ℓ_1 minimization problem (5.4), with $y = \Phi\hat{x}$. Now let $h \in N(\Phi)$ with $h \neq \Theta$ and let S be any index set such that $\#(S) = k$. Then h_S is k-sparse, hence the unique ℓ_1 minimizer for the problem $\Phi x = \Phi h_S$. Now $\Phi(h) = \Phi(h_S + h_{S^c}) = 0$, so $\Phi(-h_{S^c}) = \Phi(h_S) \neq \Theta$. By uniqueness, we must have $||h_{S^c}||_1 > ||h_S||_1$ which is the null space property.

Conversely, assume Φ satisfies the null space property of order s and S is the index set of $\hat{x}$, i.e., the set of indices i for which $\hat{x}_i \neq 0$. Then any solution x of $\Phi x = \Phi\hat{x} = y$ with $x \neq \hat{x}$ can be written as $x = \hat{x} - h$ with $h \in N(\Phi)$. Then

$$||\hat{x}||_1 = ||(\hat{x}-x_S)+x_S||_1 \leq ||(\hat{x}-x_S)||_1 + ||x_S||_1 = ||(\hat{x}_S - x_S)||_1 + ||x_S||_1 =$$

$$||h_S||_1 + ||x_S||_1 < ||h_{S^c}||_1 + ||x_S||_1 = ||-x_{S^c}||_1 + ||x_S||_1 = ||x||_1.$$

Hence $||\hat{x}||_1 < ||x||_1$. $\qquad\square$

Exercise 5.5 In the proof of Theorem 5.4 show in detail why $\Phi(-h_{S^c}) = \Phi(h_S) \neq \Theta$.

Exercise 5.6 In the proof of Theorem 5.4 show why the ℓ_1 norm rather than some other ℓ_p norm is essential.

Exercise 5.7 Show that if Φ satisfies the null space property of order k then it also satisfies the null space property for orders $1, 2, \ldots, k - 1$.

A case of special importance for the above definition and theorem occurs for real $m \times n$ matrices and real signal vectors x. Then we have

Definition 5.5 The $m \times n$ real matrix Φ acting on real vectors satisfies the null space property of order k provided for every index set S with $\#(S) = k$ we have

$$||h_S||_1 < ||h_{S^c}||_1, \quad \forall\, h \in N(\Phi),\ h \neq \Theta. \tag{5.6}$$

Theorem 5.6 *Every k-sparse vector $\hat{x}$ is the unique solution of the ℓ_1 minimization problem (5.4) if and only if Φ satisfies the null space property of order k.*

The proof of the real version of the theorem is identical to the proof of the complex version. The only difference is that the vectors x, h in the second case are restricted to be real. It is not immediately obvious for a real matrix Φ that the two definitions and results agree.

Conditions (5.5), (5.6) while difficult to verify in practice, point to an important property that the null space must possess for construction by ℓ_1 minimization. In order that the sample matrix be able to reconstruct sparse signals supported on small index sets T, the null space must counterbalance by being distributed through all indices. This is often referred to as an uncertainty principle for compressive sampling. It suggests that the components of the sampling matrix, rather than being highly structured to fit a particular type of signal, should be chosen at random. We shall see that this insight is correct.

If we can show that the conditions of Theorem 5.4 or 5.6 are satisfied then we can recover uniquely the k-sparse signal $\hat{x}$ from the sample y by solving the minimization problem

$$\hat{x} = \text{Argmin}_{x \in F(y)} ||x||_1. \tag{5.7}$$

Although, unlike ℓ_2 (least squares) minimization, this problem has no analytic solution, it can be solved by numerically efficient optimization algorithms. The complex case is equivalent to a convex second-order cone program (CSOCP), [16]. The real case is particularly easy to solve because it can be recast as a linear programming problem, and such routines are widely available.

This text is not meant to cover linear programming, but it is worth pointing out that the real ℓ_1 minimization problem for real Φ and x is equivalent to a linear programming problem. To see this we first express the $m \times n$ real sampling matrix in terms of its row vectors $r^{(k)}$:

$$\Phi = \begin{pmatrix} r^{(1)} \\ \vdots \\ r^{(m)} \end{pmatrix}, \quad r^{(k)} \in \mathbb{R}^n, \quad i = 1, \ldots, m.$$

Then we can express the ℓ_1 minimization problem (5.7) as the linear programming problem

$$\min \sum_{j=1}^{n} u_j, \text{ such that } y_k = r^{(k)} \cdot x, \quad -u_j \le x_j \le u_j. \tag{5.8}$$

Similarly the (possibly overdetermined) ℓ_1 problem

$$\min_{x \in F(y)} ||y - \Phi x||_1 = \min \sum_{i+1}^{m} |y_i - r^{(i)} \cdot x| \tag{5.9}$$

can be expressed as the linear programming problem

$$\min \sum_{k=1}^{m} u_k, \text{ such that } -u_k \le y_k - r^{(k)} \cdot x \le u_k. \tag{5.10}$$

(Note that if the x_j are complex numbers then this approach fails.) In general, a linear programming problem (LP) is a maximization or minimization problem that can be expressed entirely through linear equations and linear inequalities [16]. Such problems can be solved, for example, by the simplex method.

In the case where x and Φ are complex we write

$$x_j = x_j^r + i x_j^i, \; y_k = y_k^r + i y_k^i, \; r^{(k)} = r_r^{(k)} + i r_i^{(k)}, \quad 1 \le j \le n \; 1 \le k \le m,$$

where $x_j^r, x_j^i, y_k^r, y_k^i, r_r^{(k)}, r_i^{(k)}$ are real. Then the ℓ_1 minimization problem (5.7) can be expressed as

$$\min \sum_{j=1}^{n} u_j, \text{ such that} \tag{5.11}$$

$$u_j = \sqrt{(x_j^r)^2 + (x_j^i)^2}, \; y^r = r_r^{(k)} \cdot x^r - r^{(i)} \cdot x^i, \; y^i = r_r^{(k)} \cdot x^i + r_i^{(k)} \cdot x^r.$$

This has the form of a convex second-order cone program (CSOCP), [16].

5.3.2 A theoretical structure for deterministic compressive sampling

Instead of restricting to signals $x \in \mathbb{C}^n$ that are strictly k-sparse we will consider the larger class of compressible signals, those that can be approximated by elements of Σ_k. As a measure of the approximation of x by k-sparse signals we adopt

$$\sigma_k(x) = \inf_{z \in \Sigma_k} ||x - z||_1. \tag{5.12}$$

Exercise 5.8 We order the components of $x \in \mathbb{C}^n$ in terms of the magnitude of the absolute value, so that

$$|x_{i_1}| \geq |x_{i_2}| \geq \cdots \geq |x_{i_n}|.$$

Show that $\sigma_k(x) = \sum_{j=k+1}^{n} |x_{i_j}|$, i.e., $\sigma_k(x)$ is the sum of the absolute values of the $n - k$ smallest components of x. Show that

$$\text{Argmin} \inf_{z \in \Sigma_k} ||x - z||_1 = x^k$$

where x^k has index set $T = \{i_1, i_2, \ldots, i_k\}$. Thus, x^k is the closest k-sparse approximation of x with respect to the ℓ_1 norm.

Exercise 5.9 If

$$x = (3, -2, 6, 0, 3, -1, 2, -5) \in \mathbb{C}^8,$$

find x^k and $\sigma_k(x)$ for $k = 1, 2, \ldots, 7$.

Clearly, $\sigma_k(x) = 0 \iff x \in \Sigma_k$. We could consider x as a good candidate for compression provided $\sigma_k(x) < \epsilon$ for some suitably small ϵ. If x is compressible but not exactly k-sparse we cannot expect to reproduce it exactly, but we can try to approximate it by a k-sparse signal.

To make clearer the relation between σ_k and the null space of Φ it will prove useful to quantify the degree to which the null space property is satisfied.

Definition 5.7 Given a sampling matrix Φ, we say that Φ satisfies the null space property (NSP) of order k with constant $\rho \in (0, 1)$ provided

$$||h_T||_1 \leq \rho ||h_{T^c}||_1$$

for all index sets T with $\#T \leq k$ and all $h \in N(\Phi)$.

This definition applies either to the complex case or to the real case where both Φ and $N(\Phi)$ are required to be real.

Exercise 5.10 Show that if Φ satisfies the null space property of order k, Definition 5.3, then there exists a $\rho \in (0,1)$ such that Φ satisfies the null space property of order k with constant ρ.

Now we come to the essence of our problem. Suppose $x \in \mathbb{C}^n$ is a signal that we consider is "nearly" k-sparse and let $y = \Phi x \in \mathbb{C}^m$ be its corresponding sample. We define a decoder

$$\Delta(y) = \mathrm{Argmin}_{z, \Phi z = y} ||z||_1 = \hat{x}.$$

We want our decoder to capture the closest k-sparse approximation x^k to x and ignore the smallest $n - k$ components.. Thus we want to guarantee that the decoder will yield a unique m-sparse solution $\hat{x}$ that is as close as possible to x. In particular, if $x \in \Sigma_k$ we want $\hat{x} = x$. Recall from Exercise 5.8 that $\sigma_k(x) = ||x - x^k||_1$.

Theorem 5.8 *Suppose Φ has the null space property of order k with constant ρ. Let $x \in \mathbb{C}^n$ with $y = \Phi x$ and let $\hat{x} = \Delta(y)$. Then*

$$||\hat{x} - x||_1 \le 2\frac{1+\rho}{1-\rho}||x - x^k||_1. \tag{5.13}$$

If x is k-sparse then $\hat{x} = x$.

Proof Since $y = \Phi\hat{x}$, there is a unique $h \in N(\Phi)$ such that $\hat{x} = x + h$. Since $\hat{x}$ is a ℓ_1 minimum solution we must have

$$||\hat{x}||_1 = ||x + h||_1 \le ||x||_1. \tag{5.14}$$

Now let T_0 be the set of indices corresponding to the k largest components of x in absolute value. (Thus T_0 is the index set of x^k.) Recall from the triangle inequality $|a| = |(a+b) + (-b)| \le |a+b| + |-b|$, so $|a+b| \ge |a| - |b|$ and, by symmetry, $|a+b| \ge |b| - |a|$. Also, for any index set T and n-tuple z, let z_T be the vector with components $(z_T)_i = z_i$ for $i \in T$ and all other components zero. Then

$$||x+h||_1 = \sum_{i \in T_0}(|x_i+h_i|) + \sum_{i \in T_0^c}(|x_i+h_i|) \ge \sum_{i \in T_0}(|x_i|-|h_i|) + \sum_{i \in T_0^c}(|h_i|-|x_i|) \tag{5.15}$$

$$= (||x||_1 - ||x_{T_0^c}||_1 - ||h_{T_0}||_1) + (||h_{T_0^c}||_1 - ||x_{T_0^c}||_1)$$

$$= (||x||_1 + ||h_{T_0^c}||_1) - (2||x_{T_0^c}||_1 + ||h_{T_0}||_1).$$

Comparing the inequalities (5.14) and (5.15) and noting that $||x_{T_0^c}||_1 = ||x - x^k||_1$, we obtain

$$2||x - x^k||_1 \ge ||h_{T_0^c}||_1 - ||h_{T_0}||_1 \ge (1 - \rho)||h_{T_0^c}||_1. \tag{5.16}$$

Since $\rho < 1$ we obtain

$$||h_{T_0^c}||_1 \leq \frac{2}{1-\rho}||x - x^k||_1.$$

Since $\hat{x} = x + h$, it follows that

$$||\hat{x}-x||_1 = ||h||_1 = ||h_{T_0}||_1 + ||h_{T_0^c}||_1 \leq (\rho+1)||h_{T_0^c}||_1 \leq 2\frac{1+\rho}{1-\rho}||x-x^k||_1.$$

$\square$

The proof of the theorem shows that the k-sparse solution $\hat{x}$ of the minimization problem is unique, and allows us to get a bound on the error of reconstruction of the nearly k-sparse signal x by $\hat{x}$. It is still possible that there is more than one k-sparse signal with sample y even though any other solution doesn't satisfy the ℓ_1 minimum property. In the next section we shall see how that possibility can be eliminated.

5.3.3 Restricted Isometry Property

Theorem 5.8 is an important theoretical result but it will not be of much practical use unless we find methods for determining ρ and for designing encoder/decoder pairs with $\rho < 1$ for k small with respect to n. The null space property doesn't have direct intuitive meaning for us. A more intuitive concept is the Uniform Uncertainty Principle or Restricted Isometry Property (RIP). Let Φ be our usual $m \times n$ sampling matrix. For each k we define the restricted isometry constant δ_k as the smallest nonnegative number such that

$$(1 - \delta_k)||x||_2^2 \leq ||\Phi(x)||_2^2 \leq (1 + \delta_k)||x||_2^2 \qquad (5.17)$$

for all k-sparse signals x, i.e., $x \in \Sigma_k$. (Note that here we are using the ℓ_2 norm.) Expression (5.17) captures in an efficient and easily understandable manner how well the k-sparse signal x is captured by the sample $y = \Phi x \in \mathbb{R}^m$. Indeed if $\delta_k \geq 1$ so that the left-hand term is ≤ 0 then we could have a nonzero signal that produces a zero sample, so that we would have no capability of recapturing it. If the constant on the right-hand side of (5.17) is large then a small change in the sample $y = \Phi x$ could correspond to a large change in x. This would introduce numerical instabilities in our computation of x from y. Thus a desirable property for an encoder is $\delta_k < 1$.

Definition 5.9 Φ has the Restricted Isometry Property (RIP) for k if it has a restricted isometry constant such that $0 < \delta_k < 1$.

Another way to understand RIP is in terms of eigenvalues. If x is k-sparse with index set T then

$$||\Phi x||_2^2 = < \Phi_T x, \Phi_T x > = < \Phi_T^* \Phi_T x, x > .$$

As pointed out in the proof of Theorem 5.1, $\Phi^* \Phi$ is self-adjoint and nonnegative, Thus all eigenvalues $\lambda_i(T)$ of this square matrix are real and nonnegative, and $\lambda_{\min}(T)||x||_2^2 \leq < \Phi_T^{tr} \Phi_T x, x > \leq \lambda_{\max}(T)||x||_2^2$. It follows that

$$1 - \delta_k = \lambda_{\min} \leq \lambda_{\max} = 1 + \delta_k, \tag{5.18}$$

where the maximum and minimum are taken over all index sets T with $\leq k$ indices.

Exercise 5.11 Give the details of the derivation of inequalities (5.18).

If we have two signals $x, x' \in \Sigma_k$ then $x - x'$ may not be k-sparse, but it is $2k$-sparse. Thus $(x - x') \in \Sigma_{2k}$ and

$$(1 - \delta_{2k})||x - x'||_2^2 \leq ||\Phi(x - x')||_2^2 \leq (1 + \delta_{2k})||x - x'||_2^2. \tag{5.19}$$

If $x \neq x'$ then, to distinguish the signals, we must have $\delta_{2k} < 1$, i.e., Φ must satisfy RIP for $2k$. Moreover, for x very close to x' in norm we want the samples $y = \Phi x$ and $y' = \Phi x'$ to be very close in norm, and this would be implied by $||\Phi(x - x')||_2^2 \leq (1 + \delta_{2k})||x - x'||_2^2$.

It is worth pointing out here that the basic idea behind RIP is pervasive in signal processing theory. For signals in Hilbert spaces the concept is a frame which is discussed in Section 8.2. These concepts are not identical because frames refer to vector spaces and Σ_k is not a vector space.

Now we will show that the null space property can be implied by the more intuitive RIP. Since RIP is expressed in terms of ℓ_2 whereas the null space property is expressed in terms of ℓ_1, we need the result from Section 1.3 that the norms of a k-sparse signal x satisfy the relation $||x||_2 \leq ||x||_1 \leq \sqrt{k}||x||_2$. Now suppose Φ satisfies RIP for some fixed $2k$. This will imply that for each sample $y \in \mathbb{C}^m$ there will be at most one k-sparse solution x to $y = \Phi x$. In particular $N(\Phi)$ will contain no k-sparse vectors.

Now let us examine what RIP implies about the null space property. We want to guarantee that for all k-index sets T the inequality $||h_T||_1 < \rho||h_{T^c}||_1$ holds for a constant $\rho < 1$ and all $h \in N(\Phi)$. We start by taking any $h \in N(\Phi)$. Our approach will be to take the worst case possible for h: $T = T_0$ is the index set of the k largest components of h, in absolute value. We want to guarantee that $||h_{T_0}||_1 < \rho||h_{T_0^c}||_1$. Note that RIP applies only to k-sparse signals and, in general, the h is not k-sparse. In order to apply RIP we will decompose $h_{T_0^c}$ as a sum of signals h_{T_i}, each of which is at most k-sparse. Noting that h is an n-tuple, we use the Euclidean algorithm to write $n = ak + r$ where a is a positive integer and the integer r is the remainder, $0 \leq r < k$. We divide the indices of h into $a+1$ disjoint sets. The first set T_0 contains the k indices i for which $|h_i|$ is maximal, the set T_1 contains the indices of the next k maximal components, and so forth. The last index set T_a contains the indices i of the remaining r components, for which $|h_i|$ is the smallest. Now let h_{T_ℓ} be the n-tuple such that the component $(h_{T_\ell})_i = h_i$ for $i \in T_\ell$ and $(h_{T_\ell^c})_i = 0$ for $i \in T_\ell^c$.

Exercise 5.12 Show that (1) $h_{T_\ell} \in \Sigma_k$, (2) $h_{T_\ell} \cdot h_{T_{\ell'}} = 0$ for $\ell \neq \ell'$ and (3)

$$h = \sum_{\ell=0}^{a} h_{T_\ell}. \tag{5.20}$$

The following material is quite technical, but it leads to the easily understood result: Theorem 5.12 with an upper bound for the constant ρ in the null space property as a function of the index δ_{2k} in RIP. Let $h_{T_0 \cup T_1}$ be the n-tuple such that $(h_{T_0 \cup T_1})_i = h_i$ for $i \in T_0 \cup T_1$ with all other components zero. Thus $h_{T_0 \cup T_1} \in \Sigma_{2k}$. We can apply RIP to each k-sparse signal h_{T_ℓ} and from (5.20) we have

$$\Phi(h) = \Phi(h_{T_0 \cup T_1}) + \sum_{\ell=2}^{a} \Phi(h_{T_\ell}) = \Theta. \tag{5.21}$$

Thus,

$$||\Phi h_{T_0 \cup T_1}||_2^2 = < \Phi h_{T_0 \cup T_1}, -\sum_{j=2}^{a} \Phi h_{T_j} > = -\sum_{j=2}^{a} < \Phi h_{T_0}, \Phi h_{T_j} >$$

$$-\sum_{j=2}^{a} < \Phi h_{T_1}, \Phi h_{T_j} >$$

so

$$||\Phi h_{T_0 \cup T_1}||_2^2 \leq \sum_{j=2}^{a} \left(| < \Phi h_{T_0}, \Phi h_{T_j} > | + | < \Phi h_{T_1}, \Phi h_{T_j} > | \right). \tag{5.22}$$

To obtain an upper bound for the right-hand side we need the following result.

Lemma 5.10 *Suppose $x, x' \in \Sigma_k$ with k-index sets S, S' respectively and such that S and S' don't intersect. Let $< \cdot, \cdot >$ be the ℓ_2 inner product on $\mathbb{R}^n$. Then $| < \Phi x, \Phi x' > | \leq 2\delta_{2k}||x||_2||x'||_2$.*

Proof Let

$$\kappa = \max_{||z||_2 = ||z'||_2 = 1} | < \Phi z, \Phi z' > |$$

where the maximum is taken for all k-sparse unit vectors z with index set S and k-sparse unit vectors z' with index set S'. Then, by renormalizing, it follows that $| < \Phi x, \Phi x' > | \leq \kappa ||x||_2 ||x'||_2$. Since $z + z'$ and $z - z'$ are $2k$-sparse and $||z \pm z'||_2^2 = ||z||_2^2 + ||z'||_2^2 = 2$, we have the RIP inequalities

$$2(1 - \delta_{2k}) \leq ||\Phi(z + z')||_2^2 \leq 2(1 + \delta_{2k}),$$

$$2(1 - \delta_{2k}) \leq ||\Phi(z - z')||_2^2 \leq 2(1 + \delta_{2k}).$$

From the parallelogram law for ℓ_2, see Exercise 1.23, we have

$$< \Phi z, \Phi z' > = \frac{1}{4}(||\Phi(z + z')||_2^2 - ||\Phi(z - z')||_2^2),$$

so the RIP inequalities imply $| < \Phi z, \Phi z' > | \leq \delta_{2k}$. Thus $\kappa \leq \delta_{2k}$. □

Exercise 5.13 Verify the details of the proof of Lemma 5.10.

Applying Lemma 5.10 to the right-hand side of (5.22) we find

$$||\Phi h_{T_0 \cup T_1}||_2^2 \leq \delta_{2k}(||h_{T_0}||_2 + ||h_{T_1}||_2) \left(\sum_{j=2}^{a} ||h_{T_j}||_2 \right). \tag{5.23}$$

On the right we use the inequality $||h_{T_0}||_2 + ||h_{T_1}||_2 \leq \sqrt{2}||h_{T_0 \cup T_1}||_2$, see Exercise 1.22, and on the left-hand side we use RIP for $2k$ to obtain

$$(1 - \delta_{2k})||h_{T_0 \cup T_1}||_2^2 \leq ||\Phi h_{T_0 \cup T_1}||_2^2 \leq \sqrt{2}\delta_{2k}||h_{T_0 \cup T_1}||_2 \left(\sum_{j=2}^{a} ||h_{T_j}||_2 \right),$$

so

$$||h_{T_0 \cup T_1}||_2 \leq \frac{\sqrt{2}\delta_{2k}}{1 - \delta_{2k}} \sum_{\ell=2}^{a} ||h_{T_\ell}||_2. \tag{5.24}$$

Now for some tricky parts and the reason that we have chosen the T_ℓ by putting the indices of h in decreasing order of magnitude.

Lemma 5.11

$$||h_{T_\ell}||_2 \leq \frac{1}{\sqrt{k}}||h_{T_{\ell-1}}||_1, \qquad \ell = 2,\ldots,a.$$

Proof For any $j \in T_\ell$, $i \in T_{\ell-1}$, we have $|h_j| \leq |h_i|$, so $|h_j|$ is also bounded by the average of the $|h_i|$ as i ranges over $T_{\ell-1}$:

$$|h_j| \leq \frac{1}{k} \sum_{i \in T_{\ell-1}} |h_i|.$$

Thus

$$||h_{T_\ell}||_2^2 = \sum_{j \in T_\ell} |h_j|^2 \leq k \left(\frac{\sum_{i \in T_{\ell-1}} |h_i|}{k} \right)^2.$$

Taking the square root, we have $||h_{T_\ell}||_2 \leq ||h_{T_{\ell-1}}||_1/\sqrt{k}$. $\square$

From the lemma we have

$$\sum_{\ell=2}^{a} ||h_{T_\ell}||_2 \leq (k)^{-1/2} \sum_{\ell=1}^{a-1} ||h_{T_\ell}||_1 \leq (k)^{-1/2} \sum_{\ell=1}^{a} ||h_{T_\ell}||_1 = (k)^{-1/2}||h_{T_0^c}||_1,$$

which gives an upper bound for the right-hand side of (5.24). From $||h_{T_\ell}||_1/\sqrt{k} \leq ||h_{T_\ell}||_2$ we get a lower bound for the left-hand side of (5.24): $||h_{T_0}||_1/\sqrt{k} \leq ||h_{T_0}||_2 \leq ||h_{T_0 \cup T_1}||_2$. Putting this all together we find

$$||h_{T_0}||_1 \leq \frac{\sqrt{2}\delta_{2k}}{1-\delta_{2k}}||h_{T_0^c}||_1 = \rho||h_{T_0^c}||_1. \tag{5.25}$$

To guarantee the null space property we must have $\rho < 1$.

Theorem 5.12 *A sufficient condition for the null space property to hold and for each class $F(y)$ to contain at most one k-sparse signal is that RIP holds for $2k$-sparse signals with*

$$\rho = \frac{\sqrt{2}\delta_{2k}}{1-\delta_{2k}} < 1.$$

It follows that if $\delta_{2k} < 1/(1+\sqrt{2}) \approx 0.4142$ then the null space property is satisfied.

Exercise 5.14 If $\delta_{2k} = 1/4$ for some k and the sample matrix Φ, verify from Theorem 5.13 that the estimate $||\hat{x} - x||_1 \leq 5.5673||x - x^k||_1$ holds for the approximation of a signal x by a k-sparse signal $\hat{x}$.

Exercise 5.15 Show that even though RIP implies the null space property, the converse is false. Hint: Given an $m \times n$ sample matrix Φ let $\Xi = A\Phi$ be a new sample matrix, where A is an invertible $m \times m$ matrix. If Φ satisfies the null space property, then so does Ξ. However, if Φ satisfies RIP we can choose A so that Ξ violates RIP.

Many different bounds for ρ can be derived from RIP using modifications of the preceding argument, but the result of Theorem 5.12, due to Candés and Tao, is sufficient for our purposes. The point is that for a given $m \times n$ sample matrix Φ we would like to be able to guarantee the null space property for k as large as possible, and for a given k we would like to design appropriate sample matrices with m as small as possible.

There are also important results on ℓ_2 minimization with ℓ_1 recovery. In particular, suppose one wishes to recover the signal x from the sample y if $y = \Phi x + z$ where z is either deterministic or a noise term, and such that $||z||_2 < \epsilon$. It follows from results in [27] that if $\delta_{2k} < \sqrt{2} - 1$ then

$$||\hat{x} - x||_2 \leq C_1 \frac{||x - x^k||_1}{\sqrt{k}} + C_2\epsilon.$$

In practice the constants are small (on the order of 10). The proof of this result is technical, but similar to that of Theorem 5.8. See also [63, 97].

For more results concerning the deterministic theory of compressive sampling see [6, 21, 60, 83].

5.4 Probabilistic theory of compressive sampling

For practical computation of sampling matrices Φ for compressive sensing we employ a probabilistic approach. Rather than designing a sampling matrix especially adapted for a given type of signal, we chose a general purpose sampling matrix at random. To motivate this approach we recall that all eigenvalues $\lambda_i(T)$ of the self-adjoint matrix $\Phi_T^*\Phi_T$ are strictly positive and

$$1 - \delta_k = \lambda_{\min} \leq \lambda_{\max} = 1 + \delta_k$$

where the maximum and minimum are taken over all index sets T with $\#(T) = k$. Thus RIP is satisfied provided $\lambda_i(T) \approx 1$. This means that $\Phi_T^*\Phi_T$ must be close to the $k \times k$ identity matrix for every k-index set T. Thus the column vectors $c^{(j)}$ of the sampling matrix $\Phi = (c^{(1)}, \dots, c^{(m)})$

should satisfy $c^{(j_1)} \cdot c^{(j_2)} \approx 0$ for $j_1 \neq j_2$ and $c^{(j)} \cdot c^{(j)} \approx 1$ to obtain RIP. We want each column vector to be approximately an ℓ_2 unit vector but distinct columns should be nearly orthogonal, i.e., uncorrelated. One way to try to achieve this is to choose the elements of Φ as independent samples from a probability space with mean 0 and standard deviation 1.

Now suppose we choose each of the mn matrix elements of $\Phi = (\Phi_{ij}) = (t_{ij})$, $1 \leq i \leq m$, $1 \leq j \leq n$, independently from the population with normal distribution $N(0, 1/\sqrt{m})$. Then $\mathbf{t} = (t_{ij})$ is a matrix random variable on $\mathbb{R}^{mn}$ with probability density function

$$\Delta(\mathbf{t}) = (\frac{m}{2\pi})^{mn/2} \exp\left(-\frac{m}{2}\Sigma_{i=1}^m \Sigma_{j=1}^n t_{ij}^2\right). \tag{5.26}$$

Given any function $f(\mathbf{t}) \in L_1(\mathbb{R}^{mn}, \Delta)$ we define its expectation by

$$\bar{f}(\mathbf{t}) = E(f(\mathbf{t})) = \int_{\mathbb{R}^{mn}} f(\mathbf{t})\Delta(\mathbf{t})\Pi_{ij}\, dt_{ij}.$$

Note that E is linear, i.e.,

$$E(\alpha f_1(\mathbf{t}) + \beta f_2(\mathbf{t})) = \alpha E(f_1(\mathbf{t})) + \beta E(f_2(\mathbf{t})) \qquad \diagup$$

for real parameters α, β. Further we have the properties

$$E(1) = 1, \quad E(t_{i_1 j_1} t_{i_2 j_2}) = \delta_{i_1 i_2} \delta_{j_1 j_2}/m. \tag{5.27}$$

The second identity is a consequence of the property that two distinct matrix elements of Φ are uncorrelated.

Now let x be an n-tuple and choose the matrix elements of Φ independently from the normal distribution $N(0, 1/\sqrt{m})$ as just described. Then

$$E_\Delta(||\Phi x||_{\ell_2^m}^2) = E_\Delta(< \Phi_T^{\text{tr}}\Phi_T x, x >) = \Sigma_{i=1}^m \Sigma_{j_1, j_2=1}^n x_{j_1} x_{j_2} E_\Delta(t_{ij_1} t_{ij_2})$$
$$= ||x||_{\ell_2^n}^2.$$

Thus, if Φ lies sufficiently close to its mean value then it will satisfy RIP. In this chapter we will show that for n sufficiently large with respect to m, the probability that the random matrix Φ lies very close to the mean and satisfies RIP is near certainty. Such random matrices are relatively easy to construct. Even standard spreadsheets will do the job.

Another simple means of constructing random matrices that might satisfy RIP is through use of the distribution $\rho_1(t)$ on $\mathbb{R}$:

$$\rho_1(t) = \begin{cases} \sqrt{\frac{m}{12}}, & -\sqrt{\frac{3}{m}} \leq t \leq \sqrt{\frac{3}{m}} \\ 0, & \text{otherwise.} \end{cases} \tag{5.28}$$

Here,

$$E_{\rho_1}(1) = 1, \quad \bar{t} = E_{\rho_1}(t) = 0, \quad \sigma^2 = E_{\rho_1}(t^2) = \frac{1}{m}.$$

We choose $\Phi_{ij} = t_{ij}$ where $\mathbf{t} = (t_{ij})$ is a matrix random variable on $\mathbb{R}^{mn}$ with density function

$$\Delta_1(\mathbf{t}) = \begin{cases} (\frac{m}{12})^{mn/2} & \text{if } -\sqrt{\frac{3}{m}} \leq t_{ij} \leq \sqrt{\frac{3}{m}} \text{ for all } i,j \\ 0 & \text{otherwise.} \end{cases} \quad (5.29)$$

Given any function $f(\mathbf{t})$ of the mn components of $\mathbf{t}$ we define its expectation by

$$\bar{f}(\mathbf{t}) = E_{\rho_1}(f(\mathbf{t})) = \int_{\mathbb{R}^{mn}} f(\mathbf{t})\Delta_1(\mathbf{t})\Pi_{ij} \, dt.$$

Again we find $E_{\rho_1}(||\Phi x||_2^2) = 1$ so these random matrices appear to be good candidates to satisfy RIP for large n. A related construction is where each column vector $c^{(j)}$ of the sample matrix $\Phi = (\Phi_{ij})$ is chosen randomly as a vector on the unit m-sphere S_m: $\Sigma_{i=1}^m \Phi_{ij}^2 = ||c^{(j)}||_{\ell_2^m}^2 = 1$. Each unit column vector is chosen independently of the others. Here the probability density is just the area measure on the n-sphere. In this case

$$||\Phi x||_{\ell_2^m}^2 = \Sigma_{j=1}^n x_j^2 ||c^{(j)}||_{\ell_2^m}^2 + 2\Sigma_{1 \leq j < \ell \leq n} x_j x_\ell < c^{(j)}, c^{(\ell)} >_{\ell_2^m} \quad (5.30)$$

$$= ||x||_{\ell_2^n}^2 + 2\Sigma_{1 \leq j < \ell \leq n} x_j x_\ell < c^{(j)}, c^{(\ell)} >_{\ell_2^m}.$$

Since the unit vectors $c^{(j)}$ and $c^{(\ell)}$ are chosen randomly on the unit sphere and independently if $j \neq \ell$, the inner product $< c^{(j)}, c^{(\ell)} >_{\ell_2^m}$ will be a random variable with mean 0. Thus $E(||\Phi x||_{\ell_2^m}^2) = ||x||_{\ell_2^n}^2$, and this is another good candidate to satisfy RIP.

A fourth method of construction is in terms of the discrete probability measure (Bernoulli distribution)

$$\rho_2(t) = \begin{cases} \frac{1}{2}, & t = \pm 1 \\ 0 & \text{otherwise.} \end{cases} \quad (5.31)$$

We define the expectation of any polynomial $p(t)$ in t by

$$\bar{p}(t) = E_{\rho_2}(p(t)) \equiv \frac{1}{2}\Sigma_{t=\pm 1} p(t).$$

Here, $\bar{t} = 0$ and $\sigma^2 = E_{\rho_2}(t^2) = 1$. In this case we choose $\Phi_{ij} = t_{ij}/\sqrt{m}$ where each t_{ij} is obtained by a "coin flip." Given any polynomial function $p(\mathbf{t})$ of the mn components of $\mathbf{t} = \mathbf{t_{ij}}$ we define its expectation by

$$\bar{p}(\mathbf{t}) = E(p(\mathbf{t})) = \frac{1}{2^{mn}}\Sigma_{t_{ij}=\pm 1} p(\mathbf{t}). \quad (5.32)$$

Again E is linear and we have the properties

$$E(1) = 1, \quad E(t_{i_1 j_1} t_{i_2 j_2}) = \delta_{i_1 i_2} \delta_{j_1 j_2}. \quad (5.33)$$

Further, $E_{\rho_2}(||\Phi x||_2^2) = 1$ so these random matrices are again good candidates to satisfy RIP for large n.

5.4.1 Covering numbers

In the preceding section we have exhibited four families of random matrices such that for any real n-tuple x the random variable $||\Phi x||_{\ell_2^m}^2$ has expected value $E(||\Phi x||_{\ell_2^m}^2) = ||x||_{\ell_2^n}^2$. In order to prove RIP for matrices Φ so chosen we need to show that this random variable is concentrated about its expected value. For this we will make use of the concentration of measure inequality

$$\Pr\left(|\ ||\Phi x||_{\ell_2^m}^2 - ||x||_{\ell_2^n}^2| \geq \epsilon ||x||_{\ell_2^n}^2\right) \leq 2e^{-m c_0(\epsilon)}, \quad 0 < \epsilon < 1. \quad (5.34)$$

The left-hand side of the inequality is the probability that $|\ ||\Phi x||_{\ell_2^m}^2 - ||x||_{\ell_2^n}^2| \geq \epsilon ||x||_{\ell_2^n}^2$, where the probability is taken over all $m \times n$ matrices Φ in one of the families. Here $c_0(\epsilon) > 0$ depends only on ϵ. Relations (5.34) quantify the degree to which each of the probability distributions is concentrated about its expected value. We will derive (5.34) for three of the families and give explicit expressions for $c_0(\epsilon)$. Here we assume these results and proceed with the verification of RIP.

Our proof makes use of an estimate for the number of closed balls of radius η that are needed to completely cover the unit ball B_1 centered at the origin in N-dimensional Euclidean space. (The ball of radius r centered at the origin is $B_r = \{x \in \ell_2^N : ||x|| \leq r\}$.) This geometrical result is particularly interesting because it has applications to many practical problems concerning high-dimensional spaces, such as point clouds and learning theory, as well as compressive sampling.

Let S be a closed bounded subset of N dimensional Euclidean space. We define the covering number $\mathcal{N}(S, \eta)$ as the smallest number n of closed balls $D_1(\eta), \ldots, D_n(\eta)$ of radius η that are needed to cover S: $S \subseteq \cup_{i=1}^n D_i(\eta)$. There is no explicit formula for this number, even for the unit ball $S = B_1$. However, it is relatively easy to derive an upper bound for $\mathcal{N}(B_1, \eta)$ and that is our aim.

We say that m points $x^{(1)}, \ldots, x^{(m)}$ in S are η-distinguishable if $||x^{(i)} - x^{(j)}|| > \eta$ for $i \neq j$. We define $\mathcal{M}(S, \eta)$ as the maximal number of η-distinguishable points in S. Since S is closed and bounded this maximum exists, although the points $x^{(i)}$ are not unique.

Lemma 5.13

$$\mathcal{M}(S, 2\eta) \leq \mathcal{N}(S, \eta) \leq \mathcal{M}(S, \eta).$$

Proof For the right-hand inequality, note that if $m = \mathcal{M}(S, \eta)$ then there exist m points $x^{(1)}, \ldots, x^{(m)}$ that are η-distinguishable in S. By the maximality of m it follows that any $x \in S$ satisfies $||x - x^{(i)}|| \leq \eta$ for some i. Hence $S \subseteq \cup_{j=1}^m D_j(\eta)$ and $\mathcal{N}(S, \eta) \leq m$.

To prove the left hand inequality we use the pigeonhole principle Let $M = \mathcal{M}(S, 2\eta)$. Then there exist M points $x^{(1)}, \ldots, x^{(M)}$ in S that are 2η-distinguishable. If $\mathcal{N}(S, \eta) < M$ then S is covered by fewer than M balls of radius η. If so, at least two of the M distinct points $x^{(i)}, x^{(j)}$ must lie in the same ball $D_h(\eta)$ with center h. By the triangle inequality

$$||x^{(i)} - x^{(j)}|| \leq ||x^{(i)} - h|| + ||h - x^{(j)}|| \leq \eta + \eta = 2\eta.$$

However, this is impossible since $x^{(i)}$ and $x^{(j)}$ are 2η-distinguishable. Hence $M \leq \mathcal{N}(S, \eta)$. □

Theorem 5.14 *For the unit ball B_1 in N-dimensional Euclidean space and any $0 < \eta \leq 1$, we have*

$$(1/2\eta)^N \leq \mathcal{N}(B_1, \eta) \leq (3/\eta)^N.$$

Proof If $m = \mathcal{M}(B_1, \eta)$ there must exist m η-distinguishable points $x^{(1)}, \ldots, x^{(m)}$ in B_1. By the maximality of m it follows that any point $x \in B_1$ satisfies $||x - x^{(i)}|| \leq \eta$ for some i. Hence, if $D_j(\eta)$ is the ball of radius η centered at $x^{(j)}$, we have $B_1 \subseteq \cup_{j=1}^m D_j(\eta)$ and $\sum_{j=1}^{k=1} V(D_j) \geq V(B_1)$ where $V(D)$ is the volume of the ball D. Since $V(D_j) = \eta^N V(B_1)$ we find $m\eta^N \geq 1$. This implies $(1/\eta)^N \leq \mathcal{M}(B_1, \eta)$.

On the other hand, we can construct balls $D_j(\eta/2)$ about each $x^{(j)}$. If $x \in D_j(\eta/2)$ then, since $x = (x - x^{(j)}) + x^{(j)}$, we have

$$||x|| \leq ||x - x^{(j)}|| + ||x^{(j)}|| \leq \frac{\eta}{2} + 1 \leq \frac{3}{2}.$$

Thus each ball $D_j(\eta/2)$ is contained in the ball $B_{3/2}$. Further, since the m points $x^{(i)}$ are η-distinguishable, no two of these balls overlap. Thus, $\sum_{j=1}^{k=1} V(D_j) \leq V(B_{3/2})$ where $V(D)$ is the volume of the ball D. Since $V(D_j) = (\eta/2)^N V(B_1)$ and $V(B_{3/2}) = (3/2)^N V(B_1)$ we have $m(\eta/2)^N \leq (3/2)^N$. This implies $\mathcal{M}(B_1, \eta) \leq (3/\eta)^N$. The theorem is now an immediate consequence of Lemma 5.13. □

Lemma 5.15 *Let Φ be an $m \times n$ matrix whose matrix elements Φ_{ij} are drawn randomly and independently from a probability distribution* *that satisfies the concentration of measure inequality (5.34). Let T be an index set with $\#(T) = k < m$, and X_T be the set of all n-tuples x with index set T. Then for any $0 < \delta < 1$ and all $x \in X_T$ we have*

$$(1 - \delta)\|x\|_{\ell_2^n} \leq \|\Phi x\|_{\ell_2^m} \leq (1 - \delta)\|x\|_{\ell_2^n} \tag{5.35}$$

with probability at least $1 - 2(12/\delta)^k e^{-c_0(\delta/2)m}$.

Remark Although this result appears to be RIP, it is not. First of all, the lemma applies only to k-sparse signals x with a specific index set T, not to all k-sparse signals. Secondly the inequalities (5.35) do not hold in general, but only with a guaranteed probability. The lemma says that the probability of failure of (5.35) to hold is bounded above by $2(12/\delta)^k e^{-c_0(\delta/2)m}$. This will be a practical method for constructing sample matrices only if we can show that this bound is so close to 0 that failure virtually never occurs.

The proof of the lemma and results to follow depend on a simple but basic result from probability theory, the union bound. We sketch the derivation. Suppose $p(A^{(1)})$ is the probability that inequalities (5.35) fail to hold for $x = x^{(1)}$, and $p(A^{(2)})$ is the probability of failure for $x = x^{(2)}$. Then

$$p(A^{(1)}) = p(A^{(1)} \cap A^{(2)^c}) + p(A^{(1)} \cap A^{(2)}),$$

i.e., the probability of failure for $x^{(1)}$ is the probability of simultaneous failure for $x^{(1)}$ and success for $x^{(2)}$ plus the probability of simultaneous failure for both $x^{(1)}$ and $x^{(2)}$. Here, $0 \leq p(C) \leq 1$ for all these probabilities. Similarly we have the decomposition

$$p(A^{(2)}) = p(A^{(1)^c} \cap A^{(2)}) + p(A^{(1)} \cap A^{(2)}).$$

The probability that there is failure for at least one of $x^{(1)}, x^{(2)}$ is denoted $p(A^{(1)} \cup A^{(2)})$ and it has the decomposition

$$p(A^{(1)} \cup A^{(2)}) = p(A^{(1)} \cap A^{(2)^c}) + p(A^{(1)^c} \cap A^{(2)}) + p(A^{(1)} \cap A^{(2)}),$$

i.e., the probability of simultaneous failure for $x^{(1)}$ and success for $x^{(2)}$, plus the probability of simultaneous success for $x^{(1)}$ and failure for $x^{(2)}$, plus the probability of simultaneous failure for both $x^{(1)}$ and $x^{(2)}$. Comparing these identities and using the fact that $p(A^{(1)} \cap A^{(2)}) \geq 0$ we obtain the inequality

$$p(A^{(1)} \cup A^{(2)}) \leq p(A^{(1)}) + p(A^{(2)}).$$

By a simple induction argument we can establish

Lemma 5.16 *(union bound)* $p(A^{(1)} \cup A^{(2)} \cdots \cup A^{(h)}) \leq \sum_{i=1}^{h} p(A^{(i)})$.

Thus the probability that at least one of h events occurs is bounded above by the sum of the probabilities of occurrence of each of the events separately.

Proof of Lemma 5.15 We can assume $||x||_{\ell_2^n} = 1$ since this can be achieved for any nonzero x by multiplying (5.35) by $1/||x||_{\ell_2^n}$. Now consider a finite set Q_T of n-tuples $q^{(1)}, \ldots, q^{(K)}$, such that Q_T is contained in the unit ball in X_T, i.e., $q^{(i)} \in X_T$ and $||q^{(i)}||_{\ell_2^n} \leq 1$. We choose these vectors such that the unit ball in X_T is covered by the closed balls $D_1, \ldots, D_K$ where D_i is centered at $q^{(i)}$ and has radius $\delta/4$. Thus, if $x \in X_T$ with $||x||_{\ell_2^n} \leq 1$ then there is some point $q^{(j)} \in Q_T$ such that

$$||x - q^{(j)}||_{\ell_2^n} \leq \delta/4.$$

From Theorem 5.14, $\mathcal{N}(B_1, \delta/4) \leq (12/\delta)^k$, so we can require $\#(Q_T) \leq (12/\delta)^k$.

The concentration of measure inequality (5.34) can be written in the form

$$(1 - \epsilon)||x||_{\ell_2^n}^2 \leq ||\Phi x||_{\ell_2^m}^2 \leq (1 + \epsilon)||x||_{\ell_2^n}^2, \quad 0 < \epsilon < 1, \tag{5.36}$$

with probability of failure bounded above by $2e^{-mc_0(\epsilon)}$. Now we set $\epsilon = \delta/2$ and use the union bound for $\#(Q_T)$ events to obtain

$$(1 - \delta/2)||x||_{\ell_2^n}^2 \leq ||\Phi x||_{\ell_2^m}^2 \leq (1 + \delta/2)||x||_{\ell_2^n}^2, \text{ for all } q \in Q_T$$

with probability of failure bounded above by $2(12/\delta)^k e^{-mc_0(\delta/2)}$. We can take square roots on each side of the inequality and use the facts that $1 - \delta/2 \leq \sqrt{1 - \delta/2}$ and $\sqrt{1 + \delta/2} \leq 1 + \delta/2$ for $0 < \delta/2 < 1$ to obtain

$$(1 - \delta/2)||x||_{\ell_2^n} \leq ||\Phi x||_{\ell_2^m} \leq (1 + \delta/2)||x||_{\ell_2^n}, \text{ for all } q \in Q_T$$

with the same probability of failure as before.

We have verified (5.35) for $x \in Q_T$. We extend the result to any x in the unit sphere of X_T by using the fact that $||x - q||_{\ell_2^n} \leq \delta/4$ for some $q \in Q_T$. First we prove the right-hand inequality. Let A be the smallest number such that

$$||\Phi x||_{\ell_2^m} \leq (1 + A)||x||_{\ell_2^m} \tag{5.37}$$

for all x in the unit sphere of X_T. (Since this sphere is closed and bounded, A must exist and be finite.) Choosing any such x in the unit sphere, and approximating it by q as above, we find

$$||\Phi x||_{\ell_2^m} \leq ||\Phi q||_{\ell_2^m} + ||\Phi(x-q)||_{\ell_2^m} \leq (1+\delta/2)||q||_{\ell_2^n} + (1+A)||x-q||_{\ell_2^n}$$

$$\leq 1 + \delta/2 + (1+A)\delta/4.$$

Since A is the smallest number such that (5.37) holds we have $1 + A \leq 1 + \delta/2 + (1+A)\delta/4$ or $A \leq (3\delta/4)/(1 - \delta/4) \leq \delta$. (This last inequality follows from a simple calculus argument and verifies the right-hand side of (5.35).) The left-hand side of (5.35) follows from

$$||\Phi x||_{\ell_2^m} \geq ||\Phi q||_{\ell_2^m} - ||\Phi(x-q)||_{\ell_2^m} \geq (1-\delta/2) - (1+\delta)\delta/4 \geq 1 - \delta.$$

$\square$

Theorem 5.17 *Let Φ be an $m \times n$ matrix whose matrix elements Φ_{ij} are drawn randomly and independently from a probability distribution that satisfies the concentration of measure inequality (5.34). Let $0 < \delta < 1$. Then there exist constants $c_1, c_2 > 0$, depending only on δ, such that for all k-sparse signals x with $1 \leq k < m$ and $k \leq c_1 m/\ln(n/k)$ we have*

$$(1-\delta)||x||_{\ell_2^n} \leq ||\Phi x||_{\ell_2^m} \leq (1-\delta)||x||_{\ell_2^n} \tag{5.38}$$

with probability of failure bounded above by e^{-mc_2}.

Proof From Lemma 5.15 we have established (5.38) for all k-sparse signals $x \in X_T$, with failure probability bounded by $2(12/\delta)^k e^{-mc_0(\delta/2)}$. We employ the union bound to establish the result for all k-sparse signals. There are

$$\binom{n}{k} = \frac{n!}{(n-k)!k!} \leq (\frac{en}{k})^k \tag{5.39}$$

possible index sets T, where the last inequality follows from Stirling's formula, see Exercise 5.16. Therefore, (5.38) will fail to hold for all k-sparse signals with probability bounded by

$$2(\frac{en}{k})^k (\frac{12}{\delta})^k e^{-mc_0(\frac{\delta}{2})} = \exp\left(-mc_0(\frac{\delta}{2}) + k[\ln(\frac{en}{k}) + \ln(\frac{12}{\delta})] + \ln 2\right).$$

To finish the proof we need to find a constant c_2 such that

$$0 < c_2 \leq c_0(\frac{\delta}{2}) - \frac{k}{m}[\ln(\frac{n}{k}) + 1 + \ln(12/\delta)] - \frac{\ln 2}{m}. \tag{5.40}$$

If we limit k to values such that $k \leq c_1 m/\ln(n/k)$ for some positive constant c_1 then (5.40) will hold if $0 < c_2 \leq c_0(\frac{\delta}{2}) - c_1(1 + \frac{(2+\ln(12/\delta))}{\ln(n/k)})$. (Here we have used the bound $\ln 2 \leq k$.) By choosing the positive constant c_1 sufficiently small we can guarantee that $c_2 > 0$. $\quad\Box$

Exercise 5.16 Verify the inequality (5.39) via the following elementary argument.

1. Using upper and lower Darboux sums for $\int_1^n \ln x \, dx$, establish the inequality

$$\sum_{j=1}^{n-1} \ln j \leq \int_1^n \ln x \, dx \leq \sum_{j=2}^n \ln j.$$

2. Show that

$$(n-1)! \leq e(\frac{n}{e})^n \leq n!.$$

3. Verify that

$$\left(\begin{array}{c} n \\ k \end{array} \right) \equiv \frac{n!}{k!(n-k)!} \leq \frac{n^k}{k!} \leq \left(\frac{ne}{k} \right)^k.$$

For more properties of covering numbers see [61, 79]. For relations to learning theory see [47].

5.4.2 Digging deeper: concentration of measure inequalities

The normal distribution
We proceed with the derivation of the concentration of measure inequality (5.34) for sample matrices chosen by means of the normal distribution $N(0, 1/\sqrt{m})$. Then $\Phi = \mathbf{t} = (t_{ij})$ is a matrix random variable on $\mathbb{R}^{mn}$ with distribution function

$$\Delta(\mathbf{t}) = (\frac{m}{2\pi})^{mn/2} \exp\left(-\frac{m}{2}\Sigma_{i=1}^m \Sigma_{j=1}^n t_{ij}^2\right). \tag{5.41}$$

We chose an n-tuple x with $||x||_{\ell_2^n} = 1$ as signal and determine the probability distribution function for the random variable $||\Phi x||_{\ell_2^m}^2 = \Sigma_{i=1}^m (\Sigma_{j=1}^n t_{ij}x_j)^2$. First we determine the cumulative probability function for $||\Phi x||_{\ell_2^m}$, i.e., $P(\tau) =$

$$\Pr\left(||\Phi x||_{\ell_2^m} \leq \tau\right) = (\frac{m}{2\pi})^{mn/2} \int_{||\Phi x||_{\ell_2^m} \leq \tau} \exp\left(-\frac{m}{2}\Sigma_{i=1}^m \Sigma_{j=1}^n t_{ij}^2\right) \, d\mathbf{t}$$

$$\tag{5.42}$$

where $\tau \geq 0$. This multiple integral appears difficult to evaluate because of the dependence on the vector x. However, there is rotational symmetry here so that $P(\tau)$ is the same for all unit vectors x. To see this, note that we can use the Gram–Schmidt process to construct an orthonormal basis $\{O^{(1)}, O^{(2)}, \ldots, O^{(n)}\}$ for ℓ_2^n such that $O^{(1)} = x$. Then the $n \times n$ matrix O whose jth row is $O^{(j)}$ will be orthogonal, i.e.,

$$O = \begin{pmatrix} O^{(1)} \\ O^{(2)} \\ \vdots \\ O^{(n)} \end{pmatrix} = (O_{ij}), \quad OO^{\text{tr}} = I$$

where I is the $n \times n$ identity matrix. Using well-known properties of the determinant $\det(A)$ of the matrix A we have

$$1 = \det(I) = \det(OO^{\text{tr}}) = \det(O)\det(O^{\text{tr}}) = (\det(O))^2,$$

so $\det(O) = \pm 1$. Now we make the orthogonal change of variables,

$$T_{ij} = \sum_{k=1}^{n} t_{ik} O_{jk}, \quad i = 1, \ldots, m, \; j = 1, \ldots, n,$$

so that, in particular, $\|\Phi x\|_{\ell_2^m}^2 = \Sigma_{i=1}^m T_{i1}^2$. Since O is orthogonal this coordinate transformation preserves the sum of squares

$$\Sigma_{i=1}^m \Sigma_{j=1}^n T_{ij}^2 = \Sigma_{i=1}^m \Sigma_{j=1}^n t_{ij}^2. \tag{5.43}$$

Exercise 5.17 Verify the identity (5.43).

Furthermore, the Jacobian of the coordinate transformation has determinant $[\det(O)]^m$, so its absolute value is 1. Thus via the standard rules for change of variables in multiple integrals we have

$$P(\tau) = \left(\frac{m}{2\pi}\right)^{mn/2} \int_{\Sigma_{i=1}^m T_{i1}^2 \leq \tau} \exp\left(-\frac{m}{2} \Sigma_{i=1}^m \Sigma_{j=1}^n T_{ij}^2\right) \, d\mathbf{T}.$$

Here the region of integration is $-\infty < T_{ij} < \infty$ for $j \neq 1$, so we can carry out $m(n-1)$ integrations immediately. For the remaining integrations we introduce spherical coordinates $T_{i1} = r\omega_i$ where $\sum_{i=1}^m \omega_i^2 = 1$, i.e. r is a radial coordinate and $(\omega_1, \ldots, \omega_m)$ ranges over the unit sphere in m-space. Then $d\mathbf{T} = r^{m-1} \, dr \, d\Omega$, where $d\Omega$ is the area measure on the sphere, and $\Sigma_{i=1}^m T_{i1}^2 = r^2$. Thus

$$P(\tau) = c \int_0^{\sqrt{\tau}} e^{-mr^2/2} r^{m-1} \, dr = c' \int_0^{m\tau/2} e^{-R} R^{(m/2-1)} \, dR$$

where c' is a positive constant. The easiest way to compute c' is to note that $P(+\infty) = 1$, so $c' = 1/\Gamma(m/2)$ by the definition of the Gamma function. Now $P(\tau) = \int_0^\tau p(\alpha) \, d\alpha$ where $p(\tau)$ is the probability distribution function for $||\Phi x||_{\ell_2^n}^2$, so

$$p(\tau) = \frac{(m/2)^{m/2}}{\Gamma(m/2)} \tau^{m/2-1} e^{-m\tau/2}. \tag{5.44}$$

This is a famous distribution in probability theory, the Chi-square (or Gamma) distribution [5, 78].

Exercise 5.18 Show that $E(||\Phi x||_{\ell_2^n}) = \int_0^\infty \tau p(\tau) \, d\tau = 1$ by evaluating the integral explicitly.

Now we are ready to verify the concentration of measure inequalities for the probability distribution $p(\tau)$ of the random variable $X = ||\Phi x||_{\ell_2^m}^2$ with $E(X) = 1$. For any $\epsilon > 0$ us first consider

$$\Pr(X - E(X) > \epsilon) = \int_{E(X)+\epsilon}^\infty p(\tau) \, d\tau.$$

We will use the idea that leads to the famous Chernoff inequality [70]. Let $u > 0$ and note that

$$\Pr(X - E(X) > \epsilon) = \Pr(X > E(X) + \epsilon) = \Pr(e^{uX} > e^{u(E(X)+\epsilon)})$$

$$= \Pr(e^{u(X-E(X)-\epsilon)} > 1).$$

This is because the exponential function is 1-1 and order preserving. Thus, applying the Markov inequality (Theorem 4.24) to the random variable $Y = e^{u(X-E(X)-\epsilon)}$ with $d = 1$ and $E(X) = 1$ we have

$$\Pr(X - E(X) > \epsilon) \leq E(e^{u(X-E(X)-\epsilon)}) =$$

$$\int_0^\infty e^{u(\tau-1-\epsilon)} p(\tau) \, d\tau = e^{-u(1+\epsilon)} E(e^{uX}),$$

for $u > 0$. Using the explicit formula (5.44) for $p(\tau)$ and assuming $0 < u < m/2$ we find

$$E(e^{uX}) = \frac{(m/2)^{m/2}}{(m/2 - u)^{m/2}}. \tag{5.45}$$

Thus

$$\Pr(X - 1 > \epsilon) \leq \frac{(m/2)^{m/2} e^{-u(1+\epsilon)}}{(m/2 - u)^{m/2}}$$

for all $0 < u < m/2$. This gives us a range of inequalities. The strongest is obtained by minimizing the right-hand side in u. A first year calculus computation shows that the minimum occurs for $u = m\epsilon/2(1+\epsilon)$. Thus

$$\Pr(X - 1 > \epsilon) \leq [e^{-\epsilon}(1+\epsilon)]^{m/2} \leq e^{-m(\epsilon^2/4 - \epsilon^3/6)}. \tag{5.46}$$

Exercise 5.19 Verify the right-hand inequality in (5.46) by showing that the maximum value of the function $f(\epsilon) = (1+\epsilon)\exp(-\epsilon + \epsilon^2/2 - \epsilon^3/3)$ on the interval $0 \leq \epsilon$ is 1.

For the other inequality we reason in a similar manner.

$$\Pr(X - E(X) < -\epsilon) = \Pr(e^{u(E(X)-X-\epsilon)} > 1) < e^{u(1-\epsilon)}E(e^{-uX}),$$

for $u > 0$, so from (5.45),

$$\Pr(X - E(X) < -\epsilon) < e^{u(1-\epsilon)}\frac{(m/2)^{m/2}}{(m/2 + u)^{m/2}}.$$

For $0 < \epsilon < 1$ the minimum of the right-hand side occurs at $u = m\epsilon/2(1-\epsilon)$ and equals $[(1-\epsilon)e^\epsilon]^{m/2}$, smaller than the right-hand side of (5.46). Indeed

$$\Pr(X - 1 < -\epsilon) < [(1-\epsilon)e^\epsilon]^{m/2} \leq e^{-m(\epsilon^2/4 - \epsilon^3/6)}. \tag{5.47}$$

We conclude from (5.46), (5.47) and the union bound that the concentration of measure inequality (5.34) holds with $c_0 = \epsilon^2/4 - \epsilon^3/6$.

Exercise 5.20 Verify the right-hand inequality in (5.47) by showing that $g(\epsilon) = (1-\epsilon)e^{\epsilon + \epsilon^2/2 - \epsilon^3/3} \leq 1$ for $0 < \epsilon < 1$.

The Bernoulli distribution

For the coin flip (Bernoulli) distribution we choose the sample matrix via $\Phi_{ij} = t_{ij}/\sqrt{m}$ where the mn random variables t_{ij} take the values ± 1 independently and with equal probability $1/2$. The expectation is now given by the sum (5.32) and has properties (5.33). If x is an n-tuple with $\|x\|_{\ell_2^n} = 1$ then

$$\|\Phi x\|_{\ell_2^m}^2 = \Sigma_{i=1}^m Q_i(x)^2, \quad Q_i(x) = \Sigma_{j=1}^n x_j \Phi_{ij}. \tag{5.48}$$

Note that each of the random variables $Q_i(x)$ has mean 0, i.e., $E(Q_i(x)) = 0$ and variance $E((Q_i^2(x)) = 1/m$. One consequence of this and (5.48) is $E(\|\Phi x\|^2) = E(1) = 1$. Since for any $\epsilon > 0$ we have

$$\Pr(\|\Phi x\|_{\ell_2^m}^2 - E(\|\Phi x\|^2) \geq \epsilon) = \Pr(\Sigma_{i=1}^m Q_i^2(x) - 1 \geq \epsilon)$$

the concentration of measure inequalities for the Bernoulli distribution reduce to properties of sums of independent, identically distributed, random variables.

Just as with the normal distribution, we make use of the Markov inequality and introduce a parameter $u > 0$ that we can adjust to get an optimal inequality. Now $\Pr(\Sigma_{i=1}^m Q_i^2(x) - 1 \geq \epsilon) =$

$$\Pr(e^{u\Sigma Q_i^2(x)} \geq e^{u(1+\epsilon)}) = \Pr(\Pi_{i=1}^m e^{uQ_i^2(x)} \geq e^{u(1+\epsilon)})$$
$$= \Pr(e^{-u(1+\epsilon)}\Pi_{i=1}^m e^{uQ_i(x)} \geq 1) \leq e^{-u(1+\epsilon)}\Pi_{i=1}^m E(e^{uQ_i^2(x)}),$$

where the last step is the Markov inequality. Finally we have

$$\Pr(\Sigma_{i=1}^m Q_i^2(x) - 1 \geq \epsilon) \leq e^{-u(1+\epsilon)}[E(e^{uQ_1^2(x)})]^m, \tag{5.49}$$

since the $Q_i(x)$ are identically distributed. Similarly we have

$$\Pr(\Sigma_{i=1}^m Q_i^2(x) - 1 \leq -\epsilon) \leq e^{u(1-\epsilon)}[E(e^{-uQ_1^2(x)})]^m. \tag{5.50}$$

Returning to inequality (5.49) note that

$$E(e^{uQ_1^2(x)}) = E(\sum_{k=0}^\infty \frac{u^k Q_1(x)^{2k}}{k!}) = \sum_{k=0}^\infty \frac{u^k}{k!} E(Q_1(x)^{2k}), \tag{5.51}$$

where the interchange in order of summation and integration can be justified by the monotone convergence theorem of Lebesgue integration theory [92]. Thus, to bound (5.49) it is sufficient to bound $E(Q_1(x)^{2k})$ for all k. However, in distinction to the case of the normal distribution, the bounds for $E(e^{uQ_1^2(x)})$ and $E(Q_1(x)^{2k})$ depend on x. Thus to obtain a concentration of measure inequality valid uniformly for all signals we need to find the "worst case," i.e., to determine the vector $x = w$ on the unit n-sphere such that $E(e^{uQ_1^2(w)})$ is maximal. We will show that this worst case is achieved for $w = (1, 1, \ldots, 1)/\sqrt{n}$, [1].

We focus our attention on two components of the unit vector x. By relabeling we can assume they are the first two components $x_1 = a, x_2 = b$, so that $Q_1(x) = (at_{11} + bt_{12} + U)/\sqrt{m}$ where $U = \Sigma_{j=3}^n x_j t_{1j}$. Let $\hat{x} = (c, c, x_3, \ldots, x_n)$ where $c = \sqrt{(a^2 + b^2)/2}$. Note that $||\hat{x}||_{\ell_2^n} = 1$, i.e., $\hat{x}$ is also a unit vector.

Lemma 5.18 For $k = 1, 2, \ldots$ we have $E(Q_1(x)^{2k}) \leq E(Q_1(\hat{x})^{2k})$. Thus $E(e^{uQ_1^2(x)}) \leq E(e^{uQ_1^2(\hat{x})})$.

Proof We fix U and average $Q_1(\hat{x})^{2k} - Q_1(x)^{2k}$ over the four possibilities $t_{11} = \pm 1, t_{12} = \pm 1$. This average is $S_k/4m^k$ where

$$S_k = (U + 2c)^{2k} + 2U^{2k} + (U - 2c)^{2k} - (U + a + b)^{2k} - (U + a - b)^{2k}$$

$$-(U - a + b)^{2k} - (U - a - b)^{2k}.$$

We will show that $S_k \geq 0$. To see this we use the binomial theorem to expand each term of S_k except $2U^{2k}$ and regroup to obtain

$$S_k = 2U^{2k} + \Sigma_{i=0}^{2k} \begin{pmatrix} 2k \\ i \end{pmatrix} U^{2k-i} D_i,$$

$$D_i = (2c)^i + (-2c)^i - (a + b)^i - (a - b)^i - (-a + b)^i - (-a - b)^i.$$

If i is odd it is clear that $D_i = 0$. If $i = 2j$ is even then since $2c^2 = a^2 + b^2$ and $2(a^2 + b^2) = (a + b)^2 + (a - b)^2$ we have

$$D_{2j} = 2(2a^2 + 2b^2)^j - 2(a + b)^{2j} - 2(a - b)^{2j} = 2[(X + Y)^j - X^j - Y^j],$$

where $X = (a + b)^2, Y = (a - b)^2$. It is an immediate consequence of the binomial theorem that $(X + Y)^j \geq X^j + Y^j$ for $X, Y \geq 0$. Thus $D_{2j} \geq 0$ and

$$S_k = 2U^{2k} + \Sigma_{j=0}^{k} \begin{pmatrix} 2k \\ 2j \end{pmatrix} U^{2(k-j)} D_{2j} \geq 0.$$

To compute the expectations $E(Q_1(x)^{2k}), E(Q_1(\hat{x})^{2k})$ we can first average over t_{11}, t_{12} as above and then average over U. This process will preserve the inequality $S_k \geq 0$ so we have $E(Q_1(x)^{2k}) \leq E(Q_1(\hat{x})^{2k})$. □

If x is any unit vector with two components that are not equal, say $x_1 = a, x_2 = b$ with $a \neq b$ then we can use Lemma 5.18 to obtain a new unit vector $\hat{x} = (c, c, x_3, \ldots, x_n)$ such that $E(e^{uQ_1^2(x)}) \leq E(e^{uQ_1^2(\hat{x})})$. Proceeding in this way it will take at most $n - 1$ steps to construct the unit vector $w = (C, \ldots, C)$ where $C = 1/\sqrt{n}$ and $E(e^{uQ_1^2(x)}) \leq E(e^{uQ_1^2(w)})$. If all components of x are equal then $x = \pm w$ and $Q_1^2(\pm w) = Q_1^2(w)$. Thus the "worst case" is achieved for $x = w$.

For the worst case we have $Q_1(w) \equiv Q = (\tau_1 + \cdots + \tau_n)/\sqrt{n}$ where the τ_j are independent random variables each taking the values $\pm 1/\sqrt{m}$ with probability $1/2$. Substituting this expression for Q_1 in (5.51) and using the multinomial expansion we can obtain an expression of the form

$$E(e^{uQ^2}) = \Sigma_{k_1, \cdots, k_n} K_{k_1, \cdots, k_n} u^{k_1 + \cdots + k_n} E(\tau_1^{k_1} \cdots \tau_n^{k_n})$$

where the k_j range over the nonnegative integers and $K_{k_1, \cdots, k_n} \geq 0$. Moreover $E(\tau_1^{k_1} \cdots \tau_n^{k_n}) = 0$ if any of the k_j are odd. Thus the only nonzero terms are of the form

$$E(\tau_1^{2\ell_1} \cdots \tau_n^{2\ell_n}) = \Pi_{j=1, \cdots, n} E(\tau_j^{2\ell_j}) = \Pi_{j=1, \ldots, n} E(\tau^{2\ell_j})$$

where the ℓ_j range over the nonnegative integers and τ is a single random variable taking the values $\pm 1/\sqrt{m}$, each with probability $1/2$. This is because the τ_j are independently distributed. (Note that τ has mean 0 and standard deviation $1/\sqrt{m}$.) We conclude that the value of $E(e^{uQ^2})$ is uniquely determined by a power series in the values $E(\tau^{2k})$ for $k = 1, 2, \ldots$, such that all terms are nonnegative:

$$E(e^{uQ^2}) = \Sigma_{\ell_1, \cdots, \ell_n} K_{2\ell_1, \cdots, 2\ell_n} u^{2(\ell_1 + \cdots + \ell_n)} E(\tau^{2\ell_1}) \cdots E(\tau^{2\ell_n}). \quad (5.52)$$

We will not attempt to evaluate this sum! What we need is an upper bound and we can get it by comparing this computation with corresponding computations for the normal distribution. There we had $X = ||\Phi x||^2_{\ell^m_2} = \Sigma^m_{i=1} Q^2_i(x)$ with $Q_i(x) = \Sigma^n_{j=1} x_j t_{ij}$, where the t_{ij} are independently distributed normal random variables with distribution $N(0, 1/\sqrt{m})$. In that case we found, using spherical symmetry, that $E(e^{uQ_i(x)})$ was independent of the unit vector x. Thus we could choose $x = w$ without changing the result, and we could write $Q_i = Q = (t_1 + \cdots + t_n)/\sqrt{n}$ where the t_j are independently distributed normal random variables each with distribution $N(0, 1/\sqrt{m})$. Then the expansion for $E(e^{uQ^2})$ would take exactly the form (5.52), with the same coefficients $K_{2\ell_1, \cdots, 2\ell_n}$ and with $E(\tau^{2\ell}/m^\ell)$ replaced by $E(t^{2\ell})$. A straightforward computation for the normal distribution yields

$$E(t^{2k}) = \frac{(2k)!}{k!(2m)^k} \geq \frac{1}{m^k}, \quad (5.53)$$

whereas, since $\tau^2 = 1$ we have $E(\tau^{2k}/m^k) = 1/m^k$. Thus $E(\tau^{2k}/m^k) \leq E(t^{2k})$, so $[E(e^{uQ^2_1(x)})]^m$ from (5.49) is bounded by the corresponding expression (5.45) for the normal distribution

$$[E(e^{uQ^2_1(x)})]^m \leq \frac{(m/2)^{m/2}}{(m/2 - u)^{m/2}}. \quad (5.54)$$

A similar argument gives $E(Q^{2k}_i(x)) \leq (2k)!/k!(2m)^k$.

Using the techniques for the Chernoff inequality we achieve the same estimate as (5.46) for the normal distribution,

$$\Pr(||\Phi x||^2_{\ell^m_2} - 1 > \epsilon) \leq [e^{-\epsilon}(1 + \epsilon)]^{m/2} \leq e^{-m(\epsilon^2/4 - \epsilon^3/6)}. \quad (5.55)$$

Rather than follow the exact same method for $\Pr(||\Phi x||^2_{\ell^m_2} - 1 < -\epsilon)$, (5.50), we recast this inequality in the form

$$\Pr(||\Phi x||^2_{\ell^m_2} - 1 < -\epsilon) \leq e^{u(1-\epsilon)}[E(e^{-uQ^2_1(x)})]^m \quad (5.56)$$

$$\leq e^{u(1-\epsilon)}[E(1 - uQ^2_1(x) + \frac{u^2}{2!}Q^4_1(x)/2!)]^m$$

$$= e^{u(1-\epsilon)}[1 - uE(Q^2_1(x)) + \frac{u^2}{2}E(Q^4_1(x))]^m,$$

taking advantage of the fact that we are dealing with a convergent alternating series. Now always $E(Q_i^2(x)) = 1/m$ and $E(Q_i^{2k}(x)) \leq (2k)!/k!(2m)^k$, so a bound for the right-hand side of (5.56) is

$$\Pr(||\Phi x||_{\ell_2^m}^2 - 1 < -\epsilon) \leq e^{u(1-\epsilon)}[1 - \frac{u}{m} + \frac{3u^2}{2m^2}]^m, \quad 0 \leq u < m/2. \quad (5.57)$$

Just as in the derivation of (5.46) we make the substitution $u = m\epsilon/2(1+\epsilon)$, not optimal in this case but good enough. The result, after some manipulation, is

$$\Pr(||\Phi x||_{\ell_2^m}^2 - 1 < -\epsilon) \leq e^{-m(\epsilon^2/4 - \epsilon^3/6)}. \quad (5.58)$$

It follows directly from (5.55), (5.58) and the union bound that the concentration of measure inequality (5.34) holds with $c_0 = \epsilon^2/4 - \epsilon^3/6$, the same as for the normal distribution case.

The uniform distribution
A modification of the method applied to the Bernoulli distribution will also work for the uniform distribution $\rho_1(t)$ on $\mathbb{R}$:

$$\rho_1(t) = \begin{cases} \sqrt{\frac{m}{12}}, & \text{if } -\sqrt{\frac{3}{m}} \leq t \leq \sqrt{\frac{3}{m}} \\ 0, & \text{otherwise.} \end{cases} \quad (5.59)$$

Here, the sample matrices are defined by $\Phi_{ij} = t_{ij}$ where $\mathbf{t} = (t_{ij})$ is a matrix random variable on $\mathbb{R}^{mn}$ with density function

$$\Delta_1(\mathbf{t}) = \begin{cases} (\frac{m}{12})^{mn/2}, & \text{if } -\sqrt{\frac{3}{m}} \leq t_{ij} \leq \sqrt{\frac{3}{m}} \text{ for all } i, j \\ 0, & \text{otherwise.} \end{cases} \quad (5.60)$$

A function $f(\mathbf{t})$ has expectation

$$\bar{f}(\mathbf{t}) = E_{\rho_1}(f(\mathbf{t})) = \int_{\mathbb{R}^{mn}} f(\mathbf{t})\Delta_1(\mathbf{t}) \, d\mathbf{t}.$$

Here $Q_i(x) = \Sigma_{j=1}^n x_j t_{ij}$ and we take x to be a unit vector. Much of the derivation is word for word the same as in the Bernoulli case. Just as in that case the bounds for $E(e^{uQ_1^2(x)})$ and $E(Q_1(x)^{2k})$ depend on x. To obtain a concentration of measure inequality valid uniformly for all signals we need to find a "worst case" vector $x = w$ on the unit n-sphere such that $E(e^{uQ_1^2(w)})$ is maximal.

We focus our attention on two components of the unit vector x. By relabeling we can assume they are the first two components $x_1 = a, x_2 =$

b, so that $Q_1(x) = (at_{11} + bt_{12} + U)$ where $U = \Sigma_{j=3}^n x_j t_{1j}$. Let $\hat{x} = (c, c, x_3, \ldots, x_n)$ where $c = \sqrt{(a^2 + b^2)/2}$. Note that $||\hat{x}||_{\ell_2^n} = 1$, i.e., $\hat{x}$ is also a unit vector.

Lemma 5.19 *Suppose $ab < 0$. Then for $k = 1, 2, \ldots$ we have $E(Q_1(x)^{2k}) \leq E(Q_1(\hat{x})^{2k})$, and $E(e^{uQ_1^2(x)}) \leq E(e^{uQ_1^2(\hat{x})})$ if $u > 0$.*

Proof We fix U and average $Q_1(\hat{x}))^{2k} - Q_1(x)^{2k}$ over the range $-\sqrt{3/m} \leq t_{11}, t_{12} \leq \sqrt{3/m}$, i.e., we integrate over this square with respect to the measure $12dt_{11}dt_{12}/m$. This average is $4S_k K^{2k}/(2k+1)(2k+2)$ where $K = \sqrt{3/m}$, $V = U/K$ and

$$S_k = \frac{1}{c^2}[(V+2c)^{2k+2} - 2V^{2k+2} + (V-2c)^{2k+2}]$$

$$-\frac{1}{ab}[(V+a+b)^{2k+2} - (V+a-b)^{2k+2} - (V-a+b)^{2k+2} + (V-a-b)^{2k+2}].$$

We will show that $S_k \geq 0$. To see this we use the binomial theorem to expand each term of S_k except $2V^{2k}$ and regroup to obtain

$$S_k = \Sigma_{i=1}^{2k+2} \binom{2k+2}{i} V^{2k+2-i} D_i,$$

$$D_i = \frac{1}{c^2}[(2c)^i + (-2c)^i] - \frac{1}{ab}[(a+b)^i - (a-b)^i - (-a+b)^i + (-a-b)^i].$$

If i is odd it is clear that $D_i = 0$. If $i = 2j$ is even then since $2c^2 = a^2 + b^2$ and $2(a^2 + b^2) = (a+b)^2 + (a-b)^2$ we have

$$D_{2j} = 4(2a^2 + 2b^2)^{j-1} - \frac{1}{ab}[2(a+b)^{2j} - 2(a-b)^{2j}] = 2[(X+Y)^j - X^j - Y^j],$$

where $X = (a+b)^2, Y = (a-b)^2$. If $ab < 0$ then $-\frac{1}{ab} > 0$ and $(a+b)^2 > (a-b)^2$, so $D_{2j} > 0$ and

$$S_k = \Sigma_{j=1}^k \binom{2k+2}{2j} V^{2(k+1-j)} D_{2j} > 0.$$

To compute the expectations $E(Q_1(x)^{2k}), E(Q_1(\hat{x})^{2k})$ we can first average over t_{11}, t_{12} as above and then average over U. This process will preserve the inequality $S_k \geq 0$ so we have $E(Q_1(x)^{2k}) \leq E(Q_1(\hat{x})^{2k})$. $\square$

Since $||\Phi x||^2 = ||\Phi(-x)||^2$ we can always assume that the unit vector x has at least one component $x_\ell > 0$. The lemma shows that the "worst case" must occur for all $x_j \geq 0$. Now, assume the first two components are $x_1 = a > 0, x_2 = b > 0$, so $Q_1(x) = (at_{11} + bt_{12} + U)$ where $U = \Sigma_{j=3}^n x_j t_{1j}$. Then we can write $a = r\cos\theta$, $b = r\sin\theta$ where $r > 0$

and $0 < \theta < \pi/2$. We investigate the dependence of $E(Q_1^{2k}(x_\theta))$ on θ for fixed r. Here x_θ is a unit vector for all θ.

Lemma 5.20 *Let* $f_k(\theta) = E(Q_1^{2k}(x_\theta))$. *Then*

$$
f_k'(\theta) \begin{cases} > 0 & \text{for } 0 < \theta < \pi/4 \\ = 0 & \text{for } \theta = \pi/4 \\ < 0 & \text{for } \pi/4 < \theta < \pi/2. \end{cases}
$$

It follows from this result and the mean value theorem of calculus that the "worst case" is again $w = (1, 1, \ldots, 1)/\sqrt{n}$.

Proof We fix U and average $Q_1(x_\theta)^{2k}$ over the range $-\sqrt{3/m} \le t_{11}, t_{12} \le \sqrt{3/m}$, with respect to the measure $12 dt_{11} dt_{12}/m$. This average is $4 S_k (rK)^{2k}/(2k + 1)(2k + 2)^2$ where $K = \sqrt{3/m}$, $V = U/rK$ and

$$
S_k = \frac{1}{\sin\theta \cos\theta} [(V + \cos\theta + \sin\theta)^{2k+2} - (V + \cos\theta - \sin\theta)^{2k+2} \\ - (V - \cos\theta + \sin\theta)^{2k+2} + (V - \cos\theta - \sin\theta)^{2k+2}].
$$

We use the binomial theorem to expand each term of S_k and regroup to obtain

$$
S_k = \Sigma_{i=1}^{2k+2} \binom{2k+2}{i} V^{2k+2-i} D_i,
$$

$$
D_i = \frac{1}{\sin\theta \cos\theta} [(\cos\theta + \sin\theta)^i - (\cos\theta - \sin\theta)^i \\ - (-\cos\theta + \sin\theta)^i + (-\cos\theta - \sin\theta)^i].
$$

If i is odd it is clear that $D_i = 0$. If $i = 2j$ is even we have $D_0 = 0$ and

$$
D_{2j}(\theta) = \frac{2}{\cos\theta \sin\theta} [(\cos\theta + \sin\theta)^{2j} - (\cos\theta - \sin\theta)^{2j}], \quad j \ge 1,
$$

where

$$
S_k = \Sigma_{j=1}^{k} \binom{2k+2}{2j} V^{2(k+1-j)} D_{2j}(\theta). \tag{5.61}
$$

We investigate the dependence of D_{2j} on θ by differentiating:

$$
\frac{d}{d\theta} D_{2j}(\theta) = (\cos^2\theta - \sin^2\theta) \left[(\cos\theta + \sin\theta)^{2j} \left(\frac{2j \cos\theta \sin\theta}{(\cos\theta + \sin\theta)^2} - 1 \right) \right.
$$

$$
\left. + (\cos\theta - \sin\theta)^{2j} \left(\frac{2j \cos\theta \sin\theta}{(\cos\theta - \sin\theta)^2} + 1 \right) \right] =
$$

$$
(\cos^2\theta - \sin^2\theta) \Sigma_{\ell=0}^{j-1} \left[4j \binom{2j-2}{2\ell} - 2 \binom{2j}{2\ell+1} \right] \cos^{2\ell+1}\theta \sin^{2j-2\ell-1}\theta.
$$

Now

$$4j \binom{2j-2}{2\ell} - 2 \binom{2j}{2\ell+1} > 0, \quad \ell = 0, 1, \dots, j-1, \; j = 1, 2, \dots,$$

(5.62)

so

$$\frac{d}{d\theta} D_{2j}(\theta) \begin{cases} > 0 & \text{for } 0 < \theta < \pi/4 \\ = 0 & \text{for } \theta = \pi/4 \\ < 0 & \text{for } \pi/4 < \theta < \pi/2. \end{cases}$$

Now we average $4S_k(rK)^{2k}/(2k+1)(2k+2)^2$ over U to get $E(Q_1(x_\theta)^{2k})$ From (5.61) we obtain the same result for $f_k'(\theta)$ as we got uniformly for each term $D_{2j}'(\theta)$). $\qquad\qquad\qquad\qquad\qquad\qquad\qquad\qquad\qquad\square$

Exercise 5.21 Using Lemmas 5.19, 5.20 and the mean value theorem, show that for the uniform probability distribution ρ_1 the "worst case" is $w = (1, \dots, 1)/\sqrt{n}$.

Now that we know that $w = (1, \dots, 1)/\sqrt{n}$ is the "worst case" we can parallel the treatment for the Bernoulli distribution virtually word for word to obtain the concentration of measure inequality for the constant probability distribution ρ_1. To show that the "worst case" is dominated by the normal distribution we have to verify

$$E_{N(0,\frac{1}{\sqrt{m}})}(t^{2k}) = \frac{(2k)!}{k!(2m)^k} \ge E_{\rho_1}(t^{2k}) = \sqrt{\frac{m}{12}} \int_{-\sqrt{\frac{3}{m}}}^{\sqrt{\frac{3}{m}}} t^2 k \, dt = \frac{3^k}{(2k+1)m^k},$$

(5.63)

for all $k = 1, 2, \dots$, and this is straightforward.

Thus the concentration of measure inequality (5.34) again holds with $c_0 = \epsilon^2/4 - \epsilon^3/6$, the same as for the normal distribution and Bernoulli cases.

5.5 Discussion and practical implementation

Theorem 5.17 provides important justification for the method of compressive sampling and qualitative information about the relationship between δ, n, m and k for successful sampling. To understand this relationship let us choose δ, n, k with $0 < \delta < 1$ and $n \gg k$. We see from the proof of Theorem 5.17 that we cannot guarantee that the method will work at all unless expression (5.40) is nonnegative. Then, gradually increasing m, there will be a threshold $m \approx Ck\ln(n/k)$ where $C^{-1} = c_0(\delta/2)/[1 + (2 + \ln(12/\delta))/\ln(n/k)]$ such that the probability of failure is finite for all larger m. The probability of failure drops exponentially in m as m increases past the threshold. If the threshold value of m is $\leq n$ then the method will work. Otherwise we have no guarantee of success. In summary, our analysis says that there is a constant C such that the probability of failure is finite for

$$m \geq Ck\ln(\frac{n}{k}). \tag{5.64}$$

Here, C depends only on the ratio n/k, so is unchanged under a scaling $(k, n) \to (pk, pn)$ for p a positive integer. However the probability of failure decreases under scaling: $\exp(-mc_2) \to \exp(-pmc_2) = [\exp(-mc_2)]^p$.

Our analysis, leading to (5.64), says that we can achieve logarithmic compression for signals with very large n (for fixed k) and this prediction holds up well in practice. However, Theorem 5.17 is not practical in determining the smallest values of C that will work and the resulting probability of failure. The estimates are overly conservative. Compressive sensing works well over a much wider range of values of k, n, m than is guaranteed by the theorem. For example, let us choose $\delta \approx 0.4142$ in accordance with the deterministic theory and $c_0(\epsilon) = \epsilon^2/4 - \epsilon^3/6$ as we have computed for each of our methods of sample matrix generation. If $n = 100\,000, k = 6, m = 11\,000$ then the estimate (5.64) guarantees that the method will work with a probability of failure less than 8×10^{-8}. If m is decreased to $10\,000$, the method still works but the probability of failure is only guaranteed to be less than 8×10^{-4}. If m is further decreased to $9\,000$, then (5.64) becomes negative, and there is no guarantee that the method will work. Furthermore, we have set $\delta = \delta_{2k} = 1/(1 + \sqrt{2})$, and in accordance with the deterministic theory, this only guarantees success for $k = 3$ in this case. Thus, according to our theory we can only guarantee success in identifying an arbitrary 3-sparse signal of length $100\,000$ if we take about $11\,000$ samples, a nontrivial but not impressive result! In fact, the method performs impressively, as a little experimentation shows. In practice, the constant C is less than 3. Moreover, with a little more work we could sharpen our bounds and get more precise estimates for C. The important thing, however, was to establish that the method works, and ways of analyzing its viability, not to compute optimal values of C.

What follows is a MATLAB computer experiment that generates sparse signals and sample matrices randomly, samples the signals, reconstructs the signals from the samples, via ℓ_1 minimization and linear programming, graphs the results and checks for errors. The optimization package used is CVX, designed by Michael Grant and Stephen Boyd, with input from Yinyu Ye, which is available, free for downloading via the Internet. Compression ratios of about 10 to 1 are achievable. The first example shows perfect reconstruction for a 10-sparse signal in a 1024 length real signal vector with 110 samples. However, the parameters can easily be varied. Another issue to keep in mind is that actual signals are seldom random. They have some structure and compressive sampling may work much better for these cases. For example, if the sparse data is clumped, rather than being randomly scattered, phenomenal compression ratios of 100 to 1 can sometimes be achieved.

```
% Step 1: Define relative parameters
n=1024              % n is total length of signal vector
k=10                % k is number of sparse signals
m=110               % m*n is size of sample matrix
good=0;             % Used to check reconstructed signal
successful=0;       % Used to show if method is successful
t=(0:1:(n-1));      % This parameter is for graphing


% Step 2: Create sparse signal
support=randsample(n,k);  %Determines positions of non-zero
   %signals by selecting k distinct integers from 1 to n
x0=zeros(n,1);            % Creates n*1 vector of zeros
x0(support)=randn(k,1);   %Fills in selected positions with
                          %number from normal distribution
% Step 3: Create sample matrix using uniform distribution
A=unifrnd(-1,1,m,n)/sqrt(2*m/3); %Creates m*n matrix A

% Step4: Compress & reconstruct signal
b=A*x0;             % Multiplies signal by sample matrix
                    % b is compressed signal of length m


cvx_begin        % "cvx" is program used for reconstruction
variable x(n)    % This solves for x by minimizing L1 norm
minimize(norm(x,1)) % 'A'&'b' and x is reconstructed signal
subject to
```

```
A*x==b;
cvx_end

% Step 5: Check difference between original/recovered signals
z=x-x0;          % Calculates difference. when
for i=1:n        %difference is < 10^(-8), then it's good
    if (abs(z(i,1))<(10^-8))
        good=good+1;
    else
    end
end
if good==n       % If all points on reconstructed signal
    successful=successful+1 %are good then "successful=1"
else             % Otherwise, " successful=0 "
    successful=successful
end

% Last Step: Graph signals and compare them
plot(t,x0,'r:p') % Original signal represented in red
hold on
plot(t,x,'b')    % Reconstructed signal represented in blue
hold off
```

The second example illustrates modifications needed in the MATLAB program for ℓ_1 minimization of a complex signal, where again we have used a real sample matrix. CVX switches automatically to CSOCP to perform the minimization.

```
n=1024;
k=10;
m=110;
good=0;
OK=0;
successful=0;
success=0;
t=(0:1:(n-1));
support=randsample(n,k);

x0=zeros(n,1);
```

```
x0(support)=randn(k,1);
y0=zeros(n,1);
y0(support)=randn(k,1);
z0=complex(x0,y0);

A=randn(m,n)/sqrt(m);


b=A*z0;
br=real(b);

cvx_begin
variable x(n) complex;
minimize(norm(x,1))
subject to
A*x==b;
cvx_end


z=x-z0;
for i=1:n
    if (abs(z(i,1))<(10^-8))
        good=good+1;
    else
    end
end
if good==n
    successful=successful+1
else
    successful=successful
end


subplot(1,2,1)
plot(t,x0,'r:p'),axis([0,1024,-2.5,2.5])
hold on
plot(t,real(x),'b')
title ('Real part of signal (red pentagram) vs real part of
    reconst. (blue line)')
hold off
```

```
subplot(1,2,2)
plot(t,y0,'m:p'),axis([0,1024,-2.5,2.5])
hold on
plot(t,imag(x),'g')
title ('Imag. part of signal (magenta) vs. imag. part of
    reconst. (green)')
hold off
```

Exercise 5.22 Demonstrate the roughly logarithmic dependence of required sample size on n for fixed k by verifying that if $n = 10^{10}, k = 6, m = 22\,000$ then the estimate (5.64) guarantees that the method will work with a probability of failure less than 6×10^{-22}. Thus squaring n requires only that m be doubled for success.

5.6 Additional exercises

Exercise 5.23 Work out explicitly the action of the Vandermonde coder/decoder Example 5.2 for (a) $k = 1$, and (b) $k = 2$.

Exercise 5.24 (a) Show that the minimization problems (5.7) and (5.8) are equivalent. (b) Show that the minimization problems (5.9) and (5.10) are equivalent.

Exercise 5.25 Show that the ℓ_∞ minimization problems analogous to (5.7) and (5.9) can be expressed as problems in linear programming.

Exercise 5.26 Let Φ be a sample matrix with identical columns $\Phi = (c, c, \ldots, c)$ where $||c||_2 = 1$. Show that $\delta_1 = 0$, so this matrix does not satisfy RIP.

Exercise 5.27 Use the eigenvalue property (5.18) to show that for any sample matrix Φ it is not possible for $\delta_k = 0$ if $k \geq 2$.

Exercise 5.28 Use the $m \times n$ sample matrix Φ with constant elements $\Phi_{ij} = 1/\sqrt{m}$ and k-sparse signals x such that $x_j = 1$ for $x \in T$ to show that this sample matrix is not RIP for $k \geq 2$.

Exercise 5.29 Derive (5.53).

Exercise 5.30 Verify the right-hand inequality in (5.58).

Exercise 5.31 Verify the inequality (5.62).

Exercise 5.32 Verify the inequality (5.63).

To probe deeper into the theory and application of compressive sensing see [14, 26, 27, 28, 29, 30, 38, 62, 82, 83]. For an application of compressive sensing to radar see [67, 68]. For more advanced topics in compressive sensing, sparsity, wavelets, curvelets, edgelets, ridgelets, shearlets, sparse approximation, approximation using sparsity, sparse representations via dictionaries, robust principal component analysis, recovery of functions in high dimensions, learning, linear approximation and nonlinear approximation, the curse of dimensionality, see [6, 7, 9, 12, 14, 18, 20, 21, 22, 23, 24, 25, 28, 29, 30, 35, 36, 37, 38, 39, 40, 41, 42, 43, 47, 52, 54, 58, 59, 60, 61] and the references cited therein.

6

Discrete transforms

In this chapter, we study discrete linear transforms such as Z transforms, the Discrete Fourier transform and the Discrete Cosine transform.

6.1 Z transforms

Thus far, we have studied transforms of functions. Now we transform sequences. Let $\{x_k\}_0^\infty$ or just $x := \{x_k\} = \{x_0, \ldots\}$ denote a sequence of complex numbers.

Definition 6.1 The Z transform of x is the function

$$X[z] := \sum_0^\infty x_k z^{-k},$$

where z is a complex variable.

If x has a finite number of components, $X[z]$ is a polynomial in z^{-1}, otherwise we have a formal power series.

Example 6.2 Let $a \in \mathbb{C}$ and $x = \{a^k\}_0^\infty$. Find $X[z]$.

Solution:

$$X[z] = \sum_0^\infty a^k/z^k = \sum_0^\infty (a/z)^k = \frac{z}{z-a},$$

where for pointwise convergence we require $|a/z| < 1$. The next theorem deals with properties of Z transforms considered as formal power series.

Theorem 6.3 (i) **Linearity:** For $a, b \in \mathbb{R}$ and $w = x + y$,

$$W[z] = aX[z] + bY[z].$$

(ii) **Uniqueness**:

$$X[z] = Y[z] \implies x = y.$$

(iii) **Derivative property**:

$$(d/dz)X[z] = \frac{1}{z}\tilde{X}[z], \quad \tilde{x} = \{-kx_k\}.$$

Proof (i) and (iii) are formally self evident. (ii) follows formally from uniqueness of power series.

Exercise 6.1 Use the theorems above to show the following:

(i) If $x = \{ka^k\}$ then $X[z] = \frac{az}{(z-a)^2}$.

(ii) If $x = \{ka^{k-4}\}$ then $X[z] = \frac{a^{-3}z}{(z-a)^2}$.

Shift properties of Z transforms are important and are contained in the next theorem.

Theorem 6.4 (i) **Delayed shift property**: *Fix $l \geq 0$. Suppose we replace x by*

$$x^{(-l)} = \{0, 0, ..., 0, x_0, x_1, x_2...\},$$

adding l zeros on the left. If we then set $x_j := 0$, $j < 0$, we can write this as

$$x^{(-l)} = \{x_{-l}, x_{-l+1}, \ldots x_{-1}, x_0, x_1, \ldots\} = \{x_{k-l}\}_0^\infty.$$

Then,

$$X^{(-l)}[z] = \frac{1}{z^l}X[z].$$

(ii) **Forward shift property**: *Fix $l \geq 0$. Suppose we replace x by $x^{(l)} = \{x_l, x_{l+1}...\}$. Thus we shift forward the sequence and drop the first l terms. Then we have*

$$X^{(l)}[z] = z^l\left(X[z] - \sum_{j=0}^{l-1} x_j z^j\right).$$

Exercise 6.2 Verify Theorem 6.4.

Exercise 6.3 Let $x_k = a^k$, $k \geq 0, a \in \mathbb{R}$. Recalling that

$$Z[\{a^k\}] = \frac{z}{z-a}$$

show that

$$X^{(-2)}[z] = \frac{1}{z(z-a)}$$

and that

$$X^{(3)}[z] = \frac{z^3(a/z)^3}{1-a/z}.$$

The following theorem gives **Multiplication properties** of the Z transform.

Theorem 6.5   (i) *Multiplication by a^k. Fix $a \in \mathbb{C}$, and let $y = \{a^k x_k\}_0^\infty$, i.e., $y_k = a^k x_k$. Then*

$$Y[z] = X[\frac{z}{a}].$$

(ii) *Multiplication by k^n. Fix $n \geq 1$ and replace x by y where $y_k = k^n x_k$. Then*

$$Y[z] = (-zd/dz)^n X[z].$$

Proof

(i)

$$Y[z] = \sum_{k=0}^\infty \frac{a^k x_k}{z^k} = \sum_{k=0}^\infty \frac{x_k}{(z/a)^k} = X[\frac{z}{a}].$$

(ii) By induction, we have

$$(-zd/dz)^n(z^{-k}) = k^n z^{-k}, \; n \geq 1.$$

Thus, formally,

$$(-zd/dz)^n X[z] = \sum_{k=0}^\infty x_k k^n z^{-k} = Y[z].$$

$\square$

In analogy to continuous convolution $*$, we may now introduce convolution of sequences:

Definition 6.6 Let $x = \{x_k\}_0^\infty$ and $y = \{y_k\}_0^\infty$ be sequences. We define the convolution or Cauchy product of x and y by

$$x * y = w = \{x_k * y_k\}_0^\infty = \left\{ \sum_{j=0}^k x_j y_{k-j} \right\}_0^\infty,$$

i.e. $w_k = \sum_{j=0}^k x_j y_{k-j}$, a finite sum for each $k \geq 0$.

This definition is motivated by the theorem below which should be interpreted as the formal manipulation of power series.

Theorem 6.7 *We have* $W[z] = X[z]Y[z]$, *or the Z transform of a convolution of* x *and* y *is the product of the Z transforms of* x *and* y.

Proof

$$X[z]Y[z] = \left(\sum_{j=0}^{\infty} x_j z^{-j}\right)\left(\sum_{k=0}^{\infty} y_k z^{-k}\right) = \sum_{j,k=0}^{\infty} x_j x_k z^{-(j+k)}$$

$$= \sum_{l=0}^{\infty} z^{-l} \sum_{j,k \geq 0,\, j+k=l} x_j x_k = \sum_{l=0}^{\infty} z^{-l} \sum_{j=0}^{l} x_j y_{l-j}$$

$$= \sum_{k=0}^{\infty} z^{-k} \sum_{j=0}^{k} x_j y_{k-j} = W[z].$$

□

Note that if x, y have only a finite number of nonzero components then the Z transforms in the theorem are polynomials in z^{-1} and converge pointwise for all $z \neq 0$.

Example 6.8 Use the above theorem to show that

$$\sum_{j=0}^{k} \frac{(-1)^j}{j!(k-j)!} = 0,\ k \geq 1.$$

Solution Let $x_j = (-1)^j/j!$ and $y_j = 1/j!$. Thus $X[z] = \sum_{j=0}^{\infty} \frac{(-1)^j}{j!z^j} = \exp(-1/z)$. Similarly, $Y[z] = \exp(1/z)$. Thus by the theorem, $X[z]Y[z] = W[z] = 1$. However, 1 is the Z transform of the vector $u = (1, 0, \ldots)$. By uniqueness of Z transforms, $w = u$ so $\sum_{j=0}^{k} x_j y_{k-j} = 0,\ k \geq 1$ or $\sum_{j=0}^{k} \frac{(-1)^j}{j!(k-j)!} = 0,\ k \geq 1$.

6.2 Inverse Z transforms

As in the previous section, we work with formal power series.

Definition 6.9 Given the formal power series, $X(z) = \sum_{k=0}^{\infty} x_k z^{-k}$ where $x = \{x_k\}$, we define, formally, the *inverse Z transform* of $X(z)$, $X^{[-1]}$ by $X^{[-1]}(X[z]) = x$.

Lemma 6.10 *Let $a, b \in \mathbb{C}$ and $X[z]$, $Y[z]$ be formal power series. Then for $W[z] = aX[z] + bY[z]$ we have*

$$W^{[-1]}(W[z]) = aX^{[-1]}(X) + bY^{[-1]}(Y) = ax + by.$$

Example 6.11 Find (1) $X^{[-1]}$ for $X[z] = z/(z-a)$ and (2) $Y^{[-1]}$ for $Y[z] = z/(z-a)(z-b)$ with $a \neq b$. (1) is easily seen to be $x = \{a^k\}_0^\infty$ from uniqueness of the Z transform and the fact that $Z\left[\{a^k\}_0^\infty\right] = z/(z-a)$. To deal with (2) we use partial fractions and (i) to deduce that

$$Y^{[-1]}\left(\frac{z}{(z-a)(z-b)}\right) = Y^{[-1]}\left(\frac{1}{b-a}\left(\frac{z}{(z-a)} - \frac{z}{(z-b)}\right)\right)$$
$$= \left\{\frac{b^k - a^k}{b-a}\right\}_0^\infty.$$

Example 6.12 Find (i) $X^{[-1]}\left(\frac{z}{z^2+\alpha^2}\right)$, (ii) $Y^{[-1]}\left(\frac{z}{z^3+4z^2+5z+2}\right)$, (iii) $Z^{[-1]}\left(\frac{1}{(z-a)^2}\right)$ and (iv) $W^{[-1]}\left(\frac{z^2}{(z-a)^2}\right)$, $\alpha, a \in \mathbb{R}$.

Solution (i) We write

$$X^{[-1]}\left(\frac{z}{(z-i\alpha)(z+i\alpha)}\right) = \left\{\frac{(i\alpha)^k - (-i\alpha)^k}{2i\alpha}\right\}_0^\infty$$
$$= \left\{\alpha^{k-1}\sin\left(\frac{k\pi}{2}\right)\right\}_0^\infty = \begin{cases} 0, & k \text{ even} \\ (-1)^{(k-1)/2} & k \text{ odd.} \end{cases}$$

To see (ii), we write,

$$Y^{[-1]}\left[\frac{1}{z^3 + 4z^2 + 5z + 2}\right] = Y^{[-1]}\left(-\frac{z}{z+1}\frac{z}{(z+1)^2} - \frac{z}{z+2}\right).$$

Now,

$$(d/dz)\left(\frac{z}{(z-a)}\right) = \frac{-a}{(z-a)^2}$$

and

$$(d/dz)\sum_{k=0}^\infty a^k z^{-k} = \frac{-a}{(z-a)^2}.$$

Thus, the inverse Z transform of $z/(z-a)^2$ is $\{ka^{k-1}\}_{k=0}^\infty$. It then follows easily that (ii) is

$$Y^{[-1]}\left(\frac{z}{z^3 + 4z^2 + 5z + 2}\right) = \{(-1)^{k-1}(k+1) - (-2)^k\}_0^\infty.$$

Next, we write using (ii) and dividing by z,

$$\frac{1}{(z-a)^2} = \sum_{j=1}^{\infty} \frac{(j-1)a^{j-2}}{z^j}.$$

Thus

$$Z^{[-1]}\left(\frac{1}{(z-a)^2}\right) = \{(j-1)a^{j-2}\}_0^{\infty}$$

except that for $j = 0$, we must set $(j-1)a^{j-2} = 0$. To see (iv), we multiply by z and observe that

$$\frac{z^2}{(z-a)^2} = \sum_{j=0}^{\infty} \frac{(j+1)a^j}{z^j}$$

and so $W^{[-1]}\left(\frac{z^2}{(z-a)^2}\right) = \{(j+1)a^j\}_{j=0}^{\infty}$.

6.3 Difference equations

In this section we we show how to use Z transforms to solve difference equations which often arise in step signal analysis.

Example 6.13 Solve the difference equations:

(i)

$$y_{k+1} - cy_k = 0, \ k \geq 0.$$

(ii)

$$y_{k+2} + y_{k+1} - 2y_k = 1, \ k \geq 0$$

subject to $y_0 = 0$ and $y_1 = 1$.

(iii)

$$y_{k+2} + 2y_k = 1$$

subject to $y_0 = 1$ and $y_1 = 1$. Here, c is real and nonzero.

Solution First we look at (i): Set $y = \{y_k\}_{k=0}^{\infty}$. Taking Z transforms and using the forward shifts, we have

$$z\left(Y[z] - y_0\right) - cY[z] = 0$$

so

$$Y[z] = y_0 \left(\frac{z}{z-c}\right).$$

Thus,

$$y = Y^{[-1]}\left(y_0\left(\frac{z}{z-c}\right)\right) = y_0\left\{c^k\right\}_0^\infty.$$

So $y_k = y_0 c^k$, $k \geq 0$. To see (ii), we take Z transforms and use shift properties to deduce that

$$z^2\left(Y[z] - y_0 - \frac{y_1}{z}\right) + z(Y[z] - y_0) - 2Y[z] = \frac{z}{z-1}.$$

Thus, we have

$$\frac{Y[z]}{z} = \frac{z}{(z-1)^2(z+2)} = \frac{2}{9(z-1)} + \frac{1}{3(z-1)^2} - \frac{2}{9(z+2)}$$

from which we can now deduce that

$$y_k = 2/9 + 1/3k - 2/9(-2)^k, \quad k \geq 0.$$

Finally for (iii), taking Z transforms and using shifts, we have

$$Y[z](z^2 + 2) - z^2 - z = \frac{z}{z-1}$$

so that

$$\frac{Y[z]}{z} = \frac{z^2}{(z+\sqrt{2}i)(z-\sqrt{2}i)(z-1)}$$

$$= \frac{2+\sqrt{2}i}{6}\left(\frac{1}{z+\sqrt{2}i}\right) + \frac{2-\sqrt{2}i}{6}\left(\frac{z}{z-\sqrt{2}i}\right) + \frac{z}{3(z-1)}$$

from which we deduce that

$$y_k = \begin{cases} 1/3\left[2^{k/2+1}(-1)^{k/2} + 1\right], & k \text{ even} \\ 1/3\left[2^{k/2}(-1)^{(k+1)/2+1} + 1\right], & k \text{ odd}. \end{cases}$$

6.4 Discrete Fourier transform and relations to Fourier series

Suppose that f is a square integrable function on the interval $[0, 2\pi]$, periodic with period 2π, and that the Fourier series expansion converges pointwise to f everywhere:

$$f(t) = \sum_n \hat{f}(n)e^{int}, \quad 0 \leq t \leq 2\pi. \tag{6.1}$$

What is the effect of sampling the signal at a finite number of equally spaced points? For an integer $N > 1$ we sample the signal at points $2\pi m/N$, $m = 0, 1, \ldots, N - 1$:

$$f\left(\frac{2\pi m}{N}\right) = \sum_n \hat{f}(n)e^{2\pi inm/N}, \quad 0 \le m < N.$$

From the Euclidean algorithm we have $n = a + bN$ where $0 \le a < N$ and a, b are integers. Thus

$$f\left(\frac{2\pi m}{N}\right) = \sum_{a=0}^{N-1}\left[\sum_b \hat{f}(a+bN)\right]e^{2\pi ima/N}, \quad 0 \le m < N. \quad (6.2)$$

Note that the quantity in brackets is the projection of $\hat{f}$ at integer points to a periodic function of period N. Furthermore, the expansion (6.2) is essentially the *finite Fourier expansion*, as we shall see. However, simply sampling the signal at the points $2\pi m/N$ tells us only $\sum_b \hat{f}(a+bN)$, not (in general) $\hat{f}(a)$. This is known as **aliasing error**. If f is sufficiently smooth and N sufficiently large that all of the Fourier coefficients $\hat{f}(n)$ for $n > N$ can be neglected, then this gives a good approximation of the Fourier series.

To further motivate the discrete Fourier transform (DFT) it is helpful to consider the periodic function $f(t)$ above as a function on the unit circle: $f(t) = g(e^{it})$. Thus t corresponds to the point $(x, y) = (\cos t, \sin t)$ on the unit circle, and the points with coordinates t and $t + 2\pi n$ are identified, for any integer n. In the complex plane the points on the unit circle would just be e^{it}. Given an integer $N > 0$, let us sample f at N points $e^{\frac{2\pi i}{N}n}$, $n = 0, 1, \ldots, N - 1$, evenly spaced around the unit circle. We denote the value of f at the nth point by $f[n]$, and the full set of values by the column vector

$$f = (f[0], f[1], \ldots f[N-1]). \quad (6.3)$$

We can extend the definition of $f[n]$ for all integers n by the periodicity requirement $f[n] = f[n + kN]$ for all integers k, i.e., $f[n] = f[m]$ if $n = m \bmod N$. (This is precisely what we should do if we consider these values to be samples of a function on the unit circle.)

We will consider the vectors (6.3) as belonging to an N-dimensional inner product space P_N and expand f in terms of a specially chosen ON basis. To get basis functions for P_N we sample the Fourier basis functions $e^{(k)}(t) = e^{ikt}$ around the unit circle:

$$e^{(k)}[n] = e^{\frac{2\pi i}{N}nk} = \omega^{-nk}$$

or as a column vector

$$e^{(k)} = (e^{(k)}[0], e^{(k)}[1], \ldots e^{(k)}[N-1]) = (1, \omega^{-k}, \omega^{-2k}, \ldots \omega^{-(N-1)k}),$$
(6.4)

where ω is the primitive Nth root of unity $\omega = e^{-\frac{2\pi i}{N}}$.

Lemma 6.14

$$\sum_{n=0}^{N-1} \omega^{kn} = \begin{cases} 0 & \text{if } k = 1, 2, \ldots, N-1, \mathrm{mod} N \\ N & \text{if } k = 0, \mathrm{mod} N. \end{cases}$$

Proof Since $\omega^N = 1$ and $\omega \neq 1$ we have

$$1 - \omega^N = 0 = (1 - \omega)(1 + \omega + \omega^2 + \cdots + \omega^{N-1}).$$

Thus

$$\sum_{n=0}^{N-1} \omega^n = 1 + \omega + \omega^2 + \cdots + \omega^{N-1} = 0.$$

Since ω^k is also an Nth root of unity and $\omega^k \neq 1$ for $k = 1, \ldots, N-1$, the same argument shows that $\sum_{n=0}^{N-1} \omega^{kn} = 0$. However, if $k=0$ the sum is N. $\quad\square$

We define an inner product on P_N by

$$(f, g)_N = \frac{1}{N} \sum_{n=0}^{N-1} f[n]\bar{g}[n], \qquad f[n], g[n] \in P_N.$$

Lemma 6.15 *The functions* $e^{(k)}$, $k = 0, 1, \ldots, N-1$ *form an ON basis for* P_N.

Proof $(e^{(j)}, e^{(k)})_N =$

$$\frac{1}{N} \sum_{n=0}^{N-1} e^{(j)}[n]\bar{e}^{(k)}[n] = \frac{1}{N} \sum_{n=0}^{N-1} \omega^{(k-j)n} = \begin{cases} 0 & \text{if } j \neq k \bmod N \\ 1 & \text{if } j = k \bmod N. \end{cases}$$

$\quad\square$

Thus $(e^{(j)}, e^{(k)})_N = \delta_{jk}$ where the result is understood mod N. Now we can expand $f \sim \{f[n]\}$ in terms of this ON basis:

$$f = \frac{1}{N} \sum_{k=0}^{N-1} F[k]e^{(k)},$$

or in terms of components,

$$f[n] = \frac{1}{N} \sum_{k=0}^{N-1} F[k]e^{2\pi ikn/N} = \frac{1}{N} \sum_{k=0}^{N-1} F[k]\omega^{-nk}.$$
(6.5)

The Fourier coefficients $F[k]$ of f are computed in the standard way:
$F[k]/N = (f, e^{(k)})_N$ or

$$F[k] = \sum_{n=0}^{N-1} f[n]e^{(k)}[-n] = \sum_{n=0}^{N-1} f[n]\omega^{kn}. \tag{6.6}$$

The Parseval (Plancherel) equality reads

$$\sum_{n=0}^{N-1} f[n]\overline{g}[n] = \sum_{k=0}^{N-1} F[k]\overline{G}[k]$$

for $f, g \in P_N$.

The column vector

$$F = (F[0], F[1], F[2], \ldots, F[N-1])$$

is the *discrete Fourier transform* (DFT) of $f = \{f[n]\}$. It is illuminating to express the discrete Fourier transform and its inverse in matrix notation. The DFT is given by the matrix equation $F = \mathcal{F}_N f$ or

$$
\begin{pmatrix} F[0] \\ F[1] \\ F[2] \\ \vdots \\ F[N-1] \end{pmatrix} =
\begin{pmatrix}
1 & 1 & 1 & \cdots & 1 \\
1 & \omega & \omega^2 & \cdots & \omega^{N-1} \\
1 & \omega^2 & \omega^4 & \cdots & \omega^{2(N-1)} \\
\vdots & \vdots & \vdots & \ddots & \vdots \\
1 & \omega^{N-1} & \omega^{2(N-1)} & \cdots & \omega^{(N-1)(N-1)}
\end{pmatrix}
\begin{pmatrix} f[0] \\ f[1] \\ f[2] \\ \vdots \\ f[N-1] \end{pmatrix}.
$$
$$\tag{6.7}$$

Here $\mathcal{F}_N$ is an $N \times N$ matrix. The inverse relation is the matrix equation $f = \mathcal{F}_N^{-1} F$ or

$$
\begin{pmatrix} f[0] \\ f[1] \\ f[2] \\ \vdots \\ f[N-1] \end{pmatrix} = \frac{1}{N}
\begin{pmatrix}
1 & 1 & 1 & \cdots & 1 \\
1 & \bar{\omega} & \bar{\omega}^2 & \cdots & \bar{\omega}^{N-1} \\
1 & \bar{\omega}^2 & \bar{\omega}^4 & \cdots & \bar{\omega}^{2(N-1)} \\
\vdots & \vdots & \vdots & \ddots & \vdots \\
1 & \bar{\omega}^{N-1} & \bar{\omega}^{2(N-1)} & \cdots & \bar{\omega}^{(N-1)(N-1)}
\end{pmatrix}
\begin{pmatrix} F[0] \\ F[1] \\ F[2] \\ \vdots \\ F[N-1] \end{pmatrix},
$$
$$\tag{6.8}$$

where $\omega = e^{-2\pi i/N}, \bar{\omega} = \omega^{-1} = e^{2\pi i/N}$.

NOTE: At this point we can drop any connection with the sampling of values of a function on the unit circle. The DFT provides us with a method of analyzing any N-tuple of values f in terms of Fourier components. However, the association with functions on the unit circle is a good guide to our intuition concerning when the DFT is an appropriate tool.

Examples 6.16

1.
$$f[n] = \delta[n] = \begin{cases} 1 & \text{if } n = 0 \\ 0 & \text{otherwise.} \end{cases}$$

Here, $F[k] = 1$.

2. $f[n] = 1$ for all n. Then $F[k] = N\delta[k]$.

3. $f[n] = r^n$ for $n = 0, 1, \cdots, N - 1$ and $r \in \mathbb{C}$. Here

$$F[k] = \begin{cases} N & \text{if } r = e^{2\pi i k/N} \\ \frac{r^N - 1}{(re^{-2\pi i k/N} - 1)} & \text{otherwise.} \end{cases}$$

4. Upsampling. Given $f \in P_{N'}$ where $N' = N/2$ we define $g \in P_N$ by

$$g[n] = \begin{cases} f[\frac{n}{2}] & \text{if } n = 0, \pm 2, \pm 4, \ldots \\ 0 & \text{otherwise.} \end{cases}$$

Then $G[k] = F[k]$ where $F[k]$ is periodic with period $N/2$.

5. Downsampling. Given $f \in P_{2N}$ we define $g \in P_N$ by $g[n] = f[2n]$, $n = 0, 1, \cdots, N$. Then $G[k] = \frac{1}{2}(F[k] + F[k + N])$.

6.4.1 More properties of the DFT

Note that if $f[n]$ is defined for all integers n by the periodicity property, $f[n+jN] = f[n]$ for all integers j, the transform $F[k]$ has the same property. Indeed $F[k] = \sum_{n=0}^{N-1} f[n]\omega^{kn}$, so $F[k+jN] = \sum_{n=0}^{N-1} f[n]\omega^{(k+jN)n} = F[k]$, since $\omega^N = 1$.

Here are some other properties of the DFT. Most are just observations about what we have already shown. A few take some proof.

Lemma 6.17

- *Symmetry:* $\mathcal{F}_N^{tr} = \mathcal{F}_N$.
- *Unitarity:* $\mathcal{F}_N^{-1} = \frac{1}{N}\overline{\mathcal{F}}_N$.

   Set $\mathcal{G}_N = \frac{1}{\sqrt{N}}\mathcal{F}_N$. Then $\mathcal{G}_N$ is unitary. That is $\mathcal{G}_N^{-1} = \overline{\mathcal{G}}_N^{tr}$. Thus the row vectors of $\mathcal{G}_N$ are mutually orthogonal and of length 1.
- *Let $S : P_N \to P_N$ be the shift operator on P_N. That is $Sf[n] = f[n-1]$ for any integer n. Then $SF[k] = \omega^k F[k]$ for any integer k. Further, $S^{-1}f[n] = f[n+1]$ and $S^{-1}F[k] = \omega^{-k}F[k]$.*
- *Let $M : P_N \to P_N$ be an operator on P_N such that $Mf[n] = \omega^{-n}f[n]$ for any integer n. Then $MF[k] = F[k+1]$ for any integer k.*
- *If $f = \{f[n]\}$ is a real vector then $F[N - k] = \overline{F}[k]$.*

- *For $f, g \in P_N$ define the convolution $f * g \in P_N$ by*

$$f * g[n] = \sum_{m=0}^{N-1} f[m]g[n-m].$$

Then $f * g[n] = g * f[n]$ and $f * g[n + jN] = f * g[n]$.
- *Let $h[n] = f * g[n]$. Then $H[k] = F[k]G[k]$.*

Here are some simple examples of basic transformations applied to the 4-vector $(f[0], f[1], f[2], f[3])$ with DFT $(F[0], F[1], F[2], F[3])$ and $\omega = e^{-i\pi/2}$: We deal with Left shift, Right shift, Upsampling and Downsampling.

Operation	Data vector	DFT
Left shift	$(f[1], f[2], f[3], f[0])$	$(F[0], \omega^{-1}F[1], \omega^{-2}$ $F[2], \omega^{-3}F[3])$
Right shift	$(f[3], f[0], f[1], f[2])$	$(F[0], \omega^{1}F[1], \omega^{2}F[2],$ $\omega^{3}F[3])$
Upsampling	$(f[0], 0, f[1], 0, f[2], 0, f[3], 0)$	$(F[0], F[1], F[2], F[3],$ $F[0], F[1], F[2], F[3])$
Downsampling	$(f[0], f[2])$	$\frac{1}{2}(F[0] + F[2], F[1]$ $+F[3]).$

$$(6.9)$$

6.4.2 The discrete cosine transform

One feature that the discrete Fourier transform shares with the complex Fourier transform and the complex Fourier integral transform is that, even if the original signal or data is real, the transformed signal will be complex. It is often useful, however, to have a transform that takes real data to real data. Then, for example, one can plot both the input signal and the transformed signal graphically in two dimensions. The Fourier cosine series has this property, see expression (3.6). If the input signal is a real function the coefficients in the Fourier cosine series are real. Recall that this series was constructed for a function $f(t)$ on the interval $[0, \pi]$ by extending it to the interval $[-\pi, \pi]$ through the requirement $f(t) = f(-t)$. A similar construction works for the DFT, with some complications. One starts with the N-tuple $f = \{f[n]\}$ and constructs from it the $2N$-tuple

$$Rf = (f[N-1], f[N-2], \ldots, f[0], f[0], \ldots, f[N-2], f[N-1]) \quad (6.10)$$

by requiring symmetry with respect to reflection in the midpoint of the vector. Then one uses the DFT for $2N$-tuples on the extended vector and restricts the results to N-tuples again. The result is called the discrete cosine transform (DCT). We will work out the details in a series of exercises.

Exercise 6.4 Consider the space of $2N$-tuples $f' = (f'[0], \ldots, f'$ $[2N-1])$. Let $\mu = e^{-2\pi i/2N}$, define the appropriate inner product $(f', g')_{2N}$ for this space and verify that the vectors $e^{(\ell)'} = (e^{(\ell)'}[0], \ldots, e^{(\ell)'}[2N-1])$, $\ell = 0, \ldots, 2N-1$, form an ON basis for the space, where $e^{(\ell)'}[n] = \mu^{-\ell n}$.

Exercise 6.5 1. Show that a $2N$-tuple f' satisfies the relation $f'[n] = f'[2N-n-1]$ for all integers n if and only if $f' = Rf$, (6.10), for some N-tuple f.

2. Show that $f' = Rf$ for f an N-tuple if and only if f' is a linear combination of $2N$-tuples $E^{(k)}$, $k = 0, 1, \ldots, N-1$ where

$$E^{(k)}[n] = \frac{1}{2}(\mu^{-k/2}e^{(k)'}[n] + \mu^{k/2}e^{(-k)'}[n]) = \cos[\frac{\pi}{N}(n+1/2)k],$$

and $n = 0, 1, \ldots, 2N-1$. Recall that $e^{(-k)'}[n] = e^{(2N-k)'}[n]$.

Exercise 6.6 Verify the orthogonality relations

$$< E^{(k)}, E^{(h)} >_{2N} = \begin{cases} \frac{1}{2} & \text{if } h = k \neq 0 \\ 1 & \text{if } h = k = 0 \\ 0 & \text{if } h \neq k \end{cases}, \quad 0 \leq h, k \leq N-1. \quad (6.11)$$

Exercise 6.7 Using the results of the preceding exercises and the fact that $f[n] = f[N+n]$ for $0 \leq n \leq N-1$, establish the discrete cosine transform and its inverse:

$$F[k] = \omega[k] \sum_{n=0}^{N-1} f[n] \cos[\frac{\pi}{N}(n+1/2)k], \quad k = 0, \ldots, N-1, \quad (6.12)$$

$$f[n] = \sum_{k=0}^{N-1} \omega[k] F[k] \cos[\frac{\pi}{N}(n+1/2)k], \quad n = 0, \ldots, N-1,$$

where $\omega[0] = 1/\sqrt{N}$ and $\omega[k] = \sqrt{2/N}$ for $1 \leq k \leq N-1$.

Note that the N-tuple f is real if and only if its cosine transform F is real. The DCT is implemented in MATLAB.

6.4.3 An application of the DFT to finding the roots of polynomials

It is somewhat of a surprise that there is a simple application of the DFT to find the Cardan formulas for the roots r_0, r_1, r_2 of a third-order polynomial:

$$P_3(x) = x^3 + ax^2 + bx + c = (x - r_0)(x - r_1)(x - r_2). \qquad (6.13)$$

Let ω be a primitive cube root of unity and define the 3-vector

$$(F[0], F[1], F[2]) = (r_0, r_1, r_2).$$

Then

$$F[k] = r_k = f[0] + f[1]\omega^k + f[2]\omega^{2k}, \quad k = 0, 1, 2 \qquad (6.14)$$

where

$$f[n] = z_n = \frac{1}{3}\left(F[0] + F[1]\omega^{-n} + F[2]\omega^{-2n}\right), \quad n = 0, 1, 2.$$

Substituting relations (6.14) for the r_k into (6.13), we can expand the resulting expression in terms of the transforms z_n and powers of ω (remembering that $\omega^3 = 1$):

$$P_3(x) = A(x, z_n) + B(x, z_n)\omega + C(x, z_n)\omega^2.$$

This expression would appear to involve powers of ω in a nontrivial manner, but in fact they cancel out. To see this, note that if we make the replacement $\omega \to \omega^2$ in (6.14), then $r_0 \to r_0$, $r_1 \to r_2$, $r_2 \to r_1$. Thus the effect of this replacement is merely to permute the roots r_1, r_2 of the expression $P_3(x) = (x - r_0)(x - r_1)(x - r_2)$, hence to leave $P_3(x)$ invariant. This means that $B = C$, and $P_3(x) = A + B(\omega + \omega^2)$. However, $1 + \omega + \omega^2 = 0$ so $P_3(x) = A - B$. Working out the details we find

$$P_3(x) = x^3 - 3z_0x^2 + 3(z_0^2 - z_1z_2)x + (3z_0z_1z_2 - z_0^3 - z_1^3 - z_2^3)$$
$$= x^3 + ax^2 + bx + c.$$

Comparing coefficients of powers of x we obtain the identities

$$z_0 = -\frac{a}{3}, \quad z_1z_2 = \frac{a^2 - 3b}{9}, \quad z_1^3 + z_2^3 = \frac{-2a^3 + 9ab - 27c}{27},$$

or

$$z_1^3 z_2^3 = \left(\frac{a^2 - 3b}{9}\right)^3, \quad z_1^3 + z_2^3 = \frac{-2a^3 + 9ab - 27c}{27}.$$

It is simple algebra to solve these two equations for the unknowns z_1^3, z_2^3:

$$z_1^3 = \frac{-2a^3 + 9ab - 27c}{54} + D, \quad z_2^3 = \frac{-2a^3 + 9ab - 27c}{54} - D,$$

where

$$D^2 = \left(\frac{-2a^3 + 9ab - 27c}{54}\right)^2 - \left(\frac{a^2 - 3b}{9}\right)^3.$$

Taking cube roots, we can obtain z_1, z_2 and plug the solutions for z_0, z_1, z_2 back into (6.14) to arrive at the Cardan formulas for r_0, r_1, r_2, [98].

This method also works for finding the roots of second-order polynomials (where it is trivial) and fourth-order polynomials (where it is much more complicated). Of course it, and all such explicit methods, must fail for fifth- and higher-order polynomials.

6.5 Fast Fourier transform (FFT)

We have shown that the DFT for the column N-vector $\{f[n]\}$ is determined by the equation $F = \mathcal{F}_N f$, or in detail,

$$F[k] = \sum_{n=0}^{N-1} f[n] e^{(k)}[-n] = \sum_{n=0}^{N-1} f[n] \omega^{kn}, \qquad \omega = e^{-2\pi i/N}.$$

From this equation we see that each computation of the N-vector F requires N^2 multiplications of complex numbers. However, due to the special structure of the matrix $\mathcal{F}_N$ we can greatly reduce the number of multiplications and speed up the calculation. (We will ignore the number of additions, since they can be done much faster and with much less computer memory.) The procedure for doing this is the fast Fourier transform (FFT). It reduces the number of multiplications to about $N \log_2 N$.

The algorithm requires $N = 2^M$ for some integer M, so $F = \mathcal{F}_{2^M} f$. We split f into its even and odd components:

$$F[k] = \sum_{n=0}^{N/2-1} f[2n] \omega^{2kn} + \omega^k \sum_{n=0}^{N/2-1} f[2n+1] \omega^{2kn}.$$

Note that each of the sums has period $N/2 = 2^{M-1}$ in k, so

$$F[k+N/2] = \sum_{n=0}^{N/2-1} f[2n] \omega^{2kn} - \omega^k \sum_{n=0}^{N/2-1} f[2n+1] \omega^{2kn}.$$

Thus by computing the sums for $k = 0, 1 \ldots, N/2 - 1$, hence computing $F[k]$ we get the $F[k + N/2]$ virtually for free. Note that the first sum is the DFT of the downsampled f and the second sum is the DFT of the data vector obtained from f by first left shifting and then downsampling.

Let's rewrite this result in matrix notation. We split the $N = 2^M$ component vector f into its even and odd parts, the 2^{M-1}-vectors

$$f_e = (f[0], f[2], \ldots, f[N-2]), \qquad f_o = (f[1], f[3], \ldots, f[N-1]),$$

and divide the N-vector F into halves, the $N/2$-vectors

$$F_- = (F[0], F[1], \ldots, F[N/2 - 1]),$$

$$F_+ = (F[0 + N/2], F[1 + N/2], \ldots, F[N - 1]).$$

We also introduce the $N/2 \times N/2$ diagonal matrix $D_{N/2}$ with matrix elements $(D_{N/2})_{jk} = \omega^k \delta_{jk}$, and the $N/2 \times N/2$ zero matrix $(O_{N/2})_{jk} = 0$ and identity matrix $(I_{N/2})_{jk} = \delta_{jk}$. The above two equations become

$$F_- = \mathcal{F}_{N/2} f_e + D_{N/2} \mathcal{F}_{N/2} f_o, \qquad F_+ = \mathcal{F}_{N/2} f_e - D_{N/2} \mathcal{F}_{N/2} f_o,$$

or

$$\begin{pmatrix} F_- \\ F_+ \end{pmatrix} = \begin{pmatrix} I_{N/2} & D_{N/2} \\ I_{N/2} & -D_{N/2} \end{pmatrix} \begin{pmatrix} \mathcal{F}_{N/2} & O_{N/2} \\ O_{N/2} & \mathcal{F}_{N/2} \end{pmatrix} \begin{pmatrix} f_e \\ f_o \end{pmatrix} \qquad (6.15)$$

Note that this factorization of the transform matrix has reduced the number of multiplications for the DFT from 2^{2M} to $2^{2M-1} + 2^M$, i.e., cut them about in half, for large M. We can apply the same factorization technique to $\mathcal{F}_{N/2}$, $\mathcal{F}_{N/4}$, and so on, iterating M times. Each iteration involves 2^M multiplications, so the total number of FFT multiplications is about $M2^M$. Thus an $N = 2^{10} = 1024$ point DFT which originally involved $N^2 = 2^{20} = 1\,048\,576$ complex multiplications, can be computed via the FFT with $10 \times 2^{10} = 10\,240$ multiplications. In addition to the very impressive speed-up in the computation, there is an improvement in accuracy. Fewer multiplications leads to a smaller roundoff error. The FFT is implemented in MATLAB.

For more properties and applications of the FFT see [88].

6.6 Approximation to the Fourier transform

One of the most important applications of the FFT is to the approximation of Fourier transforms. We will indicate how to set up the calculation of the FFT to calculate the Fourier transform of a continuous function $f(t)$ with support on the interval $\alpha \le t \le \beta$ of the real line. We first map this interval on the line onto the unit circle $0 \le \phi \le 2\pi$, mod 2π via the affine transformation $t = a\phi + b$. Clearly $b = \alpha, a = \frac{\beta - \alpha}{2\pi}$. Since normally $f(\beta) \ne f(\alpha)$ when we transfer f as a function on the unit circle there will usually be a jump discontinuity at β.

We want to approximate

$$\hat{f}(\lambda) = \int_{-\infty}^{\infty} f(t)e^{-it\lambda}dt = \int_{\alpha}^{\beta} f(t)e^{-it\lambda}dt$$

$$= \frac{\beta - \alpha}{2\pi}e^{-i\lambda\alpha}\int_0^{2\pi} g(\phi)e^{-i\frac{\beta-\alpha}{2\pi}\lambda\phi}d\phi$$

$$= \frac{\beta - \alpha}{4\pi^2}e^{-i\lambda\alpha}\hat{g}(\frac{\beta - \alpha}{2\pi}\lambda)$$

where $g(\phi) = f(\frac{\beta-\alpha}{2\pi}\phi + \alpha)$.

For an N-vector DFT we will choose our sample points at $\phi = 2\pi n/N$, $n = 0, 1, \ldots, N - 1$. Thus

$$g = (g[0], g[1], \ldots, g[N-1]), \qquad g[n] = f(\frac{\beta - \alpha}{N}n + \alpha).$$

Now the Fourier coefficients

$$G[k] = \sum_{n=0}^{N-1} g[n]e_k[-n] = \sum_{n=0}^{N-1} g[n]\omega^{kn}, \qquad \omega = e^{-2\pi i/N}$$

are approximations of the coefficients $\hat{g}(k)$. Indeed $\hat{g}(k) \sim G[k]/N$. Thus

$$\hat{f}(\frac{2\pi k}{\beta - \alpha}) \sim \frac{\beta - \alpha}{4\pi^2 N}e^{-\frac{2\pi i k\alpha}{\beta - \alpha}}G[k].$$

Note that this approach is closely related to the ideas behind the Shannon sampling theorem, except here it is the signal $f(t)$ that is assumed to have compact support. Thus $f(t)$ can be expanded in a Fourier series on the interval $[\alpha, \beta]$ and the DFT allows us to approximate the Fourier series coefficients from a sampling of $f(t)$. (This approximation is more or less accurate, depending on the aliasing error.) Then we notice that the Fourier series coefficients are proportional to an evaluation of the Fourier transform $\hat{f}(\lambda)$ at the discrete points $\lambda = \frac{2\pi k}{\beta - \alpha}$ for $k = 0, \ldots, N - 1$.

6.7 Additional exercises

Exercise 6.8 Find the Z transforms of the following sequences:

(i) $\{(1/4)^k\}_0^\infty$.

(ii) $\{(-7i)^k\}_0^\infty$.

(iii) $\{8(-7i)^k + 4(1/4)^k\}_0^\infty$.

(iv) $\{5k\}_0^\infty$.

(v) $\{k^2 3^k\}_0^\infty$.

(vi) $\{2^k/k!\}_0^\infty$.

(vii) $\{x_j\}_0^\infty$ where

$$x_j = \begin{cases} (-1)^{(j-1)/2}, & j \text{ odd} \\ 0, & j \text{ even.} \end{cases}$$

(viii) $\{\cos(k\theta)\}_0^\infty$, $\theta \in \mathbb{R}$.

Exercise 6.9 Let $x = \{x_k\}_0^\infty, y, z$ be sequences. Show that

(i) $x * y = y * x$.

(ii) $(ax + by) * z = a(x * z) + b(y * z)$, $a, b \in \mathbb{R}$.

Exercise 6.10 Let $x(t)$ be the signal defined on the integers t and such that

$$x(t) = \begin{cases} 1 & \text{if } t = 0, 1 \\ 0 & \text{otherwise.} \end{cases}$$

1. Compute the convolution $x_2(t) = x * x(t)$.
2. Compute the convolution $x_3(t) = x * (x * x)(t) = x * x_2(t)$.
3. Evaluate the convolution $x_n(t) = x * x_{n-1}(t)$ for $n = 2, 3, \ldots$. Hint: Use Pascal's triangle relation for binomial coefficients.

Exercise 6.11 Find the inverse Z transforms of the the following:

(i) $\frac{z}{z-i}$.

(ii) $\frac{z}{3z+1}$.

(iii) $\frac{4z}{3z+1}$.

(iv) $\frac{1}{3z+1}$.

(v) $\frac{4z+7}{3z+1}$.

(vi) $\frac{1}{z(z-i)}$.

(vii) $\frac{3z^2+4z-5/z-z^{-2}}{z^3}$.

Exercise 6.12 Prove that if

$$X[z] = \sum_{k=0}^\infty x_k z^{-k}$$

is a power series in $1/z$ and ℓ is a positive integer, then the inverse Z transform of $z^{-\ell} X[z]$ is the sequence $\{x_{k-\ell}\}_0^\infty$.

Exercise 6.13 Using partial fractions, find the inverse Z transforms of the following functions:

(i) $\frac{z}{(z-1)(z-2)}$.

(ii) $\frac{z}{z^2-z+1}$.

(iii) $\frac{2z+1}{(z+1)(z-3)}$.

(iv) $\frac{z^2}{(2z+1)(z-1)}$.

(v) $\frac{z^2}{(z-1)^2(z^2-z+1)}$.

(vi) $\frac{2z^2-7z}{(z-1)^2(z-3)}$.

Exercise 6.14 Solve the following difference equations, using Z transform methods:

(i) $8y_{k+2} - 6y_{k+1} + y_k = 9$, $y_0 = 1$, $y_1 = 3/2$.

(ii) $y_{k+2} - 2y_{k+1} + y_k = 0$, y_0, $y_1 = 1$.

(iii) $y_{k+2} - 5y_{k+1} - 6y_k = (1/2)^k$, $y_0 = 0$, $y_1 = 0$.

(iv) $2y_{k+2} - 3y_{k+1} - 2y_k = 6k + 1$, $y_0 = 1$, $y_1 = 2$.

(v) $y_{k+2} - 4y_k = 3k - 5$, $y_0 = 0$, $y_1 = 0$.

Exercise 6.15 (i) Let a, b, c, d, e be real numbers. Consider the difference equation:

$$ay_{k+3} + by_{k+2} + cy_{k+1} + dy_k = e.$$

(ii) Let $Y[z]$ be as usual. Show that

$$Y[z](az^3 + bz^2 + cz + d) = z^3(ay_0) + z^2(ay_1 + by_0)$$
$$+ z(ay_2 + by_1 + cy_0) + \frac{ez}{z-1}.$$

Deduce that when $y_0 = y_1 = 0$,

$$Y[z] = \frac{z(ay_2) + \frac{ez}{z-1}}{az^3 + bz^2 + cz + d}.$$

(iii) Use (i)–(ii) to solve the difference equation

$$y_{k+3} + 9y_{k+2} + 26y_{k+1} + 24y_k = 60$$

subject to $y_0 = y_1 = 0$, $y_2 = 60$.

Exercise 6.16 Let $N = 6$ and $f[n] = n$, $n = 0, \ldots, 5$. Compute the DFT $F[n]$ explicitly and verify the recovery of f from its transform.

Exercise 6.17 (i) Let $N = 4$ and $f[n] = a_n$, $n = 0, \ldots, 3$. Compute the DFT $F[n]$ explicitly and verify the recovery of f from its transform.

(ii) Let $N = 4$ and $f[n] = a_n$, $n = 0, \ldots, 3$. Compute the DCT $F[n]$ explicitly and verify the recovery of f from its transform.

Exercise 6.18 This problem investigates the use of the FFT in denoising. Let

$$f = \exp(-t^2/10)\left(\sin(t) + 2\cos(5t) + 5t^2\right) \quad 0 < t \leq 2\pi.$$

Discretize f by setting $f_k = f(2k\pi/256), k = 1, \ldots, 256$. Suppose the signal has been corrupted by noise, modeled by the addition of the vector

```
x=2*rand(1,2^8)-0.5
```

in MATLAB, i.e., by random addition of a number between -2 and 2 at each value of k, so that the corrupted signal is $g = f + x$. Use MATLAB's fft command to compute $\hat{g}_k$ for $0 \leq k \leq 255$. (Note that $g_{n-k} = \overline{g_k}$. Thus the low-frequency coefficients are $\hat{g}_0, \ldots, \hat{g}_m$ and $\hat{g}_{256-m}, \ldots, \hat{g}_{256}$ for some small integer m). Filter out the high-frequency terms by setting $\hat{g}_k = 0$ for $m \leq k \leq 255 - m$ with $m = 10$. Then apply the inverse FFT to these filtered $\hat{g}_k$ to compute the filtered g_k. Plot the results and compare with the original unfiltered signal. Experiment with several different values of m.

The following sample MATLAB program is an implementation of the denoising procedure in Exercise 6.18.

```
t=linspace(0,2*pi,2^8);
x=2*rand(1,2^8)-0.5
f=exp(-t.^2/10).*(sin(t)+2*cos(5*t)+5*t.^2);
g=f+x;
m=10;  % Filter parameter.
hatg=fft(g);
%If $hatg$ is FFT of $g$, can filter out high frequency
%components from $hatg$ with command such as

denoisehatg=[hatg(1:m) zeros(1,2^8-2*m) hatg(2^8-m+1:2^8)];
subplot(2,2,1)
plot(t,f) % The signal.
subplot(2,2,2)
plot(t,x) % The noise.
subplot(2,2,3)
plot(t,g) % Signal plus noise.
subplot(2,2,4)
plot(t,ifft(denoisehatg)) % Denoised signal.
```

Exercise 6.19 Consider the signal

$$f(t) = \exp(-t^2/10)\left(\sin(t) + 2\cos(2t) + \sin(5*t) + \cos(8t) + 2t^2\right),$$

$0 < t \le 2\pi$. Discretize f by setting $f_k = f(2k\pi/256), k = 1, \ldots, 256$, as in Exercise 6.18. Thus the information in f is given by prescribing 256 ordered numbers, one for each value of k. The problem here is to compress the information in the signal f by expressing it in terms of only $256c$ ordered numbers, where $0 < c < 1$ is the compression ratio. This will involve throwing away some information about the original signal, but we want to use the compressed data to reconstitute a facsimile of f that retains as much information about f as possible. The strategy is to compute the FFT $\hat{f}$ of f and compress the transformed signal by zeroing out the fraction $1 - c$ of the terms $\hat{f}_k$ with smallest absolute value. Apply the inverse FFT to the compressed transformed signal to get a compressed signal fc_k. Plot the results and compare with the original uncompressed signal. Experiment with several different values of c. A measure of the information lost from the original signal is L2error $= \|f - fc\|_2 / \|f\|_2$, so one would like to have c as small as possible while at the same time keeping the $L2$ error as small as possible. Experiment with c and with other signals, and also use your inspection of the graphs of the original and compressed signals to judge the effectiveness of the compression.

The following sample MATLAB program is an implementation of the compression procedure in Exercise 6.19.

```
% Input: time vector t, signal vector f, compression rate c,
%(between 0 and 1)

t=linspace(0,2*pi,2^8);
f=exp(-t.^2/10).*(sin(t)+2*cos(2*t)+sin(5*t)+cos(8*t)+2*t.^2);
c=.5;
hatf=fft(f)

% Input vector f and ratio c: 0<= c <=1.
% Output is vector fc in which  smallest
% 100c% of terms f_k, in absolute value, are set
% equal to zero.
N=length(f); Nc=floor(N*c);
ff=sort(abs(hatf));
cutoff=abs(ff(Nc+1));
```

```
hatfc=(abs(hatf)>=cutoff).*hatf;

fc=ifft(hatfc);
L2error=norm(f-fc,2)/norm(f) %Relative L2 information loss

subplot(1,2,1)
plot(t,f) %Graph of f
subplot(1,2,2)
plot(t, fc,'r') % Graph of compression of f
```

7

Linear filters

In this chapter we will introduce and develop those parts of linear filter theory that are most closely related to the mathematics of wavelets, in particular perfect reconstruction filter banks. The primary emphasis will be on discrete filters. Mathematically, the setup is analogous to that of Chapter 5 where we considered the equation $\Phi x = y$ with x as a signal and y the result of sampling x by the matrix Φ. However, there are important differences in order that Φ be considered a filter.

7.1 Discrete linear filters

A *discrete-time signal* is a sequence of numbers (real or complex). The signal takes the form of an ∞-tuple x where

$$x^{\mathrm{tr}} = (\ldots, x_{-1}, x_0, x_1, x_2, \ldots).$$

Intuitively, we think of $x(n){=}x_n$ as the value of the signal at time nT where T is the time interval between successive values. Here x could be a digital sampling of a continuous analog signal or simply a discrete data stream. In general, these signals are of infinite length, in distinction to Chapter 5 where all signals were of fixed finite length and Chapter 6 where they could be semi-infinite. Usually, but not always, we will require that the signals belong to ℓ_2, i.e., that they have finite energy: $\sum_{n=-\infty}^{\infty} |x_n|^2 < \infty$. Recall that ℓ_2 is a Hilbert space with inner product

$$(x^{(1)}, x^{(2)}) = \sum_{n=-\infty}^{\infty} x_n^{(1)} \bar{x}_n^{(2)}.$$

The impulses $e^{(j)}, j = 0, \pm 1, \pm 2, \ldots$, defined by $e_n^{(j)} = \delta_{jn}$ form an ON basis for the signal space: $(e^{(j)}, e^{(k)}) = \delta_{jk}$ and $x = \sum_{n=-\infty}^{\infty} x_n e^{(n)}$. In

$$
\begin{bmatrix}
\vdots \\
y_{-1} \\
y_0 \\
y_1 \\
\vdots
\end{bmatrix}
=
\begin{bmatrix}
\cdot & \cdot & & & \cdot & \cdot \\
\cdot & \cdot & h_0 & h_{-1} & h_{-2} & \cdot & \cdot \\
\cdot & \cdot & h_1 & h_0 & h_{-1} & \cdot & \cdot \\
\cdot & \cdot & h_2 & h_1 & h_0 & \cdot & \cdot \\
\cdot & \cdot & & & \cdot & \cdot
\end{bmatrix}
\begin{bmatrix}
\vdots \\
x_{-1} \\
x_0 \\
x_1 \\
\vdots
\end{bmatrix}.
$$

Figure 7.1 Matrix filter action.

particular the impulse at time 0 is called the *unit impulse* $\delta = e^0$. The right *shift* or *delay* operator $\mathbf{R} : \ell_2 \to \ell_2$ is defined by $\mathbf{R}x(n) = x(n-1)$. Note that the action of this bounded operator is to delay the signal by one unit. Similarly the inverse operator $\mathbf{L}x(n) = x(n+1)$ (advances) shifts the signal to the left by one time unit. Note that $\mathbf{L} = \mathbf{R}^{-1}$

A *digital filter* $\mathbf{\Phi}$ is a bounded linear operator $\mathbf{\Phi} : \ell_2 \to \ell_2$ that is time invariant. The filter processes each input x and gives an output $\mathbf{\Phi}x = y \in \ell_2$. Since $\mathbf{\Phi}$ is linear, its action is completely determined by the outputs $\mathbf{\Phi}e^{(j)}$. Time invariance means that $\mathbf{\Phi}x^k = y^k$, $k = 0, \pm 1, \ldots$ whenever $\mathbf{\Phi}x = y$. (Here, $x_n^k = x_{n-k}$.) Thus, the effect of delaying the input by k units of time is just to delay the output by k units. (Another way to put this is $\mathbf{\Phi R} = \mathbf{R \Phi}$, the filter commutes with shifts.) We can associate an infinite matrix with $\mathbf{\Phi}$.

$$
\Phi = [\Phi_{ij}], \qquad \text{where } \Phi_{ij} = (\mathbf{\Phi}e^{(j)}, e^{(i)}).
$$

Thus, $\mathbf{\Phi}e^{(j)} = \sum_{i=-\infty}^{\infty} \Phi_{ij}e^{(i)}$ and $y_i = \sum_{j=-\infty}^{\infty} \Phi_{ij}x_j$. In terms of matrices, time invariance means $\Phi_{ij} = (\mathbf{\Phi}e^{(j)}, e^{(i)}) = (\mathbf{\Phi}e^{(j+k)}, e^{(i+k)}) = \Phi_{m+k,n+k}$ for all k. Hence the matrix elements Φ_{ij} depend only on the difference $i - j$: $\Phi_{ij} = h_{i-j}$ and $\mathbf{\Phi}$ is completely determined by its coefficients h_ℓ. The matrix filter action looks like Figure 7.1. Note that the matrix has diagonal bands. Here h_0 appears down the main diagonal, h_{-1} on the first superdiagonal, h_{-2} on the next superdiagonal, etc. Similarly h_1 on the first subdiagonal, etc. A matrix Φ_{ij} whose matrix elements depend only on $i - j$ is called a *Toeplitz matrix*.

Another way that we can express the action of the filter is in terms of the shift operator:

$$
\mathbf{\Phi} = \sum_{\ell=-\infty}^{\infty} h_\ell \mathbf{R}^\ell. \tag{7.1}
$$

Thus $\mathbf{R}^\ell e^{(j)} = e^{(j+\ell)}$ and

$$\Phi(\sum_j x_j e^{(j)}) = \sum_{\ell,j} x_j h_\ell e^{(j+\ell)} = \sum_{i,j} x_j h_{i-j} e^{(i)} = \sum_i y_i e^{(i)}.$$

We have $y_i = \sum_j x_j h_{i-j}$. If only a finite number of the coefficients h_n are nonzero we say that we have a *Finite Impulse Response* (FIR) filter. Otherwise we have an *Infinite Impulse Response* (IIR) filter. We can uniquely define the action of an FIR filter on *any* sequence x, not just an element of ℓ_2 because there are only a finite number of nonzero terms in (7.1), so no convergence difficulties. For an IIR filter we have to be more careful. Note that the response to the unit impulse is $\Phi e^{(0)} = \sum_j h_j e^{(j)}$.

Finally, we say that a digital filter is *causal* if it doesn't respond to a signal until the signal is received, i.e., $h_j = 0$ for $j < 0$. To sum up, a causal FIR digital filter is completely determined by the impulse response vector

$$h = (h_0, h_1, \ldots, h_N)$$

where N is the largest nonnegative integer such that $h_N \neq 0$. We say that the filter has $N + 1$ "taps."

There are other ways to represent the filter action that will prove useful. The next is in terms of the convolution of vectors. Recall that a signal x belongs to the Banach space ℓ_1 provided $\sum_{j=-\infty}^{\infty} |x_j| < \infty$.

Definition 7.1 Let x, y be in ℓ_1. The convolution $x * y$ is given by the expression

$$x * y_j = \sum_{i=-\infty}^{\infty} x_{j-i} y_i, \qquad j = 0, \pm 1, \pm 2, \ldots.$$

This definition is identical to the definitions of convolution in Chapter 6, except that here the sequences x, y are doubly infinite.

Lemma 7.2

1. $x * y \in \ell_1$
2. $x * y = y * x$

Sketch of proof

$$\sum_j |x * y_j| \leq \sum_j \sum_i |x_{j-i}| \cdot |y_i| = \sum_i \sum_j |x_j| \cdot |y_i|$$

$$= (\sum_n |x_n|)(\sum_m |y_m|) < \infty.$$

The interchange of order of summation is justified because the series are absolutely convergent. □

Now we note that the action of the filter $\mathbf{\Phi}$ can be given by the convolution

$$\mathbf{\Phi}x = h * x.$$

For an FIR filter, this expression makes sense even if x isn't in ℓ_1, since the sum is finite.

Examples 7.3 FIR filters are commonly used to analyze and report data from Wall Street. Let $x(n)$ be the Dow Jones Industrial Average at the conclusion of trading on day n. The *moving average* $y^{(1)}(n) = \frac{1}{2}(x(n) + x(n-1))$ is the sequence giving the average closing on two successive days. The *moving difference* is the filter $y^{(2)}(n) = \frac{1}{2}(x(n) - x(n-1))$. In particular $2y^{(2)}(n)$ is the statistic used to report on how much the Dow changes from one day to the next. The *150 day moving average* $y^{(3)}(n) = \frac{1}{150}\sum_{k=0}^{149} x(n-k)$ is often used to smooth out day-to-day fluctuations in the market and detect long-term trends. The filter $y^{(4)}(n) = x(n) - y^{(3)}(n)$ is used to determine if the market is leading or trailing its long-term trend, so it involves both long-term smoothing and short-term fluctuations.

Exercise 7.1 Verify that each of these four examples $y^{(j)}(n)$ defines a time-independent, causal, FIR filter, compute the associated impulse response vector $h^{(j)}(k)$ and determine the number of taps.

7.2 Continuous filters

Our emphasis will be on discrete filters, but we will examine a few basic definitions and concepts in the theory of continuous filters.

A *continuous-time signal* is a function $x(t)$ (real or complex-valued), defined for all t on the real line. Intuitively, we think of $x(t)$ as the signal at time t, say a continuous analog signal. In general, these signals are of arbitrary length. Usually, but not always, we will require that the signals belong to $L_2(\mathbb{R})$, i.e., that they have finite energy: $\int_{-\infty}^{\infty} |x(t)|^2 dt < \infty$. Sometimes we will require that $x \in L_1(\mathbb{R})$.

The *time-shift* operator $\mathbf{T}_a : L_2 \to L_2$ is defined by $\mathbf{T}_x(t) = x(t-a)$. The action of this bounded operator is to delay the signal by the time interval a. Similarly the inverse operator $\mathbf{T}_{-a}x(t) = x(t + a)$ advances the signal by a time units.

A *continuous filter* $\mathbf{\Phi}$ is a bounded linear operator $\mathbf{\Phi} : L_2 \to L_2$ that is time invariant. The filter processes each input x and gives an output $\mathbf{\Phi}x = y$. Time invariance means that $\mathbf{\Phi}(\mathbf{T}_a x)(t) = \mathbf{T}_a y(t)$, whenever $\mathbf{\Phi}x(t) = y(t)$. Thus, the effect of delaying the input by a units of time is just to delay the output by a units. (Another way to put this is $\mathbf{\Phi}\mathbf{T}_a = \mathbf{T}_a\mathbf{\Phi}$, the filter commutes with shifts.)

Suppose that $\mathbf{\Phi}$ takes the form

$$\mathbf{\Phi}x(t) = \int_{-\infty}^{\infty} \Phi(s,t)x(s)ds$$

where $\Phi(s,t)$ is a continuous function in the (s,t) plane with bounded support. Then the time invariance requirement

$$y(t - a) = \int_{-\infty}^{\infty} \Phi(s,t)x(s - a)ds$$

whenever

$$y(t) = \int_{-\infty}^{\infty} \Phi(s,t)x(s)ds,$$

for all $x \in L_2(\mathbb{R})$ and all $a \in \mathbb{R}$ implies $\Phi(s + a, t + a) = \Phi(s,t)$ for all real s, t; hence there is a continuous function h on the real line such that $\Phi(s,t) = h(t - s)$. It follows that $\mathbf{\Phi}$ is a convolution operator, i.e.,

$$\mathbf{\Phi}x(t) = h * x(t) = x * h(t) = \int_{-\infty}^{\infty} h(s)x(t - s)ds.$$

(Note: This characterization of a continuous filter as a convolution can be proved under rather general circumstances. However, h may not be a continuous function. Indeed for the identity operator $\mathbf{\Phi} = \mathbf{I}$, $h(s) = \delta(s)$, the Dirac delta function.)

Finally, we say that a continuous filter is *causal* if it doesn't respond to a signal until the signal is received, i.e., $\mathbf{\Phi}x(t) = 0$ for $t < 0$ if $x(t) = 0$ for $t < 0$. This implies $h(s) = 0$ for $s < 0$. Thus a causal filter is completely determined by the impulse response function $h(s), s \geq 0$ and we have

$$\mathbf{\Phi}x(t) = h * x(t) = x * h(t) = \int_{0}^{\infty} h(s)x(t - s)ds = \int_{-\infty}^{t} h(t - s)x(s)ds.$$

If $x, h \in L_1(\mathbb{R})$ then $y = \mathbf{\Phi}x \in L_1(\mathbb{R})$ and, by the convolution theorem

$$\hat{y}(\lambda) = \hat{h}(\lambda)\hat{x}(\lambda).$$

7.3 Discrete filters in the frequency domain

Let $x \in \ell_2$ be a discrete-time signal.

$$x^{\mathrm{tr}} = (\ldots, x_{-1}, x_0, x_1, x_2, \ldots).$$

Definition 7.4 The discrete-time Fourier transform of **x** is

$$X(\omega) = \sum_{n=-\infty}^{\infty} x_n e^{-in\omega}.$$

Note the change in point of view. The input is the set of coefficients $\{x_n\}$ and the output is the 2π-periodic function $X(\omega) \in L_2[-\pi, \pi]$. We consider $X(\omega)$ as the frequency-domain signal. We can recover the time-domain signal from the frequency-domain signal by integrating:

$$x_n = \frac{1}{2\pi} \int_{-\pi}^{\pi} X(\omega)e^{in\omega}d\omega, \qquad n = 0, \pm 1, \ldots.$$

For discrete-time signals, $x^{(1)}$, $x^{(2)}$, the Parseval identity is

$$(x^{(1)}, x^{(2)}) = \sum_{n=-\infty}^{\infty} x_n^{(1)}\overline{x}_n^{(2)} = \frac{1}{2\pi} \int_{-\infty}^{\infty} X^{(1)}(\omega)\overline{X}^{(2)}(\omega)d\omega.$$

If x belongs to ℓ_1 then the Fourier transform X is a bounded continuous function on $[-\pi, \pi]$.

In addition to the *mathematics* notation $X(\omega)$ for the frequency-domain signal, we shall sometimes use the *signal processing* notation

$$X[e^{i\omega}] = \sum_{n=-\infty}^{\infty} x_n e^{-in\omega}, \tag{7.2}$$

and the *Z transform* notation

$$X[z] = \sum_{n=-\infty}^{\infty} x_n z^{-n}. \tag{7.3}$$

Note that the Z transform is a function of the complex variable z; it differs from the Z transform in Chapter 6 only in the fact that the signal x can now be doubly infinite. It reduces to the signal processing form for $z = e^{i\omega}$. The Fourier transform of the impulse response function h of an FIR filter is a "polynomial" in z^{-1}.

To determine what high frequency and low frequency mean in the context of discrete-time signals, we try a thought experiment. We would expect a constant signal $x_n \equiv 1$ to have zero frequency and it corresponds to $X(\omega) = \delta(\omega - 0)$ where $\delta(\omega)$ is the Dirac delta function, so $\omega = 0$ corresponds to low frequency. (See Section 11.2.1 for a discussion of this "function.") The highest possible degree of oscillation for a discrete-time signal would be $x_n = (-1)^n$, i.e., the signal changes sign in each successive time interval. This corresponds to the time-domain signal $X(\omega) = \delta(\omega - \pi)$. Thus $\pm \pi$, and not 2π, correspond to high frequency.

In analogy with the properties of convolution for the Fourier transform on $\mathbb{R}$ we have

Lemma 7.5 *Let $x^{(1)}$, $x^{(2)}$ be in ℓ_1 with frequency-domain transforms $X^{(1)}(\omega), X^{(2)}(\omega)$, respectively. The frequency-domain transform of the convolution $x^{(1)} * x^{(2)}$ is $X^{(1)}(\omega)X^{(2)}(\omega)$.*

Proof

$$\sum_n x^{(1)} * x^{(2)}(n)e^{-in\omega} = \sum_n \sum_m x_{n-m}^{(1)} x_m^{(2)} e^{-i(n-m)\omega} e^{-im\omega}$$

$$= \sum_m \left(\sum_n x_{n-m}^{(1)} e^{-i(n-m)\omega} \right) x_m^{(2)} e^{-im\omega}$$

$$= \sum_m \left(\sum_n x_n^{(1)} e^{-in\omega} \right) x_m^{(2)} e^{-im\omega} = X^{(1)}(\omega)X^{(2)}(\omega).$$

The interchange of order of summation is justified because the series converge absolutely. □

NOTE: If $x \in \ell_2$ (but not ℓ_1), and h has only a *finite* number of nonzero terms, then the interchange in order of summation is still justified in the above computation, and the transform of $h * x$ is $H(\omega)X(\omega)$.

Let Φ be a digital filter: $\Phi x = h * x = y$. If $h \in \ell_1$ then we have that the action of Φ in the frequency domain is given by

$$H(\omega)X(\omega) = Y(\omega)$$

where $H(\omega)$ is the frequency transform of h. If Φ is a FIR filter then

$$H[z]X[z] = Y[z]$$

where $H[z]$ is a *polynomial* in z^{-1}. Note further that if $\Psi y = k * y$ is another FIR filter then the Z transform of $\Psi \Phi x$ is

$$K[z]H[z]X[z] \tag{7.4}$$

where $K[z]$ is the Z transform of k.

$$\begin{bmatrix} \cdot \\ y_{-1} \\ y_0 \\ y_1 \\ \cdot \end{bmatrix} = \begin{bmatrix} \cdot & \cdot & & & & & \cdot & \cdot \\ \cdot & \frac{1}{2} & \frac{1}{2} & 0 & 0 & \cdot & \cdot \\ \cdot & \cdot & \frac{1}{2} & \frac{1}{2} & 0 & \cdot & \cdot \\ \cdot & \cdot & 0 & \frac{1}{2} & \frac{1}{2} & \cdot & \cdot \\ \cdot & \cdot & & & & & \cdot & \cdot \end{bmatrix} \begin{bmatrix} \cdot \\ x_{-1} \\ x_0 \\ x_1 \\ \cdot \end{bmatrix} .$$

Figure 7.2 Moving average filter action.

One of the practical functions of a filter is to select a band of frequencies to pass, and to reject other frequencies. In the *pass band*, $|H(\omega)|$ is maximal (or very close to its maximum value). We shall frequently normalize the filter so that this maximum value is 1. In the *stop band*, $|H(\omega)|$ is 0, or very close to 0. Mathematically ideal filters can divide the spectrum into pass band and stop band; for realizable, non-ideal, filters there is a transition band where $|H(\omega)|$ changes from near 1 to near 0. A *low pass* filter is a filter whose pass band is a band of frequencies around $\omega = 0$. (Indeed we shall additionally require $H(0) = 1$ and $H(\pi) = 0$ for a low pass filter. Thus, if $\mathbf{H}$ is an FIR low pass filter we have $H(0) = \sum_{n=0}^{N} h_n = 1$.) A *high pass* filter is a filter whose pass band is a range of frequencies around $\omega = \pi$ (and we shall additionally require $|H(\pi)| = 1$ and $H(0) = 0$ for a high pass filter).

Examples 7.6 1. A simple low pass filter (moving average). Here, $\mathbf{\Phi} x = y$ where $y_n = \frac{1}{2}x_n + \frac{1}{2}x_{n-1}$. We have $N = 1$ and the filter coefficients are $h = (h_0, h_1)) = (\frac{1}{2}, \frac{1}{2})$. The frequency response is $H(\omega) = \frac{1}{2} + \frac{1}{2}e^{-i\omega} = |H(\omega)|e^{i\phi(\omega)}$ where

$$|H(\omega)| = \cos\frac{\omega}{2}, \qquad \phi(\omega) = -\frac{\omega}{2}.$$

Note that $|H(\omega)|$ is 1 for $\omega = 0$ and 0 for $\omega = \pi$, so this is a low pass filter. The Z transform is $H[z] = \frac{1}{2} + \frac{1}{2}z^{-1}$. The matrix form of the action in the time domain is given in Figure 7.2. This is a moving average filter action. For future use it will be convenient to renormalize the moving average filter so that the filter coefficients have norm 1. Thus we will often use the filter $\mathbf{C} = \sqrt{2}\mathbf{\Phi}$ with filter coefficients $c = (\frac{1}{\sqrt{2}}, \frac{1}{\sqrt{2}})$.

2. A simple high pass filter (moving difference). $\mathbf{\Phi} x = y$ where $y_n = \frac{1}{2}x_n - \frac{1}{2}x_{n-1}$ and $N = 1$. The filter coefficients are $h = (h_0, h_1) = (\frac{1}{2}, -\frac{1}{2})$. The frequency response is $H(\omega) = \frac{1}{2} - \frac{1}{2}e^{-i\omega} = |H(\omega)|e^{i\phi(\omega)}$

$$
\begin{bmatrix} \vdots \\ y_{-1} \\ y_0 \\ y_1 \\ \vdots \end{bmatrix} = \begin{bmatrix} \cdot & \cdot & & & & \cdot & \cdot \\ \cdot & -\frac{1}{2} & \frac{1}{2} & 0 & 0 & \cdot & \cdot \\ \cdot & \cdot & -\frac{1}{2} & \frac{1}{2} & 0 & \cdot & \cdot \\ \cdot & \cdot & 0 & -\frac{1}{2} & \frac{1}{2} & \cdot & \cdot \\ & & & & \cdot & \cdot \end{bmatrix} \begin{bmatrix} \vdots \\ x_{-1} \\ x_0 \\ x_1 \\ \vdots \end{bmatrix}.
$$

Figure 7.3 Moving difference filter action.

where

$$
|H(\omega)| = \sin\frac{\omega}{2}, \qquad \phi(\omega) = \frac{\pi}{2} - \frac{\omega}{2}.
$$

Note that $|H(\omega)|$ is 1 for $\omega = \pi$ and 0 for $\omega = 0$, so this is a high pass filter. The Z transform is $H[z] = \frac{1}{2} - \frac{1}{2}z^{-1}$. The matrix form of the action in the time domain is given in Figure 7.3. This is the moving difference filter action. In the following we will often renormalize the moving difference filter so that the filter coefficients have norm 1. Thus we have the filter $\mathbf{D} = \sqrt{2}\boldsymbol{\Phi}$ with filter coefficients $d = (\frac{1}{\sqrt{2}}, -\frac{1}{\sqrt{2}})$.

7.4 Other operations on discrete signals

We have already examined the action of a digital filter $\boldsymbol{\Phi}$ in both the time and frequency domains. We will now do the same for some other useful operators. Let $\{x_n : n = 0, \pm 1, \ldots\}$ be a discrete-time signal in ℓ_2 with Z transform $X[z] = \sum_{n=-\infty}^{\infty} x_n z^{-n}$ in the time domain.

- Delay. $\mathbf{R}x_n = x_{n-1}$. In the frequency domain $\mathbf{R} : X[z] \to z^{-1}X[z]$, because

$$
\sum_{n=-\infty}^{\infty} \mathbf{R}x_n z^{-n} = \sum_{n=-\infty}^{\infty} x_{n-1} z^{-n} = \sum_{m=-\infty}^{\infty} x_m z^{-m-1} = z^{-1}X[z].
$$

- Advance. $\mathbf{L}x_n = x_{n+1}$. In the frequency domain $\mathbf{L} : X[z] \to zX[z]$.
- Downsampling. $(\downarrow 2)x_n = x_{2n}$, i.e.,

$$
(\downarrow 2)x = (\ldots, x_{-2}, x_0, x_2, \ldots).
$$

In terms of matrix notation, the action of downsampling in the time domain is given by Figure 7.4. In terms of the Z transform, the

$$
\begin{bmatrix} \cdot \\ \cdot \\ x_{-2} \\ x_0 \\ x_2 \\ \cdot \\ \cdot \end{bmatrix} = \begin{bmatrix} \cdot & \cdot & & & & \cdot & \cdot \\ \cdot & 1 & 0 & 0 & 0 & 0 & \cdot \\ \cdot & 0 & 0 & 1 & 0 & 0 & \cdot \\ \cdot & 0 & 0 & 0 & 0 & 1 & \cdot \\ \cdot & \cdot & & & & \cdot & \cdot \end{bmatrix} \begin{bmatrix} \cdot \\ x_{-2} \\ x_{-1} \\ x_0 \\ x_1 \\ x_2 \\ \cdot \end{bmatrix}. \tag{7.5}
$$

Figure 7.4 Downsampling matrix action.

$$
\begin{bmatrix} \cdot \\ x_{-1} \\ 0 \\ x_0 \\ 0 \\ x_1 \\ \cdot \end{bmatrix} = \begin{bmatrix} \cdot & \cdot & & \cdot & \cdot \\ \cdot & 1 & 0 & 0 & \cdot \\ \cdot & 0 & 0 & 0 & \cdot \\ \cdot & 0 & 1 & 0 & \cdot \\ \cdot & 0 & 0 & 0 & \cdot \\ \cdot & 0 & 0 & 1 & \cdot \\ & \cdot & & \cdot & \cdot \end{bmatrix} \begin{bmatrix} \cdot \\ x_{-1} \\ x_0 \\ x_1 \\ \cdot \end{bmatrix}. \tag{7.6}
$$

Figure 7.5 Upsampling matrix action.

action is

$$
\sum_n (\downarrow 2) x_n z^{-n} = \sum_n x_{2n} z^{-n} = \frac{1}{2} \sum_n x_n (z^{\frac{1}{2}})^{-n} + \frac{1}{2} \sum_n x_n (-z^{\frac{1}{2}})^{-n}
$$

$$
= \frac{1}{2}(X[z^{\frac{1}{2}}] + X[-z^{\frac{1}{2}}]).
$$

• Upsampling.

$$
(\uparrow 2) x_n = \begin{cases} x_{\frac{n}{2}} & n \text{ even} \\ 0 & n \text{ odd} \end{cases},
$$

i.e., $(\uparrow 2)x = (\ldots, x_{-1}, 0, x_0, 0, x_1, 0, \ldots)$. In terms of matrix notation, the action of upsampling in the time domain is given by Figure 7.5. In terms of the Z transform, the action is

$$
\sum_n (\uparrow 2) x_n z^{-n} = \sum_m x_m z^{-2m} = X[z^2].
$$

• Upsampling, then downsampling. $(\downarrow 2)(\uparrow 2)x_n = x_n$, the identity operator. Note that the matrices (7.5), (7.6) give $(\downarrow 2)(\uparrow 2) = I$, the infinite identity matrix. This shows that the upsampling matrix is the right inverse of the downsampling matrix. Furthermore the upsampling matrix is just the transpose of the downsampling matrix: $(\downarrow 2)^{\text{tr}} = (\uparrow 2)$.

- Downsampling, then upsampling. $(\uparrow 2)(\downarrow 2)x_n = \begin{cases} x_n & n \text{ even} \\ 0 & n \text{ odd} \end{cases}$, i.e.,

$$(\uparrow 2)(\downarrow 2)x = (\ldots, x_{-2}), 0, x_0, 0, x_2, 0, \ldots).$$

Note that the matrices (7.6), (7.5), give $(\uparrow 2)(\downarrow 2) \neq I$. This shows that the upsampling matrix is *not* the left inverse of the downsampling matrix. The action in the frequency domain is

$$(\uparrow 2)(\downarrow 2) : X[z] \rightarrow \frac{1}{2}(X[z] + X[-z]).$$

- Flip about $N/2$. The action in the time domain is $\mathbf{F}_{N/2}x_n = x_{N-n}$, i.e., reflect x about $N/2$. If N is even then the point $x_{N/2}$ is fixed. In the frequency domain we have

$$\mathbf{F}_{N/2} : X[z] \rightarrow z^{-N} X[z^{-1}].$$

- Alternate signs. $\mathbf{A}x_n = (-1)^n x_n$ or

$$\mathbf{A}x = (\ldots, -x_{-1}, x_0, -x_1, x_2, \ldots).$$

Here

$$\mathbf{A} : X[z] \rightarrow X[-z].$$

- Alternating flip about $N/2$. The action in the time domain is $\mathbf{A}\mathbf{lF}_{N/2}x_n = (-1)^n x_{N-n}$. In the frequency domain

$$\mathbf{A}\mathbf{lF}_{N/2} : X[z] \rightarrow (-z)^{-N} X[-z^{-1}].$$

- Conjugate alternating flip about $N/2$. The action in the time domain is $\mathbf{CalF}_{N/2}x_n = (-1)^n \overline{x}_{N-n}$. In the frequency domain

$$\mathbf{CalF}_{N/2} : X[z] \rightarrow (-z)^{-N} \overline{X[-\overline{z}^{-1}]}.$$

Note that this is not a linear operator.

7.5 Additional exercises

Exercise 7.2 Suppose the only nonzero components of the input vector x and the impulse response vector h are $x_0 = 1$, $x_1 = 2$ and $h_0 = 1/4$, $h_1 = 1/2$, $h_2 = 1/4$. Compute the outputs $y_n = (x * h)_n$. Verify in the frequency domain that $Y(\omega) = H(\omega)X(\omega)$.

Exercise 7.3 Iterate the averaging filter $\mathbf{H}$ of Exercise 7.2 four times to get $\mathbf{K} = \mathbf{H}^4$. What is $K(\omega)$ and what is the impulse response k_n?

Exercise 7.4 Consider a filter with finite impulse response vector h. The problem is to verify the convolution rule $Y = HX$ in the special case that $h_1 = 1$ and all other $h_n = 0$. Thus

$$h = (\cdots, 0, 0, 0, 1, 0, \cdots).$$

It will be accomplished in the following steps:

1. What is $y = h * x = h * (\ldots, x_{-1}, x_0, x_1, \ldots)$?
2. What is $H(\omega)$?
3. Verify that $\sum y_n e^{-in\omega}$ agrees with $H(\omega)X(\omega)$.

Exercise 7.5

- Write down the infinite matrix $(\downarrow 3)$ that executes the downsampling: $(\downarrow 3)x_n = x_{3n}$.
- Write down the infinite matrix $(\uparrow 3)$ that executes the upsampling:

$$(\uparrow 3)y_n = \begin{cases} y_{\frac{n}{3}} & \text{if 3 divides } n \\ 0 & \text{otherwise} \end{cases}.$$

- Multiply the matrices $(\uparrow 3)(\downarrow 3)$ and $(\downarrow 3)(\uparrow 3)$. Describe the output for each of these product operations.

Exercise 7.6 Show that if $x \in \ell_1$ then $x \in \ell_2$.

Exercise 7.7 Verify the property (7.4) for the Z transform of the composition of two FIR filters.

Exercise 7.8 For each of the operators $\mathbf{R}, \mathbf{L}, (\downarrow 2), (\uparrow 2), \mathbf{A}, \mathbf{F}, \mathbf{AlF}$ determine which of the properties of time invariance, causality and finite impulse response are satisfied.

Exercise 7.9 A direct approach to the convolution rule $Y = HX$. What is the coefficient of z^{-n} in $\left(\sum h_k z^{-k}\right)\left(x_\ell z^{-\ell}\right)$? Show that your answer agrees with $\sum h_k x_{n-k}$.

Exercise 7.10 Let $\mathbf{H}$ be a causal filter with six taps, i.e., only coefficients $h_0, h_1, \ldots, h_5$ can be nonzero. We say that such a filter is **antisymmetric** if the reflection of h about its midpoint (5/2 in this case) takes h to its negative: $h_k = -h_{5-k}$. Can such an antisymmetric filter ever be low pass? Either give an example or show that no example exists.

8

Windowed Fourier and continuous wavelet transforms. Frames

In this chapter we study two different procedures for the analysis of time-dependent signals, locally in both frequency and time. These methods preceded the general discrete wavelet method and we shall see how they led to discrete wavelets. The first procedure, the "windowed Fourier transform" is associated with classical Fourier analysis while the second, is associated with scaling concepts related to discrete wavelets. Both of these procedures yield information about a time signal $f(t)$ that is over-complete. To understand this it is useful to return to our basic paradigm $y = \Phi x$ where x is a signal, Φ is a sample matrix and y is the sample vector. Our problem is to recover the signal x from the samples y. In this chapter $x = f(t)$ and Φ is an integral operator. However, for the moment let us consider the finite-dimensional case where x is an n-tuple, Φ is an $m \times n$ matrix and y is an m-tuple. If $m = n$ and Φ is invertible then we can obtain a unique solution $x = \Phi^{-1}y$. In the case of compressive sampling, however, $m < n$ and the problem is underdetermined. It is no longer possible to obtain x from y in general, without special assumptions on x and Φ. Now suppose $m > n$. The problem is now overdetermined. In this case one can always find m-tuples y for which there is no x such that $y = \Phi x$. However if the rectangular matrix Φ has maximal rank n and if there exists a signal x with $y = \Phi x$ then we can recover x uniquely from y. Indeed $m - n$ of the components of y are redundant and we can recover x from just n components of y that are associated to n linearly independent row vectors of Φ. On the other hand the redundancy of the information leads to $m - n$ consistency checks on the data, which could be of value for numerical computation.

We will see that the windowed Fourier transform and the continuous wavelet transform are overcomplete, i.e., infinite-dimensional analogs of overdetermined systems. Thus, some of the samples produced by these transforms can be discarded and the remaining samples may still suffice to recompute x. We will see that we can often find a countable subset of the continuous data produced by these transforms that will suffice to reproduce the signal. This idea will lead us to windowed frames in the case of the windowed Fourier transform and to discrete wavelets in the case of the continuous wavelet transform.

8.1 The windowed Fourier transform

The windowed Fourier transform approach is designed to overcome the disadvantage that the Fourier transform $\hat{f}(\omega)$ of a signal $f(t)$ requires knowledge of the signal for all times and does not reflect local properties in time. If we think of the Fourier transform operator as an infinite-dimensional analog of a sampling matrix, then each sample is the Fourier transform function at some particular frequency. Each sample gives precise information about the signal in frequency, but no information about its localization in time. We get around this by choosing a "window function" $g \in L_2(\mathbb{R})$ with $||g|| = 1$ and defining the time-frequency translation of g by

$$g^{[x_1,x_2]}(t) = e^{2\pi i t x_2} g(t + x_1). \tag{8.1}$$

Note that $|g(t)|^2$ determines a probability distribution function in t, whereas the Fourier transform $|\tilde{g}(\omega)|^2$ determines a probability distribution in ω. Now suppose g is centered about the point (t_0, ω_0) in phase (time–frequency) space, i.e., suppose

$$\int_{-\infty}^{\infty} t|g(t)|^2 dt = t_0, \quad \int_{-\infty}^{\infty} \omega|\tilde{g}(\omega)|^2 d\omega = \omega_0$$

where $\tilde{g}(\omega) = \int_{-\infty}^{\infty} g(t)e^{-2\pi i \omega t} dt = \hat{g}(2\pi\omega)$ is the Fourier transform of $g(t)$. (Note the change in normalization of the Fourier transform for this chapter only, so that $\int_{-\infty}^{\infty} |\tilde{g}(\omega)|^2 \, d\omega = \int_{-\infty}^{\infty} |g(t)|^2 \, dt$.) Then

$$\int_{-\infty}^{\infty} t|g^{[x_1,x_2]}(t)|^2 dt = t_0 - x_1, \quad \int_{-\infty}^{\infty} \omega|\tilde{g}^{[x_1,x_2]}(\omega)|^2 d\omega = \omega_0 + x_2 \tag{8.2}$$

so $g^{[x_1,x_2]}$ is centered about $(t_0 - x_1, \omega_0 + x_2)$ in phase space. To analyze an arbitrary function $f(t)$ in $L_2(\mathbb{R})$ we compute the inner product

$$F(x_1, x_2) = \langle f, g^{[x_1,x_2]} \rangle = \int_{-\infty}^{\infty} f(t)\overline{g^{[x_1,x_2]}(t)}dt$$

with the idea that $F(x_1, x_2)$ is sampling the behavior of f in a neighborhood of the point $(t_0 - x_1, \omega_0 + x_2)$ in phase space. Intuitively, we think of an ideal window $g(t)$ as a function that is nearly zero, except in a small neighborhood of $t = t_0$, and whose Fourier transform is nearly zero except in a small neighborhood of $\omega = \omega_0$.

As x_1, x_2 range over all real numbers the samples $F(x_1, x_2)$ give us enough information to reconstruct $f(t)$. It is easy to show this directly for functions f such that $f(t)\overline{g}(t - s) \in L_2(\mathbb{R})$ for all s. Indeed let's relate the windowed Fourier transform to the usual Fourier transform of f (rescaled for this chapter):

$$\tilde{f}(\omega) = \int_{-\infty}^{\infty} f(t)e^{-2\pi i \omega t} dt, \qquad f(t) = \int_{-\infty}^{\infty} \tilde{f}(\omega)e^{2\pi i \omega t} d\omega. \qquad (8.3)$$

Thus since

$$F(x_1, x_2) = \int_{-\infty}^{\infty} f(t)\overline{g(t + x_1)}e^{-2\pi i t x_2} dt,$$

we have

$$f(t)\overline{g(t + x_1)} = \int_{-\infty}^{\infty} F(x_1, x_2)e^{2\pi i t x_2} dx_2.$$

Multiplying both sides of this equation by $g(t + x_1)$ and integrating over x_1 we obtain

$$f(t) = \frac{1}{||g||^2} \int_{-\infty}^{\infty} \int_{-\infty}^{\infty} F(x_1, x_2)g(t + x_1)e^{2\pi i t x_2} dx_1 dx_2. \qquad (8.4)$$

This shows us how to recover $f(t)$ from the windowed Fourier transform, if f and g decay sufficiently rapidly at ∞. A deeper but more difficult result is the following theorem.

Theorem 8.1 *The functions $g^{[x_1, x_2]}$ are dense in $L_2(\mathbb{R})$ as $[x_1, x_2]$ runs over $\mathbb{R}^2$.*

We won't give the technical proof of this result but note that it means precisely that if $f \in L_2(\mathbb{R})$ and the inner product $(f, g^{[x_1, x_2]}) = 0$ for all x_1, x_2 then $f \equiv 0$, [85].

We see that a general $f \in L_2(\mathbb{R})$ is uniquely determined by the inner products $(f, g^{[x_1, x_2]})$, $-\infty < x_1, x_2 < \infty$. Indeed, suppose $(f_1, g^{[x_1, x_2]}) = (f_2, g^{[x_1, x_2]})$ for $f_1, f_2 \in L_2(\mathbb{R})$ and all x_1, x_2. Then with $f = f_1 - f_2$ we have $(f, g^{[x_1, x_2]}) \equiv 0$, so f is orthogonal to the closed subspace generated by the $g^{[x_1, x_2]}$. This closed subspace is $L_2(\mathbb{R})$ itself. Hence $f = 0$ and $f_1 = f_2$.

The power of this method is that with proper choice of the window g each of the samples $F(x_1, x_2)$ gives local information about f in a neighborhood of $t = t_0 - x_1, \omega = \omega_0 - x_2$ of phase space. Then the function f can be recovered from the samples. The method is very flexible, since the window function can be chosen as we wish. However, there are difficulties, the first of which is the impossibility of localizing g and $\tilde{g}$ simultaneously in phase space with arbitrary accuracy. Indeed from the Heisenberg inequality, Theorem 4.21, we find

$$\int_{-\infty}^{\infty} (t - t_0)^2 |g(t)|^2 dt \int_{-\infty}^{\infty} (\omega - \omega_0)^2 |\tilde{g}(\omega)|^2 d\omega \geq \frac{1}{16\pi^2}. \tag{8.5}$$

There will always be a trade-off: we can localize in time with arbitrary precision, but at the expense of determining local frequency information, and vice versa.

Another issue, as we shall see, is that the set of basis states $g^{[x_1, x_2]}$ is overcomplete: the coefficients $(f, g^{[x_1, x_2]})$ are not independent of one another, i.e., in general there is no $f \in L_2(\mathbb{R})$ such that $(f, g^{[x_1, x_2]}) = F(x_1, x_2)$ for an arbitrary $F \in L_2(\mathbb{R}^2)$. The $g^{[x_1, x_2]}$ are examples of *coherent states*, continuous overcomplete Hilbert space bases which are of interest in quantum optics, quantum field theory, group representation theory, etc. (The situation is analogous to the case of bandlimited signals. As the Shannon sampling theorem shows, the representation of a bandlimited signal $f(t)$ in the time domain is overcomplete. The discrete samples $f(\frac{n\pi}{\Omega})$ for n running over the integers are enough to determine f uniquely.)

Example 8.2 Consider the case $g = \pi^{-1/4} e^{-t^2/2}$. Here g is essentially its own Fourier transform, so we see that g is centered about $(t_0, \omega_0) = (0, 0)$ in phase space. Thus

$$g^{[x_1, x_2]}(t) = \pi^{-1/4} e^{2\pi i t x_2} e^{-(t + x_1)^2/2}$$

is centered about $(-x_1, x_2)$. This example is known as the *Gabor window*. It is optimal in the sense that the Heisenberg inequality becomes an equality for this case.

Since coherent states are always overcomplete it isn't necessary to compute the inner products $(f, g^{[x_1, x_2]}) = F(x_1, x_2)$ for every point in phase space. In the windowed Fourier approach one typically samples F at the lattice points $(x_1, x_2) = (ma, nb)$ where a, b are fixed positive numbers and m, n range over the integers. Here, a, b and $g(t)$ must be chosen so that the map $f \to \{F(ma, nb)\}$ is 1-1; then f can be recovered from the lattice point values $F(ma, nb)$.

8.1.1 Digging deeper: the lattice Hilbert space

There is a new Hilbert space that we shall find particularly useful in the study of windowed Fourier transforms: the *lattice Hilbert space*. This is the space V' of complex-valued functions $\varphi(x_1, x_2)$ in the plane $\mathbb{R}^2$ that satisfy the periodicity condition

$$\varphi(a_1 + x_1, a_2 + x_2) = e^{-2\pi i a_1 x_2} \varphi(x_1, x_2) \tag{8.6}$$

for $a_1, a_2 = 0, \pm 1, \ldots$ and are square integrable over the unit square:

$$\int_0^1 \int_0^1 |\varphi(x_1, x_2)|^2 dx_1 dx_2 < \infty.$$

The inner product is

$$\langle \varphi_1, \varphi_2 \rangle = \int_0^1 \int_0^1 \varphi_1(x_1, x_2) \overline{\varphi}_2(x_1, x_2) dx_1 dx_2.$$

Note that each function $\varphi(x_1, x_2)$ is uniquely determined by its values in the square $\{(x_1, x_2) : 0 \le x_1, x_2 < 1\}$. It is periodic in x_2 with period 1 and satisfies the "twist" property $\varphi(x_1 + 1, x_2) = e^{-2\pi i x_2} \varphi(x_1, x_2)$.

We can relate this space to $L_2(\mathbb{R})$ via the periodizing operator (Weil–Brezin–Zak transform)

$$\mathbf{P}f(x_1, x_2) = \sum_{n=-\infty}^{\infty} e^{2\pi i n x_2} f(n + x_1) \tag{8.7}$$

which is well defined for any $f \in L_2(\mathbb{R})$ which belongs to the Schwartz space. It is straightforward to verify that $\mathbf{P}f$ satisfies the periodicity condition (8.6), hence $\mathbf{P}f$ belongs to V'. Now

$$\langle \mathbf{P}f(\cdot, \cdot), \mathbf{P}f'(\cdot, \cdot) \rangle$$
$$= \int_0^1 dx_1 \int_0^1 dx_2 \sum_{m,n=-\infty}^{\infty} e^{2\pi i (n-m) x_2} f(n + x_1) \overline{f'(m + x_1)}$$
$$= \int_0^1 dx_1 \sum_{n=-\infty}^{\infty} f(n + x_1) \overline{f'(n + x_1)} = \int_{-\infty}^{\infty} f(t_1) \overline{f'(t)} \, dt$$
$$= (f, f')$$

so $\mathbf{P}$ can be extended to an inner product preserving mapping of $L_2(\mathbb{R})$ into V.

It is clear from the transform that if $\varphi(x_1, x_2) = \mathbf{P}f(x_1, x_2)$ then we recover $f(x_1)$ by integrating with respect to x_2 : $f(x_1) = \int_0^1 \varphi(x_1, y) dy$.

Thus we define the mapping $\mathbf{P}^*$ of V' into $L_2(\mathbb{R})$ by

$$\mathbf{P}^*\varphi(t) = \int_0^1 \varphi(t,y)dy, \quad \varphi \in V'. \tag{8.8}$$

Since $\varphi \in V'$ we have

$$\mathbf{P}^*\varphi(t+a) = \int_0^1 \varphi(t,y)e^{-2\pi iay}dy = \hat{\varphi}_{-a}(t)$$

for a an integer. (Here $\hat{\varphi}_n(t)$ is the nth Fourier coefficient of $\varphi(t,y)$.)
The Parseval formula then yields

$$\int_0^1 |\varphi(t,y)|^2 dy = \sum_{a=-\infty}^{\infty} |\mathbf{P}^*\varphi(t+a)|^2$$

so

$$\langle \varphi, \varphi \rangle = \int_0^1 \int_0^1 |\varphi(t,y)|^2 dt dy = \int_0^1 \sum_{a=-\infty}^{\infty} |\mathbf{P}^*\varphi(t+a)|^2 dt$$

$$= \int_{-\infty}^{\infty} |\mathbf{P}^*\varphi(t)|^2 dt = (\mathbf{P}^*\varphi, \mathbf{P}^*\varphi)$$

and $\mathbf{P}^*$ is an inner product preserving mapping of V' into $L_2(\mathbb{R})$. More-over, it is easy to verify that

$$\langle \mathbf{P}g, \varphi \rangle = (g, \mathbf{P}^*\varphi)$$

for $g \in L_2(\mathbb{R})$, $\varphi \in V'$, i.e., $\mathbf{P}^*$ is the adjoint of $\mathbf{P}$. Since $\mathbf{P}^*\mathbf{P} = \mathbf{I}$ on $L_2(\mathbb{R})$ it follows that $\mathbf{P}$ is a unitary operator mapping $L_2(\mathbb{R})$ **onto** V' and $\mathbf{P}^* = \mathbf{P}^{-1}$ is a unitary operator mapping V' **onto** $L_2(\mathbb{R})$.

Exercise 8.1 Show directly that $\mathbf{P}\mathbf{P}^* = \mathbf{I}$ on V'.

8.1.2 Digging deeper: more on the Zak transform

The Weil–Brezin transform (earlier used in radar theory by Zak, so also called the Zak transform) is very useful in studying the lattice sampling problem for $(f, g^{[x_1, x_2]})$, at the points $(x_1, x_2) = (ma, nb)$ where a, b are fixed positive numbers and m, n range over the integers. Restricting to the case $a = b = 1$ for the time being, we let $f \in L_2(\mathbb{R})$. Then

$$f_{\mathbf{P}}(x_1, x_2) = \mathbf{P}f(x_1, x_2) = \sum_{k=-\infty}^{\infty} e^{2\pi ikx_2} f(x_1 + k) \tag{8.9}$$

satisfies

$$f_{\mathbf{P}}(k_1 + x_1, k_2 + x_2) = e^{-2\pi i k_1 x_2} f_{\mathbf{P}}(x_1, x_2)$$

for integers k_1, k_2. (Here (8.9) is meaningful if f belongs to, say, the Schwartz class. Otherwise $\mathbf{P}f = \lim_{n \to s} \mathbf{P}f_n$ where $f = \lim_{n \to s} f_n$ and the f_n are Schwartz class functions. The limit is taken with respect to the Hilbert space norm.) If $f = g^{[m,n]}(t) = e^{2\pi i t n} g(t + m)$ we have

$$g_{\mathbf{P}}^{[m,n]}(x_1, x_2) = e^{2\pi i (n x_1 - m x_2)} g_{\mathbf{P}}(x_1, x_2).$$

Thus in the lattice Hilbert space, the functions $g_{\mathbf{P}}^{[m,n]}$ differ from $g_{\mathbf{P}}$ simply by the multiplicative factor $e^{2\pi i (n x_1 - m x_2)} = E_{n,m}(x_1, x_2)$, and as n, m range over the integers the $E_{n,m}$ form an ON basis for the lattice Hilbert space:

$$\int_0^1 \int_0^1 E_{n_1,m_1}(x_1, x_2) \overline{E_{n_2,m_2}(x_1, x_2)} dx_1 dx_2 = \delta_{n_1,n_2} \delta_{m_1,m_2}.$$

Definition 8.3 Let $f(t)$ be a function defined on the real line and let $\Xi(t)$ be the characteristic function of the set on which f vanishes:

$$\Xi(t) = \begin{cases} 1 & \text{if } f(t) = 0 \\ 0 & \text{if } f(t) \neq 0. \end{cases}$$

We say that f is nonzero **almost everywhere** (a.e.) if the L_2 norm of Ξ is 0, i.e., $||\Xi|| = 0$.

Thus f is nonzero a.e. provided the Lebesgue integral $\int_{-\infty}^{\infty} |\Xi(t)|^2 dt = 0$, so Ξ belongs to the equivalence class of the zero function. If the support of Ξ is contained in a countable set it will certainly have norm zero; it is also possible that the support is a noncountable set (such as the Cantor set). We will not go into these details here.

Theorem 8.4 *For* $(a, b) = (1, 1)$ *and* $g \in L_2(\mathbb{R})$ *the transforms* $\{g^{[m,n]} : m, n = 0 \pm 1, \pm 2, \ldots\}$ *span* $L_2(\mathbb{R})$ *iff* $\mathbf{P}g(x_1, x_2) = g_{\mathbf{P}}(x_1, x_2) \neq 0$ *a.e.*

Proof Let $\mathcal{M}$ be the closed linear subspace of $L_2(\mathbb{R})$ spanned by $\{g^{[m,n]}\}$. Clearly $\mathcal{M} = L_2(\mathbb{R})$ iff the only solution of $(f, g^{[m,n]}) = 0$ for all integers m and n is $f = 0$ a.e. Applying the Weyl–Brezin–Zak isomorphism $\mathbf{P}$ we have

$$(f, g^{[m,n]}) = \; < \mathbf{P}f, E_{n,m} \mathbf{P}g > \qquad (8.10)$$
$$= \; < [\mathbf{P}f][\overline{\mathbf{P}g}], E_{n,m} > = \; < f_{\mathbf{P}} \bar{g}_{\mathbf{P}}, E_{n,m} > .$$

Since the functions $E_{n,m}$ form an ON basis for the lattice Hilbert space it follows that $(f, g^{[m,n]}) = 0$ for all integers m, n iff $f_{\mathbf{P}}(x_1, x_2)g_{\mathbf{P}}(x_1, x_2) = 0$, a.e. If this product vanishes a.e. but $g_{\mathbf{P}} \neq 0$, a.e. then $f_{\mathbf{P}} = 0$ a.e. so $f \equiv 0$ and $\mathcal{M} = L_2(\mathbb{R})$. If $g_{\mathbf{P}} = 0$ on a set S of positive measure on the unit square, and we choose f such that the characteristic function $\chi_S = \mathbf{P}f = f_{\mathbf{P}}$ then $f_{\mathbf{P}}g_{\mathbf{P}} = \chi_S g_{\mathbf{P}} = 0$ a.e., hence $(f, g^{[m,n]}) = 0$ and $\mathcal{M} \neq L_2(\mathbb{R})$. $\qquad\square$

Example 8.5 In the case $g(t) = \pi^{-1/4}e^{-t^2/2}$ one finds that

$$g_{\mathbf{P}}(x_1, x_2) = \pi^{-1/4} \sum_{k=-\infty}^{\infty} e^{2\pi i k x_2 - (x_1+k)^2/2}.$$

This series defines a Jacobi Theta function. Using complex variable techniques it can be shown that this function vanishes at the single point $(1/2, 1/2)$ in the square $0 \leq x_1 < 1$, $0 \leq x_2 < 1$, [100]. Thus $g_{\mathbf{P}} \neq 0$ a.e. and the functions $\{g^{[m,n]}\}$ span $L_2(\mathbb{R})$. However, the expansion of an $L_2(\mathbb{R})$ function in terms of this set is not unique.

Corollary 8.6 *For $(a, b) = (1, 1)$ and $g \in L_2(\mathbb{R})$ the transforms $\{g^{[m,n]} : m, n = 0, \pm 1, \ldots\}$ form an ON basis for $L_2(\mathbb{R})$ iff $|g_{\mathbf{P}}(x_1, x_2)| = 1$, a.e.*

Proof We have

$$\delta_{mm'}\delta_{nn'} = (g^{[m,n]}, g^{[m',n']}) = < E_{n,m}g_{\mathbf{P}}, E_{n',m'}g_{\mathbf{P}} > \qquad (8.11)$$
$$= < |g_{\mathbf{P}}|^2, E_{n'-n,m'-m} > \qquad (8.12)$$

iff $|g_{\mathbf{P}}|^2 = 1$, a.e. $\qquad\square$

Example 8.7 Let $g = \phi(t)$, the Haar scaling function, (9.1). It is easy to see that $|g_{\mathbf{P}}(x_1, x_2)| \equiv 1$. Thus $\{g^{[m,n]}\}$ is an ON basis for $L_2(\mathbb{R})$.

Theorem 8.8 *For $(a, b) = (1, 1)$ and $g \in L_2(\mathbb{R})$, suppose there are constants A, B such that*

$$0 < A \leq |g_{\mathbf{P}}(x_1, x_2)|^2 \leq B < \infty$$

*almost everywhere in the square $0 \leq x_1, x_2 < 1$. Then $\{g^{[m,n]}\}$ is a basis for $L_2(\mathbb{R})$, i.e., each $f \in L_2(\mathbb{R})$ can be expanded **uniquely** in the form $f = \sum_{m,n} a_{mn}g^{[m,n]}$. Indeed,*

$$a_{mn} = \left\langle f_{\mathbf{P}}, \frac{g_{\mathbf{P}}^{[m,n]}}{|g_{\mathbf{P}}|^2} \right\rangle = \left\langle \frac{f_{\mathbf{P}}}{g_{\mathbf{P}}}, E_{n,m} \right\rangle$$

Proof By hypothesis $|g_{\mathbf{P}}|^{-1}$ is a bounded nonvanishing function on the domain $0 \leq x_1, x_2 < 1$. Hence $f_{\mathbf{P}}/g_{\mathbf{P}}$ is square integrable on this domain and, from the periodicity properties of elements in the lattice Hilbert space, $\frac{f_{\mathbf{P}}}{g_{\mathbf{P}}}(x_1 + n, x_2 + m) = \frac{f_{\mathbf{P}}}{g_{\mathbf{P}}}(x_1, x_2)$. It follows that

$$\frac{f_{\mathbf{P}}}{g_{\mathbf{P}}} = \sum a_{mn} E_{n,m}$$

where $a_{mn} = (f_{\mathbf{P}}/g_{\mathbf{P}}, E_{n,m})$, so $f_{\mathbf{P}} = \sum a_{mn} E_{n,m} g_{\mathbf{P}}$. This last expression implies $f = \sum a_{mn} g^{[m,n]}$. Conversely, given $f = \sum a_{mn} g^{[m,n]}$ we can reverse the steps in the preceding argument to obtain $a_{mn} = (f_{\mathbf{P}}/g_{\mathbf{P}}, E_{n,m})$. $\square$

Note that, in general, the basis $\{g^{[m,n]}\}$ will not be ON.

8.1.3 Digging deeper: windowed transforms

The expansion $f = \sum a_{mn} g^{[m,n]}$ is equivalent to the lattice Hilbert space expansion $f_{\mathbf{P}} = \sum a_{mn} E_{n,m} g_{\mathbf{P}}$ or

$$f_{\mathbf{P}} \bar{g}_{\mathbf{P}} = \sum (a_{mn} E_{n,m}) |g_{\mathbf{P}}|^2. \tag{8.13}$$

Now if $g_{\mathbf{P}}$ is a bounded function then $f_{\mathbf{P}} \bar{g}_{\mathbf{P}}(x_1, x_2)$ and $|g_{\mathbf{P}}|^2$ both belong to the lattice Hilbert space and are periodic functions in x_1 and x_2 with period 1. Hence,

$$f_{\mathbf{P}} \bar{g}_{\mathbf{P}} = \sum b_{mn} E_{n,m} \tag{8.14}$$

$$|g_{\mathbf{P}}|^2 = \sum c_{mn} E_{n,m} \tag{8.15}$$

with

$$b_{mn} = < f_{\mathbf{P}} \bar{g}_{\mathbf{P}}, E_{n,m} > = < f_{\mathbf{P}}, g_{\mathbf{P}} E_{n,m} > = (f, g^{[m,n]}), \tag{8.16}$$

$$c_{mn} = < g_{\mathbf{P}} \bar{g}_{\mathbf{P}}, E_{n,m} > = (g, g^{[m,n]}). \tag{8.17}$$

This gives the Fourier series expansion for $f_{\mathbf{P}} \bar{g}_{\mathbf{P}}$ as the product of two other Fourier series expansions. (We consider the functions f, g, hence $f_{\mathbf{P}}, g_{\mathbf{P}}$ as known.) If $|g_{\mathbf{P}}|^2$ never vanishes we can solve for the a_{mn} directly:

$$\sum a_{mn} E_{n,m} = (\sum b_{mn} E_{n,m})(\sum c'_{mn} E_{n,m}),$$

where the c'_{mn} are the Fourier coefficients of $|g_{\mathbf{P}}|^{-2}$. However, if $|g_{\mathbf{P}}|^2$ vanishes at some point then the best we can do is obtain the convolution equations $b = a * c$, i.e.,

$$b_{mn} = \sum_{k+k'=m, \ell+\ell'=n} a_{k\ell} c'_{k'\ell'}.$$

(We can approximate the coefficients $a_{k\ell}$ even in the cases where $|g_{\mathbf{P}}|^2$ vanishes at some points. The basic idea is to truncate $\sum a_{mn} E_{n,m}$ to a finite number of nonzero terms and to sample equation (8.13), making sure that $|g_{\mathbf{P}}|(x_1, x_2)$ is nonzero at each sample point. The a_{mn} can then be computed by using the inverse finite Fourier transform.)

The problem of $|g_{\mathbf{P}}|$ vanishing at a point is not confined to an isolated example. Indeed it can be shown that if $g_{\mathbf{P}}$ is an everywhere continuous function in the lattice Hilbert space then it must vanish at one point at least.

8.2 Bases and frames, windowed frames

8.2.1 Frames

To understand the nature of the complete sets $\{g^{[m,n]}\}$ it is useful to broaden our perspective and introduce the idea of a **frame** in an arbitrary Hilbert space $\mathcal{H}$. From this more general point of view we are given a sequence $\{f^{(j)}\}$ of vectors of $\mathcal{H}$ and we want to find conditions on $\{f^{(j)}\}$ so that we can recover an arbitrary $f \in \mathcal{H}$ from the sample inner products $(f, f^{(j)})$ on $\mathcal{H}$. (Note that this fits our paradigm $y = \Phi x$, where now the signal is $x = f$. The jth component of the sample vector y is $y_j = (f, f^{(j)})$ and the action of Φ on x is given by the inner products.)

Let ℓ_2 be the Hilbert space of countable sequences $\{\xi_j\}$ with inner product $< \xi, \eta > = \sum_j \xi_j \bar{\eta}_j$. (A sequence $\{\xi_j\}$ belongs to ℓ_2 provided $\sum_j \xi_j \bar{\xi}_j < \infty$.) Now let $\mathbf{T} : \mathcal{H} \to \ell_2$ be the linear mapping defined by

$$(\mathbf{T}f)_j = (f, f^{(j)}).$$

We require that $\mathbf{T}$ is a bounded operator from $\mathcal{H}$ to ℓ_2, i.e., that there is a finite $B > 0$ such that $\sum_j |(f, f^{(j)})|^2 \leq B||f||^2$. In order to recover f from the samples $(f, f^{(j)})$ we want $\mathbf{T}$ to be invertible with $\mathbf{T}^{-1} : \mathcal{R}_{\mathbf{T}} \to \mathcal{H}$ where $\mathcal{R}_{\mathbf{T}}$ is the range $\mathbf{T}\mathcal{H}$ of $\mathbf{T}$ in ℓ_2. Moreover, for numerical stability in the computation of f from the samples we want $\mathbf{T}^{-1}$ to be bounded. (In other words we want to require that a "small" change in the sample data $(f, f^{(j)})$ leads to a "small" change in f.) This means that there is a finite $A > 0$ such that $\sum_j |(f, f^{(j)})|^2 \geq A||f||^2$. (Note that $\mathbf{T}^{-1}\xi = f$ if $\xi_j = (f, f^{(j)})$.) If these conditions are satisfied, i.e., if there exist positive constants A, B such that

$$A||f||^2 \leq \sum_j |(f, f^{(j)})|^2 \leq B||f||^2 \qquad (8.18)$$

for all $f \in \mathcal{H}$, we say that the sequence $\{f^{(j)}\}$ is a *frame* for $\mathcal{H}$ and that A and B are *frame bounds*. In general, a frame gives completeness (so that we can recover f from the samples), but also redundancy. There are more terms than the minimal needed to determine f. However, if the set $\{f^{(j)}\}$ is linearly independent, then it forms a basis, called a *Riesz basis*, and there is no redundancy.

The *adjoint* $\mathbf{T}^*$ of $\mathbf{T}$ is the linear mapping $\mathbf{T}^* : \ell_2 \to \mathcal{H}$ defined by

$$(\mathbf{T}^*\xi, f) = <\xi, \mathbf{T}f>$$

for all $\xi \in \ell_2$, $f \in \mathcal{H}$. A simple computation yields

$$\mathbf{T}^*\xi = \sum_j \xi_j f^{(j)}. \qquad (8.19)$$

(Since $\mathbf{T}$ is bounded, so is $\mathbf{T}^*$ and the right-hand side is well-defined for all $\xi \in \ell_2$.)

Now the bounded self-adjoint operator $\mathbf{S} = \mathbf{T}^*\mathbf{T} : \mathcal{H} \to \mathcal{H}$ is given by

$$\mathbf{S}f = \mathbf{T}^*\mathbf{T}f = \sum_j (f, f^{(j)})f^{(j)}, \qquad (8.20)$$

and we can rewrite the defining inequality for the frame as

$$A||f||^2 \leq (\mathbf{T}^*\mathbf{T}f, f) \leq B||f||^2.$$

Since $A > 0$, if $\mathbf{T}^*\mathbf{T}f = \Theta$ then $f = \Theta$, so $\mathbf{S}$ is 1-1, hence invertible. Furthermore, the range $\mathbf{S}\mathcal{H}$ of $\mathbf{S}$ is $\mathcal{H}$. Indeed, if $\mathbf{S}\mathcal{H}$ is a proper subspace of $\mathcal{H}$ then we can find a nonzero vector g in $(\mathbf{S}\mathcal{H})^\perp : (\mathbf{S}f, g) = 0$ for all $f \in \mathcal{H}$. However, $(\mathbf{S}f, g) = (\mathbf{T}^*\mathbf{T}f, g) = <\mathbf{T}f, \mathbf{T}g> = \sum_j (f, f^{(j)})(f^{(j)}, g)$. Setting $f = g$ we obtain

$$\sum_j |(g, f^{(j)})|^2 = 0.$$

But then we have $g = \Theta$, a contradiction. Thus $\mathbf{S}\mathcal{H} = \mathcal{H}$ and the inverse operator $\mathbf{S}^{-1}$ exists and has domain $\mathcal{H}$.

Since $\mathbf{S}\mathbf{S}^{-1}f = \mathbf{S}^{-1}\mathbf{S}f = f$ for all $f \in \mathcal{H}$, we immediately obtain two expansions for f from (8.20):

$$(a) \ f = \sum_j (\mathbf{S}^{-1}f, f^{(j)})f^{(j)} = \sum_j (f, \mathbf{S}^{-1}f^{(j)})f^{(j)}$$

$$(b) \quad f = \sum_j (f, f^{(j)}) \mathbf{S}^{-1} f^{(j)}.$$

(The second equality in the first expression follows from the identity $(\mathbf{S}^{-1}f, f^{(j)}) = (f, \mathbf{S}^{-1}f^{(j)})$, which holds since $\mathbf{S}^{-1}$ is self-adjoint.)

Recall, that for a *positive* operator $\mathbf{S}$, i.e., an operator such that $(\mathbf{S}f, f) \geq 0$ for all $f \in \mathcal{H}$ the inequalities

$$A\|f\|^2 \leq (\mathbf{S}f, f) \leq B\|f\|^2$$

for $A, B > 0$ are equivalent to the inequalities

$$A\|f\| \leq \|\mathbf{S}f\| \leq B\|f\|.$$

This suggests that if the $\{f^{(j)}\}$ form a frame then so do the $\{\mathbf{S}^{-1}f^{(j)}\}$.

Theorem 8.9 *Suppose* $\{f^{(j)}\}$ *is a frame with frame bounds* A, B *and let* $\mathbf{S} = \mathbf{T}^*\mathbf{T}$. *Then* $\{\mathbf{S}^{-1}f^{(j)}\}$ *is also a frame, called the dual frame of* $\{f^{(j)}\}$, *with frame bounds* B^{-1}, A^{-1}.

Proof Setting $f = \mathbf{S}^{-1}g$ we have $B^{-1}\|g\| \leq \|\mathbf{S}^{-1}g\| \leq A^{-1}\|g\|$. Since $\mathbf{S}^{-1}$ is self-adjoint, this implies $B^{-1}\|g\|^2 \leq (\mathbf{S}^{-1}g, g) \leq A^{-1}\|g\|^2$. Then we have $\mathbf{S}^{-1}g = \sum_j (\mathbf{S}^{-1}g, f^{(j)})\mathbf{S}^{-1}f^{(j)}$ so

$$(\mathbf{S}^{-1}g, g) = \sum_j (\mathbf{S}^{-1}g, f^{(j)})(\mathbf{S}^{-1}f^{(j)}, g) = \sum_j |(g, \mathbf{S}^{-1}f^{(j)})|^2.$$

Hence $\{\mathbf{S}^{-1}f^{(j)}\}$ is a frame with frame bounds B^{-1}, A^{-1}. $\qquad\square$

We say that $\{f^{(j)}\}$ is a *tight frame* if $A = B$.

Corollary 8.10 *If* $\{f^{(j)}\}$ *is a tight frame then every* $f \in \mathcal{H}$ *can be expanded in the form*

$$f = A^{-1} \sum_j (f, f^{(j)})f^{(j)}.$$

Proof Since $\{f^{(j)}\}$ is a tight frame we have $A\|f\|^2 = (\mathbf{S}f, f)$ or $((\mathbf{S} - A\mathbf{E})f, f) = 0$ where $\mathbf{E}$ is the identity operator $\mathbf{E}f = f$. Since $\mathbf{S} - A\mathbf{E}$ is a self-adjoint operator we have $\|(\mathbf{S} - A\mathbf{E})f\| = 0$ for all $f \in \mathcal{H}$. Thus $\mathbf{S} = A\mathbf{E}$. However, from (8.20), $\mathbf{S}f = \sum_j (f, f^{(j)})f^{(j)}$. $\qquad\square$

Consider the special case where $\mathcal{H}$ is finite dimensional, $\mathcal{H} = \mathcal{V}_N$, an N-dimensional complex inner product space. Then all of these constructions can be understood in terms of matrices. Let $\{e^{(1)}, \ldots, e^{(N)}\}$ be an ON basis for $\mathcal{V}_N$, so that each vector $f \in \mathcal{V}_N$ can be expressed uniquely in terms of this basis by the N-tuple ϕ where $f = \sum_{i=1}^N \phi_i e^{(i)}$.

Let $\{f^{(1)}, \ldots, f^{(M)}\}$ be a frame for $\mathcal{V}_N$. We can represent the frame by the $N \times M$ matrix Ω where

$$f^{(j)} = \sum_{i=1}^{N} \Omega_{ij} e^{(i)}. \tag{8.21}$$

In terms of these bases the operator $\mathbf{T}$ is represented by the $M \times N$ matrix $\bar{\Omega}^{\mathrm{tr}} = \Omega^*$ so that $\mathbf{T}f = \xi$ becomes $\Omega^* \phi = \xi$. The operator $\mathbf{T}^*$ is represented by the $M \times N$ matrix Ω. The condition that $\{f^{(j)}\}$ be a frame is that there exist real numbers $0 < A \le B$ such that

$$A\bar{\phi}^{\mathrm{tr}} \phi \le \bar{\phi}^{\mathrm{tr}} \Omega \Omega^* \phi \le B\bar{\phi}^{\mathrm{tr}} \phi,$$

for all N-tuples ϕ. The operator $\mathbf{S}$ is represented by the $N \times N$ positive symmetric matrix $\Omega\Omega^*$. From the spectral theorem for self-adjoint matrices we see that the frame condition says exactly that the ordered eigenvalues λ_i of $\Omega\Omega^*$ satisfy

$$A \le \lambda_N \le \lambda_{N-1} \le \cdots \le \lambda_1 \le B.$$

Since the eigenvalues of $\Omega\Omega^*$ are all nonnegative, we see that the frame requirement is that the smallest eigenvalue λ_N is strictly positive. Thus $\Omega\Omega^*$ is an invertible matrix. If we have a frame then the frame vectors will span the space but in general will not be linearly independent. There will be N element spanning subsets of the frame that will form a basis, but no unique way to determine these. Thus expansions of vectors in terms of the frame will not necessarily be unique. The condition for a tight frame in a finite-dimensional inner product space is that $A = B$, i.e.,

$$A = \lambda_1 = \lambda_2 = \cdots = \lambda_N = B.$$

This means that $\Omega\Omega^* = AI_N$ where I_N is the $N \times N$ identity matrix. Thus we have $\phi = \frac{1}{A}\Omega\Omega^*\phi$ for any M-tuple ϕ, which is equivalent to the vector expansion $f = \frac{1}{A}\sum_{j=1}^{M}(f, f^{(j)})f^{(j)}$.

Exercise 8.2 Using the results of Exercise 4.15, show that the Shannon–Whittaker sampling procedure can be interpreted to define a tight frame. What is the frame bound?

8.2.2 Riesz bases

In this section we will investigate the conditions that a frame must satisfy in order for it to define a Riesz basis, i.e., in order that the set $\{f^{(j)}\}$

be linearly independent. Crucial to this question is the adjoint operator. Recall that the **adjoint** $\mathbf{T}^*$ of $\mathbf{T}$ is the linear mapping $\mathbf{T}^* : \ell_2 \to \mathcal{H}$ defined by

$$(\mathbf{T}^*\xi, f) = <\xi, \mathbf{T}f> \tag{8.22}$$

for all $\xi \in \ell_2$, $f \in \mathcal{H}$, so that

$$\mathbf{T}^*\xi = \sum_j \xi_j f^{(j)}.$$

Since $\mathbf{T}$ is bounded, it is a straightforward exercise in functional analysis to show that $\mathbf{T}^*$ is also bounded and that if $||\mathbf{T}||^2 = B$ then $||\mathbf{T}||^2 = ||\mathbf{T}^*||^2 = ||\mathbf{T}\mathbf{T}^*|| = ||\mathbf{T}^*\mathbf{T}|| = B$.

Exercise 8.3 Using the definition (8.22) of $\mathbf{T}^*$ show that if $||\mathbf{T}|| \leq K$ for some constant K then $||\mathbf{T}^*|| \leq K$. Hint: Choose $f = \mathbf{T}^*\xi$.

Furthermore, we know that $\mathbf{T}$ is invertible, as is $\mathbf{T}^*\mathbf{T}$, and if $||\mathbf{T}^{-1}||^2 = A^{-1}$ then $||(\mathbf{T}^*\mathbf{T})^{-1}|| = A^{-1}$. However, this doesn't necessarily imply that $\mathbf{T}^*$ is invertible (though it is invertible when restricted to the range of $\mathbf{T}$). $\mathbf{T}^*$ will fail to be invertible if there is a nonzero $\xi \in \ell_2$ such that $\mathbf{T}^*\xi = \sum_j \xi_j f^{(j)} = 0$. This can happen if and only if the set $\{f^{(j)}\}$ is linearly dependent. If $\mathbf{T}^*$ is invertible, then it follows easily that the inverse is bounded and $||(\mathbf{T}^*)^{-1}||^2 = A^{-1}$.

The key to all of this is the operator $\mathbf{T}\mathbf{T}^* : \ell_2 \to \mathcal{H}$ with the action

$$(\mathbf{T}\mathbf{T}^*\xi)_n = \sum_m \xi_m (f^{(m)}, f^{(n)}), \qquad \xi \in \ell_2.$$

Then matrix elements of the infinite matrix corresponding to the operator $\mathbf{T}\mathbf{T}^*$ are the inner products $(\mathbf{T}\mathbf{T}^*)_{nm} = \overline{(f^{(n)}, f^{(m)})}$. This is a self-adjoint matrix. If its eigenvalues are all positive and bounded away from zero, with lower bound $A > 0$, then it follows that $\mathbf{T}^*$ is invertible and $||\mathbf{T}^{*-1}||^2 = A^{-1}$. In this case the $f^{(j)}$ form a Riesz basis with frame bounds A, B.

We will return to this issue when we study biorthogonal wavelets.

8.2.3 Digging deeper: frames of W–H type

We can now relate frames to the lattice Hilbert space construction.

Theorem 8.11 *For* $(a, b) = (1, 1)$ *and* $g \in L_2(\mathbb{R})$, *we have*

$$0 < A \leq |g_{\mathbf{P}}(x_1, x_2)|^2 \leq B < \infty \tag{8.23}$$

almost everywhere in the square $0 \leq x_1, x_2 < 1$ *iff* $\{g^{[m,n]}\}$ *is a frame for* $L_2(\mathbb{R})$ *with frame bounds* A, B. *(By Theorem 8.8 this frame is actually a Riesz basis for* $L_2(\mathbb{R})$.*)*

Proof If (8.23) holds then $g_{\mathbf{P}}$ is a bounded function on the square. Hence for any $f \in L_2(\mathbb{R})$, $f_{\mathbf{P}}\bar{g}_{\mathbf{P}}$ is a periodic function, in x_1, x_2 on the square. Thus

$$\sum_{m,n=-\infty}^{\infty} |\langle f, g^{[m,n]}\rangle|^2 = \sum_{m,n=-\infty}^{\infty} |(f_{\mathbf{P}}, E_{n,m}g_{\mathbf{P}})|^2$$

$$= \sum_{m,n=-\infty}^{\infty} |(f_{\mathbf{P}}\bar{g}_{\mathbf{P}}, E_{n,m})|^2 = ||f_{\mathbf{P}}\bar{g}_{\mathbf{P}}||^2 \tag{8.24}$$

$$= \int_0^1 \int_0^1 |f_{\mathbf{P}}|^2 |g_{\mathbf{P}}|^2 dx_1 dx_2.$$

(Here we have used the Plancherel theorem for the exponentials $E_{n,m}$.) It follows from (8.23) that

$$A||f||^2 \leq \sum_{m,n=-\infty}^{\infty} |(f, g^{[m,n]})|^2 \leq B||f||^2, \tag{8.25}$$

so $\{g^{[m,n]}\}$ is a frame.

Conversely, if $\{g^{[m,n]}\}$ is a frame with frame bounds A, B, it follows from (8.25) and the computation (8.24) that

$$A||f_{\mathbf{P}}||^2 \leq \int_0^1 \int_0^1 |f_{\mathbf{P}}|^2 |g_{\mathbf{P}}|^2 dx_1 dx_2 \leq B||f_{\mathbf{P}}||^2$$

for an **arbitrary** $f_{\mathbf{P}}$ in the lattice Hilbert space. (Here we have used the fact that $||f|| = ||f_{\mathbf{P}}||$, since $\mathbf{P}$ is a unitary transformation.) Thus the inequalities (8.23) hold almost everywhere. □

Frames of the form $\{g^{[ma,nb]}\}$ are called *Weyl–Heisenberg* (or *W–H frames*). The Weyl–Brezin–Zak transform is not so useful for the study of W–H frames with general frame parameters (a, b). (Note that it is only the product ab that is of significance for the W–H frame parameters. Indeed, the change of variable $t' = t/a$ in (8.1) converts the frame parameters (a, b) to $(a', b') = (1, ab)$.) An easy consequence of the general definition of frames is the following:

Theorem 8.12 *Let* $g \in L_2(\mathbb{R})$ *and* $a, b, A, B > 0$ *such that*

1. $0 < A \leq \sum_m |g(x + ma)|^2 \leq B < \infty$, *a.e.,*

2. *g has support contained in an interval I where I has length b^{-1}.*

Then the $\{g^{[ma,nb]}\}$ are a W–H frame for $L_2(\mathbb{R})$ with frame bounds $b^{-1}A, b^{-1}B$.

Proof For fixed m and arbitrary $f \in L_2(\mathbb{R})$ the function $F_m(t) = f(t)\overline{g(t+ma)}$ has support in the interval $I_m = \{t + ma : x \in I\}$ of length b^{-1}. Thus $F_m(t)$ can be expanded in a Fourier series with respect to the basis exponentials $E_{nb}(t) = e^{2\pi i bnt}$ on I_m. Using the Plancherel formula for this expansion we have

$$\sum_{m,n} |\langle f, g^{[ma,nb]}\rangle|^2 = \sum_{m,n} |\langle F_m, E_{nb}\rangle|^2 \qquad (8.26)$$

$$= \frac{1}{b}\sum_m |\langle F_m, F_m\rangle| = \frac{1}{b}\sum_m \int_{I_m} |f(t)|^2 |g(t+ma)|^2 dt$$

$$= \frac{1}{b}\int_{-\infty}^{\infty} |f(t)|^2 \sum_m |g(t+ma)|^2 dt.$$

From property 1 we have then

$$\frac{A}{b}||f||^2 \le \sum_{m,n} |\langle f, g^{[ma,nb]}\rangle|^2 \le \frac{B}{b}||f||^2,$$

so $\{g^{[ma,nb]}\}$ is a W–H frame. $\qquad\square$

It can be shown that there are no W–H frames with frame parameters (a, b) such that $ab > 1$. For some insight into this case we consider the example $(a, b) = (N, 1)$, $N > 1$, N an integer. Let $g \in L_2(\mathbb{R})$. There are two distinct possibilities:

1. There is a constant $A > 0$ such that $A \le |g_\mathbf{P}(x_1, x_2)|$ almost everywhere.
2. There is no such $A > 0$.

Let $\mathcal{M}$ be the closed subspace of $L_2(\mathbb{R})$ spanned by functions $\{g^{[mN,n]}, m, n = 0 \pm 1, \pm 2, \ldots\}$ and suppose $f \in L_2(\mathbb{R})$. Then

$$\langle f, g^{[mN,n]}\rangle = (f_\mathbf{P}, E_{n,mN} g_\mathbf{P}) = (f_\mathbf{P}\bar{g}_\mathbf{P}, E_{n,mN}).$$

If possibility 1 holds, we set $f_\mathbf{P} = \bar{g}_\mathbf{P}^{-1} E_{n_0,1}$. Then $f_\mathbf{P}$ belongs to the lattice Hilbert space and $0 = (E_{n_0,1}, E_{n,mN}) = (f_\mathbf{P}\bar{g}_\mathbf{P}, E_{n,mN}) = \langle f, g^{[mN,n]}\rangle$ so $f \in \mathcal{M}^\perp$ and $\{g^{[mN,n]}\}$ is not a frame. Now suppose possibility 2 holds. Then according to the proof of Theorem 8.11, g cannot generate a frame $\{g^{[m,n]}\}$ with frame parameters $(1, 1)$ because there is no $A > 0$ such that $A||f||^2 < \sum_{m,n} |\langle f, g^{[m,n]}\rangle|^2$. Since the $\{g^{[mN,n]}\}$

corresponding to frame parameters $(1, N)$ is a proper subset of $\{g^{[m,n]}\}$, it follows that $\{g^{[mN,n]}\}$ cannot be a frame either.

For frame parameters (a, b) with $0 < ab < 1$ it is not difficult to construct W–H frames $\{g^{[ma,nb]}\}$ such that $g \in L_2(\mathbb{R})$ is a smooth function. Taking the case $a = 1, b = 1/2$, for example, let v be an infinitely differentiable function on $\mathbb{R}$ such that

$$v(x) = \begin{cases} 0 & \text{if } x \leq 0 \\ 1 & \text{if } x \geq 1 \end{cases} \tag{8.27}$$

and $0 < v(x) < 1$ if $0 < x < 1$. Set

$$g(x) = \begin{cases} 0, & x \leq 0 \\ v(x), & 0 < x < 1 \\ \sqrt{1 - v^2(x-1)}, & 1 \leq x \leq 2 \\ 0, & 2 < x. \end{cases}$$

Then $g \in L_2(\mathbb{R})$ is infinitely differentiable and with support contained in the interval $[0, 2]$. Moreover, $||g||^2 = 1$ and $\sum_n |g(x+m)|^2 \equiv 1$. It follows immediately from Theorem 8.12 that $\{g^{[m,n/2]}\}$ is a W–H frame with frame bounds $A = B = 2$.

Theorem 8.13 *Let $f, g \in L_2(\mathbb{R})$ such that $|f_{\mathbf{P}}(x_1, x_2)|$, $|g_{\mathbf{P}}(x_1, x_2)|$ are each bounded almost everywhere. Then*

$$\sum_{m,n} |\langle f, g^{[m,n]} \rangle|^2 = \sum_{m,n} \langle f, f^{[m,n]} \rangle \langle g^{[m,n]}, g \rangle.$$

Proof Since $\langle f, g^{[m,n]} \rangle = (f_{\mathbf{P}}, E_{n,m}g_{\mathbf{P}}) = (f_{\mathbf{P}}\bar{g}_{\mathbf{P}}, E_{n,m})$ we have the Fourier series expansion

$$f_P(x_1, x_2)\overline{g_P(x_1, x_2)} = \sum_{m,n} \langle f, g^{[m,n]} \rangle E_{n,m}(x_1, x_2). \tag{8.28}$$

Since $|f_P|, |g_P|$ are bounded, $f_P\bar{g}_P$ is square integrable with respect to the measure $dx_1 dx_2$ on the square $0 \leq x_1, x_2 < 1$. From the Plancherel formula for double Fourier series, we obtain the identity

$$\int_0^1 \int_0^1 |f_{\mathbf{P}}|^2 |g_{\mathbf{P}}|^2 dx_1 dx_2 = \sum_{m,n} |\langle f, g^{[m,n]} \rangle|^2.$$

Similarly, we can obtain expansions of the form (8.28) for $f_{\mathbf{P}}\bar{f}_{\mathbf{P}}$ and $g_{\mathbf{P}}\bar{g}_{\mathbf{P}}$. Applying the Plancherel formula to these two functions we find

$$\int_0^1 \int_0^1 |f_{\mathbf{P}}|^2 |g_{\mathbf{P}}|^2 dx_1 dx_2 = \sum_{m,n} \langle f, f^{[m,n]} \rangle \langle g^{[m,n]}, g \rangle.$$

$\square$

8.2.4 Continuous wavelets

Recall that if g is the sampling function for the windowed Fourier transform and g is centered about the point (t_0, ω_0) in time–frequency (phase) space, then $g^{[x_1, x_2]}$ samples a signal f in a "window" centered about the point $(t_0 - x_1, \omega_0 + x_2)$. As x_1, x_2 are varied this window is translated around the phase space, but its size and shape remain unchanged. The continuous wavelet transform employs a new type of sampling behavior that exploits a scaling of the time parameter. This changes the shape of the window but, due to the Heisenberg inequality, the area of the window is unchanged. Thus, for example, we will be able to modify the window to obtain more and more precise localization of the signal in the time domain, but at the expense of the degree of localization in the frequency domain.

Let $w \in L_2(\mathbb{R})$ with $||w|| = 1$ and define the affine translation of w by

$$w^{(a,b)}(t) = |a|^{-1/2} w \left(\frac{t-b}{a} \right) \tag{8.29}$$

where $a > 0$. Let $f(t) \in L_2(\mathbb{R})$. The *integral wavelet transform* of f is the function

$$F_w(a, b) = |a|^{-1/2} \int_{-\infty}^{\infty} f(t) \overline{w} \left(\frac{t-b}{a} \right) dt = (f, w^{(a,b)}). \tag{8.30}$$

(Note that we can also write the transform as

$$F_w(a, b) = |a|^{1/2} \int_{-\infty}^{\infty} f(au + b) \overline{w}(u) du,$$

which is well-defined as $a \to 0$.) This transform is defined in the time-scale plane. The parameter b is associated with time samples, but frequency sampling has been replaced by the scaling parameter a.

In analogy with the windowed Fourier transform, one might expect that the functions $w^{(a,b)}(t)$ span $L_2(\mathbb{R})$ as a, b range over all possible values. However, in general this is not the case. Indeed $L_2(\mathbb{R}) = \mathcal{H}^+ \oplus \mathcal{H}^-$ where $\mathcal{H}^+$ consists of the functions f_+ such that the Fourier transform $\mathcal{F}f_+(\lambda)$ has support on the positive λ-axis and the functions f_- in $\mathcal{H}^-$ have Fourier transform with support on the negative λ-axis. If the support of $\hat{w}(\lambda)$ is contained on the positive λ-axis then the same will be true of $\hat{w}^{(a,b)}(\lambda)$ for all $a > 0, b$ as one can easily check. Thus the functions $\{w^{(a,b)}\}$ will not necessarily span $L_2(\mathbb{R})$, though for some choices of w

we will still get a spanning set. However, if we choose two nonzero functions $w_\pm \in \mathcal{H}^\pm$ then the (orthogonal) functions $\{w_+^{(a,b)}, w_-^{(a,b)} : a > 0\}$ *will* span $L_2(\mathbb{R})$.

An alternative way to proceed, and the way that we shall follow first, is to compute the samples for all (a, b) such that $a \neq 0$, i.e., also to compute (8.30) for $a < 0$. Now, for example, if the Fourier transform of w has support on the positive λ-axis, we see that, for $a < 0$, $\hat{w}^{(a,b)}(\lambda)$ has support on the negative λ-axis. Then it isn't difficult to show that, indeed, the functions $w^{(a,b)}(t)$ span $L_2(\mathbb{R})$ as a, b range over all possible values (including negative a.). However, to get a convenient inversion formula, we will further require the condition (8.31) to follow (which is just $\hat{w}(0) = 0$).

We will soon see that in order to invert (8.30) and synthesize f from the transform of a single mother wavelet w we shall need to impose the condition

$$\int_{-\infty}^{\infty} w(t)dt = 0. \tag{8.31}$$

Further, we require that $w(t)$ has exponential decay at ∞, i.e., $|w(t)| \leq Ke^{-k|t|}$ for some $k, K > 0$ and all t. Among other things this implies that $|\hat{w}(\lambda)|$ is uniformly bounded in λ. Then there is a Plancherel formula.

Theorem 8.14 *Let $f, g \in L_2(\mathbb{R})$ and $C = \int |\hat{w}(\lambda)|^2 \frac{d\lambda}{|\lambda|}$. Then*

$$C \int_{-\infty}^{\infty} f(t)\overline{g}(t)dt = \int_{-\infty}^{\infty} \int_{-\infty}^{\infty} F_w(a,b)\overline{G}_w(a,b)\frac{da\,db}{a^2}. \tag{8.32}$$

Proof Assume first that $\hat{f}$ and $\hat{g}$ have their support contained in $|\lambda| \leq \Lambda$ for some finite Λ. Note that the right-hand side of (8.30), considered as a function of b, is the convolution of $f(t)$ and $|a|^{-1/2}\overline{w}(-t/a)$. Thus the Fourier transform of $F_w(a, \cdot)$ is $|a|^{1/2}\hat{f}(\lambda)\hat{w}(a\lambda)$. Similarly the Fourier transform of $\overline{G}_w(a, \cdot)$ is $|a|^{1/2}\hat{\overline{g}}(\lambda)\overline{\hat{w}}(a\lambda)$. The standard Plancherel identity gives

$$\int_{-\infty}^{\infty} F_w(a,b)\overline{G}_w(a,b)db = \frac{1}{2\pi} \int_{-\infty}^{\infty} |a|\hat{f}(\lambda)\hat{\overline{g}}(\lambda)|\hat{w}(a\lambda)|^2 d\lambda.$$

Note that the integral on the right-hand side converges, because f, g are bandlimited functions. Multiplying both sides by $da/|a|^2$, integrating with respect to a and switching the order of integration on the right (justified because the functions are absolutely integrable) we obtain

$$\int_{-\infty}^{\infty} \int_{-\infty}^{\infty} F_w(a,b)\overline{G}_w(a,b)\frac{da\,db}{a^2} = \frac{C}{2\pi} \int_{-\infty}^{\infty} \hat{f}(\lambda)\hat{\overline{g}}(\lambda)d\lambda.$$

Using the Parseval–Plancherel formula for Fourier transforms, we have the stated result for bandlimited functions.

The rest of the proof is standard. We need to bound the bandlimited restriction on f and g. Let f, g be arbitrary L_2 functions and let

$$\hat{f}_N(\lambda) = \begin{cases} \hat{f}(\lambda) & \text{if } |\lambda| \leq N \\ 0 & \text{otherwise,} \end{cases}$$

where N is a positive integer. Then $\hat{f}_N \to \hat{f}$ (in the frequency domain L_2 norm) as $N \to +\infty$, with a similar statement for $\hat{g}_N$. From the Parseval–Plancherel identity we then have $f_n \to f, g_N \to g$ (in the time domain L_2 norm). Since

$$C \int_{-\infty}^{\infty} |f_N(t) - f_M(t)|^2 dt = \int_{-\infty}^{\infty} \int_{-\infty}^{\infty} |F_{w,N}(a,b) - F_{w,M}(a,b)|^2 \frac{da\ db}{a^2}$$

it follows easily that $\{F_{w,N}\}$ is a Cauchy sequence in the Hilbert space of square integrable functions in $\mathbb{R}^2$ with measure $da\ db/|a|^2$, and $F_{w,N} \to F_w$ in the norm, as $N \to \infty$. Since the inner products are continuous with respect to this limit, we get the general result. □

Exercise 8.4 Verify that the requirement $\int_{-\infty}^{\infty} w(t)dt = 0$ ensures that C is finite, where $C = \int |\hat{w}(\lambda)|^2 \frac{d\lambda}{|\lambda|}$ in the Plancherel formula for continuous wavelets.

Exercise 8.5 Consider the Haar mother wavelet

$$w(t) = \begin{cases} 1, & 0 \leq t < \frac{1}{2} \\ -1, & -\frac{1}{2} \leq t < 0 \\ 0 & \text{otherwise.} \end{cases}$$

Compute the explicit expression for the wavelet transform. Find the explicit integral expression for C. Can you determine its exact value?

The synthesis equation for continuous wavelets is as follows.

Theorem 8.15

$$f(t) = \frac{1}{C} \int_{-\infty}^{\infty} \int_{-\infty}^{\infty} F_w(a,b)|a|^{-1/2}\overline{w}(\frac{t-b}{a})\frac{db\ da}{a^2}. \tag{8.33}$$

Proof Consider the b-integral on the right-hand side of Equation (8.33). By the Plancherel formula this can be recast as $\frac{1}{2\pi} \int_{-\infty}^{\infty} \hat{F}_w(\lambda)\overline{\hat{w}}^{(a,\cdot)}(\lambda)d\lambda$ where the Fourier transform of $F_w(a,\cdot)$ is $|a|^{1/2}\hat{f}(\lambda)\hat{w}(a\lambda)$, and the

Fourier transform of $|a|^{-1/2}w(\frac{t-\cdot}{a})$ is $|a|^{1/2}e^{-it\lambda}\hat{w}(a\lambda)$. Thus the expression on the right-hand side of (8.33) becomes

$$\frac{1}{2\pi C}\int_{-\infty}^{\infty}\frac{da}{|a|}\int_{-\infty}^{\infty}\hat{f}(\lambda)|\hat{w}(a\lambda)|^2 e^{it\lambda}d\lambda$$

$$=\frac{1}{2\pi C}\int_{-\infty}^{\infty}\hat{f}(\lambda)e^{it\lambda}d\lambda\int_{-\infty}^{\infty}|\hat{w}(a\lambda)|^2\frac{da}{|a|}.$$

The a-integral is just C, so from the inverse Fourier transform, we see that the expression equals $f(t)$ (provided it meets conditions for pointwise convergence of the inverse Fourier transform). $\qquad\square$

Now let's see how to modify these results for the case where we require $a > 0$. We choose two nonzero functions $w_\pm \in \mathcal{H}^\pm$, i.e., w_+ is a positive frequency probe function and w_- is a negative frequency probe. To get a convenient inversion formula, we require the condition (8.34) (which is just $\hat{w}_+(0) = 0, \hat{w}_-(0) = 0$):

$$\int_{-\infty}^{\infty}w_+(t)dt = \int_{-\infty}^{\infty}w_-(t)dt = 0. \tag{8.34}$$

Further, we require that $w_\pm(t)$ have exponential decay at ∞, i.e., $|w_\pm(t)| \le Ke^{-k|t|}$ for some $k, K > 0$ and all t. Among other things this implies that $|\hat{w}_\pm(\lambda)|$ are uniformly bounded in λ. Finally, we adjust the relative normalization of w_+ and w_- so that

$$C = \int_0^\infty |\hat{w}_+(\lambda)|^2\frac{d\lambda}{|\lambda|} = \int_{-\infty}^0 |\hat{w}_-(\lambda)|^2\frac{d\lambda}{|\lambda|}. \tag{8.35}$$

Let $f(t) \in L_2(\mathbb{R})$. Now the *integral wavelet transform* of f is the pair of functions

$$F_\pm(a,b) = |a|^{-1/2}\int_{-\infty}^{\infty}f(t)\overline{w}_\pm\left(\frac{t-b}{a}\right)dt = (f, w_\pm^{(a,b)}). \tag{8.36}$$

(Note that we can also write the transform pair as

$$F_\pm(a,b) = |a|^{1/2}\int_{-\infty}^{\infty}f(au+b)\overline{w}_\pm(u)du,$$

which is well-defined as $a \to 0$.) Then there is a Plancherel formula.

Theorem 8.16 *Let* $f, g \in L_2(\mathbb{R})$. *Then* $C\int_{-\infty}^{\infty}f(t)\overline{g}(t)dt =$

$$\int_0^\infty\int_{-\infty}^{\infty}\left(F_+(a,b)\overline{G}_+(a,b) + F_-(a,b)\overline{G}_-(a,b)\right)\frac{da\ db}{a^2}. \tag{8.37}$$

Proof A straightforward modification of our previous proof. Assume first that $\hat{f}$ and $\hat{g}$ have their support contained in $|\lambda| \leq \Lambda$ for some finite Λ. Note that the right-hand sides of (8.36), considered as functions of b, are the convolutions of $f(t)$ and $|a|^{-1/2}\overline{w}_\pm(-t/a)$. Thus the Fourier transform of $F_\pm(a, \cdot)$ is $|a|^{1/2}\hat{f}(\lambda)\hat{w}_\pm(a\lambda)$. Similarly the Fourier transforms of $\overline{G}_\pm(a, \cdot)$ are $|a|^{1/2}\hat{\overline{g}}(\lambda)\overline{\hat{w}}_\pm(a\lambda)$. The standard Plancherel identity gives

$$\int_{-\infty}^\infty F_+(a, b)\overline{G}_+(a, b)db = \frac{1}{2\pi}\int_0^\infty |a|\hat{f}(\lambda)\hat{\overline{g}}(\lambda)|\hat{w}_+(a\lambda)|^2 d\lambda,$$

$$\int_{-\infty}^\infty F_-(a, b)\overline{G}_-(a, b)db = \frac{1}{2\pi}\int_{-\infty}^0 |a|\hat{f}(\lambda)\hat{\overline{g}}(\lambda)|\hat{w}_-(a\lambda)|^2 d\lambda.$$

Note that the integrals on the right-hand side converge, because f, g are bandlimited functions. Multiplying both sides by $da/|a|^2$, integrating with respect to a (from 0 to $+\infty$) and switching the order of integration on the right we obtain

$$\int_0^\infty \int_{-\infty}^\infty \left(F_+(a, b)\overline{G}_+(a, b) + F_-(a, b)\overline{G}_-(a, b)\right)\frac{da\, db}{a^2} =$$

$$\frac{C}{2\pi}\int_0^\infty \hat{f}(\lambda)\hat{\overline{g}}(\lambda)d\lambda + \frac{C}{2\pi}\int_{-\infty}^0 \hat{f}(\lambda)\hat{\overline{g}}(\lambda)d\lambda$$

$$= \frac{C}{2\pi}\int_{-\infty}^\infty \hat{f}(\lambda)\hat{\overline{g}}(\lambda)d\lambda.$$

Using the Plancherel formula for Fourier transforms, we have the stated result for bandlimited functions. The rest of the proof is standard. □

Note that the positive frequency data is orthogonal to the negative frequency data.

Corollary 8.17

$$\int_0^\infty \int_{-\infty}^\infty F_+(a, b)\overline{G}_-(a, b)\frac{da\, db}{a^2} = 0. \tag{8.38}$$

Proof A slight modification of the proof of the theorem. The standard Plancherel identity gives

$$\int_{-\infty}^\infty F_+(a, b)\overline{G_-(a, b)}db = \frac{1}{2\pi}\int_0^\infty |a|\hat{f}(\lambda)\overline{\hat{g}(\lambda)}\hat{w}_+(a\lambda)\overline{\hat{w}_-(a\lambda)}d\lambda = 0$$

since $\hat{w}_+(\lambda)\overline{\hat{w}_-(\lambda)} \equiv 0$. □

The modified synthesis equation for continuous wavelets is as follows.

Theorem 8.18

$$f(t) = \tag{8.39}$$

$$\frac{1}{C} \int_0^\infty \int_{-\infty}^\infty |a|^{-1/2} \left(F_+(a,b)\overline{w}_+(\frac{t-b}{a}) + F_-(a,b)\overline{w}_-(\frac{t-b}{a}) \right) \frac{db\,da}{a^2}.$$

Proof Consider the b-integrals on the right-hand side of equations (8.39). By the Plancherel formula they become $\frac{1}{2\pi} \int_{-\infty}^\infty \hat{F}_\pm(\lambda)\overline{\hat{w}_\pm}^{(a,\cdot)}(\lambda)d\lambda$ where the Fourier transform of $F_\pm(a,\cdot)$ is $|a|^{1/2}\hat{f}(\lambda)\hat{w}_\pm(a\lambda)$, and the Fourier transform of $|a|^{-1/2}w_\pm(\frac{t-\cdot}{a})$ is $|a|^{1/2}e^{-it\lambda}\hat{w}_\pm(a\lambda)$. Thus the expressions on the right-hand side of (8.39) become

$$\frac{1}{2\pi C} \int_{-\infty}^\infty \frac{da}{|a|} \int_{-\infty}^\infty \hat{f}(\lambda)|\hat{w}_\pm(a\lambda)|^2 e^{it\lambda} d\lambda$$

$$= \frac{1}{2\pi C} \int_{-\infty}^\infty \hat{f}(\lambda)e^{it\lambda} d\lambda \int_{-\infty}^\infty |\hat{w}_\pm(a\lambda)|^2 \frac{da}{|a|}.$$

Each a-integral is just C, so from the inverse Fourier transform, we see that the expression equals $f(t)$ (provided it meets conditions for pointwise convergence of the inverse Fourier transform). $\square$

Can we get a continuous transform for the case $a > 0$ that uses a single wavelet? Yes, but not any wavelet will do. A convenient restriction is to require that $w(t)$ is a *real-valued* function with $||w|| = 1$. In that case it is easy to show that $\overline{\hat{w}(\lambda)} = \hat{w}(-\lambda)$, so $|\hat{w}(\lambda)| = |\hat{w}(-\lambda)|$. Now let

$$w_+(t) = \frac{1}{2\pi} \int_0^\infty \hat{w}(\lambda)e^{i\lambda t} d\lambda, \qquad w_-(t) = \frac{1}{2\pi} \int_{-\infty}^0 \hat{w}(\lambda)e^{i\lambda t} d\lambda.$$

Note that

$$w(t) = w_+(t) + w_-(t), \qquad ||w_+|| = ||w_-||$$

and that $w_\pm$ are, respectively, positive and negative frequency wavelets. We further require the zero area condition (which is just $\hat{w}(0) = \hat{w}_+(0) = 0, \hat{w}_-(0) = 0$):

$$\int_{-\infty}^\infty w_+(t)dt = \int_{-\infty}^\infty w_-(t)dt = 0, \tag{8.40}$$

and that $w(t)$ have exponential decay at ∞. Then

$$C = \int_0^\infty |\hat{w}(\lambda)|^2 \frac{d\lambda}{|\lambda|} = \int_{-\infty}^0 |\hat{w}_\pm(\lambda)|^2 \frac{d\lambda}{|\lambda|} \tag{8.41}$$

exists. Let $f(t) \in L_2(\mathbb{R})$. Here the integral wavelet transform of f is the function

$$F(a,b) = |a|^{-1/2} \int_{-\infty}^{\infty} f(t)\overline{w}\left(\frac{t-b}{a}\right)dt = (f, w^{(a,b)}). \tag{8.42}$$

Theorem 8.19 *Let $f, g \in L_2(\mathbb{R})$, and $w(t)$ a real-valued wavelet function with the properties listed above. Then*

$$\frac{C}{2}\int_{-\infty}^{\infty} f(t)\overline{g}(t)dt = \int_{0}^{\infty}\int_{-\infty}^{\infty} F(a,b)\overline{G}(a,b)\frac{da\ db}{a^2}. \tag{8.43}$$

Proof This follows immediately from Theorem 8.16 and the fact that

$$\int_{0}^{\infty}\int_{-\infty}^{\infty} F(a,b)\overline{G}(a,b)\frac{da\ db}{a^2} =$$

$$\int_{0}^{\infty}\int_{-\infty}^{\infty} \left(F_+(a,b) + F_-(a,b)\right)\left(\overline{G}_+(a,b) + \overline{G}_-(a,b)\right)\frac{da\ db}{a^2}$$

$$= \int_{0}^{\infty}\int_{-\infty}^{\infty} \left(F_+(a,b)\overline{G}_+(a,b) + F_-(a,b)\overline{G}_-(a,b)\right)\frac{da\ db}{a^2},$$

due to the orthogonality relation (8.38). □

The synthesis equation for continuous real wavelets is as follows.

Theorem 8.20

$$f(t) = \frac{2}{C}\int_{0}^{\infty}\int_{-\infty}^{\infty} |a|^{-1/2}F(a,b)\overline{w}(\frac{t-b}{a})\frac{db\ da}{a^2}. \tag{8.44}$$

Exercise 8.6 Show that the derivatives of the Gaussian distribution

$$w^{(n)}(t) = K_n\frac{d^n}{dt^n}G(t), \quad G(t) = \frac{1}{\sqrt{2\pi}\sigma}\exp(-t^2/2\sigma^2), \quad n = 1, 2, \ldots \tag{8.45}$$

each satisfy the conditions for reconstruction with a single wavelet in the case $a > 0$, provided the constant K_n is chosen so that $||w^{(n)}|| = 1$. In common use is the case $n = 2$, the Mexican hat wavelet. Graph this wavelet to explain the name.

Exercise 8.7 Show that the continuous wavelets $w^{(n)}$ are insensitive to signals t^k of order $k < n$. Of course, polynomials are not square integrable on the real line. However, it follows from this that $w^{(n)}$ will filter out most of the polynomial portion (of order $< n$) of a signal restricted to a finite interval.

Exercise 8.8 For students with access to the MATLAB wavelet tool-box. The MATLAB function cwt computes $F(a,b)$, (called $C_{a,b}$ by MAT-LAB) for a specified discrete set of values of a and for b ranging the length of the signal vector. Consider the signal $f(t) = e^{-t^2/10}(\sin t + 2\cos 2t + \sin(5t + \cos 8t + 2t^2)$, $0 \le t < 2\pi$. Discretize the signal into a 512 component vector and use the Mexican hat wavelet $gau2$ to compute the continuous wavelet transform $C_{a,b}$ for a taking integer values from 1 to 32. A simple code that works is

```
t=linspace(0,2*pi,2^8);
f=exp(-t.^2/10).*(sin(t)+2*cos(2*t)+sin(5*t)+cos(8*t)+2*t.^2);
c=cwt(f,1:32,'gau2')
```

For color graphical output an appropriate command is

```
cc=cwt(f,1:32,'gau2','plot')
```

This produces four colored graphs whose colors represent the magnitudes of the real and the imaginary parts of $C_{a,b}$, the absolute value of $C_{a,b}$ and the phase angle of $C_{a,b}$, respectively, for the matrix of values (a,b). Repeat the analysis for the Haar wavelet $'db1'$ or $'haar'$ and for $'cgau3'$. Taking Exercise 8.7 into account, what is your interpretation of the information about the signal obtained from these graphs?

The continuous wavelet transform is also used frequently for nonzero wavelets $w(t)$ that do not satisfy the condition $\int_{-\infty}^{\infty} w(t)\, dt = 0$, so that the synthesis formulas derived above are no longer valid. The critical requirement is that the affine translations $\{w^{(a,b)}(t) : 0 < a,\ -\infty < b < \infty\}$, (8.29), must span $L_2(\mathbb{R})$. A sufficient condition for this is that for any function $g \in L_2(\mathbb{R})$ the orthogonality condition $\int_{-\infty}^{\infty} g(t)\overline{w}^{(a,b)}(t)\, dt = 0$ for all $a > 0, b \in \mathbb{R}$ implies $||g|| = 0$. Then even though there is no simple inversion formula, the samples $F_w(a,b)$ uniquely characterize any $f \in L_2(\mathbb{R})$. Some of these wavelets provide very useful information about the phase space structure of functions f. Examples for w are the Gaussian function $G(t) = \frac{1}{\sqrt{2\pi}\sigma} \exp(-t^2/2\sigma^2)$ and, more generally, the complex Gaussians

$$w_\omega^{(n)}(t) = K_n \frac{d^n}{dt^n} e^{i\omega t} G(t)$$

for ω a fixed real number. The case $w_\omega^{(0)}(t)$ is the Morlet wavelet.

Exercise 8.9 Show that the affine translations $w^{(a,b)}(t)$ span $L_2(\mathbb{R})$ where $w(t) = G(t)$, the Gaussian distribution.

Exercise 8.10 Show that the set of affine translations of a complex Gaussian $w_\omega^{(n)}(t)$ span $L_2(\mathbb{R})$. Hint: We already know that the affine translations of the real Gaussian (8.45) span $L_2(\mathbb{R})$.

The continuous wavelet transform is overcomplete, just as is the windowed Fourier transform. To avoid redundancy (and for practical computation where one cannot determine the wavelet transform for a continuum of a, b values) we can restrict attention to discrete lattices in time-scale space. The question is: which lattices will lead to bases for the Hilbert space? Our work with discrete wavelets will give us many nontrivial examples of a lattice and wavelets that will work. Here, we look for other examples.

8.2.5 Lattices in time-scale space

To define a lattice in the time-scale space we choose two nonzero real numbers $a_0, b_0 > 0$ with $a_0 \neq 1$. Then the lattice points are $a = a_0^m, b = nb_0 a_0^m$, $m, n = 0, \pm 1, \ldots$, so

$$w^{mn}(t) = w^{(a_0^m, nb_0 a_0^m)}(t) = a_0^{-m/2} w(a_0^{-m} t - nb_0).$$

Note that if w has support contained in an interval of length ℓ then the support of w^{mn} is contained in an interval of length $a_0^{-m}\ell$. Similarly, if $\mathcal{F}w$ has support contained in an interval of length L then the support of $\mathcal{F}w^{mn}$ is contained in an interval of length $a_0^m L$. (Note that this behavior is very different from the behavior of the Heisenberg translates $g^{[ma,nb]}$. In the Heisenberg case the support of g in either position or momentum space is the same as the support of $g^{[ma,nb]}$. In the affine case the sampling of position–momentum space is on a logarithmic scale. There is the possibility, through the choice of m and n, of sampling in smaller and smaller neighborhoods of a fixed point in position space.)

The affine translates w^{mn} are called *discrete wavelets* and the function w is a *mother wavelet*. The map $\mathbf{T} : f \to \langle f, w_\pm^{mn} \rangle$ is the *discrete wavelet transform*.

NOTE: This should all become very familiar to you. The lattice $a_0 = 2^j, b_0 = 1$ corresponds to the multiresolution analysis that we will study in Chapter 9.

Exercise 8.11 This is a continuation of Exercise 8.5 on the Haar mother wavelet. Choosing the lattice $a_0 = 2^j, b_0 = 1$, for fixed integer $j \neq 0$, show that the discrete wavelets w^{mn} for $m, n = 0, \pm 1, \pm 2, \ldots$ form an ON set in $L_2(R)$. (In fact it is an ON basis.)

8.3 Affine frames

The general definitions and analysis of frames presented earlier clearly apply to wavelets. However, there is no affine analog of the Weil–Brezin–Zak transform which was so useful for Weyl–Heisenberg frames. Nonetheless we can prove the following result directly.

Lemma 8.21 *Let $w_+ \in L_2(\mathbb{R})$ such that the support of $\mathcal{F}w$ is contained in the interval $[\ell, L]$ where $0 < \ell < L < \infty$, and let $a_0 > 1, b_0 > 0$ with $(L - \ell)b_0 \le 1$. Suppose also that*

$$0 < A \le \sum_m |\mathcal{F}w_+(a_0^m\omega)|^2 \le B < \infty$$

for almost all $\omega \ge 0$. Then $\{w_+^{mn}\}$ is a frame for $\mathcal{H}^+$ with frame bounds $A/b_0, B/b_0$.

Proof Let $f \in \mathcal{H}^+$ and note that $w_+ \in \mathcal{H}^+$. For fixed m the support of $\mathcal{F}f(a_0^m y)\overline{\mathcal{F}w_+}(\omega)$ is contained in the interval $\ell \le \omega \le \ell + 1/b_0$ (of length $1/b_0$). Then

$$\sum_{m,n} |\langle f, w_+^{mn}\rangle|^2 = \sum_{m,n} |\langle \mathcal{F}f, \mathcal{F}w_+^{mn}\rangle|^2 \tag{8.46}$$

$$= \sum_{m,n} a_0^{-m} |\int_{-0}^{\infty} \mathcal{F}f(a_0^{-m}\omega)\overline{\mathcal{F}w_+}(\omega)e^{-inb_0\omega}d\omega|^2$$

$$= \sum_m \frac{a_0^{-m}}{b_0} \int_{\ell}^{\ell+1/b_0} |\mathcal{F}f(a_0^{-m}\omega)\mathcal{F}w_+(\omega)|^2 d\omega$$

$$= \frac{1}{b} \sum_m \int_0^{\infty} |\mathcal{F}f(\omega)\mathcal{F}w_+(a_0^m\omega)|^2 d\omega$$

$$= \frac{1}{b} \int_0^{\infty} |\mathcal{F}f(\omega)|^2 \left(\sum_m |\mathcal{F}w_+(a_0^m\omega)|^2\right) d\omega.$$

Since $||f||^2 = \int_0^{\infty} |\mathcal{F}f(\omega)|^2 d\omega$ for $f \in \mathcal{H}^+$, the result

$$A||f||^2 \le \sum_{m,n} |\langle f, w_+^{mn}\rangle|^2 \le B||f||^2$$

follows.   □

A very similar result characterizes a frame for $\mathcal{H}^-$. (Just let ω run from $-\infty$ to 0.) Furthermore, if $\{w_+^{mn}\}, \{w_-^{mn}\}$ are frames for $\mathcal{H}^+, \mathcal{H}^-$, respectively, corresponding to lattice parameters a_0, b_0, then $\{w_+^{mn}, w_-^{mn}\}$ is a frame for $L_2(\mathbb{R})$

Exercise 8.12 For lattice parameters $a_0 = 2$, $b_0 = 1$, choose $\hat{w}_+ = \chi_{[1,2)}$ and $\hat{w}_- = \chi_{(-2,-1]}$. Show that w_+ generates a tight frame for $\mathcal{H}^+$ with $A = B = 1$ and w_- generates a tight frame for $\mathcal{H}^-$ with $A = B = 1$, so $\{w_+^{mn}, w_-^{mn}\}$ is a tight frame for $L_2(\mathbb{R})$. (Indeed, one can verify directly that $\{w_\pm^{mn}\}$ is an *ON* basis for $L_2(\mathbb{R})$.)

Exercise 8.13 Let w be the function such that

$$\mathcal{F}w(\omega) = \frac{1}{\sqrt{\ln a}} \begin{cases} 0 & \text{if } \omega \leq \ell \\ \sin \frac{\pi}{2} v \left(\frac{\omega - \ell}{\ell(a-1)} \right) & \text{if } \ell < \omega \leq a\ell \\ \cos \frac{\pi}{2} v \left(\frac{\omega - a\ell}{a\ell(a-1)} \right) & \text{if } a\ell < \omega \leq a^2\ell \\ 0 & \text{if } a^2\ell < \omega \end{cases}$$

where $v(x)$ is defined as in (8.27). Show that $\{w^{mn}\}$ is a tight frame for $\mathcal{H}^+$ with $A = B = 1/b\ln a$. Furthermore, for $w_+ = w$ and $w_- = \bar{w}$ show that $\{w_\pm^{mn}\}$ is a tight frame for $L_2(\mathbb{R})$.

Suppose $w \in L_2(\mathbb{R})$ such that $\mathcal{F}w(\omega)$ is bounded almost everywhere and has support in the interval $[-1/2b, 1/2b]$. Then for any $f \in L_2(\mathbb{R})$ the function

$$a_0^{-m/2} \mathcal{F}f(a_0^{-m}\omega)\overline{\mathcal{F}w(\omega)}$$

has support in this same interval and is square integrable. Thus

$$\sum_{m,n} |\langle f, w_{mn}\rangle|^2 = \sum_{m,n} |a_0^{-m/2} \int_{-\infty}^{\infty} \mathcal{F}f(a_0^{-m}\omega)\overline{\mathcal{F}w(\omega)}e^{-2\pi i\omega b_0} d\omega|^2$$

$$= b_0^{-1} \sum_m \int_{-\infty}^{\infty} a_0^{-m} |\mathcal{F}f(a_0^{-m}\omega)\mathcal{F}w(\omega)|^2 d\omega$$

$$= \frac{1}{b_0} \int_{-\infty}^0 |\mathcal{F}f(\omega)|^2 \sum_m |\mathcal{F}w(a_0^m\omega)|^2 d\omega \qquad (8.47)$$

$$+ \frac{1}{b_0} \int_0^{\infty} |\mathcal{F}f(\omega)|^2 \sum_m |\mathcal{F}w(a_0^m\omega)|^2 d\omega.$$

It follows from the computation that if there exist constants $A, B > 0$ such that

$$A \leq \sum_m |\mathcal{F}w(a_0^m\omega)|^2 \leq B$$

for almost all ω, then the single mother wavelet w generates an affine frame.

8.4 Additional exercises

Exercise 8.14 Verify formulas (8.2), using the definitions (8.3).

Exercise 8.15 Derive the inequality (8.5).

Exercise 8.16 Given the function

$$g(t) = \phi(t) = \begin{cases} 1, & 0 \le t < 1 \\ 0, & \text{otherwise,} \end{cases}$$

i.e., the Haar scaling function, show that the set $\{g^{[m,n]}\}$ is an ON basis for $L_2(\mathbb{R})$. Here, m, n run over the integers. Thus $g^{[x_1, x_2]}$ is overcomplete.

Exercise 8.17 Verify expansion (8.19) for a frame.

Exercise 8.18 Show that the necessary and sufficient condition for a frame $\{f^{(1)}, \ldots, f^{(M)}\}$ in the finite-dimensional inner product space $\mathcal{V}_n$ is that the $N \times M$ matrix Ω, (8.21), be of rank N. In particular we must have $M \ge N$.

Exercise 8.19 Show that the necessary and sufficient condition for a frame $\{f^{(1)}, \ldots, f^{(M)}\}$ in the finite-dimensional inner product space $\mathcal{V}_n$ to be a Riesz basis is that $M = N$.

Exercise 8.20 Show that the $N \times M$ matrix defining the dual frame is $(\Omega\Omega^*)^{-1}\Omega$.

Exercise 8.21 Verify that $\lambda_j = \sigma_j^2$, for $j = 1, \ldots, N$ where the σ_j are the singular values of the matrix Ω, (8.21). Thus optimal values for A and B can be computed directly from the singular value decomposition of Ω.

Exercise 8.22 Show that expansion (8.21) for a frame in the finite-dimensional real inner product space $\mathcal{V}_N$ corresponds to the matrix identity $\phi = \Omega\Omega^*(\Omega\Omega^*)^{-1}\phi$.

Exercise 8.23 Find the matrix identity corresponding to the expansion (8.21) for a frame in the finite-dimensional real inner product space $\mathcal{V}_N$.

Exercise 8.24 By translation in t if necessary, assume that $\int_{-\infty}^{\infty} t|w_{\pm}(t)|^2 dt = 0$. Let

$$k_+ = \int_0^{\infty} y|\mathcal{F}w_+(y)|^2 dy, \quad k_- = \int_{-\infty}^0 y|\mathcal{F}w_-(y)|^2 dy.$$

Then $w_\pm$ are centered about the origin in position space and about $k_\pm$ in momentum space. Show that

$$\int_{-\infty}^{\infty} t|w_\pm^{(a,b)}(t)|^2 dt = -b, \quad \pm \int_0^{\infty} y|\mathcal{F}w_\pm^{(a,b)}(\pm y)|^2 dy = a^{-1}k_\pm,$$

so that $w_\pm^{(a,b)}$ are centered about $-b$ in time and $a^{-1}k_\pm$ in frequency.

Exercise 8.25 Show that if the support of w_+ is contained in an interval of length ℓ in the time domain then the support of $w_+^{(a,b)}$ is contained in an interval of length $|a|\ell$. Similarly, show that if the support of $\mathcal{F}w_+$ is contained in an interval of length L in the frequency domain then the support of $\mathcal{F}w_+^{(a,b)}$ is contained in an interval of length $|a|^{-1}L$. This implies that the length and width of the "window" in time–frequency space will be rescaled by a and a^{-1} but the area of the window will remain unchanged.

For a group theory approach to windowed Fourier and continuous wavelet transforms, with applications to radar, see [85].

9
Multiresolution analysis

9.1 Haar wavelets

Discrete wavelet analysis constitutes a method of signal decomposition quite distinct from Fourier analysis. The simplest wavelets are the Haar wavelets, studied by Haar more than 50 years before discrete wavelet theory came into vogue. From our point of view, however, Haar wavelets arise from the affine frame associated with the continuous Haar wavelet transform, and the lattice $(a_0, b_0) = (2, 1)$, see Exercises 8.5 and 8.11. The continuous transform functions are overcomplete but by restricting the transform to the lattice, we get an ON basis for $L_2(\mathbb{R})$. By exploiting the scaling properties of the affine frame we will find the Mallat algorithm, a novel way of organizing the expansion coefficients for a signal with respect to discrete wavelet bases that exploits the "zooming in" feature of the change of time scale by a factor of 2. A deep connection between filter banks and discrete wavelets will emerge. This simple example will illustrate some of the basic features of multiresolution analysis, an abstract structure for analyzing Riesz bases related to the lattice $(a_0, b_0) = (2, 1)$.

We start with the *father wavelet* or *scaling function*. For the Haar wavelets the scaling function is the *box function*

$$\phi(t) = \begin{cases} 1 & \text{if } 0 \leq t < 1 \\ 0 & \text{otherwise.} \end{cases} \tag{9.1}$$

We can use this function and its integer translates to construct the space V_0 of all step functions of the form

$$s(t) = a_k \qquad \text{for } k \leq t < k + 1,$$

where the a_k are complex numbers such that $\sum_{k=-\infty}^{\infty} |a_k|^2 < \infty$. Thus $s \in V_0 \subset L_2(\mathbb{R})$ if and only if

$$s(t) = \sum_k a_k \phi(t-k), \qquad \sum_{k=-\infty}^{\infty} |a_k|^2 < \infty.$$

Note that the $\{\phi(t-k) : k = 0, \pm 1, \ldots\}$ form an ON basis for V_0. Also, the area under the father wavelet is 1:

$$\int_{-\infty}^{\infty} \phi(t)dt = 1.$$

We can approximate signals $f(t) \in L_2(\mathbb{R})$ by projecting them on V_0 and then expanding the projection in terms of the translated scaling functions. Of course this would be a very crude approximation. To get more accuracy we can change the scale by a factor of 2.

Consider the functions $\phi(2t - k)$. They form a basis for the space V_1 of all step functions of the form

$$s(t) = a_k \qquad \text{for } \frac{k}{2} \le t < \frac{k+1}{2},$$

where $\sum_{k=-\infty}^{\infty} |a_k|^2 < \infty$. This is a larger space than V_0 because the intervals on which the step functions are constant are just $1/2$ the width of those for V_0. The functions $\{2^{1/2}\phi(2t-k) : k = 0, \pm 1, \ldots\}$ form an ON basis for V_1. The scaling function also belongs to V_1. Indeed we can expand it in terms of the basis as

$$\phi(t) = \phi(2t) + \phi(2t-1). \tag{9.2}$$

NOTE: In the next section we will study many new scaling functions ϕ. We will always require that these functions satisfy the *dilation equation*

$$\phi(t) = \sqrt{2} \sum_{k=0}^{N} c_k \phi(2t-k). \tag{9.3}$$

The $\sqrt{2}$ will be needed for normalization purposes. It is ubiquitous in discrete wavelet theory in the same way that π is ubiquitous in the equations of Fourier analysis. For the Haar scaling function $N = 1$ and $(c_0, c_1) = (1/\sqrt{2}, 1/\sqrt{2})$, just as for the moving average filter, see Examples 7.6. From (9.3) we can easily prove

Lemma 9.1 *If the scaling function is normalized so that*

$$\int_{-\infty}^{\infty} \phi(t)dt = 1,$$

then $\sum_{k=0}^{N} c_k = \sqrt{2}$.

Returning to Haar wavelets, we can iterate this rescaling procedure and define the space V_j of step functions at level j to be the Hilbert space spanned by the linear combinations of the functions $\phi(2^j t - k)$, $k = 0, \pm 1, \ldots$. These functions will be piecewise constant with discontinuities contained in the set

$$\{t = \frac{n}{2^j}, \qquad n = 0, \pm 1, \pm 2, \ldots\}.$$

The functions

$$\phi_{jk}(t) = 2^{\frac{j}{2}} \phi(2^j t - k), \qquad k = 0, \pm 1, \pm 2, \ldots$$

form an ON basis for V_j. Further we have

$$V_0 \subset V_1 \subset \cdots \subset V_{j-1} \subset V_j \subset V_{j+1} \subset \cdots$$

and the containment is strict. (Each V_j contains functions that are not in V_{j-1}.) Also, note that the dilation equation (9.2) implies that

$$\phi_{jk}(t) = \frac{1}{\sqrt{2}} [\phi_{j+1,2k}(t) + \phi_{j+1,2k+1}(t)]. \qquad (9.4)$$

Exercise 9.1 Verify explicitly that the functions $\phi_{jk}(t)$ form an ON basis for the space V_j and that the dilation equation (9.4) holds.

Our definition of the space V_j and functions $\phi_{jk}(t)$ also makes sense for negative integers j. Thus we have

$$\cdots V_{-2} \subset V_{-1} \subset V_0 \subset V_1 \subset \cdots$$

Here is an easy way to decide in which class a step function $s(t)$ belongs:

Lemma 9.2

1. $s(t) \in V_0 \Leftrightarrow s(2^j t) \in V_j$
2. $s(t) \in V_j \Leftrightarrow s(2^{-j} t) \in V_0$

Proof $s(t)$ is a linear combination of functions $\phi(t - k)$ if and only if $s(2^j t)$ is a linear combination of functions $\phi(2^j t - k)$. $\square$

Since $V_0 \subset V_1$, it is natural to look at the orthogonal complement of V_0 in V_1, i.e., to decompose each $s \in V_1$ in the form $s = s_0 + s_1$ where $s_0 \in V_0$ and $s_1 \in V_0^\perp$. We write

$$V_1 = V_0 \oplus W_0,$$

where $W_0 = \{s \in V_1 : (s, f) = 0 \text{ for all } f \in V_0\}$. It follows that the functions in W_0 are just those in V_1 that are orthogonal to the basis vectors $\phi(t - k)$ of V_0.

Note from the dilation equation that $\phi(t - k) = \phi(2t - 2k) + \phi(2t - 2k - 1) = 2^{-1/2} (\phi_{1,2k}(t) + \phi_{1,2k+1}(t))$. Thus

$$(\phi_{0k}, \phi_{1\ell}) = 2^{1/2} \int_{-\infty}^{\infty} \phi(t - k)\phi(2t - \ell)dt = \begin{cases} 2^{-1/2} & \text{if } \ell = 2k, 2k + 1 \\ 0 & \text{otherwise} \end{cases}$$

and

$$s_1(t) = \sum_k a_k \phi(2t - k) \in V_1$$

belongs to W_0 if and only if $a_{2k+1} = -a_{2k}$. Thus

$$s_1 = \sum_k a_{2k} [\phi(2t - 2k) - \phi(2t - 2k - 1)] = \sum_k a_{2k} w(t - k)$$

where

$$w(t) = \phi(2t) - \phi(2t - 1) \tag{9.5}$$

is the *Haar wavelet*, or *mother wavelet*. You can check that the wavelets $w(t - k)$, $k = 0 \pm 1, \ldots$, form an ON basis for W_0.

NOTE: In the next section we will require that associated with the father wavelet $\phi(t)$ there be a mother wavelet $w(t)$ satisfying the *wavelet equation*

$$w(t) = \sqrt{2} \sum_{k=0}^{N} d_k \phi(2t - k), \tag{9.6}$$

and such that w is orthogonal to all translations $\phi(t - k)$ of the father wavelet. For the Haar scaling function $N = 1$ and $(d_0, d_1) = (1/\sqrt{2}, -1/\sqrt{2})$, the normalized impulse response vector for the moving difference filter, see Examples 7.6.

We define functions $w_{jk}(t)$

$$w_{jk}(t) = 2^{\frac{j}{2}} w(2^j t - k) = 2^{\frac{j}{2}} (\phi(2^{j+1}t - 2k) - \phi(2^{j+1}t - 2(k + 1))),$$

$$k = 0, \pm 1, \pm 2, \ldots, \quad j = 1, 2, \ldots$$

It is easy to prove

Lemma 9.3 *For fixed j,*

$$(w_{jk}, w_{jk'}) = \delta_{kk'}, \qquad (\phi_{jk}, w_{jk'}) = 0 \tag{9.7}$$

where $k, k' = 0, \pm 1, \ldots$

Other properties proved above are that

$$\phi_{jk}(t) = \frac{1}{\sqrt{2}}(\phi_{j+1,2k}(t) + \phi_{j+1,2k+1}(t)),$$

$$w_{jk}(t) = \frac{1}{\sqrt{2}}(\phi_{j+1,2k}(t) - \phi_{j+1,2k+1}(t)).$$

Theorem 9.4 *Let W_j be the orthogonal complement of V_j in V_{j+1}:*

$$V_j \oplus W_j = V_{j+1}.$$

The wavelets $\{w_{jk}(t) : \quad k = 0, \pm 1, \ldots\}$ form an ON basis for W_j.

Proof From (9.7) it follows that the wavelets $\{w_{jk}\}$ form an ON set in W_j. Suppose $s \in W_j \subset V_{j+1}$. Then

$$s(t) = \sum_k a_k \phi_{j+1,k}(t)$$

and $(s, \phi_{jn}) = 0$ for all n. Now $\phi_{jn} = \frac{1}{\sqrt{2}}(\phi_{j+1,2n} + \phi_{j+1,2n+1})$, so

$$0 = (s, \phi_{jn}) = \frac{1}{\sqrt{2}}[(s, \phi_{j+1,2n}) + (s, \phi_{j+1,2n+1})] = \frac{1}{\sqrt{2}}(a_{2n} + a_{2n+1}).$$

Hence

$$s(t) = \sum_k a_{2k}\sqrt{2}w_{jk}(t),$$

and the set $\{w_{jk}\}$ is an ON basis for W_j.   □

Since $V_j \oplus W_j = V_{j+1}$ for all $j \geq 0$, we can iterate on j to get $V_{j+1} = W_j \oplus V_j = W_j \oplus W_{j-1} \oplus V_{j-1}$ and so on. Thus

$$V_{j+1} = W_j \oplus W_{j-1} \oplus \cdots \oplus W_1 \oplus W_0 \oplus V_0,$$

and any $s \in V_{j+1}$ can be written uniquely in the form

$$s = \sum_{k=0}^{j} w_k + s_0 \qquad \text{where } w_k \in W_k, \ s_0 \in V_0.$$

Remark Note that $(w_{jk}, w_{j'k'}) = 0$ if $j \neq j'$. Indeed, suppose $j > j'$ to be definite. Then $w_{j'k'} \in W_{j'} \subset V_{j'+1} \subseteq V_j$. Since $w_{jk} \in W_j$ it must be perpendicular to $w_{j'k'}$.

Lemma 9.5 $(w_{jk}, w_{j'k'}) = \delta_{jj'}\delta_{kk'}$ *for $j, j', \pm k, \pm k' = 0, 1, \ldots$*

Theorem 9.6

$$L_2(\mathbb{R}) = V_0 \oplus \sum_{\ell=0}^{\infty} \oplus W_\ell = V_0 \oplus W_0 \oplus W_1 \oplus \cdots,$$

so that each $f(t) \in L_2(\mathbb{R})$ can be written uniquely in the form

$$f = f_0 + \sum_{k=0}^{\infty} w_\ell, \qquad w_\ell \in W_\ell, \ f_0 \in V_0. \tag{9.8}$$

Proof Based on our study of Hilbert spaces, it is sufficient to show that for any $f \in L_2(\mathbb{R})$, given $\epsilon > 0$ we can find an integer $j(\epsilon)$ and a step function $s = \sum_k a_k \phi_{jk} \in V_j$ with a finite number of nonzero a_k and such that $||f - s|| < \epsilon$. This is easy. Since the space of step functions $S^2_{[-\infty,\infty]}$ is dense in $L_2(\mathbb{R})$ there is a step function $s'(t) \in S^2_{[-\infty,\infty]}$, nonzero on a finite number of bounded intervals, such that $||f - s'|| < \epsilon/2$. Then, it is clear that by choosing j sufficiently large, we can find an $s \in V_j$ with a finite number of nonzero a_k and such that $||s' - s|| < \epsilon/2$. Thus $||f - s|| \leq ||f - s'|| + ||s' - s|| < \epsilon$. $\qquad\square$

Note that for j a negative integer we can also define spaces V_j, W_j and functions ϕ_{jk}, w_{jk} in an obvious way, so that we have

$$L_2(\mathbb{R}) = V_j \oplus \sum_{\ell=j}^{\infty} \oplus W_\ell = V_j \oplus W_j \oplus W_1 \oplus \cdots, \tag{9.9}$$

even for negative j. Further we can let $j \to -\infty$ to get

Corollary 9.7

$$L_2(\mathbb{R}) = \sum_{\ell=-\infty}^{\infty} \oplus W_\ell = \cdots W_{-1} \oplus W_0 \oplus W_1 \oplus \cdots,$$

so that each $f(t) \in L_2(\mathbb{R})$ can be written uniquely in the form

$$f = \sum_{\ell=-\infty}^{\infty} w_\ell, \qquad w_\ell \in W_\ell. \tag{9.10}$$

In particular, $\{w_{jk} : j, k = 0, \pm 1, \pm 2, \ldots\}$ is an ON basis for $L_2(\mathbb{R})$.

Proof (If you understand that every function in $L_2(\mathbb{R})$ is determined up to its values on a set of measure zero.): We will show that $\{w_{jk}\}$ is an ON basis for $L_2(\mathbb{R})$. The proof will be complete if we can show that the space W' spanned by all finite linear combinations of the w_{jk} is dense in $L_2(\mathbb{R})$. This is equivalent to showing that the only $g \in L_2(\mathbb{R})$ such that

$(g, f) = 0$ for all $f \in W'$ is the zero vector $g \equiv 0$. It follows immediately from (9.9) that if $(g, w_{jk}) = 0$ for all j, k then $g \in V_\ell$ for all integers ℓ. This means that, almost everywhere, g is equal to a step function that is constant on intervals of length $2^{-\ell}$. Since we can let ℓ go to $-\infty$ we see that, almost everywhere, $g(t) = c$ where c is a constant. We can't have $c \neq 0$ for otherwise g would not be square integrable. Hence $g \equiv 0$. $\qquad\square$

This result settles an issue raised in Exercise 8.11.

We have a new ON basis for $L_2(\mathbb{R})$:

$$\{\phi_{0k}, w_{jk'} : \qquad j, \pm k, \pm k' = 0, 1, \ldots\}.$$

Let's consider the space V_j for fixed j. On one hand we have the scaling function basis

$$\{\phi_{j,k} : \qquad \pm k = 0, 1, \ldots\}.$$

Then we can expand any $f_j \in V_j$ as

$$f_j = \sum_{k=-\infty}^{\infty} a_{j,k} \phi_{j,k}. \tag{9.11}$$

On the other hand we have the wavelets basis

$$\{\phi_{j-1,k}, w_{j-1,k'} : \qquad \pm k, \pm k' = 0, 1, \ldots\}$$

associated with the direct sum decomposition

$$V_j = W_{j-1} \oplus V_{j-1}.$$

Using this basis we can expand any $f_j \in V_j$ as

$$f_j = \sum_{k'=-\infty}^{\infty} b_{j-1,k'} w_{j-1,k'} + \sum_{k=-\infty}^{\infty} a_{j-1,k} \phi_{j-1,k}. \tag{9.12}$$

If we substitute the relations $\phi_{j-1,k}(t) = \frac{1}{\sqrt{2}}(\phi_{j,2k}(t) + \phi_{j,2k+1}(t))$, $w_{j-1,k}(t) = \frac{1}{\sqrt{2}}(\phi_{j,2k}(t) - \phi_{j,2k+1}(t))$, into the expansion (9.12) and compare coefficients of $\phi_{j,\ell}$ with the expansion (9.11), we obtain the fundamental recursions

$$\text{Averages (lowpass)} \quad a_{j-1,k} = \tfrac{1}{\sqrt{2}}(a_{j,2k} + a_{j,2k+1}) \tag{9.13}$$

$$\text{Differences (highpass)} \quad b_{j-1,k} = \tfrac{1}{\sqrt{2}}(a_{j,2k} - a_{j,2k+1}). \tag{9.14}$$

These equations link the Haar wavelets with an $N = 1$ filter. Let $x_k = a_{jk}$ be a discrete signal. The result of passing this signal, backwards, through the (normalized) moving average filter $\mathbf{C}$ and then downsampling is $y_k = (\downarrow 2)C^T * x_k = a_{j-1,k}$, where $a_{j-1,k}$ is given by (9.13). Similarly, the result of passing the signal backwards through the (normalized) moving difference filter $\mathbf{D}$ and then downsampling is $z_k = (\downarrow 2)D^T * x_k = b_{j-1,k}$, where $b_{j-1,k}$ is given by (9.14).

If you compare the formulas (9.13), (9.14) with the action of the filters $\mathbf{C}$, $\mathbf{D}$ in Examples 7.6 you see that the correct filters differ by a time reversal. The correct analysis filters are the time-reversed filters $\mathbf{D}^T$, where the impulse response vector is $d_n^T = d_{-n}$, and $\mathbf{C}^T$. These filters are not causal. In general, the analysis recurrence relations for wavelet coefficients will involve the corresponding acausal filters $\mathbf{C}^T$, $\mathbf{D}^T$. However synthesis filters for recovering $a_{j,k}$ from the $a_{j-1,k}$ will turn out to be $\mathbf{C}$, $\mathbf{D}$ exactly. The picture is in expression (9.15).

$$a_{j-1,k} \quad \leftarrow \quad (\downarrow 2) \quad \mathbf{C}^T$$
$$\nwarrow$$
$$Output \qquad Analysis \qquad\qquad a_{jk} \quad . \qquad (9.15)$$
$$\swarrow \quad Input$$
$$b_{j-1,k} \quad \leftarrow \quad (\downarrow 2) \quad \mathbf{D}^T$$

This coupling of high pass filters and low pass filters with downsampling is called a filter bank, whose precise definition will be given in Section 9.3. We can iterate this process by inputting the output $a_{j-1,k}$ of the high pass filter to the filter bank again to compute $a_{j-2,k}, b_{j-2,k}$, etc. At each stage we save the wavelet coefficients $b_{j'k'}$ and input the scaling coefficients $a_{j'k'}$ for further processing, see expression (9.16).

$$a_{j-2,k} \quad \leftarrow \quad (\downarrow 2) \quad \mathbf{C}^T$$
$$\nwarrow$$
$$a_{j-1,k} \quad \leftarrow \quad (\downarrow 2) \quad \mathbf{C}^T$$
$$\swarrow \qquad\qquad\qquad \nwarrow$$
$$b_{j-2,k} \quad \leftarrow \quad (\downarrow 2) \quad \mathbf{D}^T \qquad\qquad\qquad a_{jk}$$
$$\swarrow \quad Input.$$
$$b_{j-1,k} \quad \leftarrow \quad (\downarrow 2) \quad \mathbf{D}^T$$
$$(9.16)$$

The output of the final stage is the set of scaling coefficients a_{0k}, b_{0k}. Thus our final output is the complete set of coefficients for the wavelet expansion

$$f_j = \sum_{j'=0}^{j} \sum_{k=-\infty}^{\infty} b_{j'k} w_{j'k} + \sum_{k=-\infty}^{\infty} a_{0k} \phi_{0k},$$

based on the decomposition

$$V_{j+1} = W_j \oplus W_{j-1} \oplus \cdots \oplus W_1 \oplus W_0 \oplus V_0.$$

What we have described is the Mallat algorithm (Mallat herring bone) for decomposing a signal. In practice f_j has finite support, so the sum on k is finite.

We can invert each stage of this process through the synthesis recursion:

$$a_{j,2k} = \frac{1}{\sqrt{2}}(a_{j-1,k} + b_{j-1,k})$$

$$a_{j,2k+1} = \frac{1}{\sqrt{2}}(a_{j-1,k} - b_{j-1,k}). \tag{9.17}$$

Thus, for level j the full analysis and reconstruction picture is expression (9.18).

$$
\begin{array}{ccccc}
\mathbf{C} & (\uparrow 2) & \leftarrow a_{j-1,k} \leftarrow & (\downarrow 2) & \mathbf{C}^T \\
\swarrow & & & & \nwarrow \\
a_{jk} & \textit{Synthesis} & & \textit{Analysis} & a_{jk} \\
\textit{Output} \nwarrow & & & \swarrow & \textit{Input.} \\
\mathbf{D} & (\uparrow 2) & \leftarrow b_{j-1,k} \leftarrow & (\downarrow 2) & \mathbf{D}^T
\end{array}
$$

$$\tag{9.18}$$

Comments on Haar wavelets:

1. For any $f(t) \in L_2(\mathbb{R})$ the scaling and wavelets coefficients of f are defined by

$$a_{jk} = (f, \phi_{jk}) = 2^{j/2} \int_{-\infty}^{\infty} f(t)\phi(2^j t - k)dt$$

$$= 2^{j/2} \int_{\frac{k}{2^j}}^{\frac{k}{2^j} + \frac{1}{2^j}} f(t)dt, \tag{9.19}$$

$$b_{jk} = (f, w_{jk}) = 2^{j/2} \int_{-\infty}^{\infty} f(t)\phi(2^{j+1} t - 2k)dt$$

$$- 2^{j/2} \int_{-\infty}^{\infty} f(t)\phi(2^{j+1} t - 2k - 1)dt$$

$$= 2^{j/2} \int_{\frac{k}{2^j}}^{\frac{k}{2^j} + \frac{1}{2^j}} [f(t) - f(t + \frac{1}{2^{j+1}})]dt. \tag{9.20}$$

If f is a continuous function and j is large then $a_{jk} \sim 2^{-j/2} f(\frac{k}{2^j})$. (Indeed if f has a bounded derivative we can develop an upper bound for the error of this approximation.) If f is continuously differentiable and j is large, then $b_{jk} \sim -\frac{1}{2^{1+3j/2}} f'(\frac{k}{2^j})$. Again this shows that the a_{jk} capture averages of f (low pass) and the b_{jk} capture changes in f (high pass).

2. Since the scaling function $\phi(t)$ is nonzero only for $0 \le t < 1$ it follows that $\phi_{jk}(t)$ is nonzero only for $\frac{k}{2^j} \le t < \frac{k+1}{2^j}$. Thus the coefficients a_{jk} depend only on the local behavior of $f(t)$ in that interval. Similarly for the wavelet coefficients b_{jk}. This is a dramatic difference from Fourier series or Fourier integrals where each coefficient depends on the global behavior of f. If f has compact support, then for fixed j, only a finite number of the coefficients a_{jk}, b_{jk} will be nonzero. The Haar coefficients a_{jk} enable us to track t intervals where the function becomes nonzero or large. Similarly the coefficients b_{jk} enable us to track t intervals in which f changes rapidly.

3. Given a signal f, how would we go about computing the wavelet coefficients? As a practical matter, one doesn't usually do this by evaluating the integrals (9.19) and (9.20). Suppose the signal has compact support. By translating and rescaling the time coordinate if necessary, we can assume that $f(t)$ vanishes except in the interval $[0, 1)$. Since $\phi_{jk}(t)$ is nonzero only for $\frac{k}{2^j} \le t < \frac{k+1}{2^j}$ it follows that each of the coefficients a_{jk}, b_{jk} will vanish except when $0 \le k < 2^j$. Now suppose that f is such that for a sufficiently large integer $j = J$ we have $a_{Jk} \sim 2^{-J/2} f(\frac{k}{2^J})$. If f is differentiable we can compute how large J needs to be for a given error tolerance. We would also want to exceed the Nyquist rate. Another possibility is that f takes discrete values on the grid $t = \frac{k}{2^J}$, in which case there is no error in our assumption. Thus, effectively, we are sampling the function $f(t)$ at the gridpoints.

Inputting the values $a_{Jk} = 2^{-J/2} f(\frac{k}{2^J})$ for $k = 0, 1, \ldots, 2^J - 1$ we use the recursion

$$\text{Averages (lowpass)} \quad a_{j-1,k} = \tfrac{1}{\sqrt{2}}(a_{j,2k} + a_{j,2k+1}) \quad (9.21)$$

$$\text{Differences (highpass)} \quad b_{j-1,k} = \tfrac{1}{\sqrt{2}}(a_{j,2k} - a_{j,2k+1}) \quad (9.22)$$

described above, see expression (9.16), to compute the wavelet coefficients b_{jk}, $j = 0, 1, \ldots, J - 1$, $k = 0, 1, \ldots, 2^j - 1$ and a_{00}.

The input consists of 2^J numbers. The output consists of $\sum_{j=0}^{J-1} 2^j + 1 = 2^J$ numbers. The algorithm is very efficient. Each recursion involves two multiplications by the factor $1/\sqrt{2}$. At level j there are $2 \cdot 2^j$ such recursions. Thus the total number of multiplications is

$2 \sum_{j=0}^{J-1} 2 \cdot 2^j = 4 \cdot 2^J - 4 < 4 \cdot 2^J$. See Figure 9.1 for a MATLAB implementation of the algorithm.

4. The preceding algorithm is an example of the Fast Wavelet Transform (FWT). It computes 2^J wavelet coefficients from an input of 2^J function values and does so with a number of multiplications $\sim 2^J$. Compare this with the FFT which needs $\sim J \cdot 2^J$ multiplications from an input of 2^J function values. In theory at least, the FWT is faster. The Inverse Fast Wavelet Transform is based on (9.17). (Note, however, that the FFT and the FWT compute different things, i.e., they divide the spectral band in different ways. Hence they aren't directly comparable.)

5. The FWT discussed here is based on filters with $N + 1$ taps, where $N = 1$. For wavelets based on more general $N + 1$ tap filters (such as the Daubechies filters), each recursion involves $N + 1$ multiplications, rather than two. Otherwise the same analysis goes through. Thus the FWT requires $\sim 2(N + 1)2^J$ multiplications.

6. Haar wavelets are very simple to implement. However they are poor at approximating continuous functions. By definition, any truncated Haar wavelet expansion is a step function. Most of the Daubechies wavelets to come are continuous and are much better for this type of approximation.

Exercise 9.2 For students with access to the wavelet toolbox of MATLAB. The dwt command of MATLAB accomplishes the filtering action of (9.15). If the signal vector is a then the command $[cA, cD] = dwt(a,' db2')$ produces vectors cA, cD, the result of applying the filters $(\downarrow 2)C^T$ and $(\downarrow 2)D^T$, respectively, to a, using the Haar wavelet $db2$. Practice with the signal a produced by

```
t=linspace(0,1,2^10);
x=2*rand(1,2^10)-0.5;
a=exp(-t.^2).*sin(100*t.^2);
a=f+x;
[cA,cD]=dwt(a,'db2');
subplot(2,2,1)
plot(a)
subplot(2,2,2)
plot(cA)
subplot(2,2,4)
plot(cD)
```

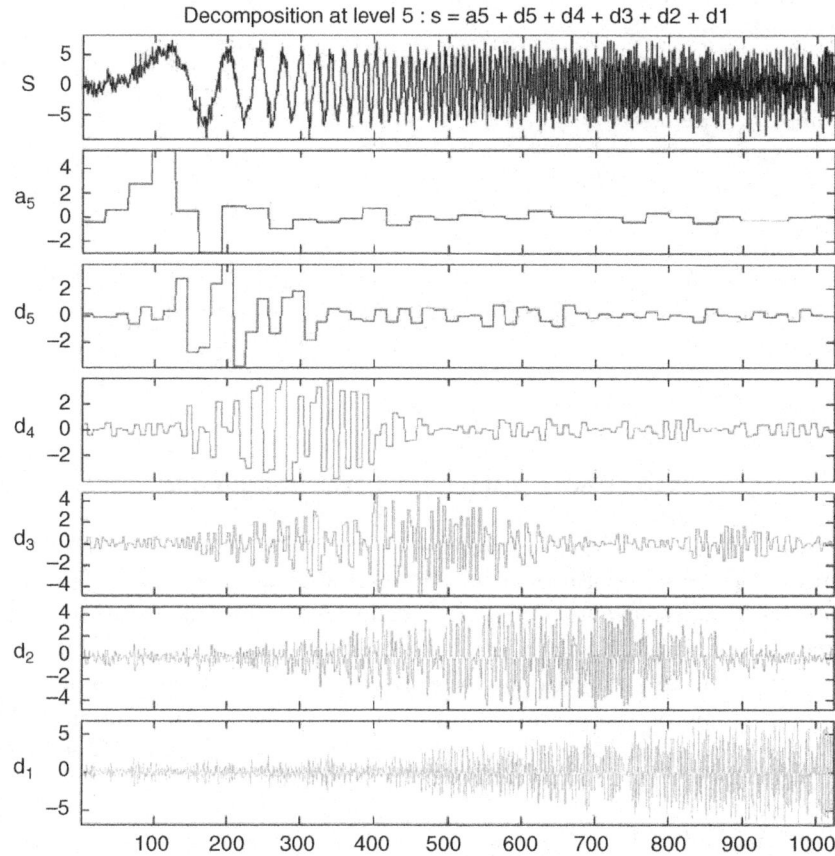

Figure 9.1 Tree structure of Haar analysis.

This is output from the Wavelet Toolbox of MATLAB. The signal $s = a_0$ is sampled at $512 = 2^9$ points, so $J = 9$ and s is assumed to be in the space V_9. The signal can be reconstructed as: $s = a_5 + d_5 + d_4 + d_3 + d_2 + d_1$. This signal is a Doppler waveform with noise superimposed. The lower-order differences contain information, but the smallest coefficients in d_1 and d_2 appear to be noise. One way of processing this signal to reduce noise and pass on the underlying information would be to set to zero all the d_1 and d_2 coefficients $b_{8,k} = b_{7,k} = 0$ whose absolute values fall below some tolerance level c and reconstruct the signal from the remaining nonzero coefficients.

The synthesizing action of (9.18) is accomplished via the inverse transformation *idwt*.

```
s=idwt(cA,cD,'db2');
subplot(1,2,1)
plot(a)
subplot(1,2,2)
plot(s)
```

Construct an iterated filter to implement the action of (9.16). Break a down into three terms, plot the results and then reconstruct the original signal.

9.2 The multiresolution structure

In this section, we deal with multiresolution structure. The Haar wavelets of the last section, with their associated nested subspaces that span L_2 are the simplest example of what is called resolution analysis. We give the full definition here. It is the main structure that we shall use for the study of wavelets, though not the only one. It allows us to focus on the important properties of wavelets, and not to get lost in the complicated details for construction of particular wavelets. Almost immediately we will see striking parallels with the study of filter banks.

Definition 9.8 Let $\{V_j : j = \ldots, -1, 0, 1, \ldots\}$ be a sequence of subspaces of $L_2(\mathbb{R})$ and $\phi \in V_0$. This is a multiresolution analysis for $L_2(\mathbb{R})$ provided the following conditions hold:

1. The subspaces are nested: $V_j \subset V_{j+1}$.
2. The union of the subspaces generates L_2 : $\overline{\cup_{j=-\infty}^{\infty} V_j} = L_2(\mathbb{R})$. (Thus, each $f \in L_2$ can be obtained as a limit of a Cauchy sequence $\{s_n : n = 1, 2, \ldots\}$ such that each $s_n \in V_{j_n}$ for some integer j_n.)
3. Separation: $\cap_{j=-\infty}^{\infty} V_j = \{0\}$, the subspace containing only the zero function. (Thus only the zero function is common to all subspaces V_j.)
4. Scale invariance: $f(t) \in V_j \iff f(2t) \in V_{j+1}$.
5. Shift invariance of V_0: $f(t) \in V_0 \iff f(t - k) \in V_0$ for all integers k.
6. ON basis: The set $\{\phi(t - k) : k = 0, \pm 1, \ldots\}$ is an ON basis for V_0.

Here, the function $\phi(t)$ is called the scaling function (or the father wavelet).

Remarks:

- Just as for the Haar wavelet, we will show that each multiresolution analysis is associated to a filter bank.
- The ON basis condition can be replaced by the (apparently weaker) condition that the translates of ϕ form a *Riesz basis*. This type of basis is most easily defined and understood from a frequency space viewpoint. We will show later that a ϕ determining a Riesz basis can be modified to a $\tilde{\phi}$ determining an ON basis.
- We can drop the ON basis condition and simply require that the integer translates of $\phi(t)$ form a basis for V_0. However, we will have to be precise about the meaning of this condition for an infinite-dimensional space. This will lead us to *biorthogonal wavelets* and non-unitary filter banks.
- The ON basis condition can be generalized in another way. It may be that there is no single function whose translates form an ON basis for V_0 but that there are m functions $\phi_1, \ldots, \phi_m$ with $m > 1$ such that the set $\{\phi_\ell(t - k) : \ell = 1, \ldots, m, k = 0, \pm 1, \ldots\}$ is an ON basis for V_0. These generate *multiwavelets* and the associated filters are *multifilters*.
- If the scaling function has finite support and satisfies the ON basis condition then it will correspond to a so-called unitary FIR filter bank. If its support is not finite, however, it will still correspond to a unitary filter bank, but one that has Infinite Impulse Response (IIR). This means that the impulse response vectors $\{c_n\}$, $\{d_n\}$ have an infinite number of nonzero components.

Examples 9.9 1. Piecewise constant functions. Here V_0 consists of the functions $f(t)$ that are constant on the unit intervals $k \le t < k + 1$:

$$f(t) = a_k \qquad \text{for } k \le t < k + 1.$$

This is exactly the Haar multiresolution analysis of the preceding section. The only change is that now we have introduced subspaces V_j for j negative. In this case the functions in V_{-n} for $n > 0$ are piecewise constant on the intervals $k \cdot 2^n \le t < (k + 1) \cdot 2^n$. Note that if $f \in V_j$ for all integers j then f must be a constant. The only square integrable constant function is identically zero, so the separation requirement is satisfied.

2. Continuous piecewise linear functions. The functions $f(t) \in V_0$ are determined by their values $f(k)$ at the integer points, and are linear between each pair of values:

$$f(t) = [f(k + 1) - f(k)](t - k) + f(k) \quad \text{for } k \le t \le k + 1.$$

Note that continuous piecewise linearity is invariant under integer shifts. Also if $f(t)$ is continuous piecewise linear on unit intervals, then $f(2t)$ is continuous piecewise linear on half-unit intervals. It isn't completely obvious, but a scaling function can be taken to be the *hat function*. The hat function $H(t)$ is the continuous piecewise linear function whose values on the integers are $H(k) = \delta_{0k}$, i.e., $H(0) = 1$ and $H(t)$ is zero on the other integers. The support of $H(t)$ is the open interval $-1 < t < 1$. Note that if $f \in V_0$ then we can write it uniquely in the form

$$f(t) = \sum_k f(k) H(t - k).$$

Although the sum could be infinite, at most two terms are nonzero for each t. Each term is linear, so the sum must be linear, and it agrees with $f(t)$ at integer times. All multiresolution analysis conditions are satisfied, except for the ON basis requirement. The integer translates of the hat function do define a basis for V_0 but it isn't ON because the inner product $(H(t), H(t-1)) \neq 0$. A scaling function does exist whose integer translates form an ON basis, but its support isn't compact.

3. Discontinuous piecewise linear functions. The functions $f(t) \in V_0$ are determined by their values and and left-hand limits $f(k), f(k-)$ at the integer points, and are linear between each pair of limit values:

$$f(t) = [f((k+1)-) - f(k)](t - k) + f(k) \quad \text{for } k \le t < k + 1.$$

Each function $f(t)$ in V_0 is determined by the two values $f(k), f((k+1)-)$ in each unit subinterval $[k, k+1)$ and two scaling functions are needed:

$$\phi_1(t) = \begin{cases} 1 & \text{if } 0 \le t < 1 \\ 0 & \text{otherwise} \end{cases} \qquad \phi_2(t) = \begin{cases} \sqrt{3}(1 - 2t) & \text{if } 0 \le t < 1 \\ 0 & \text{otherwise.} \end{cases}$$

Then

$$f(t) = \sum_k \left(\frac{f(k)}{2} [\phi_1(t - k) + \frac{1}{\sqrt{3}} \phi_2(t - k)] \right.$$
$$\left. + \frac{f((k+1)-)}{2} [\phi_1(t - k)) - \frac{1}{\sqrt{3}} \phi_2(t - k)] \right).$$

The integer translates of $\phi_1(t), \phi_2(t)$ form an ON basis for V_0. These are multiwavelets and they correspond to multifilters.

4. Shannon resolution analysis. Here V_j is the space of bandlimited signals $f(t)$ in $L_2(\mathbb{R})$ with frequency band contained in the interval $[-2^j\pi, 2^j\pi]$. The nesting property is a consequence of the fact that if $f(t)$ has Fourier transform $\hat{f}(\lambda)$ then $f(2t)$ has Fourier transform $\frac{1}{2}\hat{f}(\frac{\lambda}{2})$. The function $\phi(t) = \text{sinc }(t)$ is the scaling function. Indeed we have already shown that $||\phi|| = 1$ and the (unitary) Fourier transform of $\phi(t)$ is

$$\mathcal{F}\text{sinc }(\lambda) = \begin{cases} \frac{1}{\sqrt{2\pi}} & \text{for } |\lambda| < \pi \\ 0 & \text{for } |\lambda| > \pi. \end{cases}$$

Thus the Fourier transform of $\phi(t-k)$ is equal to $e_k(\lambda) = e^{-ik\lambda}/\sqrt{2\pi}$ in the interior of the interval $[-\pi, \pi]$ and is zero outside this interval. It follows that the integer translates of sinc (t) form an ON basis for V_0. Note that the scaling function $\phi(t)$ does not have compact support in this case.

5. The Daubechies functions D_n. This is a family of wavelets, the first of which is the Haar wavelet D_1. The remaining members of the family cannot be expressed as explicit series or integrals. We will take up their construction in the next few sections, along with the construction of associated FIR filters. We will see that each D_n corresponds to a scaling function with compact support and an ON wavelets basis.

Just as in our study of the Haar multiresolution analysis, for a general multiresolution analysis we can define the functions

$$\phi_{jk}(t) = 2^{\frac{j}{2}}\phi(2^j t - k), \qquad k = 0, \pm 1, \pm 2, \dots$$

and for fixed integer j they will form an ON basis for V_j. Since $V_0 \subset V_1$ it follows that $\phi \in V_1$ and ϕ can be expanded in terms of the ON basis $\{\phi_{1k}\}$ for V_1. Thus we have the *dilation equation*

$$\phi(t) = \sqrt{2}\sum_k c_k \phi(2t - k). \tag{9.23}$$

Since the ϕ_{jk} form an ON set, the coefficient vector c must be a unit vector in ℓ_2,

$$\sum_k |c_k|^2 = 1. \tag{9.24}$$

We will soon show that $\phi(t)$ has support in the interval $[0, N]$ if and only if the only nonvanishing coefficients of c are $(c_0, \dots, c_N)$. Scaling functions with nonbounded support correspond to coefficient vectors with infinitely many nonzero terms. Since $\phi(t) \perp \phi(t - m)$ for all nonzero m, the vector c satisfies double-shift orthogonality:

$$(\phi_{00}, \phi_{0m}) = \sum_k c_k \overline{c_{k-2m}} = \delta_{0m}. \tag{9.25}$$

Thus if we shift the vector c by a nonzero even number of places and take the inner product of the nonshifted and shifted coefficient vectors, we get 0. For FIR filters, double-shift orthogonality is associated with downsampling. For orthogonal wavelets it is associated with dilation.

From (9.23) we can easily prove

Lemma 9.10 *If the scaling function is normalized so that*

$$\int_{-\infty}^{\infty} \phi(t)dt = 1,$$

then $\sum_{k=0}^{N} c_k = \sqrt{2}$.

Also, note that the dilation equation (9.23) implies that

$$\phi_{jk}(t) = \sum_{\ell} c_{\ell-2k} \phi_{j+1,\ell}(t), \tag{9.26}$$

which is the expansion of the V_j scaling basis in terms of the V_{j+1} scaling basis. Just as in the special case of the Haar multiresolution analysis we can introduce the orthogonal complement W_j of V_j in V_{j+1}.

$$V_{j+1} = V_j \oplus W_j.$$

We start by trying to find an ON basis for the wavelet space W_0. Associated with the father wavelet $\phi(t)$ there must be a mother wavelet $w(t)$, with norm 1, and satisfying the *wavelet equation*

$$w(t) = \sqrt{2} \sum_k d_k \phi(2t - k), \tag{9.27}$$

and such that w is orthogonal to all translations $\phi(t - k)$ of the father wavelet. We will further require that w is orthogonal to integer translations of itself. For the Haar scaling function, $N = 1$ and $(d_0, d_1) = (1/\sqrt{2}, -1/\sqrt{2})$.

NOTE: In several of our examples we were able to identify the scaling subspaces, the scaling function and the mother wavelet explicitly. In general however, this won't be practical. In practice we will find conditions on the coefficient vectors c and d such that they could correspond to scaling functions and wavelets. We will then solve these conditions and demonstrate that some solutions define a multiresolution analysis, a scaling function and a mother wavelet. Virtually the entire analysis will be carried out with the coefficient vectors; we shall seldom use the scaling and wavelet functions directly. However, it is essential to show that these functions exist and have the required properties. Otherwise any results we obtain will be vacuous. Now back to our construction.

Since the ϕ_{jk} form an ON set, the coefficient vector d must be a unit vector in ℓ_2,

$$\sum_k |d_k|^2 = 1. \tag{9.28}$$

Moreover since $w(t) \perp \phi(t - m)$ for all m, the vector d satisfies double-shift orthogonality with c:

$$(w, \phi_{0m}) = \sum_k c_k \overline{d_{k-2m}} = 0. \tag{9.29}$$

The requirement that $w(t) \perp w(t - m)$ for nonzero integer m leads to double-shift orthogonality of d to itself:

$$(w(t), w(t - m)) = \sum_k d_k \overline{d_{k-2m}} = \delta_{0m}. \tag{9.30}$$

Clearly, there is no unique solution for d, but any solution will suffice. We claim that if the unit coefficient vector c is double-shift orthogonal then the coefficient vector d defined by taking the conjugate alternating flip automatically satisfies the conditions (9.29) and (9.30). Here,

$$d_n = (-1)^n \overline{c_{N-n}}. \tag{9.31}$$

This expression depends on N where the c vector for the low pass filter had $N + 1$ nonzero components. However, due to the double-shift orthogonality obeyed by c, the only thing about N that is necessary for d to exhibit double-shift orthogonality is that N be odd. Thus we will choose $N = -1$ and take $d_n = (-1)^n \overline{c_{-1-n}}$. We claim that this choice also works when $c \in \ell_2$ has an infinite number of nonzero components. Let's check for example that d is orthogonal to c:

$$S = \sum_k c_k \overline{d_{k-2m}} = \sum_k c_k (-1)^k c_{-1+2m-k}.$$

Now set $\ell = -1 + 2m - k$ and sum over ℓ:

$$S = \sum_\ell c_{1-2m-\ell}(-1)^{\ell+1} c_\ell = -S.$$

Hence $S = 0$. Thus, once the scaling function is defined through the dilation equation, the wavelet $w(t)$ is determined by the wavelet equation

$$\Omega = \begin{bmatrix} \cdot & \cdot & & & & & \cdot & \cdot \\ \cdot & c_0 & 0 & 0 & 0 & & \cdot & \cdot \\ \cdot & c_2 & c_1 & c_0 & c_{-1} & & \cdot & \cdot \\ \cdot & c_4 & c_3 & c_2 & c_1 & c_0 & \cdot & \cdot \\ \cdot & & & & & & & \cdot \\ \cdot & & & & & & & \cdot \\ \cdot & \overline{c}_{-1} & \overline{c}_0 & \overline{c}_1 & \overline{c}_2 & & \cdot & \cdot \\ \cdot & \overline{c}_{-3} & -\overline{c}_{-2} & \overline{c}_{-1} & \overline{c}_0 & & \cdot & \cdot \\ \cdot & \overline{c}_{-5} & -\overline{c}_{-4} & \overline{c}_{-3} & -\overline{c}_{-2} & \overline{c}_{-1} & \cdot & \cdot \\ \cdot & & & & & & & \cdot \\ \cdot & & & & & & \cdot & \cdot \end{bmatrix} \cdot$$

Figure 9.2 The wavelet Ω matrix.

(9.27) with $d_n = (-1)^n \overline{c_{-1-n}}$. Once w has been determined we can define functions

$$w_{jk}(t) = 2^{\frac{j}{2}} w(2^j t - k), \quad j, k = 0, \pm 1, \pm 2, \ldots.$$

It is easy to prove

Lemma 9.11

$$(w_{jk}, w_{j'k'}) = \delta_{jj'}\delta_{kk'}, \qquad (\phi_{jk}, w_{jk'}) = 0 \qquad (9.32)$$

where $j, j', k, k' = 0, \pm 1, \ldots$

The dilation equations and wavelet equations extend to:

$$\phi_{j\ell} = \sum_k c_{k-2\ell} \phi_{j+1,k}(t), \qquad (9.33)$$

$$w_{j\ell} = \sum_k d_{k-2\ell} \phi_{j+1,k}(t). \qquad (9.34)$$

Equations (9.33) and (9.34) can be understood in terms of the doubly infinite matrix Ω pictured in Figure 9.2. The rows of Ω are ON. Indeed, the ℓth upper row vector is just the coefficient vector for the expansion of $\phi_{j\ell}(t)$ as a linear combination of the ON basis vectors $\phi_{j+1,k}$. (The entry in upper row ℓ, column k is just the coefficient $c_{k-2\ell}$.) Similarly, the ℓth lower row vector is the coefficient vector for the expansion of $w_{j\ell}(t)$ as a linear combination of the basis vectors $\phi_{j+1,k}$ (and the entry in lower row ℓ, column k is the coefficient $d_{k-2\ell} = (-1)^k \overline{c_{2\ell-1-k}}$). Thus the ON property of the row vectors follows from Lemma 9.11 and the fact that ϕ_{jk}, w_{jk} have unit length.

We claim that the columns of Ω are also ON. This means that the matrix Ω is unitary and that its inverse is the transpose conjugate. A consequence of this fact is that we can solve Equations (9.33) and (9.34) explicitly to express the basis vectors $\phi_{j+1,k}$ for V_{j+1} as linear combinations of the vectors $\phi_{j,k'}$ and $w_{jk''}$. The kth column of Ω is the coefficient vector for the expansion of $\phi_{j+1,k}$ in terms of $\phi_{j,k'}$ (on the top) and $w_{jk''}$ (on the bottom).

To verify this claim, note that the columns of Ω are of two types: even (containing only terms c_{2n}, d_{2n}) and odd (containing only terms c_{2n+1}, d_{2n+1}). Thus the requirement that the column vectors of Ω are ON reduces to three types of identities:

$$\text{even–even}: \sum_\ell c_{2\ell}\overline{c_{2k+2\ell}} + \sum_\ell d_{2\ell}\overline{d_{2k+2\ell}} = \delta_{k0}, \quad (9.35)$$

$$\text{odd–odd}: \sum_\ell c_{2\ell+1}\overline{c_{2k+2\ell+1}} + \sum_\ell d_{2\ell+1}\overline{d_{2k+2\ell+1}} = \delta_{k0}, \quad (9.36)$$

$$\text{odd–even}: \sum_\ell c_{2\ell+1}\overline{c_{2k+2\ell}} + \sum_\ell d_{2\ell+1}\overline{d_{2k+2\ell}} = 0. \quad (9.37)$$

Theorem 9.12 *If c satisfies the double-shift orthogonality condition and the filter d is determined by the conjugate alternating flip*

$$d_n = (-1)^n \overline{c_{-1-n}},$$

then the columns of Ω are orthonormal.

Proof The even–even case computation is

$$\sum_\ell d_{2\ell}\overline{d_{2k+2\ell}} = \sum_\ell \overline{c_{-1-2\ell}}c_{-1-2k-2\ell} = \sum_s c_{2s+1}\overline{c_{2s+2k+1}}.$$

Thus

$$\sum_\ell c_{2\ell}\overline{c_{2k+2\ell}} + \sum_\ell d_{2\ell}\overline{d_{2k+2\ell}} = \sum_n c_n\overline{c_{n+2k}} = \delta_{k0}.$$

The other cases are similar. □

Exercise 9.3 Verify the identities (9.36), (9.37), thus finishing the proof that the columns of Ω are ON.

Now we define functions $\phi'_{j+1,k}(t)$ in V_{j+1} by

$$\phi'_{j+1,s} = \sum_h \left(\overline{c_{s-2h}}\phi_{jh} + \overline{d_{s-2h}}w_{jh} \right).$$

Substituting the expansions

$$\phi_{jh} = \sum_k c_{k-2h}\phi_{j+1,k}, \quad w_{jh} = \sum_k d_{k-2h}\phi_{j+1,k},$$

into the right-hand side of the first equation we find

$$\phi'_{j+1,s} = \sum_{hk} \left(\overline{c_{s-2h}}c_{k-2h} + \overline{d_{s-2h}}d_{k-2h} \right) \phi_{j+1,k} = \phi_{j+1,s},$$

as follows from the even–even, odd–even and odd–odd identities above. Thus

$$\phi_{j+1,s} = \sum_h \left(\overline{c_{s-2h}}\phi_{jh} + \overline{d_{s-2h}}w_{jh} \right), \tag{9.38}$$

and we have inverted the expansions

$$\phi_{j\ell} = \sum_k c_{k-2\ell}\phi_{j+1,k}(t), \tag{9.39}$$

$$w_{j\ell} = \sum_k d_{k-2\ell}\phi_{j+1,k}(t). \tag{9.40}$$

Thus the set $\{\phi_{jk}, w_{jk'}\}$ is an alternate ON basis for V_{j+1} and we have

Lemma 9.13 *The wavelets* $\{w_{jk} : k = 0, \pm 1, \ldots\}$ *form an ON basis for* W_j.

To get the wavelet expansions for functions $f \in L_2$ we can now follow the steps in the construction for the Haar wavelets. The proofs are virtually identical. Since $V_j \oplus W_j = V_{j+1}$ for all $j \geq 0$, we can iterate on j to get $V_{j+1} = W_j \oplus V_j = W_j \oplus W_{j-1} \oplus V_{j-1}$ and so on. Thus

$$V_{j+1} = W_j \oplus W_{j-1} \oplus \cdots \oplus W_1 \oplus W_0 \oplus V_0$$

and any $s \in V_{j+1}$ can be written uniquely in the form

$$s = \sum_{k=0}^{j} w_k + s_0 \qquad \text{where } w_k \in W_k, \ s_0 \in V_0.$$

Theorem 9.14

$$L_2(\mathbb{R}) = V_j \oplus \sum_{k=j}^{\infty} W_k = V_j \oplus W_j \oplus W_{j+1} \oplus \cdots,$$

so that each $f(t) \in L_2(\mathbb{R})$ *can be written uniquely in the form*

$$f = f_j + \sum_{k=j}^{\infty} w_k, \qquad w_k \in W_k, \ f_j \in V_j. \qquad (9.41)$$

We have a family of new ON bases for $L_2(\mathbb{R})$, one for each integer j:

$$\{\phi_{jk}, w_{j'k'} : \qquad j' = j, j+1, \ldots, \qquad \pm k, \pm k' = 0, 1, \ldots\}.$$

Let's consider the space V_j for fixed j. On one hand we have the scaling function basis

$$\{\phi_{j,k} : \qquad \pm k = 0, 1, \ldots\}.$$

Then we can expand any $f_j \in V_j$ as

$$f_j = \sum_{k=-\infty}^{\infty} a_{j,k} \phi_{j,k}. \qquad (9.42)$$

On the other hand we have the wavelets basis

$$\{\phi_{j-1,k}, w_{j-1,k'} : \qquad \pm k, \pm k' = 0, 1, \ldots\}$$

associated with the direct sum decomposition

$$V_j = W_{j-1} \oplus V_{j-1}.$$

Using this basis we can expand any $f_j \in V_j$ as

$$f_j = \sum_{k'=-\infty}^{\infty} b_{j-1,k'} w_{j-1,k'} + \sum_{k=-\infty}^{\infty} a_{j-1,k} \phi_{j-1,k}. \qquad (9.43)$$

If we substitute the relations

$$\phi_{j-1,\ell} = \sum_{k} c_{k-2\ell} \phi_{jk}(t), \qquad (9.44)$$

$$w_{j-1,\ell} = \sum_{k} d_{k-2\ell} \phi_{j,k}(t), \qquad (9.45)$$

into the expansion (9.42) and compare coefficients of $\phi_{j,\ell}$ with the expansion (9.43), we obtain the fundamental recursions

$$\text{Averages (lowpass)} \quad a_{j-1,k} = \sum_n c_{n-2k} a_{jn} \qquad (9.46)$$
$$\text{Differences (highpass)} \quad b_{j-1,k} = \sum_n d_{n-2k} a_{jn}. \qquad (9.47)$$

These equations link the wavelets with a unitary filter bank.

Let $x_k = a_{jk}$ be a discrete signal. The result of passing this signal through the (normalized and time-reversed) filter $\mathbf{C}^T$ and then downsampling is $y_k = ((\downarrow 2)\mathbf{C}^T * x)_k = a_{j-1,k}$, where $a_{j-1,k}$ is given by (9.46). Similarly, the result of passing the signal through the (normalized and time-reversed) filter $\mathbf{D}^T$ and then downsampling is $z_k = ((\downarrow 2)\mathbf{D}^T * x)_k = b_{j-1,k}$, where $b_{j-1,k}$ is given by (9.47). The picture, in complete analogy with that for Haar wavelets, is in expression (9.15).

We can iterate this process by inputting the output $a_{j-1,k}$ of the high pass filter to the filter bank again to compute $a_{j-2,k}, b_{j-2,k}$, etc. At each stage we save the wavelet coefficients $b_{j'k'}$ and input the scaling coefficients $a_{j'k'}$ for further processing, see expression (9.16).

The output of the final stage is the set of scaling coefficients a_{0k}, b_{0k}, assuming that we stop at $j = 0$. Thus our final output is the complete set of coefficients for the wavelet expansion

$$f_j = \sum_{j'=0}^{j-1} \sum_{k=-\infty}^{\infty} b_{j'k} w_{j'k} + \sum_{k=-\infty}^{\infty} a_{0k} \phi_{0k},$$

based on the decomposition

$$V_j = W_{j-1} \oplus W_{j-2} \oplus \cdots \oplus W_1 \oplus W_0 \oplus V_0.$$

This is the Mallat algorithm for a general multiresolution analysis.

To derive the synthesis filter bank recursion we can substitute the inverse relation

$$\phi_{j,s} = \sum_h \left(\bar{c}_{s-2h} \phi_{j-1,h} + \bar{d}_{s-2h} w_{j-1,h} \right), \tag{9.48}$$

into the expansion (9.43) and compare coefficients of $\phi_{j-1,\ell}, w_{j-1,\ell}$ with the expansion (9.42) to obtain the inverse recursion

$$a_{j,n} = \sum_k c_{2k-n} a_{j-1,k} + \sum_k d_{2k-n} b_{j-1,k}. \tag{9.49}$$

This is exactly the output of the synthesis filter bank. Thus, for level j the full analysis and reconstruction picture is expression (9.18). This entire process is known as the fast wavelet transform (FWT), and its inversion. In analogy with the Haar wavelets discussion, for any $f(t) \in L_2(\mathbb{R})$ the scaling and wavelets coefficients of f are defined by

$$a_{jk} = (f, \phi_{jk}) = 2^{j/2} \int_{-\infty}^{\infty} f(t) \overline{\phi(2^j t - k)} dt, \tag{9.50}$$

$$b_{jk} = (f, w_{jk}) = 2^{j/2} \int_{-\infty}^{\infty} f(t) \overline{w(2^j t - k)} dt.$$

9.2.1 Wavelet packets

The wavelet transform of the last section has been based on the decomposition $V_j = W_{j-1} \oplus V_{j-1}$ and its iteration. Using the symbols a_j, b_j (with the index k suppressed) for the projection of a signal f on the subspaces V_j, W_j, respectively, we have the tree structure of Figure 9.3 where we have gone down three levels in the recursion. However, a finer resolution is possible. We could also use our low pass and high pass filters to decompose the wavelet spaces W_j into a direct sum of low-frequency and high-frequency subspaces: $W_j = W_{j,0} \oplus W_{j,1}$. The new ON basis for this decomposition could be obtained from the wavelet basis $w_{jk}(t)$ for W_j exactly as the basis for the decomposition $V_j = W_{j-1} \oplus V_{j-1}$ was obtained from the scaling basis $\phi_{jk}(t)$ for V_j: $w_{j\ell,1}(t) = \sum_{k=0}^{N} d_{k-2\ell} w_{jk}(t)$ and $w_{j\ell,0}(t) = \sum_{k=0}^{N} c_{k-2\ell} w_{jk}(t)$. Similarly, the new high- and low-frequency wavelet subspaces so obtained could themselves be decomposed into a direct sum of high and low pass subspaces, and so on. The wavelet transform algorithm (and its inversion) would be exactly as before. The only difference is that the algorithm would be applied to the b_{jk} coefficients, as well as the a_{jk} coefficients. Now the picture (down three levels) is the complete (wavelet packet) Figure 9.4. With wavelet packets we have a much finer resolution of the signal and a greater variety of options for decomposing it. For example,

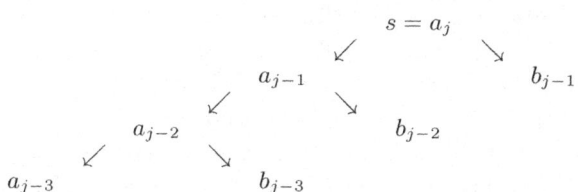

Figure 9.3 General fast wavelet transform tree.

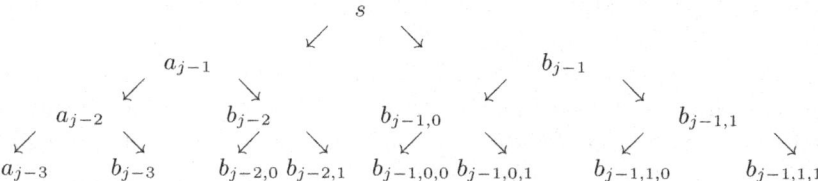

Figure 9.4 Wavelet packet tree.

we could decompose s as the sum of the eight terms at level three:
$$s = a_{j-3} + b_{j-3} + b_{j-2,0} + b_{j-2,1} + b_{j-1,0,0} + b_{j-1,0,1} + b_{j-1,1,0} + b_{j-1,1,1}.$$
A hybrid would be $s = a_{j-1} + b_{j-1,0} + b_{j-1,1,0} + b_{j-1,1,1}$. The tree structure for this algorithm is the same as for the FFT. The total number of multiplications involved in analyzing a signal at level J all the way down to level 0 is of the order $J2^J$, just as for the Fast Fourier Transform.

9.3 Filter banks and reconstruction of signals

We have seen that every multiresolution analysis is associated with an invertible linear filter system called a filter bank. In this section we will give a precise definition of a filter bank and explore some of its properties. We shall see that the determination of all possible filter banks is an algebraic problem. The partial solution of this problem is an important step for the determination of possible multiresolution structures. It is not the final step because proving the existence of wavelets is also a problem involving the tools of analysis. A given filter bank may not correspond to a multiresolution analysis, in which case the associated FWT is mathematical (and practical) nonsense. However, even then the filter bank may be of use.

Digital filters are used to analyze and process signals x_n. For purposes such as signal compression or noise reduction, it is OK to throw away some of the data generated by this filtering. However, we want to make sure that we don't (unintentionally) lose information about the original signal as we proceed with the analysis. Filter bank theory builds in the requirement that this analysis process should be invertible: We should be able to recreate (synthesize) the signal from the analysis output. Further we want this synthesis (or decoding) process to be implemented by filters. Thus, if we link the input for the synthesis filters to the output of the analysis filters we should end up with the original signal except for a fixed delay of ℓ units caused by the processing in the filters: $x_{n-\ell}$. For practical purposes we must require ℓ to be finite. This is the basic idea of *Perfect Reconstruction* of signals.

If we try to carry out the analysis with a single filter, it is essential that the filter be an invertible operator. A low pass filter would certainly fail this requirement, for example, since it would screen out the high-frequency part of the signal and lose all information about the high-frequency components of x_n. For the time being, we will consider only FIR filters and the invertibility problem is even worse for this class of

$$
\begin{bmatrix} \cdot \\ y^{(j)}_{-2} \\ y^{(j)}_{-1} \\ y^{(j)}_0 \\ y^{(j)}_1 \\ \cdot \end{bmatrix} = \begin{bmatrix} \cdot & \cdot & & & & \cdot & \cdot \\ \cdot & h^{(j)}_0 & 0 & 0 & 0 & \cdot & \cdot \\ \cdot & h^{(j)}_1 & h^{(j)}_0 & 0 & 0 & \cdot & \cdot \\ \cdot & h^{(j)}_2 & h^{(j)}_1 & h^{(j)}_0 & 0 & \cdot & \cdot \\ \cdot & h^{(j)}_3 & h^{(j)}_2 & h^{(j)}_1 & h^{(j)}_0 & \cdot & \cdot \\ \cdot & \cdot & & & & \cdot & \cdot \end{bmatrix} \begin{bmatrix} \cdot \\ x_{-2} \\ x_{-1} \\ x_0 \\ x_1 \\ \cdot \end{bmatrix},
$$

Figure 9.5 $\Phi^{(j)}$ matrix action.

filters. Recall that the Z transform $H[z]$ of an FIR filter is a polynomial in z^{-1}. Now suppose that Φ is an invertible filter with inverse Φ^{-1}. Since $\Phi\Phi^{-1} = \mathbf{I}$ where $\mathbf{I}$ is the identity filter, the convolution theorem gives us that

$$
H[z]H^{-1}[z] = 1,
$$

i.e., the Z transform of Φ^{-1} is the reciprocal of the Z transform of Φ. Except for trivial cases the Z transform of Φ^{-1} cannot be a polynomial in z^{-1}. Hence if the (nontrivial) FIR filter has an inverse, it is *not* an FIR filter. Thus for perfect reconstruction with FIR filters, we will certainly need more than one filter.

Let's try a *filter bank* with two FIR filters, $\Phi^{(0)}$ and $\Phi^{(1)}$. The input is $x = \{x_n\}$. The output of the filters is $y^{(j)} = \Phi^{(j)}x$, $j = 0, 1$. The $\Phi^{(j)}$ filter action looks like Figure 9.5 for $j = 0, 1$, where $h^{(j)}$ is the associated impulse response vector.

Note that each row of the infinite matrix $\Phi^{(j)}$ contains all zeros, except for the terms $(h^{(j)}_N, h^{(j)}_{N-1}, \ldots, h^{(j)}_0)$ which are shifted one column to the right for each successive row. (We choose N to be the largest of N_0, N_1, where $\Phi^{(0)}$ has $N_0 + 1$ taps and $\Phi^{(1)}$ has $N_1 + 1$ taps.) Thus each row vector has the same norm $||h^{(j)}||_2$.

It will turn out to be very convenient to have a filter all of whose row vectors have norm 1. Thus we will replace filters $\Phi^{(j)}$ by the normalized filters

$$
\mathbf{C} = \frac{1}{||h^{(0)}||}\Phi^{(0)}, \quad \mathbf{D} = \frac{1}{||h^{(1)}||}\Phi^{(1)}.
$$

The impulse response vector for $\mathbf{C}$ is $c = h^{(0)}/||h^{(0)}||$, so that $||c|| = 1$. Similarly, the impulse response vector for $\mathbf{D}$ is $d = h^{(1)}/||h^{(1)}||$, so that $||d|| = 1$.

$$
\begin{bmatrix} \vdots \\ y_{-2}^{(0)} \\ y_0^{(0)} \\ y_2^{(0)} \\ \vdots \end{bmatrix} = \begin{bmatrix} \cdot & \cdot & & & & \cdot & \cdot \\ \cdot & c_0 & 0 & 0 & 0 & \cdot & \cdot \\ \cdot & c_2 & c_1 & c_0 & 0 & \cdot & \cdot \\ \cdot & c_4 & c_3 & c_2 & c_1 & c_0 & \cdot \\ \cdot & \cdot & & & & \cdot & \cdot \end{bmatrix} \begin{bmatrix} \vdots \\ x_{-2} \\ x_{-1} \\ x_0 \\ x_1 \\ \vdots \end{bmatrix},
$$

Figure 9.6 $(\downarrow 2)C$ matrix action.

$$
\begin{bmatrix} \vdots \\ y_{-2}^{(1)} \\ y_0^{(1)} \\ y_2^{(1)} \\ \vdots \end{bmatrix} = \begin{bmatrix} \cdot & \cdot & & & & \cdot & \cdot \\ \cdot & d_0 & 0 & 0 & 0 & \cdot & \cdot \\ \cdot & d_2 & d_1 & d_0 & 0 & \cdot & \cdot \\ \cdot & d_4 & d_3 & d_2 & d_1 & d_0 & \cdot \\ \cdot & \cdot & & & & \cdot & \cdot \end{bmatrix} \begin{bmatrix} \vdots \\ x_{-2} \\ x_{-1} \\ x_0 \\ x_1 \\ \vdots \end{bmatrix}.
$$

Figure 9.7 $(\downarrow 2)D$ matrix action.

Now these two filters are producing twice as much output as the original input, and we want eventually to compress the output (or certainly not add to the stream of data that is transmitted). Otherwise we would have to delay the data transmission by an ever-growing amount, or we would have to replace the original one-channel transmission by a two-channel transmission. Thus we will downsample the output of filters **C** and **D**. This will effectively replace our original filters **C** and **D** by new filters

$$
(\downarrow 2)\mathbf{C} = \frac{1}{\|h_0\|}(\downarrow 2)\mathbf{\Phi}^{(0)}, \quad (\downarrow 2)\mathbf{D} = \frac{1}{\|h_1\|}(\downarrow 2)\mathbf{\Phi}^{(1)}.
$$

The $(\downarrow 2)\mathbf{C}$ filter action looks like Figure 9.6. and the $(\downarrow 2)\mathbf{D}$ filter action looks like Figure 9.7. Note that each row vector is now shifted two spaces to the right of the row vector immediately above. Schematically, we can put the $(\downarrow 2)C$ and $(\downarrow 2)D$ matrices together to display the full time-domain action as

$$
y = \tilde{\Omega}x, \quad \tilde{\Omega} = \begin{bmatrix} (\downarrow 2)C \\ (\downarrow 2)D \end{bmatrix}, \tag{9.51}
$$

where x is the signal, $\tilde{\Omega}$ is an infinite sample matrix and the sample y consists of the truncated outputs $\{y_{2n}^{(0)}\}, \{y_{2n}^{(1)}\}$. This is just the original

$$
\overline{\tilde{\Omega}^{\mathrm{tr}}} = \begin{bmatrix} (\downarrow 2)\overline{C^{\mathrm{tr}}} & (\downarrow 2)\overline{D^{\mathrm{tr}}} \end{bmatrix} = \begin{bmatrix} \cdot & \cdot & \cdot & \cdot & \cdot & \cdot & \cdot & \cdot & \cdot & \cdot \\ \cdot & \overline{c}_0 & \overline{c}_2 & \overline{c}_4 & \cdot & \cdot & \overline{d}_0 & \overline{d}_2 & \overline{d}_4 & \cdot \\ \cdot & 0 & \overline{c}_1 & \overline{c}_3 & \cdot & \cdot & 0 & \overline{d}_1 & \overline{d}_3 & \cdot \\ \cdot & 0 & \overline{c}_0 & \overline{c}_2 & \cdot & \cdot & 0 & \overline{d}_0 & \overline{d}_2 & \cdot \\ \cdot & 0 & 0 & \overline{c}_1 & \cdot & \cdot & 0 & 0 & \overline{d}_1 & \cdot \\ \cdot & 0 & 0 & \overline{c}_0 & \cdot & \cdot & 0 & 0 & \overline{d}_0 & \cdot \\ \cdot & \cdot & \cdot & \cdot & \cdot & \cdot & \cdot & \cdot & \cdot & \cdot \end{bmatrix}.
$$

Figure 9.8 $\overline{\tilde{\Omega}^{\mathrm{tr}}}$ matrix.

complete filter, with the odd-number rows removed. How can we ensure that this decimation of data from the original filters $\mathbf{C}$ and $\mathbf{D}$ still permits reconstruction of the original signal x from y? A sufficient condition is, clearly, that $\tilde{\Omega}$ should be invertible!

The invertibility requirement is very strong, and won't be satisfied in general. For example, if $\mathbf{C}$ and $\mathbf{D}$ are both low pass filters, then high-frequency information from the original signal will be permanently lost. However, if $\mathbf{C}$ is a low pass filter and $\mathbf{D}$ is a high pass filter, then there is hope that the high-frequency information from $\mathbf{D}$ and the low-frequency information from $\mathbf{C}$ will supplement one another, even after downsampling.

Initially we are going to make even a stronger requirement on $\tilde{\Omega}$ than invertibility. We are going to require that it be a *unitary* matrix. (In that case the inverse of the matrix is just the transpose conjugate and solving for the original signal x from the truncated outputs $\{y^{(0)}(2n)\},\{y^{(1)}(2n)\}$ is simple. Moreover, if the impulse response vectors c, d are real, then the matrix will be *orthogonal*.)

The transpose conjugate looks like Figure 9.8. The unitarity condition is:

$$
\overline{\tilde{\Omega}^{\mathrm{tr}}}\tilde{\Omega} = \tilde{\Omega}\overline{\tilde{\Omega}^{\mathrm{tr}}} = I, \qquad I_{ij} = \delta_{ij}.
$$

Written out in terms of the $(\downarrow 2)C$ and $(\downarrow 2)D$ matrices this is

$$
\begin{bmatrix} \overline{(\downarrow 2)C}^{\mathrm{tr}} & \overline{(\downarrow 2)D}^{\mathrm{tr}} \end{bmatrix} \begin{bmatrix} (\downarrow 2)C \\ (\downarrow 2)D \end{bmatrix} = \overline{(\downarrow 2)C}^{\mathrm{tr}}(\downarrow 2)C + \overline{(\downarrow 2)D}^{\mathrm{tr}}(\downarrow 2)D = I
$$

$$(9.52)$$

and

$$
\begin{bmatrix} (\downarrow 2)C \\ (\downarrow 2)D \end{bmatrix} \begin{bmatrix} \overline{(\downarrow 2)C}^{\mathrm{tr}} & \overline{(\downarrow 2)D}^{\mathrm{tr}} \end{bmatrix} = \begin{bmatrix} (\downarrow 2)C\overline{(\downarrow 2)C}^{\mathrm{tr}} & (\downarrow 2)C\overline{(\downarrow 2)D}^{\mathrm{tr}} \\ (\downarrow 2)D\overline{(\downarrow 2)C}^{\mathrm{tr}} & (\downarrow 2)D\overline{(\downarrow 2)D}^{\mathrm{tr}} \end{bmatrix}
$$

$$= \begin{bmatrix} I & 0 \\ 0 & I \end{bmatrix}. \tag{9.53}$$

For the filter coefficients c_k and d_k conditions (9.53) become *orthogonality to double-shifts* of the rows:

$$(\downarrow 2)C\overline{(\downarrow 2)C}^{\mathrm{tr}} = I : \qquad \sum_n c_n \bar{c}_{n-2k} = \delta_{k0} \tag{9.54}$$

$$(\downarrow 2)C\overline{(\downarrow 2)D}^{\mathrm{tr}} = 0 : \qquad \sum_n c_n \bar{d}_{n-2k} = 0 \tag{9.55}$$

$$(\downarrow 2)D\overline{(\downarrow 2)D}^{\mathrm{tr}} = I : \qquad \sum_n d_n \bar{d}_{n-2k} = \delta_{k0}. \tag{9.56}$$

Remarks • The condition says that the row vectors of $\tilde{\Omega}$ form an ON set, and that the column vectors of $\tilde{\Omega}$ also form an ON set. For a finite-dimensional matrix, only one of these requirements is needed to imply orthogonality; the other property can be proved. For infinite matrices, however, both requirements are needed.

• By normalizing the rows of $\Phi^{(0)}$, $\Phi^{(1)}$ to length 1, hence replacing these filters by the normalized filters C, D we have already gone part way to the verification of orthonormality.

• The double-shift orthogonality conditions (9.54)–(9.56) force N to be odd. For if N were even, then setting $k = N/2$ in these equations (and also $k = -N/2$ in the middle one) leads to the conditions

$$c_N \bar{c}_0 = c_N \bar{d}_0 = d_N \bar{c}_0 = d_N \bar{d}_0 = 0.$$

This violates our definition of N.

• The orthogonality condition (9.54) says that the rows of $(\downarrow 2)C$ are orthogonal, and condition (9.56) says that the rows of $(\downarrow 2)D$ are orthogonal. Condition (9.55) says that the rows of $(\downarrow 2)C$ are orthogonal to the rows of $(\downarrow 2)D$.

• If we know that the rows of $(\downarrow 2)C$ are orthogonal, then we can always construct a filter $(\downarrow 2)\mathbf{D}$, hence the impulse response vector d, such that conditions (9.55),(9.56) are satisfied. Indeed, suppose that c satisfies conditions (9.54). Then we define d by applying to c the conjugate alternating flip about $N/2$. (Recall that N *must be odd*. We are flipping the vector $c = (c_0, c_1, \ldots, c_N)$ about its midpoint, conjugating and then alternating the signs.)

$$d_n = (-1)^n \bar{c}_{N-n}, \qquad n = 0, 1, \ldots, N. \tag{9.57}$$

Thus

$$d = (d_0, d_1, \ldots, d_N) = (\bar{c}_N, -\bar{c}_{N-1}, \bar{c}_{N-2}, \ldots, -\bar{c}_0).$$

You can check by taking simple examples that this works. However in detail:

$$S \equiv \sum_n c_n \bar{d}_{n-2k} = \sum_n c_n (-1)^n c_{N-n+2k}.$$

Setting $m = N - n + 2k$ in the last sum we find

$$S = \sum_m c_{N-m+2k} c_m (-1)^{N-m} = -S,$$

since N is odd. Thus $S = 0$. Similarly,

$$T \equiv \sum_n d_n \bar{d}_{n-2k} = \sum_n (-1)^n \bar{c}_{N-n} (-1)^{n-2k} c_{N-n+2k}$$

$$= \sum_n c_{N-n+2k} \bar{c}_{N-n}.$$

Now set $m = N - n + 2k$ in the last sum:

$$T = \sum_m c_m \bar{c}_{m-2k} = \delta_{k0}.$$

This last construction is no accident. Indeed, using the facts that $c_0 c_N \neq 0$ and that the nonzero terms in a row of $(\downarrow 2)D$ overlap nonzero terms from a row of $(\downarrow 2)C$ in exactly $0, 2, 4, \ldots, N+1$ places, you can derive that d must be related to c by $\pm$ a conjugate alternating flip, in order for the rows to be ON.

Now we have to consider the remaining condition (9.52), the orthonormality of the columns of H. This is almost exactly the same as the proof of Theorem 9.12, except that there we chose $N = -1$ and here N is a given positive odd integer. The columns of H are of two types: even (containing only terms c_{2n}, d_{2n}) and odd (containing only terms c_{2n+1}, d_{2n+1}). Thus the requirement that the column vectors of H are ON reduces to three types of identities:

$$\text{even–even} : \sum_\ell c_{2\ell} \bar{c}_{2k+2\ell} + \sum_\ell d_{2\ell} \bar{d}_{2k+2\ell} = \delta_{k0}, \quad (9.58)$$

$$\text{odd–odd} : \sum_\ell c_{2\ell+1} \bar{c}_{2k+2\ell+1} + \sum_\ell d_{2\ell+1} \bar{d}_{2k+2\ell+1} = \delta_{k0}, \quad (9.59)$$

$$\text{odd–even} : \sum_\ell c_{2\ell+1} \bar{c}_{2k+2\ell} + \sum_\ell d_{2\ell+1} \bar{d}_{2k+2\ell} = 0. \quad (9.60)$$

Theorem 9.15 *If the filter* $(\downarrow 2)\mathbf{C}$ *satisfies the double-shift orthogonality condition (9.54) and the filter* $(\downarrow 2)\mathbf{D}$ *is determined by the conjugate alternating flip*

$$d_n = (-1)^n \bar{c}_{N-n}, \qquad n = 0, 1, \ldots, N, \tag{9.61}$$

then condition (9.52) holds and the columns of $\tilde{\Omega}$ *are orthonormal.*

Corollary 9.16 *If the row vectors of* $\tilde{\Omega}$ *form an ON set, then the columns are also ON and* $\tilde{\Omega}$ *is unitary.*

Exercise 9.4 Prove the even–even, odd–odd and odd–even cases of Theorem 9.15.

We can clarify the result of Theorem 9.15 and give an alternate proof by going to the frequency domain. Let's first reexamine the unitarity conditions of Section 9.3. Denote the Fourier transform of the impulse response vector c of the filter $\mathbf{C} = \frac{1}{||h_0||}\Phi^{(0)}$ by $C(\omega)$. (Recall that, by definition, $C(\omega) = \sum_{n=0}^{N} c_n e^{-n\omega}$.) Then the orthonormality of the (double-shifted) rows of $(\downarrow 2)C$ is

$$\int_{-\pi}^{\pi} e^{2ik\omega}|C(\omega)|^2 d\omega = 2\pi\delta_{k0}, \tag{9.62}$$

for integer k. Denote the Fourier transform of the impulse response vector d of the filter $\mathbf{D} = \frac{1}{||h_1||}\Phi^{(1)}$ by $D(\omega)$. Then the orthogonality of the (double-shifted) rows of $(\downarrow 2)D$ to the rows of $(\downarrow 2)C$ is expressed as

$$\int_{-\pi}^{\pi} e^{2ik\omega}C(\omega)\overline{D}(\omega)d\omega = 0 \tag{9.63}$$

for all integers k; similarly the orthonormality of the (double-shifted) rows of $(\downarrow 2)D$ is

$$\int_{-\pi}^{\pi} e^{2ik\omega}|D(\omega)|^2 d\omega = 2\pi\delta_{k0} \tag{9.64}$$

for all integers k.

Now we assume double-shift orthogonality for c and take d to be the conjugate alternating flip of c. In the frequency domain this means that Equation (9.62) holds and that

$$D(\omega) = e^{-iN\omega}\overline{C}(\pi + \omega). \tag{9.65}$$

(Note from this expression that if we choose $\mathbf{C}$ to be a low pass filter, so that $C(0) = 1, C(\pi) = 0$ then the conjugate alternating flip will have $D(0) = 0, |D(\pi)| = 1$ so that $\mathbf{D}$ will be a high pass filter.) We will require only (9.62) and (9.65) and use this to prove that $\tilde{\Omega}$ is unitary. The condition (9.63) for the orthogonality of the rows of $(\downarrow 2)C$ and $(\downarrow 2)D$ becomes

$$\int_{-\pi}^{\pi} e^{2ik\omega} C(\omega)\overline{D}(\omega)d\omega = \int_{-\pi}^{\pi} e^{i(2k+N)\omega} C(\omega)C(\pi+\omega)d\omega =$$

$$(-1)^N \int_{-\pi}^{\pi} e^{i(2k+N)\phi} C(\pi+\phi)C(\phi)d\phi = 0,$$

where $\phi = \pi + \omega$, (since N is odd and $C(\omega)$ is 2π-periodic). Similarly

$$\int_{-\pi}^{\pi} e^{2ik\omega}|D(\omega)|^2 d\omega = \int_{-\pi}^{\pi} e^{2ik\omega}|C(\pi+\omega)|^2 d\omega =$$

$$e^{-2\pi ik} \int_{-\pi}^{\pi} e^{2ik\phi}|C(\phi)|^2 d\phi = 2\pi\delta_{k0},$$

so double-shift orthogonality holds for the rows of D.

These results also enable us to demonstrate the orthonormality of the columns of $\tilde{\Omega}$. For future use we recast condition (9.62) in a new form. Since $C(\omega) = \sum_{n=0}^{N} c_n e^{-in\omega}$, this condition means that the expansion of $|C(\omega)|^2$ looks like

$$|C(\omega)|^2 = 1 + \sum_{m=1}^{N} \left[a_m \cos(2m-1)\omega + b_m \sin(2m-1)\omega \right],$$

i.e., no nonzero even powers of $e^{i\omega}$ occur in the expansion. For $N = 1$ this condition is identically satisfied. For $N = 3, 5, \dots$ it is very restrictive. An equivalent but more compact way of expressing the double-shift orthogonality in the frequency domain is

$$|C(\omega)|^2 + |C(\omega+\pi)|^2 = 2. \tag{9.66}$$

Example 9.17 Show that conditions (9.62) and (9.66) for the Fourier transform $C(\omega)$ are equivalent.

To summarize, if the filter $\mathbf{C}$ satisfies the double-shift orthogonality condition (9.54), or equivalently (9.66), then we can construct a filter $\mathbf{D}$ such that conditions (9.55), (9.56) and (9.52) hold. Thus $\tilde{\Omega}$ is unitary provided double-shift orthogonality holds for the rows of the filter matrix C.

Exercise 9.5 Show that conditions (9.58), (9.59) and (9.60) are equivalent to the equations

$$\int_{-\pi}^{\pi} e^{2ik\omega} \left(|C(\omega) + C(\pi + \omega)|^2 + |D(\omega) + D(\pi + \omega)|^2 \right) d\omega = 8\pi\delta_{k0},$$
(9.67)

$$\int_{-\pi}^{\pi} e^{2ik\omega} \left(|C(\omega) - C(\pi + \omega)|^2 + |D(\omega) - D(\pi + \omega)|^2 \right) d\omega = 8\pi\delta_{k0},$$
(9.68)

$$\int_{-\pi}^{\pi} e^{2ik\omega} \left((C(\omega) - C(\pi + \omega))\overline{(C(\omega) + C(\pi + \omega))} + \right.$$

$$\left. (D(\omega) - D(\pi + \omega))\overline{(D(\omega) + D(\pi + \omega))} \right) d\omega = 0. \qquad (9.69)$$

9.4 The unitary two-channel filter bank system

In this section, we deal with the unitary analysis–synthesis two-channel filter bank system. If $\tilde{\Omega}$ *is* unitary, then (9.52) shows us how to construct a synthesis filter bank to reconstruct the signal:

$$\overline{(\downarrow 2)C}^{\text{tr}}(\downarrow 2)C + \overline{(\downarrow 2)D}^{\text{tr}}(\downarrow 2)D = I.$$

Using the fact that the transpose of the product of two matrices is the product of the transposed matrices in the reverse order, $(EF)^{\text{tr}} = F^{\text{tr}}E^{\text{tr}}$, and that $(\downarrow 2)^{\text{tr}} = (\uparrow 2)$, see (7.5),(7.6), we have

$$\overline{(\downarrow 2)C}^{\text{tr}} = \overline{C}^{\text{tr}}(\downarrow 2)^{\text{tr}} = \overline{C}^{\text{tr}}(\uparrow 2), \quad \overline{(\downarrow 2)D}^{\text{tr}} = \overline{D}^{\text{tr}}(\downarrow 2)^{\text{tr}} = \overline{D}^{\text{tr}}(\uparrow 2).$$

Now, remembering that the order in which we apply operators in (9.52) is from right to left, we see that we have the picture of expression (9.70).

$$\begin{array}{ccccccc}
& \overline{C}^{\text{tr}} & \uparrow 2 \leftarrow & \cdots & \leftarrow \downarrow 2 & C & \\
& \nearrow & & & & & \nwarrow \\
x_{n-N} & Synthesis & & Processing & & Analysis & x_n \\
& \nwarrow & & & & & \swarrow \\
& \overline{D}^{\text{tr}} & \uparrow 2 \leftarrow & \cdots & \leftarrow \downarrow 2 & D &
\end{array}$$
(9.70)

We attach each channel of our two filter bank analysis system to a channel of a two filter bank synthesis system. On the upper channel the analysis filter **C** is applied, followed by downsampling. The output is first upsampled by the upper channel of the synthesis filter bank (which

$$\overline{C}^{\mathrm{tr}} = \begin{bmatrix} \cdot & \cdot & & & \cdot & \cdot \\ \cdot & \bar{c}_0 & \bar{c}_1 & \bar{c}_2 & \bar{c}_3 & \cdot \\ \cdot & 0 & \bar{c}_0 & \bar{c}_1 & \bar{c}_2 & \cdot \\ \cdot & 0 & 0 & \bar{c}_0 & \bar{c}_1 & \cdot \\ \cdot & 0 & 0 & 0 & \bar{c}_0 & \cdot \\ \cdot & \cdot & & & \cdot & \cdot \end{bmatrix}.$$

Figure 9.9 $\overline{C}^{\mathrm{tr}}$ matrix.

inserts zeros between successive terms of the upper analysis filter) and then filtered by $\overline{\mathbf{C}}^{\mathrm{tr}}$. (Here we mean the linear operator whose matrix with respect to the standard y-signal basis is $\overline{C}^{\mathrm{tr}}$.) On the lower channel the analysis filter $\mathbf{D}$ is applied, followed by downsampling. The output is first upsampled by the lower channel of the synthesis filter bank and then filtered by $\overline{\mathbf{D}}^{\mathrm{tr}}$. The outputs of the two channels of the synthesis filter bank are then added to reproduce the original signal.

There is still one problem. The transpose conjugate looks like Figure 9.9. The synthesis filter is *not* causal! The output of the filter at time t depends on the input at times $t + k$, $k = 0, 1, \ldots, N$. To ensure that we have causal filters we insert time delays $\mathbf{R}^N$ before the action of the synthesis filters, i.e., we replace $\overline{\mathbf{C}}^{\mathrm{tr}}$ by $\overline{\mathbf{C}}^{\mathrm{tr}}\mathbf{R}^N$ and $\overline{\mathbf{D}}^{\mathrm{tr}}$ by $\overline{\mathbf{D}}^{\mathrm{tr}}\mathbf{R}^N$. The resulting filters are causal (hence practical to build), and we have reproduced the original signal with a time delay of N, see expression (9.71).

$$\mathbf{R}^N\overline{\mathbf{C}}^{\mathrm{tr}} \quad \uparrow 2 \quad \leftarrow \quad \cdots \quad \leftarrow \quad \downarrow 2 \quad \mathbf{C}$$

$$\mathbf{R}^N x \quad = \qquad + \qquad\qquad\qquad\qquad\qquad\qquad x \qquad (9.71)$$

$$\mathbf{R}^N\overline{\mathbf{D}}^{\mathrm{tr}} \quad \uparrow 2 \quad \leftarrow \quad \cdots \quad \leftarrow \quad \downarrow 2 \quad \mathbf{D}$$

Exercise 9.6 Compute the matrices R, L of the shift operators $\mathbf{R}$ and $\mathbf{L}$.

Are there filters that actually satisfy these conditions? In the next section we will exhibit a simple solution for $N = 1$. The derivation of solutions for $N = 3, 5, \ldots$ is highly nontrivial but highly interesting, as we shall see.

9.5 A perfect reconstruction filter bank with $N = 1$

From the results of the last section, we can design a two-channel filter bank $\tilde{\Omega}$ with perfect reconstruction provided the rows of the filter $(\downarrow 2)\mathbf{C}$ are double-shift orthogonal. For general N this is a strong restriction, for $N = 1$ it is satisfied by *all* filters. Since there are only two nonzero terms in a row c_1, c_0, all double-shifts of the row are automatically orthogonal to the original row vector. It is conventional to choose $\mathbf{\Phi}^{(0)}$ to be a low pass filter, so that in the frequency domain $H^{(0)}(0) = 1, H^{(0)}(\pi) = 0$.

This uniquely determines $\mathbf{\Phi}^{(0)}$ as the moving average $\mathbf{\Phi}^{(0)} = \frac{1}{2}\mathbf{I} + \frac{1}{2}\mathbf{R}$. The frequency response is $H^{(0)}(\omega) = \frac{1}{2} + \frac{1}{2}e^{-in\omega}$ and the Z transform is $H^{(0)}[z] = \frac{1}{2} + \frac{1}{2}z^{-1}$. The norm of the impulse response vector $(1/2, 1/2)$ is $\|h^{(0)}\| = 1/\sqrt{2}$, so the normalized filter $\mathbf{C}$ is determined by $(c_0, c_1) = (h_0^{(0)}, h_1^{(0)})/\|h^{(0)}\| = (1/\sqrt{2}, 1/\sqrt{2})$. Applying the conjugate alternating flip to c we get the normalized impulse response function $\sqrt{2}(1/2, -1/2)$ of the moving difference filter, a high pass filter. Thus $\mathbf{D} = \sqrt{2}(\frac{1}{2}\mathbf{I} - \frac{1}{2}\mathbf{R})$.

Exercise 9.7 Compute the matrix forms of the action of the moving difference and moving average filters $\mathbf{C}, \mathbf{D}$ in the time domain.

The outputs of the upper and lower channels of the analysis filter bank are

$$(\downarrow 2)Cx_n = \frac{1}{\sqrt{2}}(x_{2n} + x_{2n-1}), \qquad (\downarrow 2)Dx_n = \frac{1}{\sqrt{2}}(x_{2n} - x_{2n-1}),$$
$$\tag{9.72}$$

and we see that full information about the signal is still present.

Exercise 9.8 Derive the output of the $N = 1$ upper and lower synthesis filters in analogy to expressions (9.72).

Exercise 9.9 Use the results of Exercise 9.8 and expressions (9.72), delay the output of the synthesis filters by 1 unit for causality, and add the results. Show that you get at the nth step x_{n-1}, the original signal with a delay of 1.

Exercise 9.10 Compute the matrices of the two channels of the $N = 1$ analysis filter bank. Do the same for the two channels of the synthesis filter bank.

9.6 Perfect reconstruction for two-channel filter banks

Now we are ready to investigate the general conditions for perfect reconstruction for a two-channel filter bank. The picture that we have in mind is that of expression (9.73).

$$
\begin{array}{ccccc}
\boldsymbol{\Psi}^{(0)} & \uparrow 2 & \leftarrow & \downarrow 2 & \boldsymbol{\Phi}^{(0)} \\
& & & & \nwarrow \\
\mathbf{R}^{\ell} x \quad = \quad + & & & & \quad x \qquad (9.73) \\
& & & & \swarrow \\
\boldsymbol{\Psi}^{(1)} & \uparrow 2 & \leftarrow & \downarrow 2 & \boldsymbol{\Phi}^{(1)}
\end{array}
$$

The analysis filter $\boldsymbol{\Phi}^{(0)}$ will be low pass and the analysis filter $\boldsymbol{\Phi}^{(1)}$ will be high pass, with impulse response vectors $h_n^{(0)}, h_n^{(1)}$ that have $N_0 + 1$ and $N_1 + 1$ taps, respectively. The corresponding synthesis filters are $\boldsymbol{\Psi}^{(0)}$ and $\boldsymbol{\Psi}^{(1)}$ with impulse response vectors $s^{(0)}$ and $s^{(1)}$, respectively. We will not impose unitarity, but the less restrictive condition of invertibility (with delay). This will require that the row and column vectors of

$$
\begin{pmatrix}
(\downarrow 2)\boldsymbol{\Phi}^{(0)} \\
(\downarrow 2)\boldsymbol{\Phi}^{(1)}
\end{pmatrix}
$$

are *biorthogonal*. Unitarity is a special case of this.

The operator condition for perfect reconstruction with delay ℓ is

$$
\boldsymbol{\Psi}^{(0)}(\uparrow 2)(\downarrow 2)\boldsymbol{\Phi}^{(0)} + \boldsymbol{\Psi}^{(1)}(\uparrow 2)(\downarrow 2)\boldsymbol{\Phi}^{(1)} = \mathbf{R}^{\ell}
$$

where $\mathbf{R}$ is the shift. If we apply the operators on both sides of this requirement to a signal $x = \{x_n\}$ and take the Z transform, we find

$$
\frac{1}{2} S^{(0)}[z] \left(H^{(0)}[z]X[z] + H^{(0)}[-z]X[-z] \right)
$$

$$
+ \frac{1}{2} S^{(1)}[z] \left(H^{(1)}[z]X[z] + H^{(1)}[-z]X[-z] \right) = z^{-\ell}X[z], \qquad (9.74)
$$

where $X[z]$ is the Z transform of x and $S^{(k)}[z]$ is the Z transform associated with the synthesis filter $\boldsymbol{\Psi}^{(k)}$. The coefficient of $X[-z]$ on the left-hand side of this equation is an aliasing term, due to the downsampling and upsampling. For perfect reconstruction of a general signal $X[z]$ this coefficient must vanish. Thus we have

Theorem 9.18 *A two-channel filter bank gives perfect reconstruction if and only if*

$$S^{(0)}[z]H^{(0)}[z] + S^{(1)}[z]H^{(1)}[z] = 2z^{-\ell} \tag{9.75}$$

$$S^{(0)}[z]H^{(0)}[-z] + S^{(1)}[z]H^{(1)}[-z] = 0. \tag{9.76}$$

We can satisfy requirement (9.76) by defining the synthesis filters in terms of the analysis filters:

$$S^{(0)}[z] = H^{(1)}[-z], \qquad S^{(1)}[z] = -H^{(0)}[-z]. \tag{9.77}$$

Now we focus on requirement (9.75). We introduce the *product filter*

$$P^{(0)}[z] = S^{(0)}[z]H^{(0)}[z] = H^{(1)}[-z]H^{(0)}[z].$$

Note that requirement (9.75) can now be written as

$$P^{(0)}[z] - P^{(0)}[-z] = 2z^{-\ell}, \tag{9.78}$$

where $P^{(0)}$ is a polynomial of order $2N$ in z^{-1}. Clearly, the even powers of z in $P^{(0)}[z]$ cancel out of (9.78); the restriction is only on the odd powers. This also tells us that ℓ must be an odd integer. (In particular, it can never be 0.) Further, since $P^{(0)}[z]$ is a polynomial in z^{-1} of exact order $N_0 + N_1$, necessarily $N_0 + N_1 = 2N$ is an even integer and $0 < \ell < 2N$. A further simplification involves recentering $P^{(0)}$ to factor out the delay term. Set $P[z] = z^{\ell}P^{(0)}[z]$. Then Equation (9.78) becomes the *halfband filter* equation

$$P[z] + P[-z] = 2. \tag{9.79}$$

This equation says the coefficients of the even powers of z in $P[z]$ vanish, except for the constant term, which is 1. The coefficients of the odd powers of z are undetermined design parameters for the filter bank. The highest power of z in $P[z]$ is $\ell > 0$ and the lowest power is $-(2N - \ell)$.

Given a two-channel filter bank with perfect reconstruction, we have found a halfband filter P. Now we can reverse the process. Given any halfband filter P satisfying (9.79), we can define a perfect reconstruction filter bank corresponding to any factorization $P = z^{\ell}H^{(1)}[-z]H^{(0)}[z]$, and use of the (9.76) solution to get $S^{(0)}, S^{(1)}$. To make contact with our earlier work on perfect reconstruction by unitarity, note that if we define $H^{(1)}[z]$ from $H^{(0)}[z]$ through the conjugate alternating flip (the condition for unitarity)

$$H^{(1)}[z] = z^{-N_0}\overline{H^{(0)}[-\overline{z^{-1}}]}, \tag{9.80}$$

then $N_0 = N_1 = N$, and $P^{(0)}[z] = z^{-N}\overline{H^{0)}[\overline{z}^{-1}]}H^{(0)}[z]$. Setting $z = e^{i\omega}$, substituting into (9.78) and taking the complex conjugate of both sides, we see that $\ell = N$ and

$$P(\omega) = \overline{H^{(0)}(\omega)}H^{(0)}(\omega) = |H^{(0)}(\omega)|^2. \qquad (9.81)$$

Exercise 9.11 Work out the details showing that the unitarity condition (9.80) implies $\ell = N$ and the expression (9.81) for the halfband filter.

At this point we will restrict our analysis to the symmetric case where the delay is $\ell = N$. It is this case that occurs in the study of orthogonal wavelets and biorthogonal wavelets, and, as we have seen, for unitary filter banks. Thus any trigonometric polynomial of the form

$$P[z] = 1 + \sum_{n \ \text{odd}/n=-N}^{N} a_n z^{-n}$$

will satisfy Equation (9.79). The constants a_n are design parameters that we can adjust to achieve desired performance from the filter bank. Once $P[z]$ is chosen then we have to factor it as $P[z] = z^N H^{(0)}[z]H^{(1)}[-z]$. Where $H^{(j)}$ is a polynomial in z^{-1} of order N_j, $j = 0, 1$ and $N_0 + N_1 = N$. In theory this can always be done. Indeed $z^N P[z]$ is a true polynomial in z and, by the fundamental theorem of algebra, polynomials over the complex numbers can always be factored completely: $z^N P[z] = A \prod_j (z - z_j)$. Then we can define $H^{(0)}$ and $H^{(1)}$ (but not uniquely!) by assigning N_0 of the linear factors to $H^{(0)}[z]$ and N_1 to $H^{(1)}[-z]$. If we want $H^{(0)}$ to be a low pass filter then we must require that $z = -1$ is a root of $P[z]$; if $S^{(0)}[z]$ is also to be low pass then $P[z]$ must have -1 as a double root. If $P[z]$ is to correspond to a unitary filter bank then we must have $P[e^{i\omega}] = |H^{(0)}[e^{i\omega}]|^2 \geq 0$ which is a strong restriction on the roots of $P[z]$.

9.7 Halfband filters and spectral factorization

We return to our consideration of unitary two-channel filter banks in the frequency domain. We have reduced the design problem for these filter banks to the construction of a low pass filter **C** whose rows satisfy the double-shift orthonormality requirement. In the frequency domain this takes the form

$$|C(\omega)|^2 + |C(\omega + \pi)|^2 = 2. \qquad (9.82)$$

(Recall that $C(\omega) = \sum_{n=0}^{N} c_n e^{-in\omega}$.) To determine the possible ways of constructing $C(\omega)$ we focus our attention on the halfband filter $P(\omega) = |C(\omega)|^2$, called the *power spectral response* of **C**. The frequency requirement on C can now be written as the halfband filter condition (9.79)

$$P[z] + P[-z] = 2.$$

Note also that

$$P(\omega) = \sum_{n=-N}^{N} p_n e^{-in\omega} = \left(\sum_{k=0}^{N} c_k e^{-ik\omega} \right) \left(\sum_{k=0}^{N} \overline{c}_k e^{ik\omega} \right).$$

Thus

$$p_n = \sum_{k} c_k \overline{c}_{k-n} = c * \overline{c}_n^{\mathrm{T}}, \qquad (9.83)$$

where $c_n^{\mathrm{T}} = c_{N-n}$ is the time reversal of c. Since $P(\omega) \geq 0$ we have $p_{-n} = \overline{p}_n$. The even coefficients of p can be obtained from the halfband filter condition (9.79):

$$p_{2m} = \sum_{k} c_k \overline{c}_{k-2m} = \delta_{m0}. \qquad (9.84)$$

The odd coefficients of p are undetermined. Note also that **P** is *not* a causal filter. One further comment: since **C** is a low pass filter, $C(\pi) = 0$ and $C(0) = \sqrt{2}$. Thus $P(\omega) \geq 0$ for all ω, $P(0) = 2$ and $P(\pi) = 0$.

If we find a nonnegative polynomial halfband filter $P[z]$, we are guaranteed that it can be factored as a perfect square.

Theorem 9.19 *(Fejér–Riesz) A trigonometric polynomial $p[e^{-i\omega}] = \sum_{n=-N}^{N} p_n e^{-in\omega}$, real and nonnegative for all ω, can be expressed as*

$$p[e^{-i\omega}] = |C[e^{-i\omega}]|^2$$

where $C[z] = \sum_{j=0}^{N} c_j z^{-j}$ is a polynomial in z^{-1}. The polynomial $C[z]$ can be chosen such that it has no roots outside the unit disk $|z| > 1$, in which case it is unique up to multiplication by a complex constant of modulus 1.

We will prove this shortly. First some halfband filter examples.

Example 9.20 $N = 1$

$$P[z] = 1 + \frac{z^{-1} + z}{2}, \qquad \text{or } P(\omega) = 1 + \cos \omega.$$

Here $p_0 = 1, p_1 = p_{-1} = 1/2$. This factors as

$$P(\omega) = |C(\omega)|^2 = \frac{1}{2}(1 + e^{-i\omega})(1 + e^{i\omega}) = 1 + \cos\omega$$

and leads to the moving average filter $C(\omega)$.

Example 9.21 $N = 3$ The Daubechies 4-tap filter.

$$P[z] = (1 + \frac{z^{-1} + z}{2})^2 (1 - \frac{z^{-1} + z}{4}), \text{ or } P(\omega) = (1 + \cos\omega)^2 (1 - \frac{1}{2}\cos\omega).$$

Here

$$P[z] = -\frac{1}{16}z^3 + \frac{9}{16}z + 1 + \frac{9}{16}z^{-1} - \frac{1}{16}z^{-3}.$$

Note that there are no nonzero even powers of z in $P[z]$. $P(\omega) \geq 0$ because one factor is a perfect square and $(1 - \frac{1}{2}\cos\omega) > 0$. Complete factorization of $P[z]$ isn't trivial, but is not too hard because we have already factored the term $1 + \frac{z^{-1}+z}{2}$ in our first example. Thus we have only to factor $(1 - \frac{z^{-1}+z}{4}) = (a^+ + a^- z^{-1})(a^+ + a^- z)$. The result is $a^\pm = (1 \pm \sqrt{3})/\sqrt{8}$. Finally, we get

$$C[z] = \frac{1}{4\sqrt{2}}(1 + z^{-1})^2 \left((1 + \sqrt{3}) + (1 - \sqrt{3})z^{-1}\right) \qquad (9.85)$$

$$= \frac{1}{4\sqrt{2}} \left((1 + \sqrt{3}) + (3 + \sqrt{3})z^{-1} + (3 - \sqrt{3})z^{-2} + (1 - \sqrt{3})z^{-3}\right).$$

NOTES: (1) Only the expressions a^+a^- and $(a^+)^2 + (a^-)^2$ were determined by the above calculation. We chose the solution such that all of the roots were on or inside the circle $|z| = 1$. There are four possible solutions and all lead to FIR filter banks, though not all to unitary filter banks. Instead of choosing the factors so that $P = |C|^2$ we can divide them in a different way to get $P = S^{(0)}H^{(0)}$ where $H^{(1)}$ is not the conjugate of $H^{(0)}$. This would be a biorthogonal filter bank. (2) Due to the repeated factor $(1 + z^{-1})^2$ in $C[z]$, it follows that $C(\omega)$ has a double zero at $\omega = \pi$. Thus $C(\pi) = 0$ and $C'(\pi) = 0$ and the response is *flat*. Similarly the response is flat at $\omega = 0$ where the derivative also vanishes. We shall see that it is highly desirable to maximize the number of derivatives of the low pass filter Fourier transform that vanish near $\omega = 0$ and $\omega = \pi$, both for filters and for application to wavelets. Note that the flatness property means that the filter has a relatively wide pass band and then a fast transition to a relatively wide stop band.

Proof of the Fejér–Riesz Theorem: Since $p[e^{i\omega}]$ is real we must have $p[\overline{z}^{-1}] = p[z]$. Thus, if z_j is a root of p then so is $\overline{z}_j{}^{-1}$. It follows that the roots of p not on the unit circle $|z| = 1$ must occur in pairs $z_j, \overline{z}_j{}^{-1}$ where $|z_j| < 1$. Since $p(e^{i\omega}) \geq 0$ each of the roots $w_k = e^{i\theta_k}$, $0 \leq \theta_k < 2\pi$ on the unit circle must occur with even multiplicity and the factorization must take the form

$$p[z] = \alpha^2 \Pi_{j=1}^M (1 - \frac{z_j}{z})(1 - z\overline{z}_j)\Pi_{k=1}^{N-M}(1 - \frac{w_k}{z})(1 - \frac{z}{w_k}) \qquad (9.86)$$

where $\alpha^2 = p_N / [\Pi_j \overline{z}_j \Pi_k w_k]$ and $\alpha \geq 0$. $\qquad\qquad\square$

Comments on the proof:

1. If the coefficients p_n of $p[z]$ are also real, as is the case with most of the examples in the text, then we can say more. We know that the roots of equations with real coefficients occur in complex conjugate pairs. Thus, if z_j is a root inside the unit circle, then so is $\overline{z}_j$, and then $z_j^{-1}, \overline{z}_j{}^{-1}$ must be roots outside the unit circle. Except for the special case when z_j is real, these roots will come four at a time. Furthermore, if w_k is a root on the unit circle, then so is $\overline{w}_k$, so non real roots on the unit circle also come four at a time: $w_k, w_k, \overline{w}_k, \overline{w}_k$. The roots ± 1 if they occur, will have even multiplicity.

2. From (9.86) we can set

$$C[z] = \alpha\Pi_{j=1}^M (1 - \frac{z_j}{z})\Pi_{k=1}^{N-M}(1 - \frac{w_k}{z})$$

thus uniquely defining C by the requirement that it has no roots outside the unit circle. Then $P[z] = |C[z]|^2$. On the other hand, we could factor P in different ways to get $P[z] = S[z]H[z]$. The allowable assignments of roots in the factorizations depends on the required properties of the filters S, H. For example if we want S, H to be filters with real coefficients then each complex root z_0 must be assigned to the same factor as $\overline{z}_0$.

9.8 Maxflat filters

We turn to maxflat filters. These are unitary FIR filters $\mathbf{C}$ with maximum flatness at $\omega = 0$ and $\omega = \pi$. $C(\omega)$ has exactly p zeros at $\omega = \pi$ and $N = 2p - 1$. The first member of the family, $p = 1$, is the moving average filter $(c_0, c_1) = (1/\sqrt{2}, 1/\sqrt{2})$, where $C[z] = (1 + z^{-1})/\sqrt{2}$. For general p the associated halfband filter $P(\omega) = |C(\omega)|^2$ takes the form

$$P[z] = (\frac{1 + z^{-1}}{2})^{2p}Q_{2p-2}[z], \qquad (9.87)$$

where P has degree $2N = 4p - 2$. (Note that it must have $2p$ zeros at $z = -1$.) The problem is to compute $Q_{2p-2}[z]$, where the subscript denotes that Q has exactly $2p - 2$ roots.

Since for $z = e^{i\omega}$ we have

$$\left(\frac{1 + z^{-1}}{2}\right)^2 = e^{-i\omega} \cos^2\left(\frac{\omega}{2}\right) = e^{-i\omega}\left(\frac{1 + \cos\omega}{2}\right),$$

this means that that $P(\omega)$ has the factor $(1 + \cos\omega)^p$. Now consider the case where

$$P(\omega) = \sum_{n=1-2p}^{2p-1} p_n e^{-in\omega}$$

and the p_n are *real* coefficients, i.e., the filter coefficients c_j are real. Since $p_0 = 1$, $p_2 = p_4 = \cdots = p_{2p-2} = 0$ and $p_n = p_{-n}$ is real for n odd, it follows that

$$P(\omega) = \left(\frac{1 + \cos\omega}{2}\right)^p \tilde{Q}_{p-1}(\cos\omega)$$

where $\tilde{Q}_{p-1}$ is a polynomial in $\cos\omega$ of order $p - 1$. Indeed $P(\omega)$ is a linear combination of terms in $\cos n\omega$ for n odd. For any nonnegative integer n one can express $\cos n\omega$ as a polynomial of order n in $\cos\omega$. An easy way to see that this is true is to use the formula

$$e^{in\omega} = \cos n\omega + i\sin n\omega = (e^{i\omega})^n = (\cos\omega + i\sin\omega)^n.$$

Taking the real part of these expressions and using the binomial theorem, we obtain

$$\cos n\omega = \sum_{j=0,\ldots[\frac{n}{2}]} \binom{n}{2j} (-1)^j \sin^{2j}\omega \cos^{n-2j}\omega.$$

Since $\sin^{2j}\omega = (1 - \cos^2\omega)^j$, the right-hand side of the last expression is a polynomial in $\cos\omega$ of order n.

We already have enough information to determine $\tilde{Q}_{p-1}(\cos\omega)$ uniquely! For convenience we introduce a new variable

$$y = \frac{1 - \cos\omega}{2} \quad \text{so that } 1 - y = \frac{1 + \cos\omega}{2}.$$

As ω runs over the interval $0 \leq \omega \leq \pi$, y runs over the interval $0 \leq y \leq 1$. Considered as a function of y, P will be a polynomial of order $2p - 1$ and of the form

$$P[y] = 2(1 - y)^p K_{p-1}[y]$$

where K_{p-1} is a polynomial in y of order $p-1$. Furthermore $P[0] = 2$. The halfband filter condition now reads

$$P[y] + P[1 - y] = 2,$$

so we have

$$(1 - y)^p K_{p-1}[y] = 1 - y^p K_{p-1}[1 - y]. \tag{9.88}$$

Dividing both sides of this equation by $(1 - y)^p$ we have

$$K_{p-1}[y] = (1 - y)^{-p} - y^p (1 - y)^{-p} K_{p-1}[1 - y].$$

Since the left-hand side of this identity is a polynomial in y of order $p-1$, the right-hand side must also be a polynomial. Thus we can expand both terms on the right-hand side in a power series in y and throw away all terms of order y^p or greater, since they must cancel to zero. Since all terms in the expansion of $y^p (1 - y)^{-p} K_{p-1}[1 - y]$ will be of order y^p or greater we can forget about those terms. The power series expansion of the first term is

$$(1 - y)^{-p} = \sum_{k=0}^{\infty} \binom{p + k - 1}{k} y^k,$$

and taking the terms up to order y^{p-1} we find

$$K_{p-1}[y] = \sum_{k=0}^{p-1} \binom{p + k - 1}{k} y^k.$$

Theorem 9.22 *The halfband response for the maxflat filter with p zeros is*

$$P(\omega) = 2(\frac{1 + \cos \omega}{2})^p \sum_{k=0}^{p-1} \binom{p + k - 1}{k} (\frac{1 - \cos \omega}{2})^k. \tag{9.89}$$

From this result one can use the Equations (9.83) and other facts about unitary filter banks to solve for the real coefficients c_n, at least numerically. When translated back to the Z transform, the maxflat halfband filters with $2p$ zeros at $\omega = \pi$ factor to the unitary low pass Daubechies filters C with $N = 2p - 1$. The notation for the Daubechies filter with $N = 2p - 1$ is D_{N+1}. An alternate notation is *dbp* where $2p$ is the number of zeros. We have already exhibited D_4 as an example. Of course, D_2 is the "moving average" filter.

Example 9.23 Daubechies filter coefficients are generally available in standard wavelets software packages, e.g., the MATLAB wavelet toolbox.

The command

```
F = dbwavf('dbp')
```

returns the scaling filter associated with Daubechies wavelet *dbp* (or D_{2p} where p takes values from 1 to 45). Thus, setting the printout for 4 digit accuracy we get

```
>>F = dbwavf('db1')

F =      0.5000     0.5000

>> G=dbwavf('db2')

G =      0.3415     0.5915     0.1585     -0.0915

>> H=dbwavf('db3')

H = 0.2352 0.5706 0.3252 -0.0955 -0.0604 0.0249

>> I=dbwavf('db4')
  = 0.1629 0.5055 0.4461 -0.0198 -0.1323 0.0218
    0.0233 -0.0075
```

For more accuracy, say 12 digits use

```
vpa(dbwavf('db3'),12)

ans =
[0.235233603893,0.570558457917,0.325182500264,
-0.0954672077843,-0.0604161041554,0.0249087498659]
```

Exercise 9.12 Verify that the Daubechies filter D_4, Example 9.21, satisfies the general expression given in Theorem 9.22 for the case $p = 2$.

This solution of the halfband filter equations in the maxflat case, due to Daubechies, is the most convenient one, but there are many others. As a simple example, if we replace $\cos \omega$ by $\sin \omega$ everywhere in expression (9.89) we will still satisfy the basic equation $P[z] + P[-z] = 2$. Now however, the odd n coefficients p_n will be pure imaginary, as will the c_k, d_k coefficients. The C filter will have a stop band centered on $\omega = \pi/2$

and the D filter will have a stop band centered on $\omega = 3\pi/2$, so they can not be considered as low pass and high pass. Nevertheless, the filter bank will work.

There is another expression for the Daubechies halfband filter $P(\omega)$, due to Meyer, that is also very useful. Differentiating the expressions

$$P[y] = (1 - y)^p K_{p-1}[y] = 1 - y^p K_{p-1}[1 - y]$$

with respect to y, we see that $P'[y]$ is divisible by y^{p-1} and also by $(1 - y)^{p-1}$. Since $P'[y]$ is a polynomial of order $2p - 2$ it follows that

$$P'[y] = ky^{p-1}(1 - y)^{p-1},$$

for some constant k. Now $dy = \frac{1}{2}\sin\omega \, d\omega$ so changing variables from y to ω we have

$$P'(\omega) = \frac{dy}{d\omega}P'[y] = \frac{k\sin\omega}{2^{2p-1}}\sin^{2p-2}\omega = -c\sin^{2p-1}\omega.$$

Then

$$P(\omega) = 2 - c\int_0^\omega \sin^{2p-1}\omega \, d\omega \qquad (9.90)$$

where the constant c is determined by the requirement $P(\pi) = 0$. Integration by parts yields

$$\int_0^\pi \sin^{2p-1}\omega \, d\omega = \frac{2^{2p-1}[(p-1)!]^2}{(2p-1)!} = \frac{\sqrt{\pi}\Gamma(p)}{\Gamma(p+\frac{1}{2})},$$

where $\Gamma(z)$ is the gamma function. Thus

$$c = \frac{2\Gamma(p+\frac{1}{2})}{\sqrt{\pi}\Gamma(p)}.$$

Stirling's formula says

$$\Gamma(z) \sim z^{z-\frac{1}{2}}e^{-z}\sqrt{2\pi}\left(1 + O(\frac{1}{z})\right),$$

[3], page 18, so $c \sim \sqrt{4p/\pi}$ as $p \to \infty$. Since $P'(\pi/2) = -c$, we see that the slope at the center of the maxflat filter is proportional to $\sqrt{N}$. Moreover $P(\omega)$ is monotonically decreasing as ω goes from 0 to π. One can show that the transition band gets more and more narrow.

A basic reference on maxflat filters and their relation to wavelets is [52].

9.9 Low pass iteration and the cascade algorithm

We leave aside filter bank theory temporarily and return to consideration of wavelets. We still have presented no provable examples of father wavelets other than a few that have been known for almost a century. We turn to the problem of determining new multiresolution structures. Up to now we have mostly been accumulating *necessary conditions* that must be satisfied for a multiresolution structure to exist, as well as the implications of the existence of such a structure. Our focus has been on the coefficient vectors c and d of the dilation and wavelet equations and we have used filter bank theory to find solutions of the algebraic conditions satisfied by these coefficient vectors. Now we will gradually change our point of view and search for more restrictive *sufficient conditions* that will *guarantee* the existence of a multiresolution structure. Further we will study the problem of actually computing the scaling function and wavelets from a knowledge of the coefficient vectors alone. In this section we will focus on the time domain. In the next section we will go to the frequency domain, where new insights emerge. Our work with Daubechies filter banks will prove invaluable, since these filter banks are all associated with wavelets.

Our main emphasis will be on the dilation equation

$$\phi(t) = \sqrt{2} \sum_k c_k \phi(2t - k). \tag{9.91}$$

We have already seen that if a scaling function satisfies this equation, then we can define d from c by a conjugate alternating flip and use the wavelet equation to generate the wavelet basis. Our primary interest is in scaling functions ϕ with support in a finite interval.

If ϕ has finite support then by translation in time if necessary, we can assume that the support is contained in the interval $[0, N)$ and the integer N is as small as possible. With such a $\phi(t)$ note that even though the right-hand side of (9.91) could conceivably have an infinite number of nonzero c_k, for fixed t there are only a finite number of nonzero terms. Suppose the coefficients c_k can be nonzero only for k in the interval $N_1 \leq k \leq N_2$ (where we allow for the possibility that $N_1 = -\infty$, or $N_2 = \infty$). Then the support of the function of t on the right-hand side of (9.91) is contained in $[\frac{N_1}{2}, \frac{N+N_2}{2})$. Since the support of both sides is the same, we must have $N_1 = 0, N_2 = N$. Thus c has only $N + 1$ terms that could be nonzero: $c_0, c_1, \ldots, c_N$. Further, N must be odd, in order that c satisfy the double-shift orthogonality conditions.

Lemma 9.24 *If the scaling function $\phi(t)$ (corresponding to a multiresolution analysis) has compact support on $[0, N)$, then c_k must vanish unless $0 \leq k \leq N$.*

Recall that c also must obey the double-shift orthogonality conditions $\sum_k c_k \bar{c}_{k-2m} = \delta_{0m}$ and the compatibility condition $\sum_{k=0}^{N} c_k = \sqrt{2}$ between the unit area normalization $\int \phi(t) dt = 1$ of the scaling function and the dilation equation.

One way to determine a scaling function $\phi(t)$ from the impulse response vector c is to iterate the low pass filter $\mathbf{C}$. That is, we start with an initial guess $\phi^{(0)}(t)$, say the box function on $[0, 1)$, and then iterate

$$\phi^{(i+1)}(t) = \sqrt{2} \sum_{k=0}^{N} c_k \phi^{(i)}(2t - k) \tag{9.92}$$

for $i = 1, 2, \ldots$. Note that if $\phi^{(0)}(t)$ is the box function, then $\phi^{(i)}(t)$ will be a piecewise constant function, constant on intervals of length 2^{-j}. If $\lim_{i \to \infty} \phi^{(i)}(t) \equiv \phi(t)$ exists for each t then the limit function satisfies the dilation equation (9.91). This is called the *cascade algorithm*, due to the iteration by the low pass filter. (It is analogous to Newton's method in calculus for finding a solution of the equation $f(x) = 0$ where f has a continuous derivative in an interval containing the solution. One starts with a guess $x = x_0$ and generates a series of improved approximations $x_1, x_2, \ldots, x_i, \ldots$ using the update $x_{i+1} = \mathbf{T}(x_i) = x_i - f(x_i)/f'(x_i)$. If the method converges then $f(r) = 0$ where $r = \lim_{i \to \infty} x_i$. Furthermore the solution satisfies $\mathbf{T}(r) = r$.)

Of course we don't know in general that the algorithm will converge. (We will find a sufficient condition for convergence when we look at this algorithm in the frequency domain.) For now we explore the implications of uniform convergence on $[-\infty, \infty]$ of the sequence $\phi^{(i)}(t)$ to $\phi(t)$.

We claim that the support of ϕ is contained in the interval $[0, N)$. To see this note that, first, the initial function $\phi^{(0)}$ has support in $[0, 1)$. After filtering once, we see that the new function $\phi^{(1)}$ has support in $[0, \frac{1+N}{2})$. Iterating i times, we see that $\phi^{(i)}$ has support in $[0, \frac{1+[2^i-1]N}{2^i})$. Thus if $\lim_{i \to \infty} \phi^{(i)}(t) = \phi(t)$ pointwise, then $\phi^{(0)}$ has support in $[0, N)$.

Note that at level $i = 0$ the scaling function and associated wavelets are orthonormal:

$$(w_{jk}^{(0)}, w_{j'k'}^{(0)}) = \delta_{jj'}\delta_{kk'}, \qquad (\phi_{jk}^{(0)}, w_{j'k'}^{(0)}) = 0$$

where $j, j', \pm k, \pm k' = 0, 1, \ldots$. (Of course it is *not* true in general that $\phi_{jk}^{(0)} \perp \phi_{j'k'}^{(0)}$ for $j \neq j'$.) These are just the orthogonality relations for the

Haar wavelets. This orthogonality is maintained through each iteration, and if the cascade algorithm converges uniformly, it applies to the limit function ϕ:

Theorem 9.25 *If the cascade algorithm converges uniformly in t then the limit function $\phi(t)$ and associated wavelet $w(t)$ satisfy the orthogonality relations*

$$(w_{jk}, w_{j'k'}) = \delta_{jj'}\delta_{kk'}, \qquad (\phi_{jk}, w_{jk'}) = 0, \qquad (\phi_{jk}, \phi_{jk'}) = \delta_{kk'}$$

where $j, j', \pm k, \pm k' = 0, 1, \ldots$

Proof There are only three sets of identities to prove:

$$\int_{-\infty}^{\infty} \phi(t-n)\overline{\phi(t-m)}dt = \delta_{nm} \tag{9.93}$$

$$\int_{-\infty}^{\infty} \phi(t-n)\overline{w(t-m)}dt = 0 \tag{9.94}$$

$$\int_{-\infty}^{\infty} w(t-n)\overline{w(t-m)}dt = \delta_{nm}. \tag{9.95}$$

The rest are immediate.

We will use induction. If (9.93) is true for the function $\phi^{(i)}(t)$ we will show that it is true for the function $\phi^{(i+1)}(t)$.

Clearly it is true for $\phi^{(0)}(t)$. Now

$$\int_{-\infty}^{\infty} \phi^{(i+1)}(t-n)\overline{\phi^{(i+1)}(t-m)}dt = (\phi_{0n}^{(i+1)}, \phi_{0m}^{(i+1)})$$

$$= (\sum_k c_k \phi_{1,2n+k}^{(i)}, \sum_\ell c_\ell \phi_{1,2m+\ell}^{(i)})$$

$$= \sum_{k\ell} c_k \overline{c_\ell} \, (\phi_{1,2n+k}^{(i)}, \phi_{1,2m+\ell}^{(i)}) = \sum_k c_k \overline{c_{k-2(m-n)}} = \delta_{nm}.$$

Since the convergence is uniform and $\phi(t)$ has compact support, these orthogonality relations are also valid for $\phi(t)$. The proofs of (9.94) and (9.95) are similar. □

Exercise 9.13 Use induction with the double-shift orthogonality of c and d to prove identities (9.94) and (9.95).

Note that most of the proof of the theorem doesn't depend on convergence. It simply relates properties at the ith recursion of the cascade algorithm to the same properties at the $(i+1)$-st recursion.

Corollary 9.26 *If the the orthogonality relations*

$$(w_{jk}^{(i)}, w_{j'k'}^{(i)}) = \delta_{jj'}\delta_{kk'}, \qquad (\phi_{jk}^{(i)}, w_{jk'}^{(i)}) = 0, \qquad (\phi_{jk}^{(i)}, \phi_{jk'}^{(i)}) = \delta_{kk'}$$

where $j, j', \pm k, \pm k' = 0, 1, \ldots$ are valid at the ith recursion of the cascade algorithm they are also valid at the $(i+1)$-st recursion.

9.10 Scaling functions by recursion: dyadic points

We continue our study of topics related to the cascade algorithm. We are trying to characterize multiresolution systems with scaling functions $\phi(t)$ that have support in the interval $[0, N)$ where N is an odd integer. The low pass filter that defines the system is $(c_0, \ldots, c_N)$. One of the beautiful features of the dilation equation is that it enables us to compute explicitly the values $\phi(\frac{k}{2^j})$ for all j, k, i.e., at all dyadic points. Each value can be obtained as a result of a finite (and easily determined) number of passes through the low pass filter. The dyadic points are dense in the reals, so if we know that ϕ exists and is continuous, (as we will assume in this section) we will have determined it completely. (However, if there is no scaling function associated with the filter c then our results will be nonsense!)

The hardest step in this process is the first. The dilation equation is

$$\phi(t) = \sqrt{2} \sum_{k=0}^{N} c_k \phi(2t - k). \tag{9.96}$$

If $\phi(t)$ exists, it is zero outside the interval $0 \leq t < N$, so we can restrict our attention to the values of $\phi(t)$ on $[0, N)$. We first try to compute this function on the integers $t = 0, 1, \ldots, N - 1$. Substituting these values one at a time into (9.96) we obtain the system of equations

$$
\begin{bmatrix}
\phi(0) \\
\phi(1) \\
\phi(2) \\
\phi(3) \\
\phi(4) \\
\cdots \\
\phi(N-2) \\
\phi(N-1)
\end{bmatrix}
= \sqrt{2}
\begin{bmatrix}
c_0 & 0 \\
c_2 & c_1 & c_0 & 0 & \cdots \\
c_4 & c_3 & c_2 & c_1 & c_0 & \cdots \\
c_6 & c_5 & c_4 & c_3 & c_2 & \cdots \\
c_8 & c_7 & c_6 & c_5 & c_4 & \cdots \\
& & & \cdots & & \cdots \\
& & & & \cdots & \cdots & c_{N-2} & c_{N-3} \\
& & & & & \cdots & c_N & c_{N-1}
\end{bmatrix}
\begin{bmatrix}
\phi(0) \\
\phi(1) \\
\phi(2) \\
\phi(3) \\
\phi(4) \\
\cdots \\
\phi(N-2) \\
\phi(N-1)
\end{bmatrix},
$$

or

$$\Phi(0) = M^{(0)}\Phi(0). \tag{9.97}$$

This says that $\Phi(0)$ is an eigenvector of the $N \times N$ matrix $M^{(0)}$, with eigenvalue 1. If 1 is in fact an eigenvalue of $M^{(0)}$ then the homogeneous system of equations (9.97) can be solved for $\Phi(0)$ by Gaussian elimination.

We can show that $M^{(0)}$ always has 1 as an eigenvalue, so that (9.97) always has a nonzero solution. We need to recall from linear algebra that λ is an eigenvalue of the $N \times N$ matrix $M^{(0)}$ if and only if it is a solution of the characteristic equation

$$\det[M^{(0)} - \lambda I] = 0, \tag{9.98}$$

where I is the $N \times N$ identity matrix. Since the determinant of a square matrix equals the determinant of its transpose, we have that (9.98) is true if and only if

$$\det[M^{(0)^{\mathrm{tr}}} - \lambda I] = 0.$$

Thus $M^{(0)}$ has 1 as an eigenvalue if and only if $M^{(0)^{\mathrm{tr}}}$ has 1 as an eigenvalue. We claim that the column vector $(1, 1, \ldots, 1)$ is an eigenvector of $M^{(0)^{\mathrm{tr}}}$. Note that the column sum of each of the 1st, 3rd, 5th, ... columns of $M^{(0)}$ is $\sqrt{2} \sum_k c_{2k}$, whereas the column sum of each of the even-numbered columns is $\sqrt{2} \sum_k c_{2k+1}$. However, it is a consequence of the double-shift orthogonality conditions $\sum_k c_k \bar{c}_{k-2m} = \delta_{0m}$ and the compatibility condition $\sum_{k=0}^{N} c_k = \sqrt{2}$ that each of those sums is equal to 1. Thus the column sum of each of the columns of $M^{(0)}$ is 1, which means that the row sum of each of the rows of $M^{(0)^{\mathrm{tr}}}$ is 1, which says precisely that the column vector $(1, 1, \ldots, 1)$ is an eigenvector of $M^{(0)^{\mathrm{tr}}}$ with eigenvalue 1.

The required identities

$$\sqrt{2} \sum_k c_{2k} = \sqrt{2} \sum_k c_{2k+1} = 1 \tag{9.99}$$

can be proven directly from the above conditions. However, an indirect but simple proof comes from these equations in frequency space. Then we have the Fourier transform

$$C(\omega) = \sum_k c_k e^{-ik\omega}.$$

The double-shift orthogonality condition is now expressed as

$$|C(\omega)|^2 + |C(\omega + \pi)|^2 = 2. \tag{9.100}$$

The compatibility condition says

$$C(0) = \sum_k c_k = \sqrt{2}.$$

It follows from (9.100) that $C(\pi) = 0 = \sum_k (-1)^k c_k$, so $\sum_k c_{2k} = \sum_k c_{2k+1}$ and the column sums are the same. Then from $C(0) = \sqrt{2}$ we get our desired result.

Now that we can compute the scaling function on the integers (*up to a constant multiple; we shall show how to fix the normalization constant shortly*) we can proceed to calculate $\phi(t)$ at all dyadic points $t = k/2^j$. The next step is to compute $\phi(t)$ on the half-integers $t = 1/2, 3/2, \ldots, N-1/2$. Substituting these values one at a time into (9.96) we obtain the system of equations

$$\begin{bmatrix} \phi(\frac{1}{2}) \\ \phi(\frac{3}{2}) \\ \phi(\frac{5}{2}) \\ \phi(\frac{7}{2}) \\ \phi(\frac{9}{2}) \\ \cdots \\ \phi(N-\frac{3}{2}) \\ \phi(N-\frac{1}{2}) \end{bmatrix} = \sqrt{2} \begin{bmatrix} c_1 & c_0 & & & & \\ c_3 & c_2 & c_1 & c_0 & \cdots & \\ c_5 & c_4 & c_3 & c_2 & c_1 & \cdots \\ c_7 & c_6 & c_5 & c_4 & c_3 & \cdots \\ c_9 & c_8 & c_7 & c_6 & c_5 & \cdots \\ & \cdots & & \cdots & & \\ & & \cdots & & c_{N-1} & c_{N-2} \\ & & \cdots & & 0 & c_N \end{bmatrix} \begin{bmatrix} \phi(0) \\ \phi(1) \\ \phi(2) \\ \phi(3) \\ \phi(4) \\ \cdots \\ \phi(N-2) \\ \phi(N-1) \end{bmatrix},$$

or

$$\Phi(\frac{1}{2}) = M^{(1)}\Phi(0). \tag{9.101}$$

We can continue in this way to compute $\phi(t)$ at all dyadic points. A general dyadic point will be of the form $t = n+s$ where $n = 0, 1, \ldots, N-1$ and $s < 1$ is of the form $s = k/2^j$, $k = 0, 1, \ldots, 2^j - 1$, $j = 1, 2, \ldots$ The N-rowed vector $\Phi(s)$ contains all the terms $\phi(n + s)$ whose fractional part is s:

$$\Phi(s) = \begin{bmatrix} \phi(s) \\ \phi(s + 1) \\ \phi(s + 2) \\ \phi(s + 3) \\ \phi(s + 4) \\ \cdots \\ \phi(s + N - 2) \\ \phi(s + N - 1) \end{bmatrix}.$$

Example 9.27

$$\Phi(\frac{1}{4}) = M^{(0)}\Phi(\frac{1}{2}) \qquad \Phi(\frac{3}{4}) = M^{(1)}\Phi(\frac{1}{2}).$$

There are two possibilities, depending on whether the fractional dyadic s is $< 1/2$ or $\geq 1/2$. If $2s < 1$ then we can substitute $t = s, s + 1, s + 2, \ldots, s + N - 1$, recursively, into the dilation equation and obtain the result

$$\Phi(s) = M^{(0)}\Phi(2s).$$

If $2s \geq 1$ then if we substitute $t = s, s+1, s+2, \ldots, s+N-1$, recursively, into the dilation equation we find that the lowest-order term on the right-hand side is $\phi(2s - 1)$ and that our result is

$$\Phi(s) = M^{(1)}\Phi(2s - 1).$$

If we set $\Phi(s) = 0$ for $s < 0$ or $s \geq 1$ then we have

Theorem 9.28 *The general vector recursion for evaluating the scaling function at dyadic points is*

$$\Phi(s) = M^{(0)}\Phi(2s) + M^{(1)}\Phi(2s - 1). \qquad (9.102)$$

Note that for each s at most one of the terms on the right-hand side of (9.102) is nonzero. From this recursion we can compute explicitly the value of ϕ at the dyadic point $t = n + s$. Indeed we can write s as a dyadic decimal

$$s = .s_1 s_2 s_3 \cdots = \sum_{j \geq 1} \frac{s_j}{2^j}, \qquad s_j = 0, 1.$$

If $2s < 1$ then $s_1 = 0$ and we have

$$\Phi(s) = \Phi(.0s_2 s_3 \cdots) = M^{(0)}\Phi(2s) = M^{(0)}\Phi(.s_2 s_3 s_4 \cdots).$$

If on the other hand, $2s \geq 1$ then $s_1 = 1$ and we have

$$\Phi(s) = \Phi(.1s_2 s_3 \cdots) = M^{(1)}\Phi(2s - 1) = M^{(1)}\Phi(.s_2 s_3 s_4 \cdots).$$

Iterating this process we obtain

Corollary 9.29

$$\Phi(.s_1 s_2 \cdots s_\ell) = M^{(s_1)} M^{(s_2)} \cdots M^{(s_\ell)}\Phi(0).$$

Exercise 9.14 Derive the recursion (9.102).

Remarks 1. We will show later that we can always impose the normalization condition $\sum_n \phi(n) = 1$ on our scaling functions.

2. We have shown that the $N \times N$ matrix $M^{(0)}$ has a column sum of 1, so that it always has the eigenvalue 1. If the eigenvalue 1 occurs with multiplicity one, then the matrix eigenvalue equation $\Phi(0) = M^{(0)}\Phi(0)$ will yield $N - 1$ linearly independent conditions for the N unknowns $\phi(0), \ldots, \phi(N - 1)$. These together with the normalization condition will allow us to solve (uniquely) for the N unknowns via Gaussian elimination. If $M^{(0)}$ has 1 as an eigenvalue with multiplicity $k > 1$ however, then the matrix eigenvalue equation will yield only $N - k$ linearly independent conditions for the N unknowns and these together with the normalization condition may not be sufficient to determine a unique solution for the N unknowns. The double eigenvalue 1 is not common, but not impossible.

Exercise 9.15 The Daubechies filter coefficients for D_4 ($N = 3$) are $4\sqrt{2}c = (1 + \sqrt{3}, 3 + \sqrt{3}, 3 - \sqrt{3}, 1 - \sqrt{3})$. Solve the equation, $\Phi(0) = M^{(0)}\Phi(0)$ to show that, with the normalization $\phi(0) + \phi(1) + \phi(2) = 1$, the unique solution is

$$\phi(0) = 0, \quad \phi(1) = \frac{1}{2}(1 + \sqrt{3}) \quad \phi(2) = \frac{1}{2}(1 - \sqrt{3}).$$

In analogy with the use of infinite matrices in filter theory, we can also relate the dilation equation

$$\phi(t) = \sqrt{2} \sum_{k=0}^{N} c_k \phi(2t - k)$$

to an infinite matrix M. Evaluate the equation at the values $t + i, i = 0, \pm 1, \ldots$ for any real t. Substituting these values one at a time into the dilation equation we obtain the system of equations

$$\phi(t + i) = \sum_{i,j=1}^{\infty} M_{ij}\phi(2t + j), \qquad -\infty < t < \infty. \tag{9.103}$$

We have met $\mathbf{M}$ before. The matrix elements of $\mathbf{M}$ are $M_{ij} = \sqrt{2}c_{2i-j}$. Note the characteristic double-shift of the rows.

Exercise 9.16 Show that

$$\mathbf{M} = (\downarrow 2)\sqrt{2}\mathbf{C},$$

where $\mathbf{C}$ is a low pass filter.

For any fixed t the only nonzero part of (9.103) will correspond to either $M^{(0)}$ or $M^{(1)}$. For $0 \le t < 1$ and dyadic the equation reduces to

(9.102). M shares with its finite forms $M^{(i)}$ the fact that the column sum is 1 for every column and that $\lambda = 1$ is an eigenvalue. The left eigenvector now, however, is the infinite component row vector $f^0 = (\ldots, 1, 1, 1, \ldots)$. We take the dot product of this vector only with other vectors that are finitely supported, so there is no convergence problem.

Now we are in a position to investigate some of the implications of requiring that $C(\omega)$ has a zero of order p at $\omega = \pi$ for $p > 1$. This requirement means that

$$C(\pi) = C'(\pi) = \cdots = C^{(p-1)}(\pi) = 0,$$

and since $C(\omega) = \sum_k c_k e^{-ik\omega}$, it is equivalent to

$$\sum_k (-1)^k c_k k^\ell = 0, \qquad \ell = 1, 2, \ldots, p - 1. \qquad (9.104)$$

Theorem 9.31 to follow is very important but the technical details leading up to its proof can be omitted at a first reading. We already know that

$$\sum_k c_{2k} = \sum_k c_{2k+1} = \frac{1}{\sqrt{2}} = \frac{1}{2}C(0), \qquad (9.105)$$

and that $|C(\omega)|^2 + |C(\omega + \pi)|^2 = 2$. For use in the proof of the theorem to follow, we introduce the notation

$$A_\ell = \sum_i (2i)^\ell c_{2i} = \sum_i (2i + 1)^\ell c_{2i+1}, \qquad \ell = 0, 1, \ldots, p - 1. \qquad (9.106)$$

We already know that $A_0 = 1/\sqrt{2}$.

Now M will admit a left eigenvector $f = (\ldots, \alpha_{-1}, \alpha_0, \alpha_1, \ldots)$ with eigenvalue λ, i.e., $\sum_i \alpha_i M_{ij} = \lambda \alpha_j$ provided the equations

$$\sqrt{2} \sum_{i=-\infty}^{\infty} \alpha_i c(2i - j) = \lambda \alpha_j, \qquad j = 0, \pm 1, \ldots \qquad (9.107)$$

hold where not all α_i are zero. A similar statement holds for the finite matrices $M^{(0)}, M^{(1)}$ except that i, j are restricted to the rows and columns of these finite matrices. (Indeed the finite matrices $M_{ij}^{(0)} = \sqrt{2}c_{2i-j}$ for $0 \le i, j \le N - 1$ and $M_{ij}^{(1)} = \sqrt{2}c_{2i-j+1}$ for $0 \le i, j \le N - 1$ have the property that the jth column vector of $M^{(0)}$ and the $(j + 1)$st column vector of $M^{(1)}$ each contain all of the nonzero elements in the jth column of the infinite matrix M. Thus the restriction of (9.107) to the row and column indices i, j for $M^{(0)}, M^{(1)}$ yields exactly the eigenvalue equations for these finite matrices.) We have already shown that this equation has the solution $\alpha_i = 1, \lambda = 1$, due to that fact that $C(\omega)$ has a zero of order 1 at π.

For each integer h we define the (infinity-tuple) row vector f^h by

$$f^h{}_i = i^h, \qquad i = 0, \pm 1 \ldots$$

Theorem 9.30 *If $C(\omega)$ has a zero of order $p > 1$ at $\omega = \pi$ then M (and $M^{(0)}, M^{(1)}$) have eigenvalues $\lambda_\ell = 1/2^\ell$, $\ell = 0, 1, \ldots, p-1$. The corresponding left eigenvectors y^ℓ can be expressed as*

$$(-1)^\ell y^\ell = f^\ell + \sum_{h=0}^{\ell-1} \beta_h^\ell f^h$$

where the β_h^ℓ are constants.

Proof For each $\ell = 0, 1, \ldots, p-1$ we have to verify that an identity of the following form holds:

$$\sqrt{2} \sum_i \left(i^\ell + \sum_{h=0}^{\ell-1} \beta_h^\ell i^h \right) c_{2i-j} = \frac{1}{2^\ell} \left(j^\ell + \sum_{h=0}^{\ell-1} \beta_h^\ell j^h \right),$$

for $i, j = 0, \pm 1, \pm 2, \ldots$ For $\ell = 0$ we already know this. Suppose $\ell \geq 1$.

Take first the case where $j = 2s$ is even. We must find constants β_h^ℓ such that the identity

$$\sqrt{2} \sum_i \left(i^\ell + \sum_{h=0}^{\ell-1} \beta_h^\ell i^h \right) c_{2i-2s} = \frac{1}{2^\ell} \left((2s)^\ell + \sum_{h=0}^{\ell-1} \beta_h^\ell (2s)^h \right),$$

holds for all s. Making the change of variable $i' = i - s$ on the left-hand side of this expression we obtain

$$\sqrt{2} \sum_{i'} \left((i' + s)^\ell + \sum_{h=0}^{\ell-1} \beta_h^\ell (i' + s)^h \right) c_{2i'} = \frac{1}{2^\ell} \left((2s)^\ell + \sum_{h=0}^{\ell-1} \beta_h^\ell (2s)^h \right).$$

Expanding the left-hand side via the binomial theorem and using the sums (9.106) we find

$$\sqrt{2} \sum_n s^n \left[\binom{\ell}{n} \frac{A_{\ell-n}}{2^{\ell-n}} + \sum_{h=0}^{\ell-1} \beta_h^\ell \binom{h}{n} \frac{A_{h-n}}{2^{h-n}} \right]$$

$$= \frac{1}{2^\ell} \left((2s)^\ell + \sum_{h=0}^{\ell-1} \beta_h^\ell (2s)^h \right).$$

Now we equate powers of s. The coefficient of s^ℓ on both sides is 1. Equating coefficients of $s^{\ell-1}$ we find

$$\frac{\ell A_1}{\sqrt{2}} + \beta^\ell_{\ell-1} = \frac{1}{2}\beta^\ell_{\ell-1}.$$

We can solve for $\beta^\ell_{\ell-1}$ in terms of the given sum A_1. Now the pattern becomes clear. We can solve these equations recursively for $\beta^\ell_{\ell-1}, \beta^\ell_{\ell-2}, \dots \beta^\ell_0$. Equating coefficients of $s^{\ell-j}$ allows us to express $\beta^\ell_{\ell-j}$ as a linear combination of the A_i and of $\beta^\ell_{\ell-1}, \beta^\ell_{\ell-2}, \dots \beta^\ell_{j+1}$. Indeed the equation for $\beta^\ell_{\ell-j}$ is

$$\beta^\ell_{\ell-j} = \frac{\sqrt{2}}{2^{-j} - 1}\left[\left(\begin{array}{c}\ell \\ \ell-j\end{array}\right)\frac{A_j}{2^j} + \sum_{h=\ell-j+1}^{\ell-1} \beta^\ell_h \left(\begin{array}{c}h \\ n\end{array}\right)\frac{A_{h-\ell+j}}{2^{h-\ell+j}}\right].$$

This finishes the proof for $j = 2s$. The proof for $j = 2s - 1$ follows immediately from replacing s by $s - 1/2$ in our computation above and using the fact that the A_h are the same for the sums over the even terms in c as for the sums over the odd terms. $\qquad\square$

Since $\phi(t + i) = \sum_j M_{ij}\phi(2t + j)$ and $\sum_i y^\ell_i M_{ij} = (1/2^\ell)y^\ell_j$ for $\ell = 1, \dots, p - 1$ it follows that the function

$$g_\ell(t) = \sum_i (y^\ell)_i \phi(t + i)$$

satisfies $g_\ell(t) = \frac{1}{2^\ell}g_\ell(2t)$. We will evaluate this function under the assumption that $\phi(t)$ can be obtained by pointwise convergence from the cascade algorithm. Thus we have

$$g_\ell^{(k)}(t) = \sum_i (y^\ell)_i \phi^{(k)}(t + i), \quad \phi^{(k+1)}(t + i) = \sum_j M_{ij}\phi^{(k)}(2t + j),$$

$k = 0, 1, \dots$, where we choose $\phi^{(0)}(t)$ as the Haar scaling function. Thus $g_\ell^{(k+1)}(t) = \frac{1}{2^\ell}g_\ell^{(k)}(2t)$, and

$$g_\ell^{(0)}(t) = \sum_i (y^\ell)_i \phi^{(0)}(t + i) = y^{(\ell)}_{-[2^n t]},$$

where $[2^n t]$ is the largest integer $\leq 2^n t$. Iterating this identity we have $g_\ell^{(k+n)}(t) = \frac{1}{2^{n\ell}}g_\ell^{(k)}(2^n t)$ for $n = 1, 2, \dots$ Setting $k = 0$ and going to the limit, we have

$$\sum_i (y^\ell)_i \phi(t + i) = \lim_{n \to +\infty} g_\ell^{(n)}(t) = \lim_{n \to +\infty} \frac{1}{2^{n\ell}}y^{(\ell)}_{-[2^n t]}. \qquad (9.108)$$

We first consider the simplest case: $\ell = 0$. Then $y_i^0 = 1$ for all integers i, so

$$g_0^{(0)}(t) = \sum_i \phi^{(0)}(t+i) = 1.$$

Thus the limit is 1 and we have the important formula

$$\sum_i \phi(t+i) = 1.$$

Remark This result tells us that, locally at least, we can represent constants within the multiresolution space V_0. This is related to the fact that e^0 is a left eigenvector for $M^{(0)}$ and $M^{(1)}$ which is in turn related to the fact that $C(\omega)$ has a zero at $\omega = \pi$. Next we show that if we require that $C(\omega)$ has a zero of order p at $\omega = \pi$ then we can represent the monomials $1, t, t^2, \ldots t^{p-1}$ within V_0, hence all polynomials in t of order $p - 1$. This is a highly desirable feature for wavelets and is satisfied by the Daubechies wavelets of order p.

If $\ell > 0$ we use the fact that $(y^\ell)_i = (-i)^\ell +$ lower-order terms in i and see that the limit is t^ℓ. Thus we (almost) have the

Theorem 9.31 *If $C(\omega)$ has $p \geq 1$ zeros at $\omega = \pi$ and the cascade algorithm for this filter converges in L_2 to a continuous scaling function $\phi(t)$, then*

$$\sum_k y^\ell{}_k \phi(t+k) = t^\ell, \qquad \ell = 0, 1, \ldots, p-1. \tag{9.109}$$

The reasoning we have employed to motivate this theorem is not quite convincing. In subsequent chapters we will give conditions that guarantee convergence of the cascade algorithm in the L_2 sense, and guarantee that the scaling function obtained in the limit is continuous. Here, however, we are talking about **pointwise** convergence of the cascade algorithm and that will not be guaranteed by our theoretical results. We describe the analysis that saves the theorem. An important fact from Lebesgue theory is that if $\{\phi^{(i)}\}$ is a Cauchy sequence of functions converging to ϕ in the L_2 norm then there always exists a subsequence $\{\phi^{(i_h)}\}$, $i_1 < i_2 < \cdots$ such that the pointwise limit $\lim_{h\to\infty} \phi^{(i_h)}(t) = \phi(t)$ exists almost everywhere. By restricting to an appropriate subsequence to take the pointwise limit we can establish (9.109) almost everywhere. (In Section 1.5.3 we proved this result for the case that the $\{\phi^{(i)}\}$ are step functions and that is all we need for convergence of the cascade algorithm with $\phi^{(0)}$ as the Haar scaling function.) There are only finitely many nonzero terms on the left-hand side of identity (9.109) for t in some bounded set. By assumption, each side of the equation is a continuous function of t. Two continuous functions that are equal almost everywhere must be identical. Put another way, the right-hand side minus the left-hand side is a continuous function that is 0 almost everywhere, hence it must be identically 0.

Exercise 9.17 Consider the example $g(t) = t^\ell \sin(2\pi \log_2(|t|))$ for $t \neq 0$ and $g(0) = 0$ to show that the difference equation $g(2t) = 2^\ell g(t)$ has solutions continuous for all t, other than ct^ℓ. Can you find solutions distinct from these?

Theorem 9.31 appears to suggest that if we require that $C(\omega)$ has a zero of order p at $\omega = \pi$ then we can represent the monomials $1, t, t^2, \ldots t^{p-1}$ within V_0, hence all polynomials in t of order $p - 1$. This isn't quite correct since the functions t^ℓ are not square integrable, so strictly speaking, they don't belong to V_0. However, due to the compact support of the scaling function, the series for $g^\ell(t)$ converges pointwise. Normally one needs only to represent the polynomial in a bounded domain. Then all of the coefficients y_j^ℓ that don't contribute to the sum in that bounded interval can be set equal to zero.

Exercise 9.18 Determine the order p of the zero of the filter $C(\omega)$ at $\omega = \pi$, either directly or using the sum rules (9.104) for the following filter coefficients. (In general these filters will not satisfy double-shift orthogonality but are normalized so that $C(0) = \sqrt{2}$.)

1. $c = \sqrt{2}(\frac{1}{4}, \frac{1}{2}, \frac{1}{4})$
2. $c = \frac{\sqrt{2}}{16}(1, 4, 6, 4, 1)$
3. $h = \frac{\sqrt{2}}{8}(1 - \sqrt{3}, 3 - \sqrt{3}, 3 + \sqrt{3}, 1 + \sqrt{3})$ (Daubechies in reverse)
4. $c = \frac{1}{\sqrt{2}}(\frac{1}{2}, \frac{1}{2}, \frac{1}{2}, \frac{1}{2})$

9.11 The cascade algorithm in the frequency domain

We have been studying pointwise convergence of iterations of the dilation equation in the time domain. Now we look at the dilation equation

$$\phi(t) = \sqrt{2} \sum_k c_k \phi(2t - k)$$

in the frequency domain. We again utilize the cascade algorithm

$$\phi^{(j+1)}(t) = \sqrt{2} \sum_k c_k \phi^{(j)}(2t - k), \quad j = 0, 1, \dots$$

and assume that the initial input $\phi^{(0)}(t)$ is the Haar scaling function. Taking the Fourier transform of both sides of this equation and using the fact that (for $u = 2t - k$)

$$2 \int_{-\infty}^{\infty} \phi(2t - k)e^{-i\omega t}dt = \int_{-\infty}^{\infty} \phi(u)e^{-i\omega(u+k)/2}du = e^{-i\omega k/2}\hat{\phi}(\frac{\omega}{2}),$$

we find

$$\hat{\phi}(\omega) = \frac{1}{\sqrt{2}} \left(\sum_k c_k e^{-i\omega k/2} \right) \hat{\phi}(\frac{\omega}{2}).$$

Thus the frequency domain form of the dilation equation is

$$\hat{\phi}^{(j+1)}(\omega) = 2^{-1/2}C(\frac{\omega}{2})\hat{\phi}^{(j)}(\frac{\omega}{2}). \tag{9.110}$$

Now iterate the left-hand side of the dilation equation:

$$\hat{\phi}^{(j+2)}(\omega) = \frac{1}{2}C(\frac{\omega}{2})[C(\frac{\omega}{4})\hat{\phi}^{(j)}(\frac{\omega}{4})].$$

After N steps we have

$$\hat{\phi}^{(j+N)}(\omega) = 2^{-N/2}C(\frac{\omega}{2})C(\frac{\omega}{4})\cdots C(\frac{\omega}{2^N})\hat{\phi}^{(j)}(\frac{\omega}{2^N}).$$

We want to let $N \to \infty$ in this equation. Setting $j = 0$ we recall that $\hat{\phi}^{(0)}(\omega) = (1 - e^{-i\omega})/i\omega$ so $\lim_{N\to\infty} \hat{\phi}^{(0)}(\frac{\omega}{2^N}) = \hat{\phi}^{(0)}(0) = 1$, and postulate an infinite product formula for $\hat{\phi}(\omega)$:

$$\hat{\phi}(\omega) = \lim_{N\to\infty} \hat{\phi}^{(N)}(\omega) = \Pi_{j=1}^{\infty} 2^{-1/2}C(\frac{\omega}{2^j}). \tag{9.111}$$

We studied infinite products in Section 2.4 and established the fundamental convergence conditions in Theorem 2.27 and Corollary 2.28. However formula (9.111) involves an infinite product of functions, not just numbers and it will be important to establish criteria for the infinite product to be a continuous function.

Definition 9.32 Let $\{w_j(z)\}$ be a sequence of continuous functions defined on an open connected set $\mathcal{D}$ of the complex plane, and let $\mathcal{S}$ be a closed, bounded subset of $\mathcal{D}$. The infinite product $P(z) \equiv \Pi_{j=1}^{\infty}(1 + w_j(z))$ is said to be uniformly convergent on $\mathcal{S}$ if

1. there exists a fixed $m_0 \geq 1$ such that $w_j(z) \neq -1$ for $k \geq m_0$, and every $z \in \mathcal{S}$, and

2. for any $\epsilon > 0$ there exists a fixed $N(\epsilon)$ such that for $n > N(\epsilon)$, $m \geq m_0$ and every $z \in S$ we have

$$|P_{m,n}(z)| \cdot |1 - P_{n+1,n+p}(z)| < \epsilon, p \geq 1.$$

Then from standard results in calculus,[92], we can verify the following.

Theorem 9.33 *Suppose $w_j(z)$ is continuous in $\mathcal{D}$ for each j and that the infinite product $P(z) \equiv \Pi_{j=1}^{\infty}(1+w_j(z))$ converges uniformly on every closed bounded subset of $\mathcal{D}$. Then $P(z)$ is a continuous function in $\mathcal{D}$.*

Now we return to the infinite product formula for the scaling function:

$$\hat{\phi}(\omega) = \Pi_{j=1}^{\infty} 2^{-1/2} C(\frac{\omega}{2^j}).$$

Note that this infinite product converges, uniformly and absolutely on all finite intervals. Indeed $2^{-1/2}C(0) = 1$ and the derivative of the 2π-periodic function $C'(\omega)$ is uniformly bounded: $|C'(\omega)| \leq K\sqrt{2}$. Then $C(\omega) = C(0) + \int_0^{\omega} C'(s)ds$ so

$$|2^{-1/2}C(\omega)| \leq 1 + K|\omega| \leq e^{K|\omega|}.$$

Since $K\sum_{j=1}^{\infty} \frac{|\omega|}{2^j} = K|\omega|$ converges, the infinite product converges absolutely and, from the proof of Theorem 2.27 we have the (very crude) upper bound $|\hat{\phi}(\omega)| \leq e^{K|\omega|}$.

Example 9.34 The moving average filter has filter coefficients $c_0 = c_1 = 1/\sqrt{2}$ and $C(\omega) = (1 + e^{-i\omega})/\sqrt{2}$. The product of the first N factors in the infinite product formula is

$$2^{-N/2}C^{(N)}(\omega) = \frac{1}{2^N}(1+e^{-i\omega/2})(1+e^{-i\omega/4})(1+e^{-i\omega/8})\cdots(1+e^{-i\omega/2^N}).$$

The following identities (easily proved by induction) are needed:

Lemma 9.35

$$(1 + z)(1 + z^2)(1 + z^4)\cdots(1 + z^{2^{n-1}}) = \sum_{k=0}^{2^n - 1} z^k = \frac{1 - z^{2^n}}{1 - z}.$$

Then, setting $z = e^{-i\omega/2^N}$, we have

$$2^{-N/2}C^{(N)}(\omega) = \frac{1}{2^N}\frac{1 - e^{-i\omega}}{1 - e^{-i\omega/2^N}}.$$

Now let $N \to \infty$. The numerator is constant. The denominator goes like $2^N(i\omega/2^N - \omega^2/2^{2N+1} + \cdots) \to i\omega$. Thus

$$\hat{\phi}(\omega) = \lim_{N \to \infty} 2^{-N/2} C^{(N)}(\omega) = \frac{1 - e^{-i\omega}}{i\omega}, \qquad (9.112)$$

basically the sinc function.

Although the infinite product formula for $\hat{\phi}(\omega)$ always converges pointwise and uniformly on any closed bounded interval, this does not solve our problem. We need to have $\hat{\phi}(\omega)$ decay sufficiently rapidly at infinity so that it belongs to L_2. At this point all we have is a weak solution to the problem. The corresponding $\phi(t)$ is not a function but a *generalized function* or *distribution*. One can get meaningful results only in terms of integrals of $\phi(t)$ and $\hat{\phi}(\omega)$ with functions that decay very rapidly at infinity and their Fourier transforms that also decay rapidly. Thus we can make sense of the generalized function $\phi(t)$ by defining the expression on the left-hand side of

$$2\pi \int_{-\infty}^{\infty} \phi(t)\overline{g}(t)dt = \int_{-\infty}^{\infty} \hat{\phi}(\omega)\overline{\hat{g}}(\omega)d\omega$$

by the integral on the right-hand side, for all g and $\hat{g}$ that decay sufficiently rapidly. We shall not go that route because we want $\phi(t)$ to be a true function.

Already the crude estimate $|\hat{\phi}(\omega)| < e^{K|\omega|}$ in the complex plane does give us some information. The Paley–Weiner Theorem (whose proof is beyond the scope of this book [93, 50]) says, essentially that for a function $\phi(t) \in L_2(\mathbb{R})$ the Fourier transform can be extended into the complex plane such that $|\hat{\phi}(\omega)| < K_0 e^{K|\omega|}$ if and only if $\phi(t) \equiv 0$ for $|t| > K$. It is easy to understand why this is true. If $\phi(t)$ vanishes for $|t| > K$ then $\hat{\phi}(\omega) = \int_{-K}^{K} \phi(t)e^{-i\omega t}dt$ can be extended into the complex ω plane and the above integral satisfies this estimate. If $\phi(t)$ is nonzero in an interval around t_0 then it will make a contribution to the integral whose absolute value would grow at the approximate rate $e^{|t_0\omega|}$.

Thus we know that if $\hat{\phi}$ belongs to L_2, so that $\phi(t)$ exists, then $\phi(t)$ has compact support. We also know that if $\sum_{k=0}^{N} c_k = \sqrt{2}$, our solution (if it exists) is unique.

9.12 Some technical results

We are in the process of constructing a family of *continuous* scaling functions with compact support and such that the integer translates of each scaling function form an ON set. It isn't yet clear, however, that each of these scaling functions will generate a basis for $L_2(\mathbb{R})$, i.e. that any function in L_2 can be approximated by these wavelets. Throughout this section we make the following assumptions concerning the scaling function $\phi(t)$.

Assumptions: Suppose that $\phi(t)$ is continuous with compact support on the real line and satisfies the orthogonality conditions $(\phi_{0k}, \phi_{0\ell}) = \int \phi(t-k)\overline{\phi(t-\ell)}dt = \delta_{k\ell}$ in V_0. Let V_j be the subspace of $L_2(\mathbb{R})$ with ON basis $\{\phi_{jk} : k = 0, \pm 1, \ldots\}$ where $\phi_{jk}(t) = 2^{j/2}\phi(2^j t - k)$. Suppose ϕ satisfies the normalization condition $\int \phi(t)dt = 1$ and the dilation equation

$$\phi(t) = \sqrt{2}\sum_{k=0}^{N} c_k \phi(2t-k)$$

for finite N.

We will show that such a scaling function does indeed determine a multiresolution analysis. (Note that the Haar scaling function does not satisfy the continuity condition.) These technical results, though important, are fairly routine abstract arguments in harmonic analysis rather than features unique to wavelets. Thus their proofs can be omitted in a first reading

Lemma 9.36 *There is a constant $C > 0$ such that $|f(t)| \le C\|f\|$ for all functions $f \in V_0$.*

Proof If $f \in V_0$ then we have $f(t) = \sum_k c_k \phi_{0k}(t)$. We can assume pointwise equality as well as Hilbert space equality, because for each t only a finite number of the continuous functions $\phi(t-k)$ are nonzero. Since $c_k = (f, \phi_{0k})$ we have

$$f(t) = \int h(t,s)f(s)ds, \quad \text{where } h(t,s) = \sum_k \phi(t-k)\overline{\phi(s-k)}.$$

Again, for any t, s only a finite number of terms in the sum for $h(t, s)$ are nonzero. For fixed t the kernel $h(t, s)$ belongs to the inner product space of square integrable functions in s. The norm square of h in this space is

$$\|h(t,\cdot)\|_s^2 = \sum_k |\phi(t-k)|^2 \le K^2$$

for some positive constant K. This is because only a finite number of the terms in the k-sum are nonzero and $\phi(t)$ is a bounded function. Thus by the Schwarz inequality we have

$$|f(t)| = |(k(t, \cdot), f)_s| \leq ||h(t, \cdot)|| \cdot ||f|| \leq K||f||.$$

$\square$

Theorem 9.37 *The separation property for multiresolution analysis holds:* $\cap_{j=-\infty}^{\infty} V_j = \{0\}$.

Proof Suppose $f \in V_{-j}$. This means that $f(2^j t) \in V_0$. By the lemma, we have

$$|f(2^j t)| \leq K||f(2^j \cdot)|| = K 2^{-j/2} ||f||.$$

If $f \in V_{-j}$ for all j then $f(t) \equiv 0$.

$\square$

Theorem 9.38 *The density property for multiresolution analysis holds:* $\overline{\cup_{j=-\infty}^{\infty} V_j} = L_2(\mathbb{R})$.

Proof Let $R_{ab}(t)$ be a rectangular function:

$$R_{ab}(t) = \begin{cases} 1 & \text{for } a \leq t \leq b \\ 0 & \text{otherwise,} \end{cases}$$

for $a < b$. We will show that $R_{ab} \in \overline{\cup_{j=-\infty}^{\infty} V_j}$. Since every step function is a linear combination of rectangular functions, and since the step functions are dense in $L_2(\mathbb{R})$, this will prove the theorem. Let $P_j R_{ab}$ be the orthogonal projection of R_{ab} on the space V_j. Since $\{\phi_{jk}\}$ is an ON basis for V_j we have

$$P_j R_{ab} = \sum_k \alpha_k \phi_{jk}.$$

We want to show that $||R_{ab} - P_j R_{ab}|| \to 0$ as $j \to +\infty$. Since $R_{ab} - P_j R_{ab} \perp V_j$ we have

$$||R_{ab}||^2 = ||R_{ab} - P_j R_{ab}||^2 + ||P_j R_{ab}||^2,$$

so it is sufficient to show that $||P_j R_{ab}||^2 \to ||R_{ab}||^2$ as $j \to +\infty$. Now

$$||P_j R_{ab}||^2 = \sum_k |\alpha_k|^2 = \sum_k |(R_{ab}, \phi_{jk})|^2 = 2^j \sum_k |\int_a^b \phi(2^j t - k) dt|^2$$

so

$$||P_j R_{ab}||^2 = 2^{-j} \sum_k |\int_{2^j a}^{2^j b} \phi(t - k) dt|^2.$$

The support of $\phi(t)$ is contained in some finite interval with integer endpoints $m_1 < m_2$: $m_1 \leq t \leq m_2$. For each integral in the summand there are three possibilities:

1. The intervals $[2^j a, 2^j b]$ and $[m_1, m_2]$ are disjoint. In this case the integral is 0.

2. $[m_1, m_2] \subset [2^j a, 2^j b]$. In this case the integral is 1.

3. The intervals $[2^j a, 2^j b]$ and $[m_1, m_2]$ partially overlap. As j gets larger and larger this case is more and more infrequent. Indeed if $ab \neq 0$ this case won't occur at all for sufficiently large j. It can only occur if say, $a = 0, m_1 \leq 0, m_2 > 0$. For large j the number of such terms would be fixed at $|m_1|$. In view of the fact that each integral squared is multiplied by 2^{-j} the contribution of these boundary terms goes to 0 as $j \to +\infty$.

Let $N_j(a, b)$ equal the number of integers between $2^j a$ and $2^j b$. Clearly $N_j(a, b) \sim 2^j(b - a)$ and $2^{-j} N_j(a, b) \to b - a$ as $j \to +\infty$. Hence

$$\lim_{j \to +\infty} ||P_j R_{a,b}||^2 = b - a = \int_a^b 1 dt = ||R_{ab}||^2.$$

$\square$

9.13 Additional exercises

Exercise 9.19 Let $\phi(t)$ and w be the Haar scaling and wavelet functions, respectively. Let V_j and W_j be the spaces generated by $\phi_{j,k}(t) = 2^{j/2}\phi(2^j t - k)$ and $w_{j,k}(t) = 2^{j/2}w(2^j t - k)$, $k = 0, \pm1, \ldots$, respectively. Suppose the real-valued function $f(t) = \sum_k a_k \phi_{1,k}(t)$ belongs to V_1 and $f \perp V_0$. Show that $a_{2\ell+1} = -a_{2\ell}$ for all integers ℓ. Conclude that f can be expressed as a linear combination of the functions, $w_{0,k}(t)$, hence that $f \in W_0$.

Exercise 9.20 Let $\phi(t), w(t) V_j$, and W_j be as defined in Exercise 9.19. Let $g(t)$ be defined on $0 \leq t < 1$ and given by

$$g(t) = \begin{cases} -1 & 0 \leq t < 1/4 \\ 4 & 1/4 \leq t < 1/2 \\ 2 & 1/2 \leq t < 3/4 \\ -3 & 3/4 \leq t < 1. \end{cases}$$

1. Express g in terms of the basis for V_2.
2. Decompose g into its component parts in W_1, W_0 and V_0. In other words, find the Haar wavelet decomposition for g.
3. Sketch each of the four decompositions.

Exercise 9.21 Reconstruct a signal $f \in V_3$ given that its only nonzero coefficients in the Haar wavelet decomposition are

$$\{a_{2,k}\} = \{\frac{1}{2}, 2, \frac{5}{2}, -\frac{3}{2}\}, \ \{b_{2,k}\} = \{-\frac{3}{2}, -1, \frac{1}{2}, -\frac{1}{2}\}.$$

Here the first entry in each list corresponds to $k = 0$. Sketch the graph of f.

Exercise 9.22 Reconstruct a signal $g \in V_3$ given that its only nonzero coefficients in the Haar wavelet decomposition are

$$\{a_{1,k}\} = \{\frac{3}{2}, -1\}, \ \{b_{1,k}\} = \{-1, -\frac{3}{2}\}, \ \{b_{2,k}\} = \{-\frac{3}{2}, -\frac{3}{2}, -\frac{1}{2}, -\frac{1}{2}\}.$$

Here the first entry in each list corresponds to $k = 0$. Sketch the graph of g.

Exercise 9.23 Let

$$f = \exp(-t^2/10) \left(\sin(t) + 2\cos(2t) + \sin(5t) + \cos(8t) + 2t^2 \right).$$

Discretize (sample) the function $f(t)$ over the interval $0 \le t \le 1$ computing the signal vector $f = \{a_{8,k}\}$ where $a_{8,k} = f(2k\pi/256), k = 1, \ldots, 256$, so that the discretization belongs to V_8 for the Haar decomposition. In other words, use $J = 8$ as the top level so that there are $2^8 = 256$ nodes in the discretization. Implement the FWT decomposition algorithm for Haar wavelets. Plot the resulting levels $f^{(j)} \in V_j$, $j = 0, \ldots, 7$ and compare with the original signal. Here you can use one of

1. The display facilities of the MATLAB wavelet toolbox. Note: You can prepare a data file with commands such as

   ```
   t=linspace(0,1,2^8)
   ```

   and

   ```
   f=exp(-t.^2/10).*(sin(t)+2*cos(2*t)+sin(5*t)+cos(8*t)+2*t.
   ```

   MATLAB can handle data files of length 2^j for $j \le 10$. The command

   ```
   save   signal f
   ```

will save the vectors f and t in the data file *signal.mat*. Then for the "load signal" option you simply open the file signal.mat and use the analysis and processing functions of the wavelet toolbox.

2. The MATLAB Wavelet Toolbox commands **wavedec** and **waverec**. Here **wavedec** is the MATLAB wavelet decomposition routine. Its inputs are the discretized signal vector f, the number of levels j (so that for a maximal decomposition number of nodes is 2^j = length of f = length of t) and 'dbp' where **dbp** is the corresponding Daubechies wavelet, i.e., **db1** or *'haar'* is the Haar wavelet, **db2** is the Daubechies 4-tap wavelet, etc. The appropriate command is

```
[C,L] = wavedec(f,j,'dbp')
```

Recall that f is associated to the time nodes t (e.g.,

```
t=linspace (0,1,2^8)
```

to decompose the interval $[0,1]$ into 2^8 terms). The output C is the 2^j component coefficients vector consisting of the coefficients $a_{0,k}$ and $b_{0,k}, b_{1,k}, \ldots b_{j-1,k}$, in that order. L is a bookkeeping vector, listing the lengths of the coefficient vectors for $a_{0,k}, b_{0,k}, b_{1,k}, \ldots b_{j-1,k}$, in that order, followed by the length of f. The wavelet recovery command, to obtain the signal **fr** from the wavelet coefficients, is

```
fr = waverec(C,L,'dbp')
```

This is just the inverse of the **wavedec** command. However, one can process the C output of **wavedec** and obtain a modified file of wavelet coefficients MC. Then the command

```
fr = waverec(MC,L,'dbp')
```

will give the reconstruction of the modified signal. A related useful command is

```
[NC,NL,cA] = upwlev(C,L,'dbp')
```

a one-dimensional wavelet analysis function. This command performs the single-level reconstruction of the wavelet decomposition structure $[C, L]$ giving the new one $[NC, NL]$, and extracts the last coefficients vector cA. Thus since $[C, L]$ is a decomposition at level j, $[NC, NL]$ is the same decomposition at level $j - 1$ and cA is the coefficients vector at this level. The default signal extension for these algorithms is zero padding. This makes no difference for Haar wavelets, but is necessary to take into account for all other Daubechies wavelets.

Exercise 9.24 Prove Lemma 9.10.

Exercise 9.25 Verify Equation (9.26).

Exercise 9.26 From the definition (9.30), show that d has length 1 and is double-shift orthogonal to itself.

Exercise 9.27 Let V_0 be the space of all square integrable signals $f(t)$ that are piecewise linear and continuous with possible corners occurring at only the integers $0, \pm 1, \pm 2, \ldots$ Similarly, for every integer j let V_j be the space of all square integrable signals $f(t)$ that are piecewise linear and continuous with possible corners occurring at only the dyadic points $k/2^j$ for k an integer. Let $\phi(t)$ be the hat function

$$\phi(t) = \begin{cases} t+1, & -1 \le t \le 0, \\ 1-t, & 0 < t \le 1, \\ 0, & |t| > 1. \end{cases}$$

From the text, $f(t) = \sum_k f(k)\phi(t-k)$ for every $f \in V_0$, so the hat function and its integer translates form a (nonorthogonal) basis for V_0. This defines the linear spline multiresolution analysis, though it is not orthonormal.

1. Verify that the spaces V_j are nested, i.e., $V_j \subset V_{j+1}$.
2. Verify that $f(t) \in V_j \Longleftrightarrow f(2^{-j}t) \in V_0$.
3. What is the scaling function basis for V_1?
4. Compute the coefficients c_k in the dilation equation $\phi(t) = \sqrt{2} \sum_k c_k \phi(2t-k)$, where $\phi(t)$ is the hat function.

Exercise 9.28 Using the fact that N is odd, show that

$$\int_{-\pi}^{\pi} e^{2ik\omega} D(\omega)\overline{D(\pi+\omega)}d\omega = -\int_{-\pi}^{\pi} e^{2ik\omega} C(\omega)\overline{C(\pi+\omega)}d\omega. \quad (9.113)$$

Exercise 9.29 Use the double-shift orthogonality conditions in the frequency domain and (9.113) to verify Equations (9.67), (9.68) and (9.69).

Exercise 9.30 Show that the ideal low pass filter:

$$C(\omega) = \begin{cases} \sqrt{2}, & 0 \le |\omega| < \frac{\pi}{2} \\ 0, & \frac{\pi}{2} \le |\omega| \le \pi \end{cases}$$

satisfies condition (9.82) for double-shift orthogonality. Verify that the impulse response vector has coefficients

$$
c_n = \begin{cases} \frac{1}{\sqrt{2}}, & n = 0 \\ \pm\frac{\sqrt{2}}{\pi n}, & n \text{ odd} \\ 0, & n \text{ even}, n \neq 0. \end{cases}
$$

Determine the associated high pass filter $D(\omega)$. Why is this not a practical filter bank? What is the delay ℓ?

Exercise 9.31 What is the frequency response for the maxflat Daubechies filter with $p = 2$? Graph $P(\omega)$ and $|C(\omega)|$.

Exercise 9.32 Compute $\{c_n\}$ for the halfband Daubechies filter with $p = 5$. Verify that $C(\omega)$ has four zero derivatives at $\omega = 0$ and $\omega = \pi$.

Exercise 9.33 Find $H^{(1)}[z]$, $S^{(0)}[z]$ and $S^{(1)}[z]$ for the biorthogonal filter bank with $H^{(0)}[z] = \left(\frac{1+z^{-1}}{2}\right)^3$ and

$$
P^{(0)}[z] = \frac{1}{16}\left(-1 + 9z^{-2} + 16z^{-3} + 9z^{-4} - z^{-6}\right).
$$

Exercise 9.34 Suppose a real infinite matrix Q has the property $Q^{\mathrm{tr}}Q = I$. Show that the columns of Q are mutually orthogonal unit vectors. Does it follow that $QQ^{\mathrm{tr}} = I$?

Exercise 9.35 Let $\mathbf{C}$, $\mathbf{H}$ be real low pass filters, each satisfying the double-shift row orthogonality condition. Does the product $\mathbf{CH}$ satisfy the double-shift row orthogonality condition?

Exercise 9.36 Plot the frequency response function $|C(\omega)|$ in the interval $0 \leq \omega \leq 2\pi$ for the Daubechies 4-tap filter. Point out the features of the graph that cause this filter to belong to the maxflat class.

Exercise 9.37 Suppose the halfband filter

$$
P(\omega) = 1 + \sum_{n \text{ odd}} p(n)e^{-in\omega}
$$

satisfies $P(0) = 2$. From these facts alone, deduce that $P(\pi) = 0$, i.e., the filter is low pass.

Exercise 9.38 The reverse Daubechies 4-tap filter is related to Daubechies $D_4 = db2$ by reversing the order of the filter coefficients in the Z transform. Thus the transform of the reverse filter is

$$
C[z] = \frac{1}{4\sqrt{2}}\left((1-\sqrt{3}) + (3-\sqrt{3})z^{-1} + (3+\sqrt{3})z^{-2} + (1+\sqrt{3})z^{-3}\right).
$$

It satisfies the properties $C[1] = \sqrt{2}$, $C[-1] = 0$ (low pass) and $|C[z]|^2 + |C[-z]|^2 = 2$ (double-shift orthogonality). Is it true that the reverse filter also has a root of degree 2 at $z = -1$, i.e., that $C'(-1) = 0$? Justify your answer.

––––––––––––

For some additional references on wavelets and their applications to signal processing see [10, 18, 35, 41, 43, 45, 54, 81].

10

Discrete wavelet theory

In this chapter we will provide some solutions to the questions of the existence of wavelets with compact and continuous scaling functions, of the L_2 convergence of the cascade algorithm, and the accuracy of approximation of functions by wavelets. For orthogonal wavelets we use a normalized impulse response vector $c = (c_0, \ldots, c_N)$ that satisfies double-shift orthogonality, though for nonorthogonal wavelets we will weaken this orthogonality restriction on c. We will start the cascade algorithm at $i = 0$ with the Haar scaling function $\phi^{(0)}(t)$, and then update to get successive approximations $\phi^{(i)}(t)$, $i = 0, 1, \ldots$ The advantages of starting with the Haar scaling function in each case are simplicity and the fact that orthogonality of the scaling function to integer shifts is satisfied automatically. We already showed in Section 9.9 that if c satisfies double-shift orthogonality and $\phi^{(i)}(t)$ is orthogonal to integer shifts of itself, then $\phi^{(i+1)}(t)$ preserves this orthogonality. Our aim is to find conditions such that the $\phi(t) = \lim_{i \to \infty} \phi^{(i)}(t)$ exists, first in the L_2 sense and then pointwise. The Haar scaling function is a step function, and it is easy to see that each iteration of the cascade algorithm maps a step function to a step function. Thus our problem harkens back to the Hilbert space considerations of Section 1.5.1. If we can show that the sequence of step functions $\phi^{(0)}, \phi^{(1)}, \phi^{(2)}, \ldots$ is Cauchy in the L_2 norm then it will follow that the sequence converges. Thus, we must show that for any $\epsilon > 0$ there is an integer N_ϵ such that $||\phi^{(i)} - \phi^{(j)}|| < \epsilon$ whenever $i, j > N_\epsilon$. We will develop the machinery to establish this property.

In the proof of Theorem 9.25 in the previous chapter we related the inner products

$$\int_{-\infty}^{\infty} \phi^{(i+1)}(t)\overline{\phi}^{(i+1)}(t - k)dt$$

to the inner products $\int_{-\infty}^{\infty} \phi^{(i)}(t)\overline{\phi}^{(i)}(t-k)dt$ in successive passages through the cascade algorithm. Recall that this relationship is as follows. Let

$$a_k^{(i)} = (\phi_{00}^{(i)}, \phi_{0k}^{(i)}) = \int_{-\infty}^{\infty} \phi^{(i)}(t)\overline{\phi}^{(i)}(t-k)dt,$$

be the vector of inner products at stage i. Note that although $a^{(i)}$ is an infinite-component vector, since $\phi^{(i)}(t)$ has support limited to the interval $[0, N]$ only the $2N - 1$ components $a^{(i)}(k)$, $k = -N + 1, \ldots - 1, 0, 1 \ldots, N - 1$ can possibly be nonzero. We can use the cascade recursion to express $a_s^{(i+1)}$ as a linear combination of terms $a_k^{(i)}$:

$$a_s^{(i+1)} = \int_{-\infty}^{\infty} \phi^{(i+1)}(t)\overline{\phi}^{(i+1)}(t-s)dt = \int_{-\infty}^{\infty} \overline{\phi}^{(i+1)}(t)\phi^{(i+1)}(t+s)dt$$

$$= 2\sum_{k,\ell} \overline{c}_k c_\ell \int_{-\infty}^{\infty} \overline{\phi}^{(i)}(2t-k)\phi^{(i)}(2t+2s-\ell)dt$$

$$= \sum_{k,\ell} \overline{c}_k c_\ell \int_{-\infty}^{\infty} \overline{\phi}^{(i)}(t-k)\phi^{(i)}(t+2s-\ell)dt$$

$$= \sum_{k,j} \overline{c}_k c(2s+j) \int_{-\infty}^{\infty} \overline{\phi}^{(i)}(t)\phi^{(i)}(t+k-j)dt$$

$$= \sum_{m,j} c_{2s+j}\overline{c}_{m+j} \int_{-\infty}^{\infty} \overline{\phi}^{(i)}(t)\phi^{(i)}(t+m)dt,$$

where $\ell = 2s + j$, $m = k - j$. Thus

$$a_s^{(i+1)} = \sum_{j,\ell} c_{2s+j}\overline{c}_{\ell+j}a_\ell^{(i)}. \tag{10.1}$$

In matrix notation this is just

$$a^{(i+1)} = Ta^{(i)} = (\downarrow 2)C\overline{C}^{\mathrm{tr}}a^{(i)} \tag{10.2}$$

where the matrix elements of the T matrix (the *transition matrix*) are given by

$$T_{s\ell} = \sum_{j} c_{2s+j}\overline{c}_{\ell+j}.$$

Although T is an infinite matrix, the only matrix elements that correspond to action on vectors a (whose components a_k necessarily vanish unless $-N+1 \le k \le N-1$) are those contained in the $(2N-1) \times (2N-1)$

block $-N+1 \leq s, \ell \leq N-1$. When we discuss the eigenvalues and eigenvectors of T we are normally talking about this $(2N-1) \times (2N-1)$ submatrix. We emphasize the relation with other matrices that we have studied before:

$$T = (\downarrow 2)C\overline{C}^{\mathrm{tr}} = \frac{1}{\sqrt{2}}M\overline{C}^{\mathrm{tr}}.$$

NOTE: Since the filter C is low pass, the matrix T shares with the matrix M the property that the column sum of each column equals 1. Indeed $\sum_s c_{2s+j} = 1/\sqrt{2}$ for all j and $\sum_j c_j = \sqrt{2}$, so $\sum_s T_{s\ell} = 1$ for all ℓ. Thus, just as is the case with M, we see that T admits the left eigenvector $f^0 = (\ldots, 1, 1, 1, \ldots)$ with eigenvalue 1.

If we apply the cascade algorithm to the inner product vector of an actual scaling function corresponding to c:

$$a_k = (\phi_{00}, \phi_{0k}) = \int_{-\infty}^{\infty} \phi(t)\overline{\phi}(t-k)\,dt,$$

we just reproduce the inner product vector:

$$a_s = \sum_{j,\ell} c_{2s+j}\overline{c}_{\ell+j}a_\ell, \tag{10.3}$$

or

$$a = Ta = (\downarrow 2)C\overline{C}^{\mathrm{tr}}a. \tag{10.4}$$

Since $a_k = \delta_{0k}$ in the orthogonal case, this just says that

$$1 = \sum_j |c_j|^2,$$

which we already know to be true. Thus T always has 1 as an eigenvalue, with associated eigenvector $a_k = \delta_{0k}$.

Let's look at the shift orthogonality of the scaling function in the frequency domain (the frequency analogs of (9.93)) as well as the update of the orthogonality relations as we iterate the cascade algorithm. We consider the vector a where

$$a_k = (\phi_{00}, \phi_{0k}) = \int_{-\infty}^{\infty} \phi(t)\overline{\phi}(t-k)\,dt = (\phi_{0i}, \phi_{0j}), \tag{10.5}$$

for $k = j - i$, and its finite Fourier transform $A(\omega) = \sum_k a_k e^{-ik\omega}$. Note that the integer translates of the scaling function are orthonormal if and only if $a_k = \delta_{0k}$, i.e., $A(\omega) \equiv 1$. However, for later use in the study of biorthogonal wavelets, we shall also consider the possibility that the translates are not orthonormal.

Using the Plancherel equality and the fact that the Fourier transform of $\phi(t - k)$ is $e^{-ik\omega}\hat\phi(\omega)$ we have in the frequency domain

$$a_k = \frac{1}{2\pi} \int_{-\infty}^{\infty} \hat\phi(\omega)\overline{\hat\phi}(\omega)e^{ik\omega}d\omega = \frac{1}{2\pi} \int_0^{2\pi} \sum_{n=-\infty}^{\infty} |\hat\phi(\omega + 2\pi n)|^2 e^{ik\omega}d\omega.$$

Theorem 10.1

$$A(\omega) = \sum_{-\infty}^{\infty} |\hat\phi(\omega + 2\pi n)|^2.$$

The integer translates of $\phi(t)$ are orthonormal if and only if $A(\omega) \equiv 1$.

The function $A(\omega)$ and its transform, the vector of inner products $(\phi(t), \phi(t + k))$ will be major players in our study of the L_2 convergence of the cascade algorithm. Let's derive some of its properties with respect to the dilation equation. We will express $A(2\omega)$ in terms of $A(\omega)$ and $H(\omega)$. Since $\hat\phi(2\omega) = H(\omega)\hat\phi(\omega)$ from the dilation equation, we have

$$\hat\phi(2\omega + 2\pi n) = H(\omega + \pi n)\hat\phi(\omega + \pi n)$$

$$= \begin{cases} H(\omega)\hat\phi(\omega + 2\pi k) & n = 2k \\ H(\omega + \pi)\hat\phi(\omega + \pi + 2\pi k) & n = 2k + 1 \end{cases}$$

since $H(\omega + 2\pi) = H(\omega)$. Squaring and adding to get $A(2\omega)$ we find

$$A(2\omega) = |H(\omega)|^2 \sum_k |\hat\phi(\omega + 2\pi k)|^2 + |H(\omega + \pi)|^2 \sum_k |\hat\phi(\omega + \pi + 2\pi k)|^2$$

$$= |H(\omega)|^2 A(\omega) + |H(\omega + \pi)|^2 A(\omega + \pi). \tag{10.6}$$

Essentially the same derivation shows how $A(\omega)$ changes with each pass through the cascade algorithm. Let

$$a_k^{(i)} = (\phi_{00}^{(i)}, \phi_{0k}^{(i)}) = \int_{-\infty}^{\infty} \phi^{(i)}(t)\overline{\phi}^{(i)}(t - k)dt, \tag{10.7}$$

and its associated Fourier transform $A^{(i)}(\omega) = \sum_k a_k^{(i)} e^{-ik\omega}$ denote the information about the inner products of the functions $\phi^{(i)}(t)$ obtained from the ith passage through the cascade algorithm. Since $\hat\phi^{(i+1)}(2\omega) = H(\omega)\hat\phi^{(i)}(\omega)$ we see immediately that

$$A^{(i+1)}(2\omega) = |H(\omega)|^2 A^{(i)}(\omega) + |H(\omega + \pi)|^2 A^{(i)}(\omega + \pi). \tag{10.8}$$

10.1 L_2 convergence

Now we have reached a critical point in the convergence theory for wavelets! We will show that the necessary and sufficient condition for the cascade algorithm to converge in L_2 to a unique solution of the dilation equation is that the transition matrix T has a non-repeated eigenvalue 1 and all other eigenvalues λ are such that $|\lambda| < 1$. Since the only nonzero part of T is a $(2N - 1) \times (2N - 1)$ block with very special structure, this is something that can be checked in practice. The proof uses the Jordan canonical form of matrix algebra, [66].

Theorem 10.2 *The infinite matrix $T = (\downarrow 2)C\overline{C}^{\mathrm{tr}} = \frac{1}{\sqrt{2}}M\overline{C}^{\mathrm{tr}}$ and its finite submatrix T_{2N-1} always have $\lambda = 1$ as an eigenvalue. The cascade iteration $a^{(i+1)} = Ta^{(i)}$ converges in L_2 to the eigenvector $a = Ta$ if and only if the following condition is satisfied:*

- *All of the eigenvalues λ of T_{2N-1} satisfy $|\lambda| < 1$ except for the simple eigenvalue $\lambda = 1$.*

Proof Let λ_j be the $2N - 1$ eigenvalues of T_{2N-1}, including multiplicities. Then there is a basis for the space of $2N - 1$-tuples with respect to which T_{2N-1} takes the *Jordan canonical form*

$$\tilde{T}_{2N-1} = \begin{pmatrix} \lambda_1 & & & & & & \\ & \ddots & & & & & \\ & & \lambda_p & & & & \\ & & & A_{p+1} & & & \\ & & & & A_{p+2} & & \\ & & & & & \ddots & \\ & & & & & & A_{p+q} \end{pmatrix}$$

where the Jordan blocks look like

$$A_s = \begin{pmatrix} \lambda_s & 1 & 0 & \cdots & 0 & 0 \\ 0 & \lambda_s & 1 & \cdots & 0 & 0 \\ \cdots & & & & & \cdots \\ 0 & 0 & 0 & \cdots & \lambda_s & 1 \\ 0 & 0 & 0 & \cdots & 0 & \lambda_s \end{pmatrix}.$$

If the eigenvectors of T_{2N-1} form a basis, for example if there are $2N - 1$ distinct eigenvalues, then with respect to this basis $\tilde{T}_{2N-1}$ would be diagonal and there would be no Jordan blocks. If however, there are not enough eigenvectors to form a basis then the more general Jordan form will hold, with Jordan blocks. Now suppose we perform the cascade recursion n times. Then the action of the iteration on the base space will be

$$
\tilde{T}^n_{2N-1} =
\begin{pmatrix}
\lambda_1^n & & & & & & \\
& \ddots & & & & & \\
& & \lambda_p^n & & & & \\
& & & A^n_{p+1} & & & \\
& & & & A^n_{p+2} & & \\
& & & & & \ddots & \\
& & & & & & A^n_{p+q}
\end{pmatrix}
$$

where $A^n_s =$

$$
\begin{pmatrix}
\lambda_s^n & \binom{n}{1}\lambda_s^{n-1} & \binom{n}{2}\lambda_s^{n-2} & \cdots & \binom{n}{m_s-1}\lambda_s^{n-m_s+1} \\
0 & \lambda_s^n & & \cdots & \binom{n}{m_s-2}\lambda_s^{n-m_s+2} \\
\cdots & & & & \cdots \\
0 & 0 & 0 & \cdots & \binom{n}{1}\lambda_s^{n-1} \\
0 & 0 & 0 & \cdots & \lambda_s^n
\end{pmatrix}
$$

and A_s is an $m_s \times m_s$ matrix with m_s the multiplicity of the eigenvalue λ_s. If there is an eigenvalue with $|\lambda_j| > 1$ then the corresponding terms in the power matrix will blow up and the cascade algorithm will fail to converge. (Of course if the original input vector has zero components corresponding to the basis vectors with these eigenvalues and the computation is done with perfect accuracy, one might have convergence. However, the slightest deviation, such as due to roundoff error, would introduce a component that would blow up after repeated iteration. Thus in practice the algorithm would diverge. The same remarks apply to Theorem 9.25 and Corollary 9.26. With perfect accuracy and filter coefficients that satisfy double-shift orthogonality, one can maintain orthogonality of the shifted scaling functions at each pass of the cascade algorithm if orthogonality holds for the initial step. However, if the algorithm diverges, this theoretical result is of no practical importance. Roundoff error would lead to meaningless results in successive iterations.)

Similarly, if there is a Jordan block corresponding to an eigenvalue $|\lambda_j| = 1$ then the algorithm will diverge. If there is no such Jordan block, but there is more than one eigenvalue with $|\lambda_j| = 1$ then there may be convergence, but it won't be unique and will differ each time the algorithm is applied. If, however, all eigenvalues satisfy $|\lambda_j| < 1$ except for the single eigenvalue $\lambda_1 = 1$, then in the limit as $n \to \infty$ we have

$$\lim_{n\to\infty} \tilde{T}_{2N-1}^n = \begin{pmatrix} 1 & & & \\ & 0 & & \\ & & \ddots & \\ & & & 0 \end{pmatrix}$$

and there is convergence to a unique limit. □

In the frequency domain the action of the **T** operator is

$$\mathbf{T}X(2\omega) = \frac{1}{2}|C(\omega)|^2 X(\omega) + \frac{1}{2}|C(\omega + \pi)|^2 X(\omega + \pi). \tag{10.9}$$

Here $X(\omega) = \sum_{n=-N}^{N} x_n e^{-in\omega}$ and x_n is a $2N-1$-tuple. In the z-domain this is

$$\mathbf{T}X[z^2] = \frac{1}{2}C[z]\,\overline{C}[z^{-1}]\,X[z] + \frac{1}{2}C[-z]\,\overline{C}[-z^{-1}]\,X[-z] \tag{10.10}$$

where $C[z] = \sum_{k=0}^{N} c_k z^{-k}$ and $\overline{C}[z^{-1}] = \sum_{k=0}^{N} \overline{c_k} z^k$. Here, $x \neq 0$ is an eigenvector of **T** with eigenvalue λ if and only if $\mathbf{T}x = \lambda x$, i.e.,

$$2\lambda X[z^2] = C[z]\,\overline{C}[z^{-1}]\,X[z] + C[-z]\,\overline{C}[-z^{-1}]\,X[-z]. \tag{10.11}$$

We can gain some additional insight into the behavior of the eigenvalues of **T** through examining it in the z-domain. Of particular interest is the effect on the eigenvalues of $p > 1$ zeros at $z = -1$ for the low pass filter **C**.

We can write $C[z] = (\frac{1+z^{-1}}{2})^{p-1} C_0[z]$ where $C_0[z]$ is the Z transform of the low pass filter $\mathbf{C}_0$ with a single zero at $z = -1$. In general, $\mathbf{C}_0$ won't satisfy the double-shift orthonormality condition, but we will still have $C_0[1] = \sqrt{2}$ and $C_0[-1] = 0$. This means that the column sums of the corresponding transition matrix T_0 are equal to 1 so that in the time domain $\mathbf{T}_0$ admits the left-hand eigenvector $f^0 = (1, 1, \ldots, 1)$ with eigenvalue 1. Thus T_0 also has some right eigenvector with eigenvalue 1. Here, $\mathbf{T}_0$ is acting on a $2N_0 - 1$-dimensional space, where $N_0 = N - 2(p-1)$.

Our strategy will be to start with $C_0[z]$ and then successively multiply it by the $p - 1$ terms $(1 + z^{-1})/2$, one-at-a-time, until we reach $C[z]$. At each stage we will use Equation (10.11) to track the behavior of the eigenvalues and eigenvectors. Each time we multiply by the factor $(1 + z^{-1})/2$ we will add two dimensions to the space on which we are acting. Thus there will be two additional eigenvalues at each recursion. Suppose we have reached stage $C_s[z] = (\frac{1+z^{-1}}{2})^s C_0[z]$ in this process, with $0 \leq s < p - 1$. Let $x_{(s)}$ be an eigenvector of the corresponding operator $\mathbf{T}_s$ with eigenvalue λ_s. In the z-domain we have

$$2\lambda_s X_s[z^2] = C_s[z]\, \overline{C}_s[z^{-1}]\, X_s[z] + C_s[-z]\, \overline{C}_s[-z^{-1}]\, X_s[-z]. \quad (10.12)$$

Now let $C_{s+1}[z] = (\frac{1+z^{-1}}{2})C_s[z]$, $X_{s+1}[z] = (1 - z^{-1})(1 - z)X_s[z]$. Then, since

$$(\frac{1 + z^{-1}}{2})(\frac{1 + z}{2})(1 - z^{-1})(1 - z) = \frac{1}{4}(1 - z^{-2})(1 - z^2)$$

the eigenvalue equation transforms to

$$2(\frac{\lambda_s}{4})X_{s+1}[z^2] = C_{s+1}[z]\, \overline{C}_{s+1}[z^{-1}]\, X_{s+1}[z]$$

$$+ C_{s+1}[-z]\, \overline{C}_{s+1}[-z^{-1}]\, X_{s+1}[-z].$$

Thus, each eigenvalue λ_s of T_s transforms to an eigenvalue $\lambda_s/4$ of T_{s+1}. In the time domain, the new eigenvectors are linear combinations of shifts of the old ones. There are still two new eigenvalues and their associated eigenvectors to be accounted for. One of these is the eigenvalue 1 associated with the left-hand eigenvector $e^0 = (1, 1, \dots, 1)$. (The associated right-hand eigenvector is the all important a.) To find the last eigenvalue and eigenvector, we consider an intermediate step between C_s and C_{s+1}. Let

$$K_{s+1/2}[z] = (\frac{1 + z^{-1}}{2})\frac{1}{2}C_s[z]\, \overline{C}_s[z^{-1}], \quad X_{s+1/2}[z] = (1 - z^{-1})X_s[z].$$

Then, since

$$(\frac{1 + z^{-1}}{2})(1 - z^{-1}) = \frac{1}{2}(1 - z^{-2})$$

the eigenvalue equation transforms to

$$\frac{1}{2}\lambda_s X_{s+1/2}[z^2] = K_{s+1/2}[z]\, X_{s+1/2}[z] + K_{s+1/2}[-z]\, X_{s+1/2}[-z].$$

This equation doesn't have the same form as the original equation, but in the time domain it corresponds to

$$(\downarrow 2)K_{s+1/2}x_{(s+1/2)} = \frac{1}{2}\lambda x_{(s+1/2)}.$$

The eigenvectors of C_s transform to eigenvectors of $(\downarrow 2)K_{s+1/2}$ with halved eigenvalues. Since $K_{s+1/2}[1] = 1, K_{s+1/2}[-1] = 0$, the columns of $(\downarrow 2)K_{s+1/2}$ sum to 1, and $(\downarrow 2)K_{s+1/2}$ has a left-hand eigenvector $e^0 = (1, 1, \ldots, 1)$ with eigenvalue 1. Thus it also has a new right-hand eigenvector with eigenvalue 1. Now we repeat this process for $(\frac{1+z}{2})K_{s+1/2}[z]$, which gets us back to the eigenvalue problem for C_{s+1}. Since existing eigenvalues are halved by this process, the new eigenvalue 1 for $(\downarrow 2)K_{s+1/2}$ becomes the eigenvalue $1/2$ for T_s.

Theorem 10.3 *If $C[z]$ has a zero of order p at $z = -1$ then T_{2N-1} has eigenvalues* $1, \frac{1}{2}, \ldots (\frac{1}{2})^{2p-1}$.

Theorem 10.4 *Assume that $\phi(t) \in L_2(\mathbb{R})$. Then the cascade sequence $\phi^{(i)}(t)$ converges in L_2 to $\phi(t)$ if and only if the convergence criteria of Theorem 10.2 hold:*

• *All of the eigenvalues λ of T_{2N-1} satisfy $|\lambda| < 1$ except for the simple eigenvalue $\lambda = 1$.*

Proof Assume that the convergence criteria of Theorem 10.2 hold. We want to show that

$$||\phi^{(i)} - \phi||^2 = ||\phi^{(i)}||^2 - (\phi^{(i)}, \phi) - (\phi, \phi^{(i)}) + ||\phi||^2$$
$$= a_0^{(i)} - \overline{b}_0^{(i)} - b_0^{(i)} + a_0 \to 0$$

as $i \to \infty$, see (10.2), (10.31). Here

$$a_k^{(i)} = \int_{-\infty}^{\infty} \phi(t)\overline{\phi}^{(i)}(t - k)dt, \qquad b_k^{(i)} = \int_{-\infty}^{\infty} \phi(t)\overline{\phi}^{(i)}(t - k)dt.$$

With the conditions on $\mathbf{T}$ we know that each of the vector sequences $a^{(i)}$, $b^{(i)}$ will converge to a multiple of the vector a as $i \to \infty$. Since $a_k = \delta_{0k}$ we have $\lim_{i \to \infty} a_k^{(i)} = \mu \delta_{0k}$ and $\lim_{i \to \infty} b_k^{(i)} = \nu \delta_{0k}$. Now at each stage of the recursion we have $\sum_k a_k^{(i)} = \sum_k b_k^{(i)} = 1$, so $\mu = \nu = 1$. Thus as $i \to \infty$ we have

$$a_0^{(i)} - \overline{b}_0^{(i)} - b_0^{(i)} + a_0 \to 1 - 1 - 1 + 1 = 0.$$

(Note: This argument is given for orthogonal wavelets where $a_k = \delta_{0k}$. However, a modification of the argument works for biorthogonal wavelets as well. Indeed the normalization condition $\sum_k a_k^{(i)} = \sum_k b_k^{(i)} = 1$ holds also in the biorthogonal case.) Conversely, this sequence can only converge to zero if the iterates of $\mathbf{T}$ converge uniquely, hence only if the convergence criteria of Theorem 10.2 are satisfied. $\square$

Theorem 10.5 *If the convergence criteria of Theorem 10.2 hold, then the $\phi^{(i)}(t)$ are a Cauchy sequence in L_2, converging to $\phi(t)$.*

Proof We need to show only that the $\phi^{(i)}(t)$ are a Cauchy sequence in L_2. Indeed, since L_2 is complete, the sequence must then converge to some $\phi \in L_2$. We have

$$||\phi^{(m)} - \phi^{(i)}||^2 = ||\phi^{(m)}||^2 - (\phi^{(m)}, \phi^{(i)}) - (\phi^{(i)}, \phi^{(m)}) + ||\phi^{(i)}||^2.$$

From the proof of the preceding theorem we know that $\lim_{i \to \infty} ||\phi^{(i)}||^2 = 1$. Set $m = i + j$ for fixed $j > 0$ and define the vector

$$c^{(i)}(j)_k = (\phi_{00}^{(i+j)}, \phi_{0k}^{(i)}) = \int_{-\infty}^{\infty} \phi^{(i+j)}(t)\overline{\phi}^{(i)}(t - k)dt,$$

i.e., the vector of inner products at stage i. A straightforward computation yields the recursion

$$c^{(i+1)}(j) = \mathbf{T}c^{(i)}(j).$$

Since $\sum_k c^{(i+1)}(j)_k = 1$ at each stage i in the recursion, it follows that $\lim_{i \to \infty} c^{(i)}(j)_k = a_k$ for each j. The initial vectors for these recursions are $c^{(0)}(j)_k = \int_{-\infty}^{\infty} \phi^{(j)}(t)\overline{\phi}^{(0)}(t - k)dt$. We have $\sum_k c^{(0)}(j)_k = 1$ so $c^{(i)}(j)_k \to a_k$ as $i \to \infty$. Furthermore, by the Schwarz inequality $|c^{(0)}(j)_k| \le ||\phi^{(j)}|| \cdot ||\phi^{(0)}|| = 1$, so the components are uniformly bounded. Thus $\mathbf{T}^i c^{(0)}(j)_0 \to a_0 = 1$ as $i \to \infty$, uniformly in j. It follows that

$$||\phi^{(i+j)} - \phi^{(i)}||^2 = ||\phi^{(i+j)}||^2 - (\phi^{(i+j)}, \phi^{(i)}) - (\phi^{(i)}, \phi^{(i+j)}) + ||\phi^{(i)}||^2 \to 0$$

as $i \to \infty$, uniformly in j. $\square$

We continue our examination of the eigenvalues of $\mathbf{T}$, particularly in cases related to Daubechies wavelets. We have observed that in the frequency domain an eigenfunction X corresponding to eigenvalue λ of the $\mathbf{T}$ operator is characterized by the equation

$$2\lambda X(\omega) = |C(\frac{\omega}{2})|^2 X(\frac{\omega}{2}) + |C(\frac{\omega}{2} + \pi)|^2 X(\frac{\omega}{2} + \pi), \qquad (10.13)$$

where $X(\omega) = \sum_{n=-N}^{N} x_n e^{-in\omega}$ and x is a $2N - 1$-tuple. We normalize X by requiring that $||X|| = 1$.

Theorem 10.6 *If $C(\omega)$ satisfies the conditions*

- $|C(\omega)|^2 + |C(\omega + \pi)|^2 \equiv 2$,

- $C(0) = \sqrt{2}, \qquad C(\pi) = 0,$

- $|C(\omega)| \neq 0$ for $-\pi/2 < \omega < \pi/2,$

then **T** *has a simple eigenvalue* 1 *and all other eigenvalues satisfy* $|\lambda| < 1$.

Proof The key is the observation that for any fixed ω with $0 < |\omega| < \pi$ we have $\lambda X(\omega) = \alpha X(\frac{\omega}{2}) + \beta X(\frac{\omega}{2} + \pi)$ where $\alpha > 0, \beta \geq 0$ and $\alpha + \beta = 1$. Thus $\lambda X(\omega)$ is a weighted average of $X(\frac{\omega}{2})$ and $X(\frac{\omega}{2} + \pi)$.

- There are no eigenvalues with $|\lambda| > 1$. For, suppose X were a normalized eigenvector corresponding to λ. Also suppose $|X(\omega)|$ takes on its maximum value at ω_0, such that $0 < \omega_0 < 2\pi$. Then $|\lambda| \cdot |X(\omega_0)| \leq \alpha |X(\frac{\omega_0}{2})| + \beta |X(\frac{\omega_0}{2} + \pi)|$ so, say, $|\lambda| \cdot |X(\omega_0)| \leq |X(\frac{\omega_0}{2})|$. Since $|\lambda| > 1$ this is impossible unless $|X(\omega_0)| = 0$. On the other hand, setting $\omega = 0$ in the eigenvalue equation we find $\lambda X(0) = X(0)$, so $X(0) = 0$. Hence $X(\omega) \equiv 0$ and λ is not an eigenvalue.

- There are no eigenvalues with $|\lambda| = 1$ but $\lambda \neq 1$. For, suppose X was a normalized eigenvector corresponding to λ. Also suppose $|X(\omega)|$ takes on its maximum value at ω_0, such that $0 < \omega_0 < 2\pi$. Then $|\lambda| \cdot |X(\omega_0)| \leq \alpha |X(\frac{\omega_0}{2})| + \beta |X(\frac{\omega_0}{2} + \pi)|$, so $|X(\omega_0)| = |X(\frac{\omega_0}{2})|$. (Note: This works exactly as stated for $0 < \omega_0 < \pi$. If $\pi < \omega_0 < 2\pi$ we can replace ω_0 by $\omega_0' = \omega_0 - 2\pi$ and argue as before. The same remark applies to the cases to follow.) Furthermore, $X(\omega_0) = \lambda^{-1} X(\frac{\omega_0}{2})$. Repeating this argument n times we find that $X(\frac{\omega}{2^n}) = \lambda^n X(\omega_0)$. Since $X(\omega)$ is a continuous function, the left-hand side of this expression is approaching $X(0)$ in the limit. Further, setting $\omega = 0$ in the eigenvalue equation we find $\lambda X(0) = X(0)$, so $X(0) = 0$. Thus $|X(\omega_0)| = 0$ and λ is not an eigenvalue.

- $\lambda = 1$ is an eigenvalue, with the unique (normalized) eigenvector $X(\omega) = \frac{1}{\sqrt{2\pi}}$. Indeed, for $\lambda = 1$ we can assume that $X(\omega)$ is real, since both $X(\omega)$ and $\overline{X}(\omega)$ satisfy the eigenvalue equation. Now suppose the eigenvector X takes on its maximum positive value at ω_0, such that $0 < \omega_0 < 2\pi$. Then $X(\omega_0) \leq \alpha X(\frac{\omega_0}{2}) + \beta X(\frac{\omega_0}{2} + \pi)$, so $X(\omega_0) = X(\frac{\omega_0}{2})$. Repeating this argument n times we find that $X(\frac{\omega}{2^n}) = X(\omega_0)$. Since $X(\omega)$ is a continuous function, the left-hand side of this expression is approaching $X(0)$ in the limit. Thus $X(\omega_0) = X(0)$. Now repeat the same argument under the supposition that the eigenvector

X takes on its minimum value at ω_1, such that $0 < \omega_1 < 2\pi$. We again find that $X(\omega_1) = X(0)$. Thus, $X(\omega)$ is a constant function. We already know that this constant function is indeed an eigenvector with eigenvalue 1.

- There is no nontrivial Jordan block corresponding to the eigenvalue 1. Denote the normalized eigenvector for eigenvalue 1, as computed above, by $X_1(\omega) = \frac{1}{\sqrt{2\pi}}$. If such a block existed there would be a function $X(\omega)$, not normalized in general, such that $\mathbf{T}X(\omega) = X(\omega) + X_1(\omega)$, i.e.,

$$X(\omega) + \frac{1}{\sqrt{2\pi}} = \frac{1}{2}|C(\frac{\omega}{2})|^2 X(\frac{\omega}{2}) + \frac{1}{2}|C(\frac{\omega}{2} + \pi)|^2 X(\frac{\omega}{2} + \pi).$$

Now set $\omega = 0$. We find $X(0) + \frac{1}{\sqrt{2\pi}} = X(0)$ which is impossible. Thus there is no nontrivial Jordan block for $\lambda = 1$.

□

It can be shown that that the condition $|C(\omega)| \neq 0$ for $-\pi/2 < \omega < \pi/2$, can be relaxed just to hold for $-\pi/3 < \omega < \pi/3$.

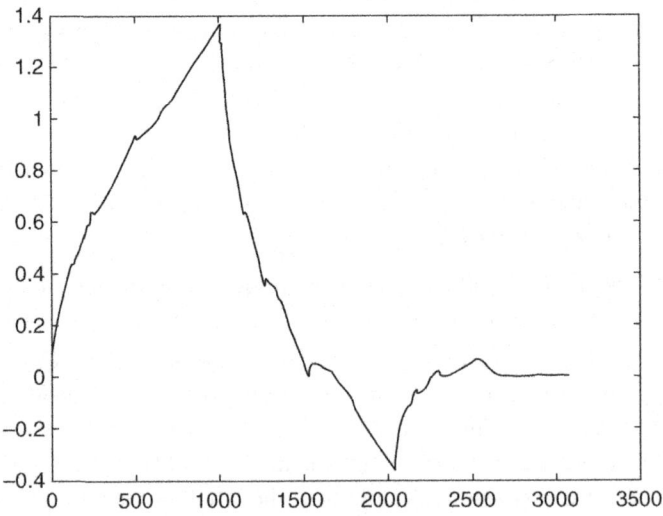

Figure 10.1 The D_4 scaling function.

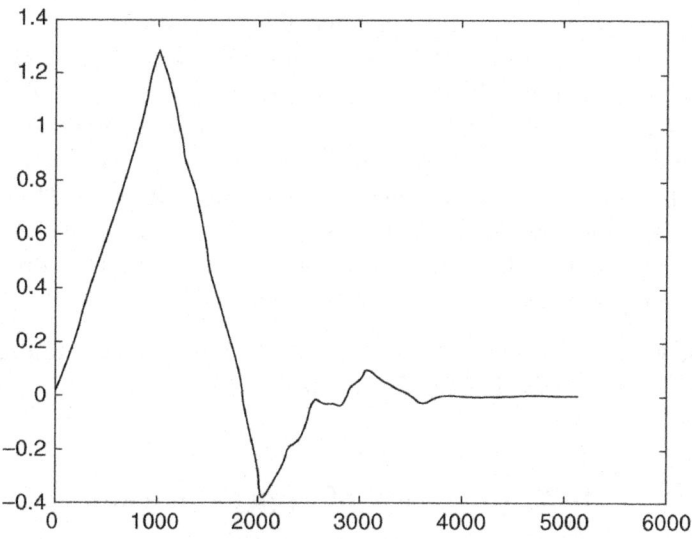

Figure 10.2 The D_6 scaling function.

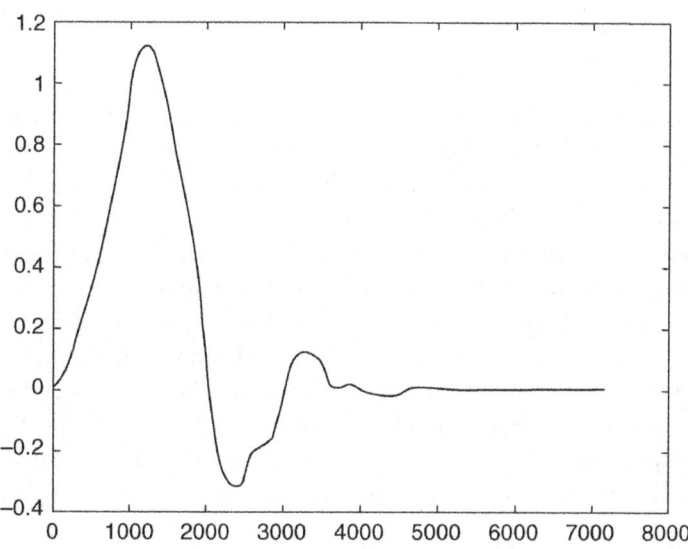

Figure 10.3 The D_8 scaling function.

From our prior work on the maxflat filters we saw that $|C(\omega)|^2$ is 2 for $\omega = 0$ and decreases (strictly) monotonically to 0 for $\omega = \pi$. In particular, $|C(\omega)| \neq 0$ for $0 \leq \omega < \pi$. Since $|C(\omega)|^2 = 2 - |C(\omega + \pi)|^2$ we also have $|C(\omega)| \neq 0$ for $0 \geq \omega > -\pi$. Thus the conditions of the preceding theorem are satisfied, and the cascade algorithm converges for each maxflat system to yield the Daubechies wavelets D_N, where $N = 2p - 1$ and p is the number of zeros of $C(\omega)$ at $\omega = \pi$. The scaling function is supported on the interval $[0, N]$. Polynomials of order $\leq p$ can be approximated with no error in the wavelet space V_0.

Exercise 10.1 For each of the Figures 10.1, 10.2, 10.3, determine the significance of the horizontal and vertical scales.

10.2 Accuracy of approximation

In this section we assume that the criteria for the eigenvalues of $\mathbf{T}$ are satisfied and that we have a multiresolution system with scaling function $\phi(t)$ supported on $[0, N]$. The related low pass filter transform function $C(\omega)$ has $p > 0$ zeros at $\omega = \pi$. We know that

$$\sum_k y^\ell{}_k \phi(t + k) = t^\ell, \qquad \ell = 0, 1, \ldots, p - 1,$$

so polynomials in t of order $\leq p - 1$ can be expressed in V_0 with no error. We assume that the scaling function $\phi(t)$ is pointwise bounded on the real line. (Soon we will be able to prove this boundedness for all Daubechies scaling functions.)

Given a function $f(t)$ we will examine how well f can be approximated pointwise by wavelets in V_j, as well as approximated in the L_2 sense. We will also look at the rate of decay of the wavelet coefficients b_{jk} as $j \to \infty$. Clearly, as j grows the accuracy of approximation of $f(t)$ by wavelets in V_j grows, but so does the computational difficulty. We will not go deeply into approximation theory, but far enough so that the basic dependence of the accuracy on j and p will emerge. We will also look at the smoothness of wavelets, particularly the relationship between smoothness and p.

Let's start with pointwise convergence. Fix $j = J$ and suppose that f has p continuous derivatives in the neighborhood $|t - t_0| \leq 1/2^J$ of t_0. Let

$$f_J(t) = \sum_k a_{Jk} \phi_{Jk}(t) = \sum_k a_{Jk} 2^{J/2} \phi(2^J t - k),$$

$$a_{Jk} = (f, \phi_{Jk}) = 2^{J/2} \int_{-\infty}^{\infty} f(t)\overline{\phi}(2^J t - k)dt,$$

be the projection of f on the scaling space V_J. We want to estimate the pointwise error $|f(t) - f_J(t)|$ in the neighborhood $|t - t_0| \leq 1/2^J$.

Recall Taylor's theorem with integral remainder, from basic calculus.

Theorem 10.7 *If $f(t)$ has p continuous derivatives on an interval containing t_0 and t, then*

$$f(t) = \sum_{k=0}^{p-1} f^{(k)}(t_0)\frac{(t-t_0)^k}{k!} + R_p(t, t_0), \quad R_p(t, t_0) = \int_{t_0}^{t} \frac{(t-x)^p}{p!} f^{(p)}(x)dx$$

where $f^{(0)}(t) = f(t)$.

Since all polynomials of order $\leq p - 1$ can be expressed exactly in V_J we can assume that the first p terms in the Taylor expansion of f have already been canceled exactly by terms a'_{Jk} in the wavelet expansion. Thus

$$|f(t) - f_J(t)| = |R_p(t, t_0) - \sum_k a''_{Jk}\phi_{Jk}(t)|,$$

where the $a''_{Jk} = (R_p(t, t_0), \phi_{Jk})$ are the remaining coefficients in the wavelet expansion $(a_{Jk} = a'_{Jk} + a''_{Jk})$. Note that for fixed t the sum in our error expression contains only N nonzero terms at most. Indeed, the support of $\phi(t)$ is contained in $[0, N)$, so the support of $\phi_{Jk}(t)$ is contained in $[\frac{k}{2^J}, \frac{k+N}{2^J})$. (That is, if $\phi_{Jk}(\tau) \neq 0$ for some τ we have $\frac{k}{2^J} \leq \tau < \frac{k+N}{2^J}$.) Thus $\phi_{Jk}(\tau) = 0$ unless $k = [2^J\tau] - \ell, \ell = 0, 1, \ldots, N-1$, where $[x]$ is the greatest integer in x. If $|f^{(p)}|$ has upper bound M_p in the interval $|t - t_0| \leq 1/2^J$ then $|R_p(t, t_0)| \leq M_p/2^{J(p+1)}(p + 1)!$ and we can derive similar upper bounds for the other N terms that contribute to the sum, to obtain

$$|f(t) - f_J(t)| \leq \frac{CM_p}{2^{J(p+1)}} \tag{10.14}$$

where C is a constant, independent of f and J. If f has only $\ell < p$ continuous derivatives in the interval $|t - t_0| \leq 1/2^J$ then the p in estimate (10.14) is replaced by ℓ:

$$|f(t) - f_J(t)| \leq \frac{CM_\ell}{2^{J(\ell+1)}}.$$

Note that this is a *local* estimate; it depends on the smoothness of f in the interval $|t - t_0| \leq 1/2^J$. Thus once the wavelets choice is fixed, the local rate of convergence can vary dramatically, depending only on the local behavior of f. This is different from Fourier series or Fourier integrals where a discontinuity of a function at one point can slow the rate of convergence at all points. Note also the dramatic improvement of convergence rate due to the p zeros of $C(\omega)$ at $\omega = \pi$.

It is interesting to note that we can investigate the pointwise convergence in a manner very similar to our approach to Fourier series and Fourier integral pointwise convergence. Since $\sum_k \phi(t+k) = 1$ and $\int \phi(t)dt = 1$ we can write (for $0 \le t \le 1/2^j$.)

$$|f(t) - f_J(t)| = |f(t) - \sum_k 2^J \int_{-\infty}^{\infty} f(x)\overline{\phi}(2^J x - k)dx \; \phi(2^J t - k)|$$

$$= |\sum_k \left(f(t) - 2^J \int_{-\infty}^{\infty} f(x)\overline{\phi}(2^J x - k)dx \right) \phi(2^J t - k)|$$

$$\le \sup |\phi| \sum_{k=-N+1}^{0} \left| \int_{-\infty}^{\infty} \left[f(t) - f(\frac{u+k}{2^J}) \right] \overline{\phi}(u)du \right|. \tag{10.15}$$

Now we can make various assumptions concerning the smoothness of f to get an upper bound for the right-hand side of (10.15). We are not taking any advantage of special features of the wavelets employed. Here we assume that f is continuous everywhere and has finite support. Then, since f is uniformly continuous on its domain, it is easy to see that the following function exists for every $h \ge 0$:

$$\omega(h) = \sup_{|t-t'| \le h, \; t,t' \text{ real}} |f(t) - f(t')|. \tag{10.16}$$

Clearly, $\omega(h) \to 0$ as $h \to 0$. We have the bound

$$|f(t) - f_J(t)| \le \sup |\phi| \cdot N\omega(\frac{N}{2^J}) \int_0^N |\phi(u)|du.$$

If we repeat this computation for $t \in [\frac{k}{2^J}, \frac{k+1}{2^J}]$ we get the same upper bound. Thus this bound is uniform for all t, and shows that $f_J(t) \to f(t)$ uniformly as $J \to \infty$.

Now we turn to the estimation of the wavelet expansion coefficients

$$b_{jk} = (f, w_{jk}) = 2^{j/2} \int_{-\infty}^{\infty} f(t)\overline{w}(2^j t - k)dt \tag{10.17}$$

where $w(t)$ is the mother wavelet. We could use Taylor's theorem for f here too, but present an alternate approach. Since $w(t)$ is orthogonal to all integer translates of the scaling function $\phi(t)$ and since all polynomials of order $\leq p - 1$ can be expressed in V_0, we have

$$\int_{-\infty}^{\infty} t^\ell w(t) dt = 0, \qquad \ell = 0, 1, \ldots, p-1. \tag{10.18}$$

Thus the first p moments of w vanish. We will investigate some of the consequences of the vanishing moments. Consider the functions

$$I_1(t) = \int_{-\infty}^{t} w(\tau_0) d\tau_0 \tag{10.19}$$

$$I_2(t) = \int_{-\infty}^{t} I_1(\tau_1) d\tau_1 = \int_{-\infty}^{t} d\tau_1 \int_{-\infty}^{\tau_1} w(\tau_0) d\tau_0$$

$$I_m(t) = \int_{-\infty}^{t} I_{m-1}(\tau_{m-1}) d\tau_{m-1}, \qquad m = 1, 2, \ldots p.$$

From Equation (10.18) with $\ell = 0$ it follows that $I_1(t)$ has support contained in $[0, N)$. Integrating by parts the integral in Equation (10.18) with $\ell = 1$, it follows that $I_2(t)$ has support contained in $[0, N)$. We can continue integrating by parts in this series of equations to show, eventually, that $I_p(t)$ has support contained in $[0, N)$. (This is as far as we can go, however.)

Now integrating by parts p times we find

$$b_{jk} = 2^{j/2} \int_{-\infty}^{\infty} f(t) \overline{w}(2^j t - k) dt = 2^{-j/2} \int_{-\infty}^{\infty} f(\frac{u+k}{2^j}) \overline{w}(u) du$$

$$= 2^{-j/2}(-1)^p \int_{-\infty}^{\infty} \frac{d^p}{du^p} f(\frac{u+k}{2^j}) \overline{I}_p(u) du$$

$$= 2^{j/2-pj}(-1)^p \int_{-\infty}^{\infty} f^{(p)}(\frac{u+k}{2^j}) \overline{I}_p(u) du$$

$$= 2^{j/2-pj}(-1)^p \int_{-\infty}^{\infty} f^{(p)}(t) \overline{I}_p(2^j t - k) dt$$

$$= 2^{j/2-pj}(-1)^p \int_{\frac{k}{2^j}}^{\frac{N+k}{2^j}} f^{(p)}(t) \overline{I}_p(2^j t - k) dt.$$

If $|f^{(p)}(t)|$ has a uniform upper bound M_p then we have the estimate

$$|b_{jk}| \leq \frac{C M_p}{2^{(p+1/2)j}} \tag{10.20}$$

where C is a constant, independent of f, j, k. If, moreover, $f(t)$ has bounded support, say within the interval $(0, K)$, then we can assume that $b_{jk} = 0$ unless $0 \leq k \leq 2^j K$. We already know that the wavelet basis is complete in $L_2(\mathbb{R})$. Let's consider the decomposition

$$L_2(\mathbb{R}) = V_J \oplus \sum_{j=J}^{\infty} W_j.$$

We want to estimate the L_2 error $||f - f_J||$ where f_J is the projection of f on V_J. From the Plancherel equality and our assumption that the support of f is bounded, this is

$$||f - f_J||^2 = \sum_{j=J}^{\infty} \sum_{k=0}^{2^j K} |b_{jk}|^2 < \frac{2C^2 M_p^2 K}{2^{2pJ}}. \tag{10.21}$$

Again, if f has $1 \leq \ell < p$ continuous derivatives then the p in (10.21) is replaced by ℓ.

Much more general and sophisticated estimates than these are known, but these provide a good guide to the convergence rate dependence on j, p and the smoothness of f.

Next we consider the estimation of the scaling function expansion coefficients

$$a_{jk} = (f, \phi_{jk}) = 2^{j/2} \int_{-\infty}^{\infty} f(t)\overline{\phi}(2^j t - k)dt. \tag{10.22}$$

In order to start the FWT recursion for a function f, particularly a continuous function, it is very common to choose a large $j = J$ and then use function samples to approximate the coefficients: $a_{Jk} \sim 2^{-J/2} f(k/2^J)$. This may not be a good policy. Let's look more closely. Since the support of $\phi(t)$ is contained in $[0, N)$ the integral for a_{jk} becomes

$$a_{jk} = 2^{j/2} \int_{\frac{k}{2^j}}^{\frac{N+k}{2^j}} f(t)\overline{\phi}(2^j t - k)dt = 2^{-j/2} \int_0^N f(\frac{u+k}{2^j})\overline{\phi}(u)du.$$

The approximation above is the replacement

$$\int_0^N f(\frac{u+k}{2^j})\overline{\phi}(u)du \sim f(\frac{k}{2^j}) \int_0^N \overline{\phi}(u)du = f(\frac{k}{2^j}).$$

If j is large and f is continuous this using of samples of f isn't a bad estimate. If f is discontinuous or only defined at dyadic values, the sampling could be wildly inaccurate. Note that if you start the FWT recursion at $j = J$ then

$$f_J(t) = \sum_k a_{Jk}\phi_{Jk}(t),$$

so it would be highly desirable for the wavelet expansion to correctly fit the sample values at the points $\ell/2^J$:

$$f(\frac{\ell}{2^J}) = f_J(\frac{\ell}{2^J}) = \sum_k a_{Jk}\phi_{Jk}(\frac{\ell}{2^J}).$$ (10.23)

However, if you use the sample values $f(k/2^J)$ for a_{Jk} then in general $f_J(\ell/2^J)$ will not reproduce the sample values! A possible fix is to prefilter the samples to ensure that (10.23) holds. For Daubechies wavelets, this amounts to replacing the integral $a_{Jk} = (f, \phi_{Jk})$ by the sum $a_{Jk}^c = \sum_\ell f(\ell/2^J)\phi_{Jk}(\ell/2^J)$. These "corrected" wavelet coefficients a_{Jk}^c will reproduce the sample values. However, there is no unique, final answer as to how to determine the initial wavelet coefficients. The issue deserves some thought, rather than a mindless use of sample values.

10.3 Smoothness of scaling functions and wavelets

Our last major issue in the construction of scaling functions and wavelets via the cascade algorithm is their smoothness. So far we have shown only that the Daubechies scaling functions are in $L_2(\mathbb{R})$. We will use the method of Theorem 10.3 to examine this. The basic result is this: The matrix T has eigenvalues $1, \frac{1}{2}, \frac{1}{4}, \ldots, \frac{1}{2^{2p-1}}$ associated with the zeros of $C(\omega)$ at $\omega = \pi$. If all other eigenvalues λ of T satisfy $|\lambda| < \frac{1}{4^s}$ then $\phi(t)$ and $w(t)$ have s derivatives. We will show this for integer s. It is also true for fractional derivatives s, although we shall not pursue this.

Recall that in the proof of Theorem 10.3 we studied the effect on $\mathbf{T}$ of multiplying $C[z]$ by factors $\frac{1+z^{-1}}{2}$, each of which adds a zero at $z = -1$. We wrote $C[z] = (\frac{1+z^{-1}}{2})^{p-1}C_0[z]$ where $C_0[z]$ is the Z transform of the low pass filter $\mathbf{C}_0$ with a single zero at $z = -1$. Our strategy was to start with $C_0[z]$ and then successively multiply it by the $p-1$ terms $(\frac{1+z^{-1}}{2})$, one-at-a-time, until we reached $C[z]$. At each stage, every eigenvalue λ_i of the preceding matrix T_i transformed to an eigenvalue $\lambda_i/4$ of T_{i+1}. There were two new eigenvalues added (1 and 1/2), associated with the new zero of $C(\omega)$.

In going from stage i to stage $i+1$ the infinite product formula for the scaling function

$$\hat{\phi}^{(i)}(\omega) = \Pi_{j=1}^\infty 2^{-1/2}C_i(\frac{\omega}{2^j}),$$ (10.24)

changes to

$$\hat{\phi}^{(i+1)}(\omega) = \Pi_{j=1}^\infty 2^{-1/2}C_{i+1}(\frac{\omega}{2^j}) =$$ (10.25)

$$\Pi_{j=1}^{\infty}\left(\frac{1}{2} + \frac{1}{2}e^{-i\omega/2^j}\right)\Pi_{j=1}^{\infty}2^{-1/2}C_i(\frac{\omega}{2^j}) = \left(\frac{1 - e^{-i\omega}}{i\omega}\right)\hat{\phi}^{(i)}(\omega).$$

The new factor is the Fourier transform of the box function. Now suppose that T_i satisfies the condition that all of its eigenvalues λ_i are < 1 in absolute value, except for the simple eigenvalue 1. Then $\hat{\phi}^{(i)}(\omega) \in L_2$, so $\int_{-\infty}^{\infty} |\omega|^2|\hat{\phi}^{(i+1)}(\omega)|^2 d\omega < \infty$. Now

$$\phi^{(i+1)}(t) = \frac{1}{2\pi}\int_{-\infty}^{\infty}\hat{\phi}^{(i+1)}(\omega)e^{i\omega t}d\omega$$

and the above inequality allows us to differentiate with respect to t under the integral sign on the right-hand side:

$$\phi'^{(i+1)}(t) = \frac{1}{2\pi}\int_{-\infty}^{\infty}i\omega\hat{\phi}^{(i+1)}(\omega)e^{i\omega t}d\omega.$$

The derivative not only exists, but by the Plancherel theorem, it is square integrable. Another way to see this (modulo a few measure-theoretic details) is in the time domain. There the scaling function $\phi^{(i+1)}(t)$ is the convolution of $\phi^{(i)}(t)$ and the box function:

$$\phi^{(i+1)}(t) = \int_0^1 \phi^{(i)}(t-x)dx = \int_{t-1}^t \phi^{(i)}(u)du.$$

Hence $\phi'^{(i+1)}(t) = \phi^{(i)}(t) - \phi^{(i)}(t-1)$. (Note: Since $\phi^{(i)}(t) \in L_2$ it is locally integrable.) Thus $\phi^{(i+1)}(t)$ is differentiable and it has one more derivative than $\phi^{(i)}(t)$. It follows that once we have (the non-special) eigenvalues of T_i less than one in absolute value, each succeeding zero of C adds a new derivative to the scaling function.

Theorem 10.8 *If all eigenvalues λ of T satisfy $|\lambda| < \frac{1}{4^s}$ (except for the special eigenvalues $1, \frac{1}{2}, \frac{1}{4}, \ldots, \frac{1}{2^{2p-1}}$, each of multiplicity one) then $\phi(t)$ and $w(t)$ have s derivatives.*

Corollary 10.9 *The convolution $\phi^{(i+1)}(t)$ has $k+1$ derivatives if and only if $\phi^{(i)}(t)$ has k derivatives.*

Proof From the proof of the theorem, if $\phi^{(i)}(t)$ has k derivatives, then $\phi^{(i+1)}(t)$ has an additional derivative. Conversely, suppose $\phi^{(i+1)}(t)$ has $k+1$ derivatives. Then $\phi'^{(i+1)}(t)$ has k derivatives, and from (10.3) we have

$$\phi'^{(i+1)}(t) = \phi^{(i)}(t) - \phi^{(i)}(t-1).$$

Thus, $\phi^{(i)}(t) - \phi^{(i)}(t-1)$ has k derivatives. Now $\phi^{(i)}(t)$ corresponds to the FIR filter C_i so $\phi^{(i)}(t)$ has support in some bounded interval $[0, M)$. Note that the function

$$\phi^{(i)}(t) - \phi^{(i)}(t-M) = \sum_{j=1}^{M} \left(\phi^{(i)}(t) - \phi^{(i)}(t-j) \right)$$

must have k derivatives, since each term on the right-hand side has k derivatives. However, the support of $\phi^{(i)}(t)$ is disjoint from the support of $\phi^{(i)}(t-M)$, so $\phi^{(i)}(t)$ itself must have k derivatives.   □

Corollary 10.10 *If $\phi(t)$ has s derivatives in L_2 then $s < p$. Thus the maximum possible smoothness is $p - 1$.*

Example 10.11 Daubechies D_4. Here $N = 3$, $p = 2$, so the T matrix is 5×5 ($2N - 1 = 5$), or T_5. Since $p = 2$ we know four roots of T_5: $1, \frac{1}{2}, \frac{1}{4}, \frac{1}{8}$. We can use MATLAB to find the remaining root. It is $\lambda = \frac{1}{4}$. This just misses permitting the scaling function to be differentiable; we would need $|\lambda| < \frac{1}{4}$ for that. Indeed by plotting the D_4 scaling function using the MATLAB toolbox you can see that there are definite corners in the graph. Even so, it is less smooth than it appears. It can be shown that, in the frequency domain, $\int_{-\infty}^{\infty} |\omega|^{2s} |\hat{\phi}(\omega)|^2 d\omega < \infty$ for $0 \leq s < 1$ (but not for $s = 1$). This implies continuity of $\phi(t)$, but not quite differentiability.

Example 10.12 General Daubechies D_M. Here $M = N + 1 = 2p$. By determining the largest eigenvalue in absolute value other than the known $2p$ eigenvalues $1, \frac{1}{2}, \ldots, \frac{1}{2^{2p-1}}$ one can compute the number of derivatives admitted by the scaling functions. The results for the smallest values of p are as follows. For $p = 1, 2$ we have $s = 0$. For $p = 3, 4$ we have $s = 1$. For $p = 5, 6, 7, 8$ we have $s = 2$. For $p = 9, 10$ we have $s = 3$. Asymptotically s grows as $0.2075p +$ constant.

We can derive additional smoothness results by considering more carefully the pointwise convergence of the cascade algorithm in the frequency domain:

$$\hat{\phi}(\omega) = \Pi_{j=1}^{\infty} 2^{-1/2} C(\frac{\omega}{2^j}).$$

Recall that we only had the crude upper bound $|\hat{\phi}(\omega)| \leq e^{C_0 |\omega|}$ where $|C'(\omega)| \leq C_0 \sqrt{2}$ and $C(\omega) = C(0) + \int_0^{\omega} C'(s) ds$, so

$$|C(\omega)| \leq \sqrt{2}(1 + C_0 |\omega|) \leq \sqrt{2} e^{C_0 |\omega|}.$$

This shows that the infinite product converges uniformly on any compact set, and converges absolutely. It is far from showing that $\hat{\phi}(\omega) \in L_2$. We clearly can find much better estimates. For example if $C(\omega)$ satisfies the double-shift orthogonality relations $|C(\omega)|^2 + |C(\omega + \pi)|^2 = 2$ then $|C(\omega)| \leq \sqrt{2}$, which implies that $|\hat{\phi}(\omega)| \leq 1$. This doesn't prove that $\hat{\phi}$ decays sufficiently rapidly at ∞ so that it is square integrable, but it suggests that we can find much sharper estimates.

The following ingenious argument by Daubechies improves the exponential upper bound on $\hat{\phi}(\omega)$ to a polynomial bound. First, let $\sqrt{2}C_1 \geq \sqrt{2}$ be an upper bound for $|C(\omega)|$,

$$|C(\omega)| \leq \sqrt{2}C_1 \quad \text{for } 0 \leq \omega < \pi.$$

Set

$$\Phi_k(\omega) = \Pi_{j=1}^k 2^{-1/2} C(\frac{\omega}{2^j}),$$

and note that $|\Phi_k(\omega)| \leq e^{C_0}$ for $|\omega| \leq 1$. Next we bound $|\Phi_k(\omega)|$ for $|\omega| > 1$. For each $|\omega| > 1$ we can uniquely determine the positive integer $K(\omega)$ so that $2^{K-1} < |\omega| \leq 2^K$. Now we derive upper bounds for $|\Phi_k(\omega)|$ in two cases: $k \leq K$ and $k > K$. For $k \leq K$ we have

$$|\Phi_k(\omega)| = \Pi_{j=1}^k 2^{-1/2} |C(\frac{\omega}{2^j})| \leq C_1^K \leq C_1^{1+\log_2 |\omega|} = C_1 |\omega|^{\log_2 C_1},$$

since $\log_2 C_1^{\log_2 |\omega|} = \log_2 |\omega|^{\log_2 C_1} = \log_2 C_1 \log_2 |\omega|$. For $k > K$ we have

$$|\Phi_k(\omega)| = \Pi_{j=1}^K 2^{-1/2} |C(\frac{\omega}{2^j})| \Pi_{j=1}^{k-K} |C(\frac{\omega}{2^{K+j}})|$$

$$\leq C_1^K |\Phi_{k-K}(\frac{\omega}{2^K})| \leq C_1 |\omega|^{\log_2 C_1} e^{C_0}.$$

Combining these estimates we obtain the uniform upper bound

$$|\Phi_k(\omega)| \leq C_1 e^{C_0} (1 + |\omega|^{\log_2 C_1})$$

for all ω and all integers k. Now when we go to the limit we have the polynomial upper bound

$$|\hat{\phi}(\omega)| \leq C_1 e^{C_0} (1 + |\omega|^{\log_2 C_1}). \tag{10.26}$$

Thus still doesn't give L_2 convergence or smoothness results, but together with information about p zeros of $C(\omega)$ at $\omega = \pi$ we can use it to obtain such results.

The following result clarifies the relationship between the smoothness of the scaling function $\phi(t)$ and the rate of decay of its Fourier transform $\hat{\phi}(\omega)$.

Lemma 10.13 *Suppose* $\int_{-\infty}^{\infty} |\hat{\phi}(\omega)|(1 + |\omega|)^{n+\alpha} d\omega < \infty$, *where* n *is a non-negative integer and* $0 \leq \alpha < 1$. *Then* $\phi(t)$ *has* n *continuous derivatives, and there exists a positive constant* A *such that* $|\phi^{(n)}(t + h) - \phi^{(n)}(t)| \leq A|h|^{\alpha}$, *uniformly in* t *and* h.

Sketch of proof From the assumption, the right-hand side of the formal inverse Fourier relation

$$\phi^{(m)}(t) = \frac{i^m}{2\pi} \int_{-\infty}^{\infty} \hat{\phi}(\omega)\omega^m e^{i\omega t} d\omega$$

converges absolutely for $m = 0, 1, \ldots, n$. It is a straightforward application of the Lebesgue dominated convergence theorem, [91], to show that $\phi^{(m)}(t)$ is a continuous function of t for all t. Now for $h > 0$

$$\frac{\phi^{(n)}(t + h) - \phi^{(n)}(t)}{h^{\alpha}} = \frac{i^n}{2\pi} \int_{-\infty}^{\infty} \hat{\phi}(\omega)\omega^n \left(\frac{e^{i\omega(t+h)} - e^{i\omega t}}{h^{\alpha}}\right) d\omega$$

$$= \frac{i^{n+1}}{\pi} \int_{-\infty}^{\infty} \hat{\phi}(\omega)\omega^{n+\alpha} e^{i\omega(t+\frac{h}{2})} \frac{\sin\frac{\omega h}{2}}{(\omega h)^{\alpha}} d\omega. \tag{10.27}$$

Note that the function $\sin\frac{\omega h}{2}/(\omega h)^{\alpha}$ is bounded for all ωh. Thus there exists a constant M such that $|\sin\frac{\omega h}{2}/(\omega h)^{\alpha}| \leq M$ for all ωh. It follows from the hypothesis that the integral on the right-hand side of (10.27) is bounded by a constant A. A slight modification of this argument goes through for $h < 0$. □

NOTE: A function $f(t)$ such that $|f(t + h) - f(t)| \leq A|h|^{\alpha}$ is said to be Hölder continuous with modulus of continuity α.

Now let's investigate the influence of p zeros at $\omega = \pi$ of the low pass filter function $C(\omega)$. In analogy with our earlier analysis of the cascade algorithm (10.24) we write $H(\omega) = (\frac{1+e^{-i\omega}}{2})^p C_{-1}(\omega)$. Thus the FIR filter $C_{-1}(\omega)$ still has $C_{-1}(0) = \sqrt{2}$, but it doesn't vanish at $\omega = \pi$. Then the infinite product formula for the scaling function

$$\hat{\phi}(\omega) = \Pi_{j=1}^{\infty} 2^{-1/2} C(\frac{\omega}{2^j}), \tag{10.28}$$

changes to

$$\hat{\phi}(\omega) = \Pi_{j=1}^{\infty} \left(\frac{1}{2} + \frac{1}{2}e^{-i\omega/2^j}\right)^p \Pi_{j=1}^{\infty} 2^{-1/2} C_{-1}(\frac{\omega}{2^j})$$

$$= \left(\frac{1 - e^{-i\omega}}{i\omega}\right)^p \hat{\phi}_{-1}(\omega). \tag{10.29}$$

The new factor is the Fourier transform of the box function, raised to the power p. From (10.26) we have the upper bound

$$|\hat{\phi}_{-1}(\omega)| \leq C_1 e^{C_0}(1 + |\omega|^{\log_2 C_1}) \tag{10.30}$$

with

$$C_1 \geq \begin{cases} \max_{\omega \in [0,2\pi]} |H_{-1}(\omega)| \\ 1. \end{cases}$$

This means that $|\hat{\phi}_{-1}(\omega)|$ decays at least as fast as $|\omega|^{\log_2 C_1}$ for $|\omega| \to \infty$, hence that $|\hat{\phi}(\omega)|$ decays at least as fast as $|\omega|^{\log_2 C_1 - p}$ for $|\omega| \to \infty$.

Example 10.14 Daubechies D_4. The low pass filter is

$$C(\omega) = \sqrt{2} \left(\frac{1 + e^{-i\omega}}{2}\right)^2 \left[\frac{1}{2}[1 + \sqrt{3}] + \frac{1}{2}[1 - \sqrt{3}]e^{-i\omega}\right]$$

$$= \left(\frac{1 + e^{-i\omega}}{2}\right)^2 C_{-1}(\omega).$$

Here $p = 2$ and the maximum value of $|C_{-1}(\omega)|$ is $\sqrt{2}C_1 = \sqrt{6}$ at $\omega = \pi$. Since $\log_2 \sqrt{3} = 0.79248\ldots$ it follows that $|\hat{\phi}(\omega)|$ decays at least as fast as $|\omega|^{-1.207\cdots}$ for $|\omega| \to \infty$. Thus we can apply Lemma 10.13 with $n = 0$ to show that the Daubechies D_4 scaling function is continuous with modulus of continuity at least $\alpha = 0.207\ldots$

We can also use the previous estimate and the computation of C_1 to get a (crude) upper bound for the eigenvalues of the T matrix associated with $C_{-1}(\omega)$, hence for the non-obvious eigenvalues of T_{2N-1} associated with $C(\omega)$. If λ is an eigenvalue of the T matrix associated with $C_{-1}(\omega)$ and $X(\omega)$ is the corresponding eigenvector, then the eigenvalue equation

$$2\lambda X(\omega) = |C_{-1}(\frac{\omega}{2})|^2 X(\frac{\omega}{2}) + |C_{-1}(\frac{\omega}{2} + \pi)|^2 X(\frac{\omega}{2} + \pi),$$

is satisfied, where $X(\omega) = \sum_{n=-N+2p}^{N-2p} x_n e^{-in\omega}$ and x_n is a $2(N - 2p) - 1$-tuple. We normalize X by requiring that $||X|| = 1$. Let $M = \max_{\omega \in [0,2\pi]} |X(\omega)| = |X(\omega_0)|$. Setting $\omega = \omega_0$ in the eigenvalue equation, and taking absolute values we obtain

$$|\lambda| \cdot M = \frac{1}{2} \left| |C_{-1}(\frac{\omega_0}{2})|^2 X(\frac{\omega_0}{2}) + |C_{-1}(\frac{\omega_0}{2} + \pi)|^2 X(\frac{\omega_0}{2} + \pi) \right|$$

$$\leq C_1^2 M + C_1^2 M = 2C_1^2 M.$$

Thus $|\lambda| \leq 2C_1^2$ for all eigenvalues associated with $C_{-1}(\omega)$. This means that the non-obvious eigenvalues of T_{2N-1} associated with $C(\omega)$ must satisfy the inequality $|\lambda| \leq 2C_1^2/4^p$. In the case of D_4 where $C_1^2 = 3$ and $p = 2$ we get the upper bound $3/8$ for the non-obvious eigenvalue. In fact, the eigenvalue is $1/4$.

10.4 Additional exercises

Exercise 10.2 Apply the cascade algorithm to the inner product vector of the ith iterate of the cascade algorithm with the scaling function itself

$$b_k^{(i)} = \int_{-\infty}^{\infty} \phi(t)\overline{\phi}^{(i)}(t-k)dt,$$

and derive, by the method leading to (10.1), the result

$$b^{(i+1)} = Tb^{(i)}. \tag{10.31}$$

Exercise 10.3 The application of Theorem 10.1 to the Haar scaling function leads to an identity for the sinc function. Derive it.

Exercise 10.4 The finite T matrix for the dilation equation associated with a prospective family of wavelets has the form

$$T = \begin{pmatrix} 0 & \frac{-1}{16} & 0 & 0 & 0 \\ 1 & \frac{9}{16} & 0 & \frac{-1}{16} & 0 \\ 0 & \frac{9}{16} & 1 & \frac{9}{16} & 0 \\ 0 & \frac{-1}{16} & 0 & \frac{9}{16} & 1 \\ 0 & 0 & 0 & \frac{-1}{16} & 0 \end{pmatrix}.$$

Designate the components of the low pass filter defining this matrix by $c_0, \ldots c_N$.

1. What is N in this case? Give a reason for your answer.
2. The vector $a_k = \int \phi(t)\overline{\phi(t-k)}dt$, for $k = -2, -1, 0, 1, 2$, is an eigenvector of the T matrix with eigenvalue 1. By looking at T can you tell if these wavelets are ON?
3. The Jordan canonical form for T is

$$J = \begin{pmatrix} 1 & 0 & 0 & 0 & 0 \\ 0 & \frac{1}{2} & 0 & 0 & 0 \\ 0 & 0 & \frac{1}{8} & 0 & 0 \\ 0 & 0 & 0 & \frac{1}{4} & 1 \\ 0 & 0 & 0 & 0 & \frac{1}{4} \end{pmatrix}.$$

What are the eigenvalues of T?

4. Does the cascade algorithm converge in this case to give an L_2 scaling function $\phi(t)$? Give a reason for your answer.

5. What polynomials $P(t)$ can be expanded by the $\phi(t-k)$ basis, with no error. (This is equivalent to determining the value of p.) Give a reason for your answer.

11

Biorthogonal filters and wavelets

11.1 Resumé of basic facts on biorthogonal filters

Previously, our main emphasis has been on orthogonal filter banks and orthogonal wavelets. Now we will focus on the more general case of biorthogonality. For filter banks this means, essentially, that the analysis filter bank is invertible (but not necessarily unitary) and the synthesis filter bank is the inverse of the analysis filter bank. We recall some of the main facts from Section 9.6, in particular, Theorem 9.18: A two-channel filter bank gives perfect reconstruction when

$$S^{(0)}[z]H^{(0)}[z] + S^{(1)}[z]H^{(1)}[z] = 2z^{-\ell}$$

$$S^{(0)}[z]H^{(0)}[-z] + S^{(1)}[z]H^{(1)}[-z] = 0. \qquad (11.1)$$

This is the mathematical expression of (11.2). (Recall that $H^{(j)}[z]$ is the Z transform of the analysis filter $\mathbf{\Phi}^{(j)}$, $S^{(j)}$ is the Z transform of the synthesis filter $\mathbf{\Psi}^{(j)}$, and $\mathbf{R}$ is the right-shift or delay operator.)

$$
\begin{array}{ccccc}
\mathbf{\Psi}^{(0)} & \uparrow 2 & \leftarrow & \downarrow 2 & \mathbf{\Phi}^{(0)} \\
& & & & \nwarrow \\
\mathbf{R}^{\ell}x \;=\; & + & & & x \;, \qquad (11.2) \\
& & & & \swarrow \\
\mathbf{\Psi}^{(1)} & \uparrow 2 & \leftarrow & \downarrow 2 & \mathbf{\Phi}^{(1)}
\end{array}
$$

We can find solutions of (11.1) by defining the synthesis filters in terms of the analysis filters:

$$S^{(0)}[z] = H^{(1)}[-z], \qquad S^{(1)}[z] = -H^{(0)}[-z],$$

and introducing the halfband filter

$$P[z] = z^{\ell}S^{(0)}[z]H^{(0)}[z]$$

such that

$$P[z] - P[-z] = 2. \tag{11.3}$$

The delay ℓ must be an odd integer. Equation (11.3) says the coefficients of the even powers of z in $P[z]$ vanish, except for the constant term, which is 1. The coefficients of the odd powers of z are undetermined design parameters for the filter bank. Once solutions for $P[z]$ are determined, the possible biorthogonal filter banks correspond to factorizations of this polynomial.

The perfect reconstruction conditions have a simple expression in terms of 2×2 matrices:

$$\left[\begin{array}{cc} S^{(0)}[z] & S^{(1)}[z] \\ S^{(0)}[-z] & S^{(1)}[-z] \end{array} \right] \left[\begin{array}{cc} \dot{H}^{(0)}[z] & \dot{H}^{(0)}[-z] \\ \dot{H}^{(1)}[z] & \dot{H}^{(1)}[-z] \end{array} \right] = 2 \left[\begin{array}{cc} 1 & 0 \\ 0 & 1 \end{array} \right],$$

or $\mathbf{S}\dot{\mathbf{H}} = 2\mathbf{I}$ where $\dot{H}^{(0)}[z] = z^\ell H^{(0)}[z]$, $\dot{H}^{(1)}[z] = z^\ell H^{(1)}[z]$ are recentered. (Note that since these are *finite matrices*, the fact that $\dot{\mathbf{H}}[z]$ has a left inverse implies that it has the same right inverse, and is invertible.)

Example 11.1 Daubechies D_4 halfband filter is

$$P[z] = \frac{1}{16} \left(-z^3 + 9z + 16 + 9z^{-1} - z^{-3} \right).$$

The shifted filter $P^{(0)}[z] = z^{-\ell}P[z]$ must be a polynomial in z^{-1} with nonzero constant term. Thus $\ell = 3$ and

$$P^{(0)}[z] = \frac{1}{16} \left(-1 + 9z^{-2} + 16z^{-3} + 9z^{-4} - z^{-6} \right).$$

Note that $p = 2$ for this filter, so $P^{(0)}[z] = (1 + z^{-1})^4 Q[z]$ where Q has only two roots, $r = 2 - \sqrt{3}$ and $r^{-1} = 2 + \sqrt{3}$. There are a variety of factorizations $P^{(0)}[z] = S^{(0)}[z]H^{(0)}[z]$, depending on which factors are assigned to $H^{(0)}$ and which to $S^{(0)}$. For the construction of filters it makes no difference if the factors are assigned to $H^{(0)}$ or to $S^{(0)}$ (there is no requirement that $H^{(0)}$ be a low pass filter, for example). In the following table we list the possible factorizations of $P^{(0)}[z]$ associated with the $D4$ filter in terms of Factor 1 and Factor 2.

Remark If we want to use these factorizations for the construction of biorthogonal or orthogonal wavelets, however, then $H^{(0)}$ and $S^{(0)}$ are *required* to be low pass. Furthermore it is not enough that conditions (11.1) hold. The T matrices corresponding to the low pass filters $H^{(0)}$ and $S^{(0)}$ must *each* have the proper eigenvalue structure to guarantee L_2 convergence of the cascade algorithm. Thus some of the factorizations listed in the table will not yield useful wavelets.

Case	Factor 1	Degree N	Factor 2	Degree N
1	1	0	$(1 + z^{-1})^4 (r - z^{-1})$ $(r^{-1} - z^{-1})$	6
2	$(1 + z^{-1})$	1	$(1 + z^{-1})^3 (r - z^{-1})$ $(r^{-1} - z^{-1})$	5
3	$(r - z^{-1})$	1	$(1 + z^{-1})^4 (r^{-1} - z^{-1})$	5
4	$(1 + z^{-1})^2$	2	$(1 + z^{-1})^2 (r - z^{-1})$ $(r^{-1} - z^{-1})$	4
5	$(1 + z^{-1})(r - z^{-1})$	2	$(1 + z^{-1})^3 (r^{-1} - z^{-1})$	4
6	$(r - z^{-1})(r^{-1} - z^{-1})$	2	$(1 + z^{-1})^4$	4
7	$(1 + z^{-1})^3$	3	$(1 + z^{-1})(r - z^{-1})$ $(r^{-1} - z^{-1})$	3
8	$(1 + z^{-1})^2 (r - z^{-1})$	3	$(1 + z^{-1})^2 (r^{-1} - z^{-1})$	3

For cases $3, 5, 8$ we can switch $r \leftrightarrow r^{-1}$ to get new possibilities. A common notation is $(N_A + 1)/(N_S + 1)$ where N_A is the degree of the analysis filter $H^{(0)}$ and N_S is the degree of the synthesis filter $S^{(0)}$, i.e. (# of analysis coefficients)/(# of synthesis coefficients). (However, this notation doesn't define the filters uniquely.) Thus from case 4 we could produce the 5/3 filters $H^{(0)}[z] = \frac{1}{8}(-1 + 2z^{-1} + 6z^{-2} + 2z^{-3} - z^{-4})$ and $S^{(0)}[z] = \frac{1}{2}(1 + 2z^{-1} + z^{-2})$, or the 3/5 filters with $H^{(0)}$ and $S^{(0)}$ interchanged. The orthonormal Daubechies D_4 filter 4/4 comes from case 7.

Let's investigate what these requirements say about the finite impulse response vectors $h_n^{(0)}, h_n^{(1)}, s_n^{(0)}, s_n^{(1)}$. (Recall that $H^{(0)}[z] = \sum_n h_n^{(0)} z^{-n}$, and so forth.) The halfband filter condition for $P[z]$, recentered , says

$$\sum_n h_{\ell+n}^{(0)} s_{2k-n}^{(0)} = \delta_{0k}, \qquad (11.4)$$

and

$$\sum_n h_{\ell+n}^{(1)} s_{2k-n}^{(1)} = -\delta_{0k}, \qquad (11.5)$$

or

$$\sum_n \breve{h}_n^{(0)} s_{2k+n}^{(0)} = \delta_{0k}, \qquad (11.6)$$

and

$$\sum_n \breve{h}_n^{(1)} s_{2k+n}^{(1)} = -\delta_{0k}, \qquad (11.7)$$

where

$$\breve{h}_n^{(j)} = h_{\ell-n}^{(j)},$$

so $s_n^{(0)} = (-1)^n h_{\ell-n}^{(1)}$, $s_n^{(1)} = (-1)^{n+1} h_{\ell-n}^{(0)}$. The conditions

$$\dot{H}^{(0)}[z]S^{(1)}[z] - \dot{H}^{(0)}[-z]S^{(1)}[-z] = 0,$$
$$\dot{H}^{(1)}[z]S^{(0)}[z] - \dot{H}^{(1)}[-z]S^{(0)}[-z] = 0,$$

imply

$$\sum_n \breve{h}_n^{(0)} s_{2k+n}^{(1)} = 0, \tag{11.8}$$

$$\sum_n \breve{h}_n^{(1)} s_{2k+n}^{(0)} = 0. \tag{11.9}$$

Expression (11.6) gives us some insight into the support of $\breve{h}_n^{(0)}$ and $s_n^{(0)}$. Since $P^{(0)}[z]$ is an even-order polynomial in z^{-1} it follows that the sum of the orders of $H^{(0)}[z]$ and $S^{(0)}[z]$ must be even. This means that $\breve{h}_n^{(0)}$ and $s_n^{(0)}$ are each nonzero for an even number of values, or each are nonzero for an odd number of values.

11.2 Biorthogonal wavelets: multiresolution structure

In this section we will introduce a multiresolution structure for biorthogonal wavelets, a generalization of what we have done for orthogonal wavelets. Again there will be striking parallels with the study of biorthogonal filter banks. We will go through this material rapidly, because it is so similar to what we have already presented.

Definition 11.2 Let $\{V_j : j = \ldots, -1, 0, 1, \ldots\}$ be a sequence of subspaces of $L_2(\mathbb{R})$ and $\phi \in V_0$. Similarly, let $\{\tilde{V}_j : j = \ldots, -1, 0, 1, \ldots\}$ be a sequence of subspaces of $L_2(\mathbb{R})$ and $\tilde{\phi} \in \tilde{V}_0$. This is a biorthogonal multiresolution analysis for $L_2(\mathbb{R})$ provided the following conditions hold:

1. The subspaces are nested: $V_j \subset V_{j+1}$ and $\tilde{V}_j \subset \tilde{V}_{j+1}$.
2. The union of the subspaces generates $L_2 : L_2(\mathbb{R}) = \overline{\cup_{j=-\infty}^{\infty} V_j} = \overline{\cup_{j=-\infty}^{\infty} \tilde{V}_j}$.
3. Separation: $\cap_{j=-\infty}^{\infty} V_j = \cap_{j=-\infty}^{\infty} \tilde{V}_j = \{0\}$, the subspace containing only the zero function. (Thus only the zero function is common to all subspaces V_j, or to all subspaces $\tilde{V}_j$.)

4. Scale invariance: $f(t) \in V_j \iff f(2t) \in V_{j+1}$, and $\tilde{f}(t) \in \tilde{V}_j \iff \tilde{f}(2t) \in \tilde{V}_{j+1}$.
5. Shift invariance of V_0 and $\tilde{V}_0$: $f(t) \in V_0 \iff f(t-k) \in V_0$ for all integers k, and $\tilde{f}(t) \in \tilde{V}_0 \iff \tilde{f}(t-k) \in \tilde{V}_0$ for all integers k.
6. Biorthogonal bases: The set $\{\phi(t-k) : k = 0, \pm 1, \ldots\}$ is a Riesz basis for V_0, the set $\{\tilde{\phi}(t-k) : k = 0, \pm 1, \ldots\}$ is a Riesz basis for $\tilde{V}_0$ and these bases are biorthogonal:

$$< \phi_{0k}, \tilde{\phi}_{0\ell} >= \int_{-\infty}^{\infty} \phi(t-k)\tilde{\phi}(t-\ell)dt = \delta_{k\ell}.$$

Now we have two scaling functions, the *synthesizing function* $\phi(t)$, and the *analyzing function* $\tilde{\phi}(t)$. The $\tilde{V}, \tilde{W}$ spaces are called the **analysis multiresolution** and the spaces V, W are called the **synthesis multiresolution**.

In analogy with orthogonal multiresolution analysis we can introduce complements W_j of V_j in V_{j+1}, and $\tilde{W}_j$ of $\tilde{V}_j$ in $\tilde{V}_{j+1}$:

$$V_{j+1} = V_j + W_j, \quad \tilde{V}_{j+1} = \tilde{V}_j + \tilde{W}_j.$$

However, these will no longer be orthogonal complements. We start by constructing a Riesz basis for the analysis wavelet space $\tilde{W}_0$. Since $\tilde{V}_0 \subset \tilde{V}_1$, the analyzing function $\tilde{\phi}(t)$ must satisfy the *analysis dilation equation*

$$\tilde{\phi}(t) = 2 \sum_k \breve{h}_k^{(0)} \tilde{\phi}(2t - k), \qquad (11.10)$$

where

$$\sum_k \breve{h}_k^{(0)} = 1$$

for compatibility with the requirement $\int_{-\infty}^{\infty} \tilde{\phi}(t)dt = 1$.

Similarly, since $V_0 \subset V_1$, the synthesis function $\phi(t)$ must satisfy the *synthesis dilation equation*

$$\phi(t) = 2 \sum_k s_k^{(0)} \phi(2t - k), \qquad (11.11)$$

where

$$\sum_k s_k^{(0)} = 1 \qquad (11.12)$$

for compatibility with the requirement $\int_{-\infty}^{\infty} \phi(t) \, dt = 1$.

Remarks 1. There is a problem here. If the filter coefficients are derived from the halfband filter $P^{(0)}[z] = H^{(0)}[z]S^{(0)}[z]$ as in the previous section, then for low pass filters we have $2 = P[1] = H^{(0)}[1]S^{(0)}[1]$ $= [\sum_k h_k^{(0)}][\sum_k s_k^{(0)}]$, so we can't have $\sum_k h_k^{(0)} = \sum_k s_k^{(0)} = 1$ simultaneously. To be definite we will always choose the $H^{(0)}$ filter such that $\sum_k h_k^{(0)} = 1$. Then we must replace the expected synthesis dilation equations (11.11) and (11.12) by

$$\phi(t) = \sum_k s_k^{(0)} \phi(2t - k), \tag{11.13}$$

where

$$\sum_k s_k^{(0)} = 2 \tag{11.14}$$

for compatibility with the requirement $\int_{-\infty}^{\infty} \phi(t)dt = 1$.

Since the $s^{(0)}$ filter coefficients are fixed multiples of the $\check{h}^{(1)}$ coefficients we will also need to alter the analysis wavelet equation by a factor of 2 in order to obtain the correct orthogonality conditions (and we have done this below).

2. We introduced the modified analysis filter coefficients $\check{h}_n^{(j)} = h_{\ell-n}^{(j)}$ in order to adapt the biorthogonal filter identities to the identities needed for wavelets, but we left the synthesis filter coefficients unchanged. We could just as well have left the analysis filter coefficients unchanged and introduced modified synthesis coefficients $\check{s}_n^{(j)} = s_{\ell-n}^{(j)}$.

Associated with the analyzing function $\tilde{\phi}(t)$ there must be an analyzing wavelet $\tilde{w}(t)$, with norm 1, and satisfying the *analysis wavelet equation*

$$\tilde{w}(t) = \sum_k \check{h}_k^{(1)} \tilde{\phi}(2t - k), \tag{11.15}$$

and such that $\tilde{w}$ is orthogonal to all translations $\phi(t-k)$ of the synthesis function.

Associated with the synthesis function $\phi(t)$ there must be an synthesis wavelet $w(t)$, with norm 1, and satisfying the *synthesis wavelet equation*

$$w(t) = 2\sum_k s_k^{(1)} \phi(2t - k), \tag{11.16}$$

and such that w is orthogonal to all translations $\tilde{\phi}(t-k)$ of the analysis function.

Since $\tilde{w}(t) \perp \phi(t - m)$ for all m, the vector $\breve{h}^{(1)}$ satisfies double-shift orthogonality with $s^{(0)}$:

$$< \phi(t - m), \tilde{w}(t) >= \sum_k \breve{h}_k^{(1)} s_{k+2m}^{(0)} = 0. \qquad (11.17)$$

The requirement that $\tilde{w}(t) \perp w(t - m)$ for nonzero integer m leads to double-shift orthogonality of $\breve{h}^{(1)}$ with $s^{(1)}$:

$$< w(t - m), \tilde{w}(t) >= \sum_k \breve{h}_k^{(1)} s_{k-2m}^{(1)} = \delta_{0m}. \qquad (11.18)$$

Since $w(t) \perp \tilde{\phi}(t - m)$ for all m, the vector $\breve{h}^{(0)}$ satisfies double-shift orthogonality with $s^{(1)}$:

$$< w(t), \tilde{\phi}(t - m) >= \sum_k \breve{h}_k^{(0)} s_{k+2m}^{(1)} = 0. \qquad (11.19)$$

The requirement that $\phi(t) \perp \tilde{\phi}(t - m)$ for nonzero integer m leads to double-shift orthogonality of $\breve{h}^{(0)}$ with $s^{(0)}$:

$$< \phi(t), \tilde{\phi}(t - m) >= \sum_k \breve{h}_k^{(0)} s_{k-2m}^{(0)} = \delta_{0m}. \qquad (11.20)$$

Thus,

$$W_0 = \mathrm{Span}\{w(t - k)\}, \quad \tilde{W}_0 = \mathrm{Span}\{\tilde{w}(t - k)\}, \quad V_0 \perp \tilde{W}_0, \quad \tilde{V}_0 \perp W_0.$$

Once $w, \tilde{w}$ have been determined we can define functions

$$w_{jk}(t) = 2^{\frac{j}{2}} w(2^j t - k), \quad \tilde{w}_{jk}(t) = 2^{\frac{j}{2}} \tilde{w}(2^j t - k),$$

$$\phi_{jk}(t) = 2^{\frac{j}{2}} \phi(2^j t - k), \quad \tilde{\phi}_{jk}(t) = 2^{\frac{j}{2}} \tilde{\phi}(2^j t - k),$$

for $j, k = 0, \pm 1, \pm 2, \ldots$ It is easy to prove the biorthogonality result.

Lemma 11.3

$$< w_{jk}, \tilde{w}_{j'k'} > = \delta_{jj'} \delta_{kk'}, \qquad < \phi_{jk}, \tilde{w}_{jk'} > = 0, \qquad (11.21)$$
$$< \phi_{jk}, \tilde{\phi}_{jk'} > = \delta_{kk'}, \qquad < w_{jk'}, \tilde{\phi}_{jk} >= 0,$$

where $j, j', k, k' = 0, \pm 1, \ldots$

The dilation and wavelet equations extend to:

$$\tilde{\phi}_{j\ell} = 2 \sum_k \breve{h}_{k-2\ell}^{(0)} \tilde{\phi}_{j+1,k}(t), \qquad (11.22)$$

$$\phi_{j\ell} = \sum_k s_{k-2\ell}^{(0)} \phi_{j+1,k}(t), \qquad (11.23)$$

$$\tilde{w}_{j\ell} = \sum_k \check{h}^{(1)}_{k-2\ell} \tilde{\phi}_{j+1,k}(t), \tag{11.24}$$

$$w_{j\ell} = 2 \sum_k s^{(1)}_{k-2\ell} \phi_{j+1,k}(t). \tag{11.25}$$

Now we have

$$V_j = \text{Span}\{\phi_{j,k}\}, \quad \tilde{V}_j = \text{Span}\{\tilde{\phi}_{j,k}\},$$

$$W_j = \text{Span}\{w_{j,k}\}, \quad \tilde{W}_j = \text{Span}\{\tilde{w}_{j,k}\}, \quad V_j \perp \tilde{W}_j, \quad \tilde{V}_j \perp W_j,$$

so we can get biorthogonal wavelet expansions for functions $f \in L_2$.

Theorem 11.4

$$L_2(\mathbb{R}) = V_j + \sum_{k=j}^{\infty} W_k = V_j + W_j + W_{j+1} + \cdots = \sum_{j=-\infty}^{\infty} W_j,$$

so that each $f(t) \in L_2(\mathbb{R})$ can be written uniquely in the form

$$f = f_j + \sum_{k=j}^{\infty} w_k, \qquad w_k \in W_k, \ f_j \in V_j. \tag{11.26}$$

Similarly

$$L_2(\mathbb{R}) = \tilde{V}_j + \sum_{k=j}^{\infty} \tilde{W}_k = \tilde{V}_j + \tilde{W}_j + \tilde{W}_{j+1} + \cdots = \sum_{j=-\infty}^{\infty} \tilde{W}_j,$$

so that each $\tilde{f}(t) \in L_2(\mathbb{R})$ can be written uniquely in the form

$$\tilde{f} = \tilde{f}_j + \sum_{k=j}^{\infty} \tilde{w}_k, \qquad \tilde{w}_k \in \tilde{W}_k, \ \tilde{f}_j \in \tilde{V}_j. \tag{11.27}$$

We have a family of new biorthogonal bases for $L_2(\mathbb{R})$, two for each integer j:

$$\{\phi_{jk}, w_{j'k'}; \ \tilde{\phi}_{jk}, \tilde{w}_{j'k'} : \ j' = j, j+1, \ldots, \ \pm k, \pm k' = 0, 1, \ldots\}.$$

Let's consider the space V_j for fixed j. On the one hand we have the scaling function basis $\{\phi_{j,k} : \ \pm k = 0, 1, \ldots\}$. Then we can expand any $f_j \in V_j$ as

$$f_j = \sum_{k=-\infty}^{\infty} \tilde{a}_{j,k} \phi_{j,k}, \quad \tilde{a}_{j,k} = <f_j, \tilde{\phi}_{j,k}> . \tag{11.28}$$

On the other hand we have the wavelets basis $\{\phi_{j-1,k}, w_{j-1,k'} : \pm k, \pm k' = 0, 1, \ldots\}$ associated with the direct sum decomposition $V_j = W_{j-1} + V_{j-1}$. Using this basis we can expand any $f_j \in V_j$ as

$$f_j = \sum_{k'=-\infty}^{\infty} \tilde{b}_{j-1,k'} w_{j-1,k'} + \sum_{k=-\infty}^{\infty} \tilde{a}_{j-1,k} \phi_{j-1,k}, \qquad (11.29)$$

where

$$\tilde{b}_{j-1,k} = < f_j, \tilde{w}_{j-1,k} >, \quad \tilde{a}_{j-1,k} = < f_j, \tilde{\phi}_{j-1,k} > .$$

There are exactly analogous expansions in terms of the $\tilde{\phi}, \tilde{w}$ basis.

If we substitute the relations

$$\tilde{\phi}_{j-1,\ell} = 2 \sum_k \breve{h}^{(0)}_{k-2\ell} \tilde{\phi}_{j,k}(t), \qquad (11.30)$$

$$\tilde{w}_{j-1,\ell} = \sum_k \breve{h}^{(1)}_{k-2\ell} \tilde{\phi}_{j,k}(t), \qquad (11.31)$$

into the expansion (11.28) and compare coefficients of $\phi_{j,\ell}$ with the expansion (11.29), we obtain the following fundamental recursions.

Theorem 11.5 *Fast Wavelet Transform.*

$$\text{Averages (lowpass)} \quad \tilde{a}_{j-1,k} = \sum_n \breve{h}^{(0)}_{n-2k} \tilde{a}_{jn} \qquad (11.32)$$

$$\text{Differences (highpass)} \quad \tilde{b}_{j-1,k} = \tfrac{1}{2} \sum_n \breve{h}^{(1)}_{n-2k} \tilde{a}_{jn}. \qquad (11.33)$$

These equations link the wavelets with the biorthogonal filter bank. Let $x_k = \tilde{a}_{jk}$ be a discrete signal. The result of passing this signal through the (time-reversed) filter $(\breve{H}^{(0)})^T$ and then downsampling is $y_k = (\downarrow 2)(\breve{H}^{(0)})^T * x_k = \tilde{a}_{j-1,k}$, where $\tilde{a}_{j-1,k}$ is given by (11.32). Similarly, the result of passing the signal through the (time-reversed) filter $(\breve{H}^{(1)})^T$ and then downsampling is $z_k = (\downarrow 2)(\breve{H}^{(1)})^T * x_k = \tilde{b}_{j-1,k}$, where $\tilde{b}_{j-1,k}$ is given by (11.33).

We can iterate this process by inputting the output $\tilde{a}_{j-1,k}$ of the low pass filter to the filter bank again to compute $\tilde{a}_{j-2,k}, \tilde{b}_{j-2,k}$, etc. At each stage we save the wavelet coefficients $\tilde{b}_{j'k'}$ and input the scaling coefficients $\tilde{a}_{j'k'}$ for further processing, see expression (11.34).

$$\tilde{a}_{j-2,k} \;\leftarrow\; (\downarrow 2) \;\; \breve{H}^{(0)})^T$$

$$\tilde{a}_{j-1,k} \;\leftarrow\; (\downarrow 2) \;\; \breve{H}^{(0)})^T$$

$$\tilde{b}_{j-2,k} \;\leftarrow\; (\downarrow 2) \;\; \breve{H}^{(1)})^T$$

$$\tilde{a}_{jk}$$
$$Input.$$

$$\tilde{b}_{j-1,k} \;\leftarrow\; (\downarrow 2) \;\; \breve{H}^{(1)})^T$$

$$(11.34)$$

The output of the final stage is the set of scaling coefficients $\tilde{a}_{0k}$, assuming that we stop at $j = 0$. Thus our final output is the complete set of coefficients for the wavelet expansion

$$f_j = \sum_{j'=0}^{j-1} \sum_{k=-\infty}^{\infty} \tilde{b}_{j'k} w_{j'k} + \sum_{k=-\infty}^{\infty} \tilde{a}_{0k} \phi_{0k},$$

based on the decomposition

$$V_j = W_{j-1} + W_{j-2} + \cdots + W_1 + W_0 + V_0.$$

To derive the synthesis filter bank recursion we can substitute the inverse relation

$$\phi_{j,\ell} = \sum_h \left(\bar{s}^{(0)}_{\ell-2h} \phi_{j-1,h} + \bar{s}^{(1)}_{\ell-2h} w_{j-1,h} \right), \tag{11.35}$$

into the expansion (11.29) and compare coefficients of $\phi_{j-1,\ell}, w_{j-1,\ell}$ with the expansion (11.28) to obtain the inverse recursion.

Theorem 11.6 *Inverse Fast Wavelet Transform.*

$$\tilde{a}_{j,\ell} = \sum_k s^{(0)}_{\ell-2k} \tilde{a}_{j-1,k} + \sum_k s^{(1)}_{\ell-2k} \tilde{b}_{j-1,k}. \tag{11.36}$$

Thus, for level j the full analysis and reconstruction picture is expression (11.37).

$$\breve{S}^{(0)} \quad (\uparrow 2) \quad \leftarrow \tilde{a}_{j-1,k} \leftarrow \quad (\downarrow 2) \quad (\breve{H}^{(0)})^T$$

$$\tilde{a}_{jk} \qquad\qquad Synthesis \qquad\qquad Analysis \qquad\qquad \tilde{a}_{jk}$$
$$Output \qquad\qquad\qquad\qquad\qquad\qquad\qquad\qquad Input.$$

$$\breve{S}^{(1)} \quad (\uparrow 2) \quad \leftarrow \tilde{b}_{j-1,k} \leftarrow \quad (\downarrow 2) \quad (\breve{H}^{(1)})^T$$

$$(11.37)$$

For any $f(t) \in L_2(\mathbb{R})$ the scaling and wavelets coefficients of f are defined by

$$\tilde{a}_{jk} = <f, \tilde{\phi}_{jk}> = 2^{j/2} \int_{-\infty}^{\infty} f(t)\tilde{\phi}(2^j t - k)dt,$$

$$\tilde{b}_{jk} = <f, \tilde{w}_{jk}> = 2^{j/2} \int_{-\infty}^{\infty} f(t)\tilde{w}(2^j t - k)dt. \qquad (11.38)$$

11.2.1 Sufficient conditions for biorthogonal multiresolution analysis

We have seen that the coefficients $\check{h}_i, s_i$ in the analysis and synthesis dilation and wavelet equations must satisfy exactly the same double-shift orthogonality properties as those that come from biorthogonal filter banks. Now we will assume that we have coefficients $\check{h}_i, s_i$ satisfying these double-shift orthogonality properties and see if we can construct analysis and synthesis functions and wavelet functions associated with a biorthogonal multiresolution analysis.

The construction will follow from the cascade algorithm, applied to functions $\phi^{(i)}(t)$ and $\tilde{\phi}^{(i)}(t)$ in parallel. We start from the Haar scaling function $\phi^{(0)}(t) = \tilde{\phi}^{(0)}(t)$ on $[0,1]$. We apply the low pass filter $\mathbf{S}^{(0)}$, recursively and with scaling, to the $\phi^{(i)}(t)$, and the low pass filter $\check{\mathbf{H}}^{(0)}$, recursively and with scaling, to the $\tilde{\phi}^{(i)}(t)$. (For each pass of the algorithm we rescale the iterate by multiplying it by $1/\sqrt{2}$ to preserve normalization.)

Theorem 11.7 *If the cascade algorithm converges uniformly in L_2 for both the analysis and synthesis functions, then the limit functions $\phi(t), \tilde{\phi}(t)$ and associated wavelets $w(t), \tilde{w}(t)$ satisfy the orthogonality relations*

$$<w_{jk}, \tilde{w}_{j'k'}> = \delta_{jj'}\delta_{kk'}, \qquad <\phi_{jk}, \tilde{\phi}_{jk'}> = \delta_{kk'},$$

$$<\phi_{jk}, \tilde{w}_{j,k'}> = 0, \qquad <w_{jk}, \tilde{\phi}_{j,k'}> = 0,$$

where $j, j', \pm k, \pm k' = 0$.

Proof There are only three sets of identities to prove:

1. $$\int_{-\infty}^{\infty} \phi(t-n)\tilde{\phi}(t-m)dt = \delta_{nm},$$

2. $$\int_{-\infty}^{\infty} \phi(t-n)\tilde{w}(t-m)dt = 0,$$

3. $$\int_{-\infty}^{\infty} w(t-n)\tilde{w}(t-m)dt = \delta_{nm}.$$

The rest are duals of these, or immediate.

1. We will use induction. If 1 is true for the functions $\phi^{(i)}(t), \tilde{\phi}^{(i)}(t)$ we will show that it is true for the functions $\phi^{(i+1)}(t), \tilde{\phi}^{(i+1)}(t)$. Clearly it is true for $\phi^{(0)}(t), \tilde{\phi}^{(0)}(t)$. Now

$$\int_{-\infty}^{\infty} \phi^{(i+1)}(t-n)\tilde{\phi}^{(i+1)}(t-m)dt = <\phi_{0n}^{(i+1)}, \tilde{\phi}_{0m}^{(i+1)}>$$

$$= <\sum_k s_k^{(0)} \phi_{1,2n+k}^{(i)}, \sum_\ell \check{h}_\ell^{(0)} \tilde{\phi}_{1,2m+\ell}^{(i)}>$$

$$= \sum_{k\ell} s_k^{(0)} \check{h}_\ell^{(0)} <\phi_{1,2n+k}^{(i)}, \tilde{\phi}_{1,2m+\ell}^{(i)}> = \sum_k s_k^{(0)} \check{h}_{k-2(m-n)}^{(0)} = \delta_{nm}.$$

Since the convergence is L_2, these orthogonality relations are also valid in the limit for $\phi(t), \tilde{\phi}(t)$.

2.

$$\int_{-\infty}^{\infty} \phi^{(i+1)}(t-n)\tilde{w}^{(i+1)}(t-m)dt = <\phi_{0n}^{(i+1)}, \tilde{w}_{0m}^{(i+1)}>$$

$$= <\sum_k s_k^{(0)} \phi_{1,2n+k}^{(i)}, \sum_\ell \check{h}_\ell^{(1)} \tilde{w}_{1,2m+\ell}^{(i)}>$$

$$= \sum_{k\ell} s_k^{(0)} \check{h}_\ell^{(1)} <\phi_{1,2n+k}^{(i)}, \tilde{w}_{1,2m+\ell}^{(i)}> = 2\sum_k s_k^{(0)} \check{h}_{k-2(m-n)}^{(1)} = 0,$$

because of the double-shift orthogonality of $s^{(0)}$ and $\check{h}^{(1)}$.

3.

$$\int_{-\infty}^{\infty} w^{(i+1)}(t-n)\tilde{w}^{(i+1)}(t-m)dt = <w_{0n}^{(i+1)}, \tilde{w}_{0m}^{(i+1)}>$$

$$= \sum_k s_k^{(1)} \check{h}_{k-2(m-n)}^{(1)} = \delta_{nm},$$

because of the double-shift orthonormality of $s^{(1)}$ and $\check{h}^{(1)}$.

□

Theorem 11.8 *Suppose the filter coefficients satisfy double-shift orthogonality conditions (11.6), (11.7), (11.8) and, (11.9), as well as the conditions of Theorem 10.2 for the matrices $T = (\downarrow 2)2S^{(0)}(\overline{S}^{(0)})^{\text{tr}}$ and $\tilde{T} = (\downarrow 2)2\check{H}^{(0)}(\overline{\check{H}}^{(0)})^{\text{tr}}$, which guarantee L_2 convergence of the cascade* *algorithm. Then the synthesis functions* $\phi(t-k), w(t-k)$ *are orthogonal to the analysis functions* $\tilde{\phi}(t-\ell), \tilde{w}(t-\ell)$, *and each scaling space is orthogonal to the dual wavelet space:*

$$V_j \perp \tilde{W}_j, \qquad W_j \perp \tilde{V}_j. \tag{11.39}$$

Also

$$V_j + W_j = V_{j+1}, \qquad \tilde{V}_j + \tilde{W}_j = \tilde{V}_{j+1}$$

where the direct sums are, in general, not orthogonal.

Corollary 11.9 *The wavelets*

$$w_{jk} = 2^{\frac{j}{2}} w(2^j - k), \qquad \tilde{w}_{jk} = 2^{\frac{j}{2}} \tilde{w}(2^j - k),$$

are biorthogonal bases for L_2:

$$\int_{-\infty}^{\infty} w_{jk}(t) \tilde{w}_{j'k'}(t) dt = \delta_{jj'} \delta_{kk'}. \tag{11.40}$$

Since the details of the construction of biorthogonal wavelets are somewhat complicated, we present a toy illustrative example. We show how it is possible to associate a scaling function $\phi(t)$ with the identity low pass filter $H_0(\omega) = H_0[z] \equiv 1$. Of course, this really isn't a low pass filter, since $H_0(\pi) \neq 0$ in the frequency domain and the scaling function will not be a function at all, but a distribution or "generalized function." If we apply the cascade algorithm construction to the identity filter $H_0(\omega) \equiv 1$ in the frequency domain, we easily obtain the Fourier transform of the scaling function as $\hat{\phi}(\omega) \equiv 1$. Since $\hat{\phi}(\omega)$ isn't square integrable, there is no true function $\phi(t)$. However there is a distribution $\phi(t) = \delta(t)$ with this transform. Distributions are linear functionals on function classes, and are informally defined by the values of integrals of the distribution with members of the function class.

Recall that $f \in L_2(\mathbb{R})$ belongs to the *Schwartz class* if f is infinitely differentiable everywhere, and there exist constants $C_{n,q}$ (depending on f) such that $|t^n \frac{d^q}{dt^q} f| \leq C_{n,q}$ on R for each $n, q = 0, 1, 2, \ldots$. One of the pleasing features of this space of functions is that f belongs to the class if and only if $\hat{f}$ belongs to the class. We will define the distribution $\phi(t)$ by its action as a linear functional on the Schwartz class. Consider the Parseval formula

$$\int_{-\infty}^{\infty} \phi(t) f(t) dt = \frac{1}{2\pi} \int_{-\infty}^{\infty} \hat{\phi}(\omega) \hat{f}(\omega) d\omega$$

where f belongs to the Schwartz class. We will *define* the integral on the left-hand side of this expression by the integral on the right-hand side. Thus

$$\int_{-\infty}^{\infty} \phi(t) f(t) dt = \frac{1}{2\pi} \int_{-\infty}^{\infty} \hat{\phi}(\omega) \hat{f}(\omega) d\omega = \frac{1}{2\pi} \int_{-\infty}^{\infty} \hat{f}(\omega) d\omega = f(0)$$

from the inverse Fourier transform. This functional is the *Dirac Delta Function*, $\phi(t) = \delta(t)$. It picks out the value of the integrand at $t = 0$.

We can use the standard change of variables formulas for integrals to see how distributions transform under variable change. Since $H_0(\omega) \equiv 1$ the corresponding filter has $h_0^{(0)} = 1$ as its only nonzero coefficient. Thus the dilation equation in the signal domain is $\phi(t) = 2\phi(2t)$. Let's show that $\phi(t) = \delta(t)$ is a solution of this equation. On one hand

$$\int_{-\infty}^{\infty} \phi(t) f(t) dt = \frac{1}{2\pi} \int_{-\infty}^{\infty} \hat{\phi}(\omega) = f(0)$$

for any function $f(t)$ in the Schwartz class. On the other hand (for $\tau = 2t$)

$$\int_{-\infty}^{\infty} 2\phi(2t) f(t) dt = \int_{-\infty}^{\infty} \phi(\tau) f(\frac{\tau}{2}) d\tau = g(0) = f(0)$$

since $g(\tau) = f(\tau/2)$ belongs to the Schwartz class. The distributions $\delta(t)$ and $2\delta(2t)$ are the same.

Now we proceed with our example and consider the biorthogonal scaling functions and wavelets determined by the biorthogonal filters

$$H^{(0)}[z] = 1, \qquad S^{(0)}[z] = \frac{1}{2} + z^{-1} + \frac{1}{2} z^{-2}.$$

Then we require $H^{(1)}[z] = S^{(0)}[-z], S^{(1)}[z] = -H^{(0)}[-z]$, so

$$H^{(1)}[z] = \frac{1}{2} - z^{-1} + \frac{1}{2} z^{-2}, \qquad S^{(1)}[z] = -1.$$

The low pass analysis and synthesis filters are related to the product filter $P^{(0)}$ by

$$P^{(0)}[z] = H^{(0)}[z] S^{(0)}[z], \text{ where } P^{(0)}[1] = 2, \ P^{(0)}[z] - P^{(0)}[-z] = 2z^{-\ell},$$

and ℓ is the delay. We find that $P^{(0)}[z] = \frac{1}{2} + z^{-1} + \frac{1}{2} z^{-2}$, so $\ell = 1$.

To pass from the biorthogonal filters to coefficient identities needed for the construction of wavelets we have to modify the filter coefficients. In Section 11.2 we modified the analysis coefficients to obtain new coefficients $\breve{h}_n^{(j)} = h_{\ell-n}^{(j)}$, $j = 0, 1$, and left the synthesis coefficients as they were. Since the choice of what is an analysis filter and what is a synthesis filter is arbitrary, we could just as well have modified the synthesis coefficients to obtain new coefficients $\check{s}_n^{(j)} = s_{\ell-n}^{(j)}$, $j = 0, 1$, and left the analysis coefficients unchanged. In this problem it is important that one of the low pass filters be $H^{(0)}[z] = 1$ (so that the delta function is the scaling function). If we want to call that an analysis filter, then we have to modify the synthesis coefficients.

Thus the nonzero analysis coefficients are

$$(h_0^{(0)}) = (1) \quad \text{and} \quad (h_0^{(1)}, h_1^{(1)}, h_2^{(1)}) = (\frac{1}{2}, -1, \frac{1}{2}).$$

The nonzero synthesis coefficients are

$$(s_0^{(1)}) = (-1) \quad \text{and} \quad (s_0^{(0)}, s_1^{(0)}, s_2^{(0)}) = (\frac{1}{2}, 1, \frac{1}{2}).$$

Since $\check{s}_n^{(j)} = s_{\ell-n}^{(j)} = s_{1-n}^{(j)}$ for $j = 0, 1$ we have nonzero components $(\check{s}_{-1}^{(0)}, \check{s}_0^{(0)}, \check{s}_1^{(0)}) = (\frac{1}{2}, 1, \frac{1}{2})$ and $(\check{s}_1^{(1)}) = (-1)$, so the modified synthesis filters are $\check{S}^{(0)}[z] = \frac{1}{2}z + 1 + \frac{1}{2}z^{-1}$, and $\check{S}^{(1)}[z] = -z^{-1}$.

The analysis dilation equation is $\tilde{\phi}(t) = 2\tilde{\phi}(2t)$ with solution $\tilde{\phi}(t) = \delta(t)$, the Dirac delta function. The synthesis dilation equation should be

$$\phi(t) = \sum_k \check{s}_k^{(0)} \phi(2t - k),$$

or

$$\phi(t) = \frac{1}{2}\phi(2t + 1) + \phi(2t) + \frac{1}{2}\phi(2t - 1).$$

It is straightforward to show that the hat function (centered at $t = 0$)

$$\phi(t) = \begin{cases} 1 + t & -1 \leq t < 0 \\ 1 - t & 0 \leq t < 1 \\ 0 & \text{otherwise} \end{cases}$$

is the proper solution to this equation.

The analysis wavelet equation is

$$\tilde{w}(t) = \sum_k h_k^{(1)} \tilde{\phi}(2t - k), \quad \text{or} \quad \tilde{w}(t) = \frac{1}{2}\tilde{\phi}(2t) - \tilde{\phi}(2t - 1) + \frac{1}{2}\tilde{\phi}(2t - 2).$$

The synthesis wavelet equation is

$$w(t) = 2\sum_k \check{s}_k^{(1)} \phi(2t - k), \quad \text{or} \quad w(t) = -2\phi(2t - 1).$$

Now it is easy to verify explicitly the biorthogonality conditions

$$\int \phi(t)\tilde{\phi}(t-k)dt = \int w(t)\tilde{w}(t-k)dt = \delta(k),$$

$$\int \phi(t)\tilde{w}(t-k)dt = \int \tilde{\phi}(t)w(t-k)dt = 0.$$

11.3 Splines

A *spline* $f(t)$ of order N on the grid of integers is a piecewise polynomial

$$f_j(t) = a_{0,j} + a_{1,j}(t-j) + \cdots a_{N,j}(t-j)^N, \qquad j \le t \le j+1,$$

such that the pieces fit together smoothly at the gridpoints:

$$f_j(j) = f_{j-1}(j), f'_j(j) = f'_{j-1}(j), \ldots, f_j^{(N-1)}(j) = f_{j-1}^{(N-1)}(j),$$

$j = 0, \pm1, \ldots$ Thus $f(t)$ has $N-1$ continuous derivatives for all t. The Nth derivative exists for all non-integer t and for $t = j$ the right- and left-hand derivatives $f^{(N)}(j+0), f^{(N)}(j-0)$ exist. Splines are widely used for approximation of functions by interpolation. That is, if $F(t)$ is a function taking values $F(j)$ at the gridpoints, one approximates F by an N-spline f that takes the same values at the gridpoints: $f(j) = F(j)$, for all j. Then by subdividing the grid (but keeping N fixed) one can show that these N-spline approximations get better and better for sufficiently smooth F. The most commonly used splines are the *cubic splines*, where $N = 3$.

Splines have a close relationship with wavelet theory. Usually the wavelets are biorthogonal, rather than orthogonal, and one set of N-splines can be associated with several sets of biorthogonal wavelets. We will look at a few of these connections as they relate to multiresolution analysis. We take our low pass space V_0 to consist of the $(N-1)$-splines on unit intervals and with compact support. The space V_1 will then contain the $(N-1)$-splines on half-intervals, etc. We will find a basis $\phi(t-k)$ for V_0 (but usually not an ON basis).

We have already seen examples of splines for the simplest cases. The 0-splines are piecewise constant on unit intervals. This is just the case of Haar wavelets. The scaling function $\phi_0(t)$ is just the box function

$$\phi_0(t) = B(t) = \begin{cases} 1, & 0 \le t < 1 \\ 0, & \text{otherwise.} \end{cases}$$

Here, $\hat{\phi}_0(\omega) = \hat{B}(t) = \frac{1}{i\omega}(1 - e^{-i\omega})$, essentially the sinc function. The set $\phi(t - k)$ is an ON basis for V_0.

The 1-splines are continuous piecewise linear functions. The functions $f(t) \in V_0$ are determined by their values $f(k)$ at the integer points, and are linear between each pair of values:

$$f(t) = [f(k + 1) - f(k)](t - k) + f(k) \quad \text{for } k \leq t \leq k + 1.$$

The scaling function ϕ_1 is the *hat function*. The hat function $H(t)$ is the continuous piecewise linear function whose values on the integers are $H(k) = \delta_{1k}$, i.e., $H(1) = 1$ and $H(t)$ is zero on the other integers. The support of $H(t)$ is the open interval $-0 < t < 2$. Furthermore, $H(t) = \phi_1(t) = (B * B)(t)$, i.e., the hat function is the convolution of two box functions. Moreover, $\hat{\phi}_1(\omega) = \hat{B}^2(t) = (\frac{1}{i\omega})^2(1 - e^{-i\omega})^2$. Note that if $f \in V_0$ then we can write it uniquely in the form

$$f(t) = \sum_k f(k)H(t - k + 1).$$

All multiresolution analysis conditions are satisfied, except for the ON basis requirement. The integer translates of the hat function do define a Riesz basis for V_0 (though we haven't completely proved it yet) but it isn't ON because the inner product $(H(t), H(t - 1)) \neq 0$. A scaling function does exist whose integer translates form an ON basis, but its support isn't compact. It is usually simpler to stick with the nonorthogonal basis, but embed it into a biorthogonal multiresolution structure, as we shall see.

Based on our two examples, it is reasonable to guess that an appropriate scaling function for the space V_0 of $(N - 1)$-splines is

$$\phi_{N-1}(t) = (B * B * \cdots * B)(t) \quad N \text{ times}, \qquad (11.41)$$

$$\hat{\phi}_{N-1}(\omega) = \hat{B}^N(t) = (\frac{1}{i\omega})^N(1 - e^{-i\omega})^N.$$

This special spline is called a *B-spline*, where the B stands for *basis*.

Let's study the properties of the B-spline. First recall the definition of the convolution:

$$f * g(t) = \int_{-\infty}^{\infty} f(t - x)g(x)dx = \int_{-\infty}^{\infty} f(x)g(t - x)dx.$$

If $f = B$, the box function, then

$$B * g(t) = \int_{t-1}^{t} g(x)dx = \int_{0}^{1} g(t - x)dx.$$

Now note that $\phi_N(t) = B * \phi_{N-1}(t)$, so

$$\phi_N(t) = \int_{t-1}^{t} \phi_{N-1}(x)dx. \qquad (11.42)$$

Using the fundamental theorem of calculus and differentiating, we find

$$\phi_N'(t) = \phi_{N-1}(t) - \phi_{N-1}(t-1). \qquad (11.43)$$

Recall that $\phi_0(t)$ is piecewise constant, has support in the interval $[0,1)$, and discontinuities 1 at $t = 0$ and -1 at $t = 1$.

Theorem 11.10 *The function $\phi_N(t)$ has the following properties:*

1. *It is a spline, i.e., it is piecewise polynomial of order N.*
2. *The support of $\phi_N(t)$ is contained in the interval $[0, N+1)$.*
3. *The jumps in the Nth derivative at $t = 0, 1, \ldots, N+1$ are the alternating binomial coefficients $(-1)^t \begin{pmatrix} N \\ t \end{pmatrix}$.*

Proof By induction on N. We observe that the theorem is true for $N = 0$. Assume that it holds for $N = M - 1$. Since $\phi_{M-1}(t)$ is piecewise polynomial of order $M - 1$ and with support in $[0, M)$, it follows from (11.42) that $\phi_M(t)$ is piecewise polynomial of order M and with support in $[0, M+1)$. Denote by $[\phi_N^{(M)}(j)] = \phi_N^{(M)}(j+0) - \phi_N^{(M)}(j-0)$ the jump in $\phi_N^{(M)}(t)$ at $t = j$. Differentiating (11.43) $M - 1$ times for $N = M$, we find

$$[\phi_N^{(M)}(j)] = [\phi_N^{(M-1)}(j)] - [\phi_N^{(M-1)}(j-1)]$$

for $j = 0, 1, \ldots M + 1$ where

$$[\phi_N^{(M)}(j)] = (-1)^j \begin{pmatrix} M-1 \\ j \end{pmatrix} - (-1)^{j-1} \begin{pmatrix} M-1 \\ j-1 \end{pmatrix}$$

$$= (-1)^j \left(\frac{(M-1)!}{j!(M-j-1)!} + \frac{(M-1)!}{(j-1)!(M-j)!} \right)$$

$$= \frac{(-1)^j M!}{j!(M-j)!} \left(\frac{M-j}{M} + \frac{j}{M} \right) = (-1)^j \begin{pmatrix} M \\ j \end{pmatrix}.$$

$\square$

Normally we start with a low pass filter **H** with finite impulse response vector h_n and determine the scaling function via the dilation equation and the cascade algorithm. Here we have a different problem. We have a candidate B-spline scaling function $\phi_{N-1}(t)$ with Fourier transform

$$\hat{\phi}_{N-1}(\omega) = (\frac{1}{i\omega})^N (1 - e^{-i\omega})^N.$$

What is the associated low pass filter function $H(\omega)$? The dilation equation in the Fourier domain is

$$\hat{\phi}_{N-1}(\omega) = H(\frac{\omega}{2})\hat{\phi}_{N-1}(\frac{\omega}{2}).$$

Thus

$$(\frac{1}{i\omega})^N(1 - e^{-i\omega})^N = H(\frac{\omega}{2})(\frac{1}{i\omega/2})^N(1 - e^{-i\omega/2})^N.$$

Solving for $H(\omega/2)$ and rescaling, we find

$$H(\omega) = \left(\frac{1 + e^{-i\omega}}{2}\right)^N$$

or

$$H[z] = \left(\frac{1 + z^{-1}}{2}\right)^N.$$

This is as nice a low pass filter as you could ever expect. *All* of its zeros are at -1! We see that the finite impulse response vector $h_n = \frac{1}{2^N}\begin{pmatrix} N \\ n \end{pmatrix}$, so that the dilation equation for a spline is

$$\phi_{N-1}(t) = 2^{1-N}\sum_{k=0}^{N}\begin{pmatrix} N \\ n \end{pmatrix}\phi_{N-1}(2t - k). \qquad (11.44)$$

For convenience we list some additional properties of B-splines. The first two properties follow directly from the fact that they are scaling functions, but we give direct proofs anyway.

Lemma 11.11 *The B-spline $\phi_N(t)$ has the properties:*

1. $\int_{-\infty}^{\infty} \phi_N(t)dt = 1.$
2. $\sum_k \phi_N(t + k) = 1.$
3. $\phi_N(t) = \phi_N(N + 1 - t),$ *for $N = 1, 2, \ldots$*
4. *Let*

$$\chi_k(t) = \begin{cases} 1, & t \geq k \\ 0, & \text{otherwise.} \end{cases}$$

$$\phi_N(t) = \sum_{k=0}^{N+1}\begin{pmatrix} N + 1 \\ k \end{pmatrix}(-1)^k \frac{(t-k)^N}{N!}\chi_k(t) \text{ for } N = 1, 2, \ldots$$

5. $\phi_N(t) \geq 0.$
6. $a_k = \int_{-\infty}^{\infty} \phi_N(t)\phi_N(t + k)dt = \phi_{2N+1}(N + k + 1).$

Proof

1. $\int_{-\infty}^{\infty} \phi_N(t)dt = \hat{\phi}_N(0) = 1.$
2.

$$\sum_k \phi_N(t+k) = \sum_k \int_{t+k-1}^{t+k} \phi_{N-1}(x)dx = \int_{-\infty}^{\infty} \phi_{N-1}(t)dt = 1.$$

3. Use induction on N. The statement is obviously true for $N = 1$. Assume it holds for $N = M - 1$. Then

$$\phi_M(M + 1 - t) = \int_{M-t}^{M+1-t} \phi_{M-1}(x)dx = \int_{t-1}^{t} \phi_{M-1}(M - u)du$$

$$= \int_{t-1}^{t} \phi_{M-1}(u)du = \phi_M(t).$$

4. From the third property in the preceding theorem, we have

$$\phi_N^{(N)}(t) = \sum_{k=0}^{N+1} \binom{N+1}{k} (-1)^k \chi_k(t).$$

Integrating this equation with respect to t, from $-\infty$ to t, N times, we obtain the desired result.
5. Follows easily, by induction on N.
6. The Fourier transform of $\phi_N(t)$ is $\hat{\phi}_N(\omega) = \hat{\phi}_0^{N+1}(\omega) = (\frac{1}{2\pi i\omega})^{N+1}(1 - e^{-i\omega})^{N+1}$ and the Fourier transform of $\phi_N(t+k)$ is $e^{ik\omega}\hat{\phi}_N(\omega)$. Thus the Plancherel formula gives

$$\int_{-\infty}^{\infty} \phi_N(t)\phi_N(t+k)dt = 2\pi \int_{-\infty}^{\infty} |\hat{\phi}_N(\omega)|^2 e^{ik\omega}d\omega =$$

$$2\pi \int_{-\infty}^{\infty} (\frac{1}{2\pi i\omega})^{2N+2}(1 - e^{-i\omega})^{2N+2} e^{i(N+k+1)\omega}d\omega = \phi_{2N+1}(N + k + 1).$$

$\square$

An N-spline $f(t)$ in V_0 is uniquely determined by the values $f(j)$ it takes at the integer gridpoints. (Similarly, functions $f(t)$ in V_J are uniquely determined by the values $f(k/2^J)$ for integer k.)

Lemma 11.12 *Let $f, \tilde{f} \in V_0$ be N-splines such that $f(j) = \tilde{f}(j)$ for all integers j. Then $f(t) \equiv \tilde{f}(t)$.*

Proof Let $g(t) = f(t) - \tilde{f}(t)$. Then $g \in V_0$ and $g(j) = 0$ for all integers j. Since $g \in V_0$, it has compact support. If g is not identically 0, then there is a least integer J such that $g_J \not\equiv 0$. Here

$$g_J(t) = a_{0,j} + a_{1,j}(t - J) + \cdots + a_{N,j}(t - J)^N, \qquad J \leq t \leq J + 1.$$

However, since $g_{J-1}(t) \equiv 0$ and since

$$g_J(J) = g_{J-1}(J), g'_J(J) = g'_{J-1}(J), \ldots, g_J^{(N-1)}(J) = g_{J-1}^{(N-1)}(J),$$

we have $g_J(t) = a_{N,J}(t - J)^N$. But $g_J(J + 1) = 0$, so $a_{N,J} = 0$ and $g_J(t) \equiv 0$, a contradiction. Thus $g(t) \equiv 0$. $\qquad\square$

Now let's focus on the case $N = 3$. For cubic splines the finite impulse response vector is $h_n = \frac{1}{16}(1, 4, 6, 4, 1)$ whereas the cubic B-spline $\phi_3(t)$ has jumps $1, -4, 6, -4, 1$ in its third derivative at the gridpoints $t = 0, 1, 2, 3, 4$, respectively. The support of $\phi_3(t)$ is contained in the interval $[0, 4)$, and from the second lemma above it follows easily that $\phi_3(0) = \phi_3(4) = 0$, $\phi_3(1) = \phi_3(3) = 1/6$. Since the sum of the values at the integer gridpoints is 1, we must have $\phi_3(2) = 4/6$. We can verify these values directly from the fourth property of the preceding lemma.

We will show directly, i.e., without using wavelet theory, that the integer translates $\phi_3(t - k)$ form a basis for the resolution space V_0 in the case $N = 3$. This means that any $f(t) \in V_0$ can be expanded uniquely in the form $f(t) = \sum_k s_k \phi_3(t - k)$ for expansion coefficients s_k and all t. Note that for fixed t, at most four terms on the right-hand side of this expression are nonzero. According to the previous lemma, if the right-hand sum agrees with $f(t)$ at the integer gridpoints then it agrees everywhere. Thus it is sufficient to show that given the input vector f where $f_j = f(j)$ we can always solve the equation

$$f(j) = \sum_k s_k \phi_3(j - k), \qquad j = 0, \pm 1, \ldots \tag{11.45}$$

for the vector s with components s_k, $k = 0, \pm 1, \ldots$ We can write (11.45) as a convolution equation $f = b * s$ where

$$b = (b_0, b_1, b_2, b_3, b_4) = (0, \frac{1}{6}, \frac{4}{6}, \frac{1}{6}, 0).$$

We need to invert this equation and solve for s in terms of f. Let $\mathbf{B}$ be the FIR filter with impulse response vector b. Passing to the frequency domain, we see that (11.45) takes the form

$$F(\omega) = B(\omega)S(\omega), \quad F(\omega) = \sum_j f_j e^{-ij\omega}, \quad S(\omega) = \sum_k s_k e^{-ik\omega},$$

and

$$B(\omega) = \sum_j b_j e^{-ij\omega} = \frac{e^{-i\omega}}{6}(1 + 4e^{-i\omega} + e^{-2i\omega}) = \frac{e^{-2i\omega}}{3}(2 + \cos\omega).$$

Note that $B(\omega)$ is bounded away from zero for all ω, hence B is invertible and has a bounded inverse B^{-1} with

$$B^{-1}(\omega) = \frac{3e^{2i\omega}}{2 + \cos\omega} = \frac{6e^{2i\omega}}{(2 + \sqrt{3})(1 + [2 - \sqrt{3}]e^{i\omega})(1 + [2 - \sqrt{3}]e^{-i\omega})}$$

$$= -\sqrt{3}e^{2i\omega}\left(\frac{[2 - \sqrt{3}]e^{i\omega}}{1 + [2 - \sqrt{3}]e^{i\omega}} - \frac{1}{1 + [2 - \sqrt{3}]e^{-i\omega}}\right) = \sum_n c_n e^{in\omega}$$

where

$$c_n = \begin{cases} \sqrt{3}(2 + \sqrt{3})^{2-n}(-1)^n & \text{if } n \geq 3 \\ \sqrt{3}(2 + \sqrt{3})^{n-2}(-1)^n & \text{if } n \leq 2. \end{cases}$$

Thus,

$$s = c * f \quad \text{or} \quad s_k = \sum_j c_{k-j} f_j.$$

Note that $\mathbf{B}^{-1}$ is an infinite impulse response (IIR) filter.

The integer translates $\phi_N(t - k)$ of the B-splines form a Riesz basis of V_0 for each N. To see this we note from Section 10.1 that it is sufficient to show that the infinite matrix of inner products

$$A_{ij} = \int_{-\infty}^{\infty} \phi_N(t - i)\phi_N(t - j) \, dt = \int_{-\infty}^{\infty} \phi_N(t)\phi_N(t + i - j) \, dt = a_{i-j}$$

has positive eigenvalues, bounded away from zero. We have studied matrices of this type many times. Here, A is a Toeplitz matrix with associated impulse vector a_k. The action of A on a column vector x, $y = Ax$, is given by the convolution $y_n = (a * x)_n$. In frequency space the action of A is given as multiplication by the function

$$A(\omega) = \sum_k a_k e^{-ik\omega} = \sum_{n=-\infty}^{\infty} (\hat{\phi}_N(\omega + 2\pi n))^2.$$

Now

$$(\hat{\phi}_N(\omega + 2\pi n))^2 = \left(\frac{4\sin^2(\frac{\omega}{2})}{(\omega + 2\pi n)^2}\right)^{N+1}. \tag{11.46}$$

On the other hand, from Lemma 11.11 we have

$$A(\omega) = \sum_k \phi_{2N+1}(N + k + 1)e^{-ik\omega}$$

$$= \phi_{2N+1}(N + 1) + 2\sum_{k>0} \phi_{2N+1}(N + k + 1)\cos(k\omega).$$

Since all of the inner products are nonnegative, and their sum is 1, the maximum value of $A(\omega)$, hence the norm of A, is $A(0) = B = 1$. It is now evident from (11.46) that $A(\omega)$ is bounded away from zero for every N. (Indeed the term (11.46) alone, with $n = 0$, is bounded away from zero on the interval $[-\pi, \pi]$.) Hence the translates always form a Riesz basis.

We can say more. Computing $A'(\omega)$ by differentiating term-by-term, we find

$$A'(\omega) = -2 \sum_{k>0} \phi_{2N+1}(N + k + 1)k \sin(k\omega).$$

Thus, $A(\omega)$ has a critical point at $\omega = 0$ and at $\omega = \pi$. Clearly, there is an absolute maximum at $\omega = 0$. It can be shown that there is also an absolute minimum at $\omega = \pi$. Thus $A_{\min} = \sum_k \phi_{2N+1}(N + k + 1)(-1)^k$.

Although the B-spline scaling function integer translates don't form an ON basis, they (along with any other Riesz basis of translates) can be "orthogonalized" by a simple construction in the frequency domain. Recall that the necessary and sufficient condition for the translates $\phi(t - k)$ of a scaling function to be an ON basis for V_0 is that

$$A(\omega) = \sum_{n=-\infty}^{\infty} |\hat{\phi}(\omega + 2\pi n)|^2 \equiv 1.$$

In general this doesn't hold for a Riesz basis. However, for a Riesz basis, we have $|A(\omega)| > 0$ for all ω. Thus, we can define a modified scaling function $\tilde{\phi}_N(t)$ by

$$\hat{\tilde{\phi}}_N(\omega) = \frac{\hat{\phi}_N(\omega)}{\sqrt{|A(\omega)|}} = \frac{\hat{\phi}_N(\omega)}{\sqrt{\sum_{n=-\infty}^{\infty} |\hat{\phi}(\omega + 2\pi n)|^2}}$$

so that $\tilde{A}(\omega) \equiv 1$. If we carry this out for the B-splines we get ON scaling functions and wavelets, but with infinite support in the time domain. Indeed, from the explicit formulas that we have derived for $A(\omega)$ we can expand $1/\sqrt{|A(\omega)|}$ in a Fourier series

$$\frac{1}{\sqrt{|A(\omega)|}} = \sum_{k=-\infty}^{\infty} e_k e^{ik\omega}$$

so that

$$\hat{\tilde{\phi}}_N(\omega) = \sum_{k=-\infty}^{\infty} e_k e^{ik\omega} \hat{\phi}_N(\omega),$$

or

$$\tilde{\phi}_N(t) = \sum_{k=-\infty}^{\infty} e_k \phi_N(t-k).$$

This expresses the scaling function generating an ON basis as a convolution of translates of the B-spline. Of course, the new scaling function does not have compact support.

Usually however, the B-spline scaling function $\phi_N(t)$ is embedded in a family of biorthogonal wavelets. There is no unique way to do this. A natural choice is to have the B-spline as the scaling function associated with the synthesis filter $\mathbf{S}^{(0)}$. Since $P^{(0)}[z] = H^{(0)}[z]S^{(0)}[z]$, the halfband filter $P^{(0)}$ must admit $\left(\frac{1+z^{-1}}{2}\right)^{N+1}$ as a factor, to produce the B-spline. If we take the halfband filter to be one from the Daubechies (maxflat) class then the factor $H^{(0)}[z]$ must be of the form $\left(\frac{1+z^{-1}}{2}\right)^{2p-N-1} Q_{2p-2}[z]$. We must then have $2p > N+1$ so that $H^{(0)}[z]$ will have a zero at $z = -1$. The smallest choice of p may not be appropriate, because the conditions for a stable basis and convergence of the cascade algorithm, and the Riesz basis condition for the integer translates of the analysis scaling function may not be satisfied. For the cubic B-spline, a choice that works is

$$S^{(0)}[z] = \left(\frac{1+z^{-1}}{2}\right)^4, \qquad H^{(0)}[z] = \left(\frac{1+z^{-1}}{2}\right)^4 Q_6[z],$$

i.e., $p = 4$. This 11/5 filter bank corresponds to Daubechies D_8 (which would be 8/8). The analysis scaling function is *not* a spline.

11.4 Generalizations of filter banks and wavelets

In this section we take a brief look at some extensions of the theory of filter banks and of wavelets. In one case we replace the integer 2 by the integer M, and in the other we extend scalars to vectors. These are, most definitely, topics of current research.

11.4.1 M channel filter banks and M band wavelets

Although two-channel filter banks are the norm, M-channel filter banks with $M > 2$ are common. There are M analysis filters and the output from each is downsampled ($\downarrow M$) to retain only $1/M$ th the information.

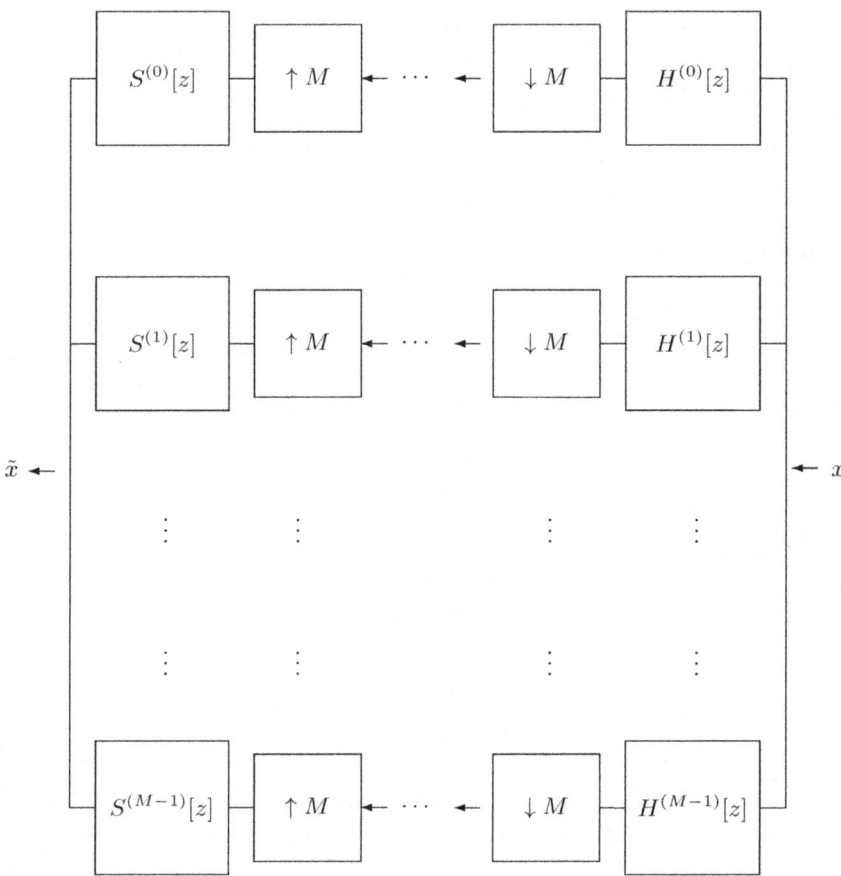

Figure 11.1 M-channel filter bank.

For perfect reconstruction, the downsampled output from each of the M analysis filters is upsampled ($\uparrow M$), passed through a synthesis filter, and then the outputs from the M synthesis filters are added to produce the original signal, with a delay. The picture, viewed from the Z transform domain is that of Figure 11.1.

We need to define $\downarrow M$ and $\uparrow M$.

Lemma 11.13 *In the time domain, $y = (\downarrow M)x$ has components $y_n = x_{Mn}$. In the frequency domain this is*

$$Y(\omega) = \frac{1}{M}\left[X(\frac{\omega}{M}) + X(\frac{\omega + 2\pi}{M}) + \cdots + X(\frac{\omega + (M-1)2\pi}{M})\right].$$

The Z transform is

$$Y[z] = \frac{1}{M}\sum_{k=0}^{M-1} X[z^{1/M}e^{2\pi ik/M}].$$

Lemma 11.14 *In the time domain, $x = (\uparrow M)y$ has components*

$$x_k = \begin{cases} y_{\frac{k}{M}}, & \text{if } M \text{ divides } k \\ 0, & \text{otherwise.} \end{cases}$$

In the frequency domain this is

$$X(\omega) = Y(M\omega).$$

The Z transform is

$$X[z] = Y[z^M].$$

The Z transform of $u = (\downarrow M)(\uparrow M)x$ is

$$U[z] = \frac{1}{M}\sum_{k=0}^{M-1} X[ze^{2\pi ik/M}].$$

Note that $(\uparrow M)(\downarrow M)$ is the identity operator, whereas $(\downarrow M)(\uparrow M)x$ leaves every Mth element of x unchanged and replaces the other elements by zeros.

The operator condition for perfect reconstruction with delay ℓ is

$$\sum_{j=0}^{M-1} \mathbf{S}^{(j)}(\uparrow M)(\downarrow M)\mathbf{H}^{(j)} = \mathbf{R}^\ell$$

where $\mathbf{R}$ is the right shift. If we apply the operators on both sides of this requirement to a signal $x=\{x_n\}$ and take the Z transform, we find

$$\frac{1}{M}\sum_{j=0}^{M-1} S^{(j)}[z]\sum_{k=0}^{M-1} H^{(j)}[zW^k]X[zW^k]$$
$$= z^{-\ell}X[z], \tag{11.47}$$

where $X[z]$ is the Z transform of x, and $W = e^{-2\pi i/M}$. The coefficients of $X[zW^k]$ for $1 < k < M-1$ on the left-hand side of this equation are aliasing terms, due to the downsampling and upsampling. For perfect reconstruction of a general signal $X[z]$ these coefficients must vanish. Thus we have

Theorem 11.15 *An M-channel filter bank gives perfect reconstruction when*

$$\sum_{j=0}^{M-1} S^{(j)}[z]H^{(j)}[z] = Mz^{-\ell}, \tag{11.48}$$

$$\sum_{j=0}^{M-1} S^{(j)}[z]H^{(j)}[zW^k] = 0, \qquad k = 1,\ldots, M-1. \tag{11.49}$$

In matrix form this reads

$$
\begin{bmatrix}
H^{(0)}[z] & H^{(1)}[z] & \cdots & H^{(M-1)}[z] \\
H^{(0)}[zW] & H^{(1)}[zW] & \cdots & H^{(M-1)}[zW] \\
\vdots & \vdots & \vdots & \vdots \\
H^{(0)}[zW^{M-1}] & H^{(1)}[zW^{M-1}] & \cdots & H^{(M-1)}[zW^{M-1}]
\end{bmatrix}
\begin{bmatrix}
S^{(0)}[z] \\
S^{(1)}[z] \\
\vdots \\
S^{(M-1)}[z]
\end{bmatrix}
$$

$$
=
\begin{bmatrix}
Mz^{-\ell} \\
0 \\
\vdots \\
0
\end{bmatrix}.
$$

In the case $M = 2$ we could find a simple solution of the requirements (11.49) by defining the synthesis filters in terms of the analysis filters. However, this is not possible for general M and the design of these filter banks is more complicated. Associated with M-channel filter banks are M-band filters. The dilation and wavelet equations are

$$\phi_\ell(t) = M \sum_k h_k^{(\ell)} \phi(Mt - k), \qquad \ell = 0, 1, \ldots, M-1.$$

Here $h_k^{(\ell)}$ is the finite impulse response vector of the FIR filter $\mathbf{H}^{(\ell)}$. Usually $\ell = 0$ is the dilation equation (for the scaling function $\phi_0(t)$ and $\ell = 1, \ldots, M-1$ are wavelet equations for $M-1$ wavelets $\phi_\ell(t)$, $\ell = 1, \ldots, M-1$. In the frequency domain the equations become

$$\hat{\phi}_\ell(\omega) = H(\frac{\omega}{M})\hat{\phi}_\ell(\frac{\omega}{M}), \qquad \ell = 0, 1, \ldots, M-1,$$

and the iteration limit is

$$\hat{\phi}_\ell(\omega) = \prod_{k=1}^{\infty} \left(H(\frac{\omega}{M^k}) \right) \hat{\phi}_\ell(0), \qquad \ell = 0, 1, \ldots, M-1,$$

assuming that the limit exists and $\hat{\phi}_\ell(0)$ is well defined. For more information about these M-band wavelets and the associated multiresolution structure see, for example, [2, 17, 40, 53, 95].

11.4.2 Multifilters and multiwavelets

Next we go back to filters corresponding to the case $M = 2$, but now we let the input vector x be an $r \times \infty$ input matrix. Thus each component $x(n)$ is an r-vector $x_j(n)$, $j = 1, \ldots, r$. Instead of analysis filters $H^{(0)}, H^{(1)}$ we have analysis multifilters $\mathbf{H}^{(0)}, \mathbf{H}^{(1)}$, each of which is an $r \times r$ matrix of filters, e.g., $\mathbf{H}^{(0)}_{(j,k)}, j, k = 1, \ldots r$. Similarly we have synthesis multifilters $\mathbf{S}^{(0)}, \mathbf{S}^{(1)}$, each of which is an $r \times r$ matrix of filters.

Formally, part of the theory looks very similar to the scalar case. Thus the Z transform of a multifilter $\mathbf{H}$ is $\mathbf{H}[z] = \sum_k h(k)z^{-k}$ where $h(k)$ is the $r \times r$ matrix of filter coefficients. In fact we have the following theorem (with the same proof).

Theorem 11.16 *A multifilter gives perfect reconstruction when*

$$\mathbf{S}^{(0)}[z]\mathbf{H}^{(0)}[z] + \mathbf{S}^{(1)}[z]\mathbf{H}^{(1)}[z] = 2z^{-\ell}\mathbf{I} \qquad (11.50)$$

$$\mathbf{S}^{(0)}[z]\mathbf{H}^{(0)}[-z] + \mathbf{S}^{(1)}[z]\mathbf{H}^{(1)}[-z] = \mathbf{0}. \qquad (11.51)$$

Here, all of the matrices are $r \times r$. We can no longer give a simple solution to Equation (11.50) because $r \times r$ matrices do not, in general, commute.

There is a corresponding theory of multiwavelets. The dilation equation is

$$\Phi(t) = 2 \sum_k h^{(0)}(k)\Phi(2t - k),$$

where each $\Phi_j(t), j = 1, \ldots, r$ is a vector of r scaling functions. The wavelet equation is

$$w(t) = 2 \sum_k h^{(1)}(k)w(2t - k),$$

where each $w_j(t), j = 1, \ldots, r$ is a vector of r wavelets.

A simple example of a multiresolution analysis is "Haar's hat." Here the space V_0 consists of piecewise linear (discontinuous) functions. Each such function is linear between integer gridpoints $t_j = j$ and right continuous at the gridpoints: $f(j) = f(j+0)$. However, in general $f(j) \neq f(j - 0)$. Each function is uniquely determined by the 2-component input vector $(f(j), f(j + 1 - 0)) = (a_j, b_j)$, $j = 0, \pm 1, \ldots$ Indeed, $f(t) = (b_j - a_j)(t - j) + a_j$ for $j \leq t < j+1$. We can write this representation as

$$f(t) = \frac{b_j + a_j}{2}\phi_1(t - j) + \frac{b_j - a_j}{2}\phi_2(t - j), \qquad j \leq t < j + 1,$$

where $\frac{b_j + a_j}{2}$ is the *average* of f in the interval $j \leq t < j + 1$ and $b_j - a_j$ is the *slope*. Here,

$$\phi_1(t) = \begin{cases} 1 & 0 \leq t < 1 \\ 0 & \text{otherwise} \end{cases} \qquad \phi_2(t) = \begin{cases} 2t - 1 & 0 \leq t < 1 \\ 0 & \text{otherwise.} \end{cases}$$

Note that $\phi_1(t)$ is just the box function. Note further that the integer translates of the two scaling functions $\phi_1(t + k)$, $\phi_2(t + \ell)$ are mutually orthogonal and form an ON basis for V_0. The same construction goes over if we halve the interval. The dilation equation for the box function is (as usual)

$$\phi_1(t) = \phi_1(2t) + \phi_1(2t - 1).$$

You can verify that the dilation equation for $\phi_2(t)$ is

$$\phi_2(t) = \frac{1}{2}\left[\phi_2(2t) + \phi_2(2t - 1) - \phi_1(2t) + \phi_1(2t - 1)\right].$$

They go together in the *matrix dilation equation*

$$\begin{bmatrix} \phi_1(t) \\ \phi_2(t) \end{bmatrix} = \begin{bmatrix} 1, & 0 \\ -\frac{1}{2}, & \frac{1}{2} \end{bmatrix} \begin{bmatrix} \phi_1(2t) \\ \phi_2(2t) \end{bmatrix} + \begin{bmatrix} 1, & 0 \\ \frac{1}{2}, & \frac{1}{2} \end{bmatrix} \begin{bmatrix} \phi_1(2t - 1) \\ \phi_2(2t - 1) \end{bmatrix}.$$
$$\tag{11.52}$$

11.5 Finite length signals

In our previous study of discrete signals in the time domain we have usually assumed that these signals were of infinite length, and we have designed filters and passed to the treatment of wavelets with this in mind. This is a good assumption for files of indefinite length such as audio files. However, some files (in particular video files) have a fixed length L. How do we modify the theory, and the design of filter banks, to process signals of fixed finite length L? We give a very brief discussion of finite length signals.

Here is the basic problem. The input to our filter bank is the finite signal $x_0, \ldots x_{L-1}$, and nothing more. What do we do when the filters call for values x_n where n lies outside that range? There are two basic approaches. One is to redesign the filters (so-called boundary filters) to process only blocks of length L. We shall not treat this approach. The other is to embed the signal of length L as part of an infinite signal (in which no additional information is transmitted) and to process the extended signal in the usual way. Here are some of the possibilities:

1. *Zero-padding*, or *constant-padding*. Set $x_n = c$ for all $n < 0$ or $n \geq L$. Here c is a constant, usually 0. If the signal is a sampling of a continuous function, then zero-padding ordinarily introduces a discontinuity.

2. Extension by periodicity (*wraparound*). We require $x_n = x_m$ if $n = m$, mod L, i.e., $x_n = x_k$ for $k = 0, 1, \ldots, L-1$ if $n = aL + k$ for some integer a. Again this ordinarily introduces a discontinuity.

3. Extension by reflection. There are two principal ways this is done. The first is called *whole-point symmetry*. We are given the finite signal $x_0, \ldots x_{L-1}$. To extend it we reflect at position 0. Thus, we define $x_{-1} = x_1, x_{-2} = x_2, \ldots, x_{-[L-2]} = x_{L-2}$. This determines x_n in the $2L-2$ strip $n = -L+2, \ldots, L-1$. Note that the values of x_0 and x_{L-1} each occur once in this strip, whereas the values of $x_1, \ldots, x_{L-2}$ each occur twice. Now x_n is defined for general n by $2L - 2$ periodicity. Thus whole-point symmetry is a special case of wraparound, but the periodicity is $2L - 2$, not L. This is sometimes referred to as a $(1,1)$ extension, since neither endpoint is repeated.

The second symmetric extension method is called *half-point symmetry*. We are given the finite signal $x_0, \ldots x_{L-1}$. To extend it we reflect at position $-1/2$, halfway between 0 and -1. Thus, we define $x_{-1} = x_0, x_{-2} = x_1, \ldots, x_{-L} = x_{L-1}$. This determines x_n in the $2L$ strip $n = -L, \ldots, L - 1$. Note that the values of x_0 and x_{L-1} each occur twice in this strip, as do the values of $x_1, \ldots, x_{L-2}$. Now x_n is defined for general n by $2L$ periodicity. Thus we have again a special case of wraparound, but the periodicity is $2L$, not L, or $2L - 2$. This is sometimes referred to as a $(2,2)$ extension, since both endpoints are repeated. We used this same extension in the derivation of the Discrete Cosine Transform (DCT), Section 6.4.2.

Finite signals that are extended by wraparound or symmetric reflection are periodic. Thus they can be analyzed and processed by tools such as the DFT that exploit this periodicity. That will be the subject of the next section. If the original signal is a sampling of a differentiable function, symmetric extension has the advantage that it maintains continuity at the boundary, only introducing a discontinuity in the first derivative. The downside to this is that extension of the signal by reflection roughly doubles its length. Thus after processing the analyzed signal we need to downsample it to obtain a signal of length L, satisfying perfect reconstruction. There are many ways to accomplish this, e.g., some of the solutions are implemented in the wavelet toolbox of MATLAB. A detailed discussion can be found in [95].

11.6 Circulant matrices

Most of the methods for treating finite length signals of length L introduced in the previous section involve extending the signal as infinite and periodic. For wraparound, the period is L; for whole-point symmetry the period is $2L - 2$; for half-point symmetry it is $2L$. To take advantage of this structure we modify the definitions of the filters so that they exhibit this same periodicity. We will call this period M, (with the understanding that this number is the period of the underlying data: L, $2L - 2$ or $2L$). Then the data can be considered as a repeating M-tuple and the filters map repeating M-tuples to repeating M-tuples. Thus for passage from the time domain to the frequency domain, we are, in effect, using the discrete Fourier transform (DFT), base M.

For infinite signals the matrices of FIR filters $\mathbf{\Phi}$ are Toeplitz, the filter action is given by convolution, and this action is diagonalized in the frequency domain as multiplication by the Fourier transform of the finite impulse response vector of $\mathbf{\Phi}$. There are perfect $M \times M$ analogies for Toeplitz matrices. These are the *circulant matrices*. Thus, the infinite signal in the time domain becomes an M-periodic signal, the filter action by Toeplitz matrices becomes action by $M \times M$ circulant matrices, and the finite Fourier transform to the frequency domain becomes the DFT, base M. Implicitly, we have worked out most of the mathematics of this action in Chapter 6. We recall some of this material to link with the concepts of filter bank theory.

Recall that the FIR filter $\mathbf{\Phi}$ can be expressed in the form $\mathbf{\Phi} = \sum_{k=0}^{N} h_k \mathbf{R}^k$ where h is the impulse response vector and $\mathbf{R}$ is the delay operator $\mathbf{R}x_n = x_{n-1}$. If $N < M$ we can define the action of $\mathbf{\Phi}$ (on data consisting of repeating M-tuples) by restriction. For the restriction we have $\mathbf{R}^M = \mathbf{I}$. Thus the delay operator action is represented by the $M \times M$ *cyclic permutation matrix* R defined by $Rx_n = x_{n-1}$, mod M. (Here and in the following discussion, all matrices are $M \times M$.) For example, for $M = 4$,

$$Rx = \begin{bmatrix} 0 & 0 & 0 & 1 \\ 1 & 0 & 0 & 0 \\ 0 & 1 & 0 & 0 \\ 0 & 0 & 1 & 0 \end{bmatrix} \begin{bmatrix} x_0 \\ x_1 \\ x_2 \\ x_3 \end{bmatrix} = \begin{bmatrix} x_3 \\ x_0 \\ x_1 \\ x_2 \end{bmatrix}.$$

The action of $\mathbf{\Phi}$ on repeating M-tuples becomes

$$\mathbf{\Phi} = \sum_{n=0}^{N} h_n \mathbf{R}^n.$$

This is an instance of a circulant matrix.

Definition 11.17 An $M \times M$ matrix $\mathbf{\Phi}$ is called a *circulant* if all of its diagonals (main, sub and super) are constant and the indices are interpreted mod M. Thus, there is an M-vector with components h_k such that $A_{\ell,k} = h_{\ell-k}$, mod M.

Example 11.18

$$\Phi = \begin{pmatrix} 1 & 5 & 3 & 2 \\ 2 & 1 & 5 & 3 \\ 3 & 2 & 1 & 5 \\ 5 & 3 & 2 & 1 \end{pmatrix}.$$

Here, $M = 4$, $a_0 = 1$, $a_1 = 2$, $a_2 = 3$, $a_3 = 5$.

Recall that the column vector

$$X = (X[0], X[1], X[2], \ldots, X[M-1])$$

is the *Discrete Fourier transform* (DFT) of $x = \{x[n],\ n = 0, 1, \ldots, M-1\}$ if it is given by the matrix equation $X = \mathcal{F}x$ or

$$\begin{pmatrix} X[0] \\ X[1] \\ X[2] \\ \vdots \\ X[M-1] \end{pmatrix} = \begin{pmatrix} 1 & 1 & 1 & \cdots & 1 \\ 1 & \omega & \omega^2 & \cdots & \omega^{M-1} \\ 1 & \omega^2 & \omega^4 & \cdots & \omega^{2(M-1)} \\ \vdots & \vdots & \vdots & \ddots & \vdots \\ 1 & \omega^{M-1} & \omega^{2(M-1)} & \cdots & \omega^{(M-1)(M-1)} \end{pmatrix} \begin{pmatrix} x[0] \\ x[1] \\ x[2] \\ \vdots \\ x[M-1] \end{pmatrix}$$

(11.53)

where $\omega = \overline{W} = e^{-2\pi i/M}$. Thus,

$$X[k] = \tilde{x}(k) = \sum_{n=0}^{M-1} x[n]\overline{W}^{nk} = \sum_{n=0}^{M-1} \mathbf{x}[n]e^{-2\pi ink/M}.$$

Here $\mathcal{F} = \overline{F}$ is again an $M \times M$ matrix. The inverse relation is the matrix equation $x = \mathcal{F}^{-1}X$ or

$$x[n] = \frac{1}{M} \sum_{k=0}^{M-1} X[k]W^{nk} = \frac{1}{M} \sum_{k=0}^{M-1} X[k]e^{2\pi ink/M},$$

where $\mathcal{F}^{-1} = \frac{1}{M}F$ and $\overline{W} = \omega = e^{-2\pi i/M}$, $W = \bar{\omega} = \omega^{-1} = e^{2\pi i/M}$.
  Note that

$$\overline{F}F = MI, \qquad \overline{F}^{-1} = \frac{1}{M}F.$$

For $x, y \in P_M$ (the space of repeating M-tuples) we define the *convolution* $x * y \in P_M$ by

$$x * y[n] = z[n] \equiv \sum_{m=0}^{M-1} x[m]y[n-m].$$

Then $Z[k] = X[k]Y[k]$.
  Now note that the DFT of $\mathbf{R}^n x[m]$ is $\hat{(\mathbf{R})}^n x[k]$ where

$$\hat{(\mathbf{R})}^n x[k] = \sum_{m=0}^{M-1} (\mathbf{R})^n x[m]e^{-2\pi imk/M} = \sum_{m=0}^{M-1} x[m-n]e^{-2\pi imk/M}$$

$$= \overline{W}^{nk} \sum_{m=0}^{M-1} x[m]e^{-2\pi imk/M} = \overline{W}^{nk}X[k] = \overline{W}^{nk}\hat{x}[k].$$

Thus,

$$(\hat{\mathbf{\Phi}})x[k] = \sum_{n=0}^{N} h_n \hat{(\mathbf{R}^n x)}[k] = \sum_{n=0}^{N} h_n \overline{W}^{nk}\hat{x}[k]$$

$$= \hat{h}_k \hat{x}[k].$$

In matrix notation, this reads

$$(\hat{\mathbf{\Phi}})x = \overline{F}\Phi x = D^\Phi \overline{F}x$$

where D^Φ is the $M \times M$ diagonal matrix

$$(D^\Phi)_{jk} = \hat{h}_k \delta_{jk}.$$

Since x is arbitrary we have $\overline{F}\Phi = D^\Phi \overline{F}$ or

$$D^\Phi = (F)^{-1}\Phi F.$$

This is the exact analog of the convolution theorem for Toeplitz matrices in frequency space. It says that circulant matrices are diagonal in the

DFT frequency space, with diagonal elements that are the DFT of the impulse response vector h.

11.7 Additional exercises

Exercise 11.1 Verify that the 2-spline scaling function $\phi_2(t)$ can be embedded in a 5/3 filter bank defined by

$$S^{(0)}[z] = \left(\frac{1 + z^{-1}}{2}\right)^2, \qquad H^{(0)}[z] = \left(\frac{1 + z^{-1}}{2}\right)^2 \left[-1 + 4z^{-1} - z^{-2}\right],$$

i.e., $p = 2$. This corresponds to Daubechies D_4 (which would be 4/4). Work out the filter coefficients for the analysis and synthesis filters. Show that the convergence criteria are satisfied, so that this defines a family of biorthogonal wavelets.

Exercise 11.2 Verify Lemma 11.14.

Exercise 11.3 Let $x = (x_0, x_1, x_2, x_3)$ be a finite signal. Write down the infinite signals obtained from x by (a) zero padding, (b) constant padding with $c = 1$, (c) wraparound, (d) whole-point reflection, (e) half-point reflection.

Exercise 11.4 Verify explicitly that the circulant matrix Φ of Example 11.18 is diagonal in the DFT frequency space and compute its diagonal elements.

Additional properties of biorthogonal wavelets can be found in [39]. For more properties and applications of multiwavelets, see [74].

12

Parsimonious representation of data

An important theme in this book is the study of systems $y = \Phi x$. We have looked carefully at the case when Φ is linear and can be taken as an $m \times n$ matrix or more generally a linear operator on a Hilbert space. In the various applications we have considered thus far, x is the signal or input, Φ is the transform and y is the sample or output. We have studied in detail the reconstruction of x from y and have developed many tools to deal with this case.

This chapter is concerned with the parsimonious representation of data x, a fundamental problem in many sciences. Here, typically x is a vector in some high-dimensional vector space (object space), and Φ is an operation (often nonlinear) we perform on x, for example dimension reduction, reconstruction, classification, etc. The parsimonious representation of data typically means obtaining accurate models y of naturally occurring sources of data, obtaining optimal representations of such models, and rapidly computing such optimal representations. Indeed, modern society is fraught with many high-dimensional highly complex data problems, critical in diverse areas such as medicine, geology, critical infrastructure, health and economics, to name just a few.

This is an introduction to some topics in this subject showing how to apply mathematical tools we have already developed: linear algebra and SVD, Fourier transforms/series, wavelets/infinite products, compressive sampling and linear filters, PCA and dimension reduction/clustering, compression methods (linear and nonlinear). The subject is vast and we aim only to give the reader an introduction which is self-contained within the framework of this book.

12.1 The nature of digital images

12.1.0 What is an image?

An image consists of an array of pixels, digitized where at each pixel there is an attribute vector. For example, a continuous image of a face may consist of N rows of numbers from $0, \ldots, 15$ and M rows from $0, \ldots, 15$. The pixel at coordinates $n = 3$ and $m = 10$ has, say, integer brightness 110. This is the attribute vector denoted by $a[m, n]$. Thus one may think of an image as a point in some high-dimensional Euclidean space. In every gray scale image, the attribute at each pixel is a number indicating the brightness level of that pixel. The number is usually an N bit binary number which lies between 0 and 2^{N-1}. Here 2^N is the full dynamic range of the image. Consider the examples below:

$$\begin{bmatrix} 1 & 2 & 2 & 1 \\ 2 & 3 & 3 & 2 \\ 1 & 2 & 2 & 1 \\ 0 & 1 & 1 & 0 \end{bmatrix}, \quad \begin{bmatrix} 1 & 2 & 2 & 1 \\ 2 & 2 & 2 & 2 \\ 1 & 2 & 2 & 1 \\ 1 & 1 & 1 & 1 \end{bmatrix}.$$

In both examples, the mean is 1.5 and the scenes are equally bright. In the first example, the variance is 0.75 and in the second, the variance is 0.25, thus the variance indicates contrast. In the case of a gray scale image, the values recorded for the pixel brightness range, are between 0 and $2^N - 1$. Thus the pixels can take on 2^N distinct brightness values in the full dynamic range. The dynamic range of the image is the actual range of values that occur. Note that in the first example above, it is 4 and in the second, it is 2. If 256 is the full dynamic range, these pictures will appear black to the human eye, even though they may have structure. In the picture below, the human eye will see two white lines fading towards the ends.

$$\begin{bmatrix} 130 & 208 & 208 & 156 \\ 0 & 0 & 0 & 0 \\ 182 & 234 & 234 & 156 \\ 0 & 0 & 0 & 0 \end{bmatrix}$$

This picture was obtained from one with a smaller dynamic range by a technique called contrast stretching. It uses the following linear transformation:

$$b[m, n] = \frac{(2^N - 1)a[m, n] - \min(m, n)}{\max(m, n) - \min(m, n)}.$$

Exercise 12.1 Consider the image immediately above. Use a contrast stretch to construct this image from the image

$$\begin{bmatrix} 5 & 8 & 8 & 6 \\ 0 & 0 & 0 & 0 \\ 7 & 9 & 9 & 6 \\ 0 & 0 & 0 & 0 \end{bmatrix}$$

Sometimes, we wish to modify the picture, so that all brightness values are equally likely. The point is to allow all brightness values to have similar contrasts. In practice, this requires the histogram of all pixel values to occur in a straight line. The technique is called histogram normalization. Consider the example below, (a) which has 16 elements and dynamic range 2^3. We wish to equalize the histogram so that all 8 values are used equally often, i.e., twice. One possible solution, (b), is shown below.

$$(a) \quad \begin{bmatrix} 7 & 7 & 0 & 1 \\ 0 & 6 & 1 & 0 \\ 7 & 1 & 0 & 1 \\ 0 & 7 & 2 & 0 \end{bmatrix}, \quad (b) \quad \begin{bmatrix} 7 & 7 & 2 & 4 \\ 2 & 5 & 4 & 1 \\ 6 & 3 & 1 & 3 \\ 0 & 6 & 5 & 0 \end{bmatrix}.$$

Human perception of color, brightness, and contrast is not uniform or even linear. Thus the format of the image does make a difference. For example, contrast stretching from the view point of a computer, is simply a rescaling and so does not affect the image at all. Histogram normalization, however, does change the statistical properties of the image. Our perception of brightness is not linear and is naturally skewed to edge detection. For example, it is easy to construct an image whose top half shows a linear increase in brightness between bands, whereas the bottom half shows an exponential increase in brightness between bands which leads to a contrast at edges.

All **colors**, may be obtained as a combination of the three RBG primary colors, Red, Green and Blue. Thus when we represent a color digital image, we have a three dimensional attribute vector (R,G,B) associated with every point. Each component, is exactly a gray scale image. **Noise:** Images of whatever type, are generated from sensors which measure light signal and return digital information and any process of this kind attracts noise. There are different types of noise. For example: (1) Instrument noise; (2) Photon noise: Sensors count photons and the quantum physics laws governing this counting are non deterministic.

Thus the noise effect is nonlinear or Gaussian and depends on the signal; (3) Thermal noise: Sensor arrays are affected by temperature, heat generates electrons as photons do; (4) Dark current: This is the signal received from the sensor in the absence of any external signal. This should be ideally zero; (5) Amplifier noise: Various amplification and transfer processes occur before the signal is digitized. All may add noise; (6) Quantization noise: This occurs when an analog signal is converted to digital numbers. Thus the error is a rounding one and may be significant in dark images. Let us consider one popular type of noise referred to as SNR (signal to noise): In this case, the signal is approximately constant in the region of interest and the variation of the region is due to noise. Moreover, the noise distribution is the same over the entire image with constant standard deviation over the region of interest (ROI).

Exercise 12.2 Assume we have a system producing a dynamic range of 256, i.e., 8 bits and the system adds two percent (of the dynamic range) random noise to the image. Suppose we measure the dark image first and then add two percent random noise to it as shown. Show that the contrast stretching has no effect on the dynamic range/noise ratio.

$$
I_{\text{signal}} = \begin{bmatrix} 5 & 8 & 8 & 6 \\ 0 & 0 & 0 & 0 \\ 7 & 9 & 9 & 6 \\ 0 & 0 & 0 & 0 \end{bmatrix}, \quad I_{\text{signal+noise}} = \begin{bmatrix} 8 & 3 & 8 & 2 \\ 5 & 2 & 5 & 2 \\ 7 & 5 & 11 & 3 \\ 4 & 3 & 4 & 1 \end{bmatrix}.
$$

See Figure 12.1.

Figure 12.1 The pixels at coordinates $n = a$ and $m = b$ have different integer brightness. This is the attribute vector.

12.1.1 Image enhancement

Image enhancement techniques are designed to improve the quality of the image. They can be subjective in the sense that they change the appearance of the image, they may be used to select features such as edge detection, for example, or they could be used to remove or reduce certain cases of noise. Typically, image enhancement techniques fall into three categories: pointwise operations, local operations and global operations. In what follows, we will present examples of these three groups of techniques and we will see they relate to machinery we have developed in earlier chapters.

12.1.2 Local/smoothing filters

We have already defined the notion of a filter and studied it in some detail. We will now use local filters for image enhancement. Local filters are local operations which take a group of neighboring pixels and use their values to generate a new value for the central pixel. This is implemented as follows: We take a matrix A, typically $n \times n$ with n odd and use it to define a moving window across the picture. The entries of the matrix are then determined by weights given to the pixels in order to calculate the new value of the central pixel. We describe this mathematically by taking convolution

$$I_{\text{new}}(x, y) = A * I_{\text{old}}(x, y).$$

Example 12.1 Consider a rectangular unweighted smoothing filter applied to an image. The bold region in the image leads to the bold number in the output image. The same process is applied to all the other entries except the edge rows and edge columns.

$$A = 1/9 \begin{bmatrix} 1 & 1 & 1 \\ 1 & 1 & 1 \\ 1 & 1 & 1 \end{bmatrix}, \qquad I_{\text{old}} = \begin{bmatrix} 0 & 1 & 2 & 1 & 0 \\ 1 & 2 & 3 & 1 & 0 \\ 2 & 3 & 1 & 0 & 2 \\ 3 & 1 & 0 & 2 & 3 \\ 1 & 0 & 2 & 3 & 3 \end{bmatrix}$$

$$I_{\text{new}} = \begin{bmatrix} 0 & 1 & 2 & 1 & 0 \\ 1 & 1.67 & \mathbf{1.56} & 1.11 & 0 \\ 2 & 1.78 & 1.44 & 1.33 & 2 \\ 3 & 1.44 & 1.33 & 1.78 & 3 \\ 1 & 0 & 2 & 3 & 3 \end{bmatrix}.$$

Our next example is a circular smoothing filter:

Example 12.2

$$A = 1/5 \begin{bmatrix} 0 & 1 & 0 \\ 1 & 1 & 1 \\ 0 & 1 & 0 \end{bmatrix}, \qquad I_{\text{old}} = \begin{bmatrix} 0 & 1 & 2 & 1 & 0 \\ 1 & 2 & 3 & 1 & 0 \\ 2 & 3 & 1 & 0 & 2 \\ 3 & 1 & 0 & 2 & 3 \\ 1 & 0 & 2 & 3 & 3 \end{bmatrix},$$

$$I_{\text{new}} = \begin{bmatrix} 0 & 1 & 2 & 1 & 0 \\ 1 & 2 & 1.8 & 1 & 0 \\ 2 & 1.8 & 1.4 & 1.2 & 2 \\ 3 & 1.4 & 1.2 & 1.6 & 3 \\ 1 & 0 & 2 & 3 & 3 \end{bmatrix}.$$

We present one more example on weighted circular filters. Observe that the central pixel is multiplied by 4, the ones to the left, to the right, above and below by 1 and the corner pixels neglected (i.e., multiplied by 0).

Example 12.3

$$A = 1/8 \begin{bmatrix} 0 & 1 & 0 \\ 1 & 4 & 1 \\ 0 & 1 & 0 \end{bmatrix}, \qquad I_{\text{old}} = \begin{bmatrix} 0 & 1 & 2 & 1 & 0 \\ 1 & 2 & 3 & 1 & 0 \\ 2 & 3 & 1 & 0 & 2 \\ 3 & 1 & 0 & 2 & 3 \\ 1 & 0 & 2 & 3 & 3 \end{bmatrix},$$

$$I_{\text{new}} = \begin{bmatrix} 0 & 1 & 2 & 1 & 0 \\ 1 & 2 & 2.25 & 1 & 0 \\ 2 & 2.25 & 1.25 & 0.75 & 2 \\ 3 & 1.25 & 0.75 & 1.75 & 3 \\ 1 & 0 & 2 & 3 & 3 \end{bmatrix}.$$

12.1.3 Edge enhancement filters

Edges may be detected in an image by looking for high values of the derivative as you move through the image along some direction. Typically we have the following rule: dark to light produces a large positive derivative, light to dark a large negative derivative, and uniform a low value of the derivative. When we think of derivative, we mean numerical directional derivative. Roughly speaking, we give meaning to this by taking the difference between successive values in the picture where successive means in the direction chosen. Edges can then be estimated in different directions, or the magnitude of the derivative can be estimated to select edges in any direction. Noise can be a problem, so we smooth at the same time. In the example, below, smoothing is implemented by filters in the horizontal directions and vertical directions, respectively.

$$A_h = 1/3 \begin{bmatrix} 1 & 0 & -1 \\ 1 & \mathbf{0} & -1 \\ 0 & 0 & -1 \end{bmatrix}, \qquad A_v = 1/3 \begin{bmatrix} 1 & 1 & 1 \\ 0 & \mathbf{0} & 0 \\ -1 & -1 & -1 \end{bmatrix}.$$

Computation of the horizontal derivative. Edges are shown in bold.

$$I_{\text{old}} = \begin{bmatrix} 0 & 1 & 5 & 4 & 6 & 3 \\ 2 & 2 & 6 & 5 & 6 & 2 \\ 1 & 2 & 7 & 6 & 5 & 0 \\ 0 & 1 & 6 & 7 & 6 & 1 \\ 1 & 2 & 7 & 6 & 5 & 2 \\ 0 & 0 & 5 & 5 & 6 & 1 \end{bmatrix}, \quad I_{\text{new}} = \begin{bmatrix} 0 & 1 & 5 & 4 & 6 & 3 \\ 2 & \mathbf{-5} & \mathbf{-3.33} & 0.3 & \mathbf{3.3} & 2 \\ 1 & \mathbf{-5.33} & \mathbf{-4.33} & 0.7 & 5 & 0 \\ 0 & \mathbf{-6} & \mathbf{-4.67} & 1.3 & \mathbf{5.3} & 1 \\ 1 & \mathbf{-5.67} & \mathbf{-5} & 0.3 & 4.7 & 2 \\ 0 & 0 & 5 & 5 & 6 & 1 \end{bmatrix}.$$

We may compute a derivative h in an angle direction θ using the well-known formula:

$$[h_\theta] = \cos(\theta).[h_x] + \sin(\theta).[h_y]$$

and its magnitude by

$$h^2 = h_x^2 + h_y^2.$$

Since high values for the magnitude will indicate a change of brightness in some direction, this is a useful edge finder. Sometimes we use the Laplace operator (second derivative) for edge enhancement/edge detection.

The Laplacian:

$$\begin{bmatrix} -1 & -1 & -1 \\ 1 & 8 & -1 \\ -1 & -1 & -1 \end{bmatrix}$$

Edge Enhancement:

$$\begin{bmatrix} -1 & -1 & -1 \\ 1 & 9 & -1 \\ 1 & -1 & -1 \end{bmatrix}$$

Note the following: The second derivative in the horizontal direction is $(1, 2, 1)$; an approximation for the Laplace operator is:

$$\begin{bmatrix} 0 & 1 & 0 \\ 1 & -4 & 1 \\ 0 & 1 & 1 \end{bmatrix}$$

The form shown above is less sensitive to rotation.

12.1.4 Mathematical morphology

Morphology is the study of shapes. In image processing, this is the processing and analysis of shapes within an image. We will study binary images in this section and also give some ideas on how to generalize to gray scale and thus color images. We will explain our ideas via set theory language. Let us mention four basic morphology operations. (1) dilation: growing the shape, filling small gaps and smoothers; (2) erosion: shrinkage of the shape; (3) opening (and closing): idempotent shape operations; (4) skeleton: this is the kernel of the shape.

Dilation: Let B be a structuring element and A be a shape in a binary image, in the sense that A and B are defined by positions of points in the underlying image. The *dilation* of A and B, which we will write as AdB, is

$$AdB = \{a + b, \, a \in A, \, \text{and } b \in B\} .$$

Here d is commutative and A is swelled by B. We also require that $0 \in B$ so that $A \subseteq AdB$.

Erosion: The *erosion* of A and B written as AeB is defined as

$$AeB = \{x \in B, x + a \in A, \, \forall a \in A\} .$$

We will assume that $0 \in A$ so that $AeB \subseteq A$. Erosion shrinks the shape. Another interesting way to visualize erosion is by the set of points where B can be centered and still be in contained in A. This is done in the following way: The shift of a shape A by a position x, A_x is defined by

$$A_x = \cup \{a + x, \, a \in A\} .$$

Then, we can write

$$AeB = \{x,\, A_x \subseteq A\}$$

and similarly we can write

$$AdB = \cup \{B_a,\, a \in A\}.$$

Openings and Closings: We define the closing of an image A by a structuring element K as $AcK = (AdK)eK$ and the opening action on A similarly by $AoK = (AeK)dK$. The opening of A is the union of all shifts of A that we can fit completely inside A.

See Figures 12.2–12.5.

Discrete example dilation:

A is the shape given by the • entries in the 6x6 image below.

B is a 2x2 structuring element given by: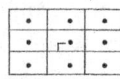

We want to dilate the shape **A** by the structuring element B. The x's in the right-hand image are the points added.

Figure 12.2 Discrete example dilation.

Continuous example dilation:

Consider the nuts-and-bolts image below on the left.

We will dilate this twice using a circular

structuring element B, where B is

The right-hand image is the result.

Figure 12.3 Continuous example dilation.

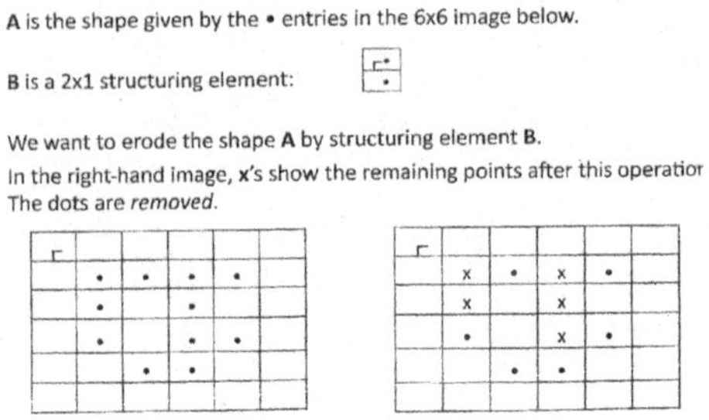

A is the shape given by the • entries in the 6x6 image below.

B is a 2x1 structuring element:

We want to erode the shape **A** by structuring element **B**.
In the right-hand image, **x**'s show the remaining points after this operation
The dots are *removed*.

Figure 12.4 Discrete example erosion.

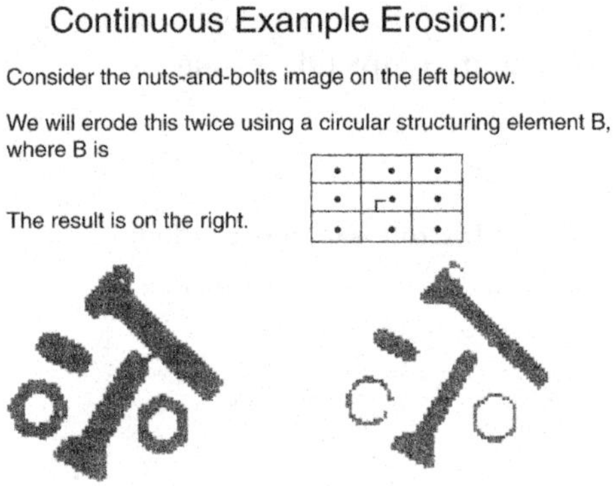

Continuous Example Erosion:

Consider the nuts-and-bolts image on the left below.

We will erode this twice using a circular structuring element B,
where B is

The result is on the right.

Figure 12.5 Continuous example erosion.

12.1.5 Morphology for shape detection

In this section, we explain how morphology operators may be used for the extraction of boundaries and skeletons from shapes in the sense of extraction of information for classification, counting objects, and enhancing edges for example. The boundary of a shape A may be defined by taking A and removing its interior. By interior, we mean taking the erosion of A by a suitable structuring element K. Thus we may write $\partial A = A \cap (AeK)^c$ where we are taking complements in the last operation.

Exercise 12.3 Consider the examples already studied. For a suitable structuring element K of your choice, calculate (1) The boundary of the element below using the structuring element shown. (2) The boundary of the nuts and bolts example using a dilation of 2. (3) Study the thickened boundary $(AdK) \cap (A^c dK)$. What do you find?

Consider now a morphology operator which will preserve the topology of the shape in the sense of distilling its essence in some natural sense. We call such an operator a skeleton. One practical construction of a skeleton may be implemented using the following heuristic algorithm:

- Set up a 3×3 moving window.
- Repeatedly set the center pixel to zero unless one of the following conditions occurs: (1) an isolated pixel is found; (2) removing a pixel would change the connectivity; (3) removing a pixel would shorten a line.

Exercise 12.4 Apply the skeletal algorithm above to the nuts and bolts example above.

The last part of this section deals briefly with representations of gray scale (and by extension, color) images. In this case, we may think of our image as a three-dimensional object. It has a rectangular base (the pixels) and the height over each pixel is the brightness of the pixel.

We are able to carry across most of the theory developed for binary image morphology by the use of the following two operators on the 3D grey scale shapes.

- Top: If A is a shape, $T[A]$ will be the top surface of the shape defined by

$$T[A](x) = \max \{y; (x, y) \in A\}.$$

Note that $T[A]$ is a function over the image.
- Umbra: A set A is called an umbra iff $(x, y) \in A$ implies that $(x, z) \in A$ for all $z \leq y$.
- If f is the gray scale intensity function over an image, then $U[f]$ is the umbra of f defined by:

$$U[f] = \{(x, y); y \leq f(x)\}, \; T[f] = f.$$

If A is an umbra, $UT[A] = A$.

The filter ideas we have discussed and introduced earlier in this section can also be implemented with morphology variants. One such filter is the open–close smoothing filter where open is understood to smooth within and close is understood to smooth without. We illustrate this with Figure 12.6.

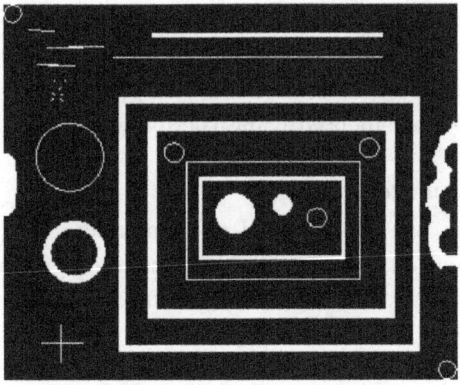

Figure 12.6 Some examples.

12.1.6 Fourier transform

The Fourier transform (FT) is an important image processing tool which is used to decompose an image into its sine and cosine components. The output of the transformation represents the image in the Fourier or frequency domain, while the input image is the spatial domain equivalent. In the Fourier domain image, each point represents a particular frequency contained in the spatial domain image. The Fourier transform is used in a wide range of applications, such as image analysis, image filtering, image reconstruction and image compression. We restrict ourselves to digital images, so we will restrict this discussion to the Discrete Fourier transform (DFT).

The DFT is the sampled FT and therefore does not contain all frequencies forming an image, but only a set of samples which is large enough to fully describe the spatial domain image. The number of frequencies corresponds to the number of pixels in the spatial domain image, i.e. the image in the spatial domain and Fourier domain are of the same size.

For a square image of size $N \times N$, the two-dimensional DFT $F(k,l)$ is given by:

$$F(k,l) = \frac{1}{N^2} \sum_{a=0}^{N-1} \sum_{b-0}^{N-1} f(a,b) \exp(-i2\pi(ka/N + lb/N))$$

where $f(a,b)$ is the image in the spatial domain and the exponential term is the basis function corresponding to each point $F(k,l)$ in the Fourier space. The equation can be interpreted as: the value of each point $F(k,l)$

is obtained by multiplying the spatial image with the corresponding base function and summing the result.

The basis functions are sine waves and cosine waves with increasing frequencies, i.e., $F(0,0)$ represents the DC (discrete cosine)-component of the image which corresponds to the average brightness and $F(N - 1, N - 1)$ represents the highest frequency.

In a similar way, the Fourier image can be re-transformed to the spatial domain. The inverse Fourier transform is given by:

$$f(a,b) = \frac{1}{N^2} \sum_{k=0}^{N-1} \sum_{l=0}^{N-1} F(k,l) \exp(i2\pi(ka/N + lb/N)).$$

To obtain the result for the above equations, a double sum has to be calculated for each image point. However, because the FT is separable, it can be written as

$$F(k,l) = \frac{1}{N} \sum_{b=0}^{N-1} P(k,b) \exp(-i2\pi lb/N)$$

where

$$P(k,b) = \frac{1}{N} \sum_{a=0}^{N-1} f(a,b) \exp(-i2\pi ka/N).$$

Using these two formulas, the spatial domain image is first transformed into an intermediate image using N one-dimensional FTs. This intermediate image is then transformed into the final image, again using N one-dimensional FTs. Expressing the two-dimensional FT in terms of a series of $2N$ one-dimensional transforms decreases the number of required computations. Even with these computational savings, the ordinary one-dimensional DFT has N^2 complexity. This can be reduced to $N \log N$ if we employ the fast Fourier transform (FFT) to compute the one-dimensional DFTs. This is a significant improvement, in particular for large images. There are various forms of the FFT and most of them restrict the size of the input image that may be transformed, often to $N = 2^n$ where n is an integer.

The FT produces a complex number valued output image which can be displayed with two images, either with the real part and imaginary part or with magnitude and phase. In image processing, often only the magnitude of FT is displayed, as it contains most of the information of the geometric structure of the spatial domain image. However, if we want to re-transform the Fourier image into the correct spatial domain after some processing in the frequency domain, we must make sure to preserve both magnitude and phase of the Fourier image. The Fourier domain image has a much greater range than the image in the spatial domain. Hence, to be sufficiently accurate, its values are usually calculated and stored in float values. The FT is used if we also want to access the geometric characteristics of a spatial domain image. Because the image in the Fourier domain is decomposed into its sinusoidal components, it is easy to examine or process certain frequencies of the image, thus influencing the geometric structure in the spatial domain.

In most implementations the Fourier image is shifted in such a way that the DC-value (i.e. the image mean) $F(0,0)$ is displayed in the center of the image. The further away from the center an image point is, the higher is its corresponding frequency.

What follows pertains to the images: FT1–FT8 in Figures 12.7–12.14.

We start off by applying the Fourier transform to the image FT1 (Figure 12.7).

The magnitude calculated from the complex result is shown in FT2 (Figure 12.8).

We can see that the DC-value is by far the largest component of the image. However, the dynamic range of the Fourier coefficients (i.e. the intensity values in the Fourier image) is too large to be displayed on the screen, therefore all other values appear as black. Now apply a logarithmic transformation to the image (FT3 (Figure 12.9)) (replace each pixel value by its log).

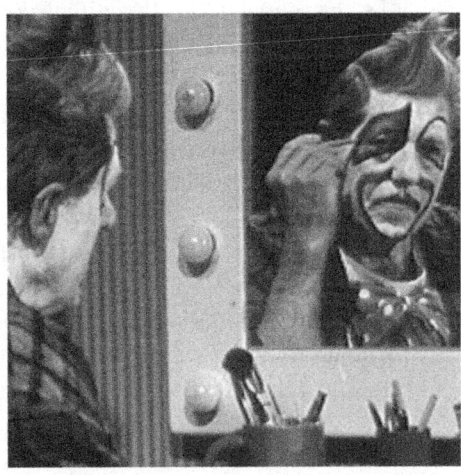

Figure 12.7 Example(Fourier transform 1).

Figure 12.8 Example(Fourier transform 2).

Figure 12.9 Example(Fourier transform 3).

The result shows that the image contains components of all frequencies, but that their magnitude gets smaller for higher frequencies. Hence, low frequencies contain more image information than the higher ones. The transform image also tells us that there are two dominating directions in the Fourier image, one passing vertically and one horizontally through the center. These originate from the regular patterns in the background of the original image.

Figure 12.10 Example(Fourier transform 4).

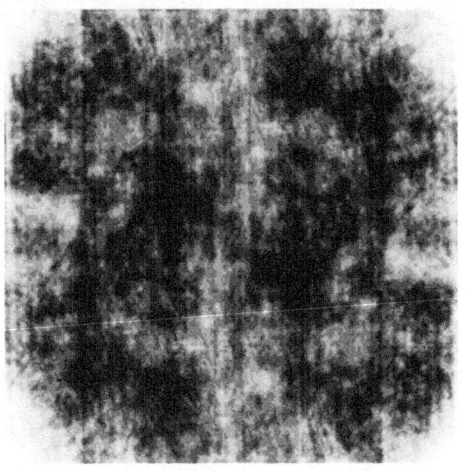

Figure 12.11 Example(Fourier transform 5).

The phase of the Fourier transform of the same image is shown in image FT4 (Figure 12.10).

The value of each point determines the phase of the corresponding frequency. As in the magnitude image, we can identify the vertical and horizontal lines corresponding to the patterns in the original image. The phase image does not yield much new information about the structure of

Figure 12.12 Example(Fourier transform 6).

Figure 12.13 Example(Fourier transform 7).

the spatial domain image; therefore, in the following examples, we will restrict ourselves to displaying only the magnitude of the FT.

Before we leave the phase image entirely, however, note that we may apply the inverse Fourier transform to the magnitude image (FT5 (Figure 12.11)) while ignoring the phase.

Although this image contains the same frequencies (and amount of frequencies) as the original input image, it is corrupted beyond recognition.

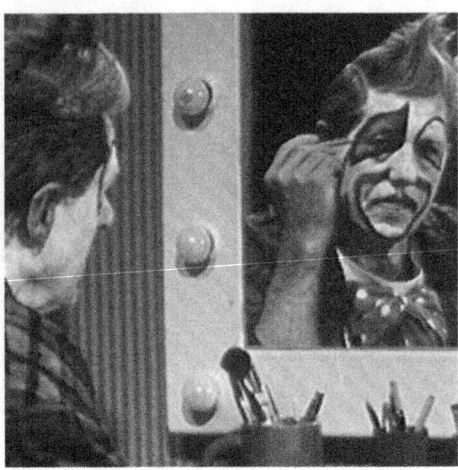

Figure 12.14 Example(Fourier transform 8).

This shows that the phase information is crucial to reconstruct the correct image in the spatial domain. See FT6–FT8 (Figure 12.12–12.14).

Exercise 12.5 Take FTs of two images. Add them using a suitable morphology operator for example a blend, and take the inverse Fourier Transform of the sum. Explain the result.

Exercise 12.6 Add different sorts of noise to two images and compare the FT's. Investigate if the Fourier Transform is distributive over multiplication.

12.2 Pattern recognition and clustering

12.2.1 Understanding images

In this section, we study the extraction of information from images. For example, we may be interested in recognizing handwritten characters, extracting and recognizing target shapes or faces, mineral or land use mapping, vegetation and human settlement, boundaries of cells obscured via noise. We may also be interested in partial information from images or extracting information with high probability, clustering or computing neural networks. Images contain a variety of information which come from color, texture, shapes and patterns. As we have already seen, we may associate with each pixel an attribute vector which then contains all the information in the image. For computer recognition, we need to design procedures to analyze patterns in the sense of constructing pattern vectors from the image which then contain relevant pattern information. Once we have obtained pattern vectors, we need to identify information from them. One way to do this is via the following three steps: (1) template matching where we match pattern vectors whose meaning is known, (2) cluster the patterns into homogeneous groups, (3) train an algorithm to assign correctly the patterns to homogeneous groups. Types of pattern problems where (1–3) may be applied are for example (A) handwritten digits (10 classes known), target detection (2 classes known, target or non target), land use (a finite number of classes known, say houses, offices, factories, parks, etc.); (B) minerals present in a drill core, vegetation types present in a satellite scene and causes of gravity anomalies in an airborne survey (in all three the classes are unknown).

12.2.2 A probability model for recognition of patterns. Bayes' rule

Consider a handwritten digit example where all the digits are equally likely. A character Q is presented and then generates a pattern or feature vector x. We want to compute probabilities that Q is correctly assigned to one of ten character classes which we list as $A_0, A_1, A_2, \ldots, A_9$. We are interested in the probability $P(A_i|x)$, that Q is a handwritten version of digit i, $0 \leq i \leq 9$. Here, $P(A_i|x)$ is the probability density function of x belonging to a specific class i.

Conditional probability and Bayes' rule. To set up a probability model for pattern recognition, we first review the concept of conditional probability. Although we will use the handwritten digit example, it should be clear that this concept and Bayes' rule to follow have wide applicability. Let S be the set of all characters Q. (For convenience we take this to be a very large but finite set.) Assume that there is a probability set function P defined on all subsets of S. This means that (1) $P(B)$ is a real number ≥ 0 for every subset B of S, (2) $P(S) = 1$ and (3) $P(B_1 \cup B_2 \cup \cdots \cup B_n) = \sum_{j=1}^{n} P(B_j)$ for any n pairwise disjoint subsets B_i of S, [5, 78]. In this example we interpret $P(B)$ as the probability that a character Q belongs to the subset B. Now suppose C is a subset of S such that $P(C) \neq 0$.

Definition 12.4 The conditional probability that a character $Q \in C$ belongs to a set $B \subseteq S$ is defined by

$$P(B|C) = \frac{P(B \cap C)}{P(C)}.$$

Exercise 12.7 Show that $P(B|C)$ is a probability set function on the subsets of C.

Let A_i, $i = 0, \ldots, 9$ be the set of all characters Q in the ith character class. Since each character Q belongs to exactly one of those mutually distinct classes we have $S = A_0 \cup \cdots \cup A_9$ and $A_i \cap A_j = \phi$ for all $i \neq j$. Now we interpret $P(A_i)$ as the probability that a character Q is a handwritten version of digit i. Clearly we must require $P(A_i) > 0$ for each i and since P is a probability set function on S, $\sum_{j=0}^{9} P(A_j) = 1$. Since the sets A_j are mutually disjoint and S is their union, it follows that for any subset B of S we have

$$P(B) = \sum_{j=0}^{9} P(A_j)P(B|A_j). \tag{12.1}$$

Exercise 12.8 Prove (12.1) and give its interpretation for the handwriting problem.

Now suppose C is a subset of S such that $P(C) \neq 0$. Bayes' rule expresses the conditional probability $P(A_i|B)$ in terms of the conditional probabilities $P(B|A_j)$:

Theorem 12.5 *Bayes' rule.*

$$P(A_i|B) = \frac{P(A_i)P(B|A_i)}{\sum_{j=0}^{9} P(A_j)P(B|A_j)}. \tag{12.2}$$

Proof Since $A_i \cap B = B \cap A_i$ we have from Definition 12.4 that $P(B \cap A_i) = P(A_i)P(B|A_i) = P(B)P(A_i|B)$. Thus $P(A_i|B) = P(A_i)P(B|A_i)/P(B)$, and substituting for $P(B)$ from (12.1) we obtain the desired result. □

Now we continue with the construction of our model. To each pattern vector x we can associate the subspace X of S consisting of all characters Q yielding x as a pattern. Thus the probability $P(A_i|x)$, that Q is a handwritten version of digit i is exactly the conditional probability $P(A_i|X)$. Next we assume that it is equally likely that a randomly chosen character Q corresponds to each of the 10 digits i: $P(A_i) = 1/10$. Finally we assume that for each of our pattern vectors x we can determine the probability set functions $P(X|A_i) \equiv P(x|A_i)$ experimentally by considering a large number of handwritten digits Q for which i is known and computing their vectors. Indeed, a perfect classifier x^j would be of the form $P(x^j|A_i) = \delta_{ij}$. Applying Bayes' rule in this case and using the fact that $P(A_j) = 1/10$ for all j we find

$$P(A_i|x) = \frac{P(x|A_i)}{\sum_{j=0}^{9} P(x|A_l)}.$$

12.2.3 Metrics

Most pattern clustering or template matching procedures rely on determining some measure of distance between the pattern vectors and so metrics arise in a natural way. The idea of metrics in collections of data is a vastly studied subject with many applications. Some applications are: weighted Euclidean metrics where the weight W is positive definite, i.e., $x^T W x > 0$ for all nonzero x; metrics which arise from linear-nonlinear dimension reduction processes on finite-dimensional spaces, for example diffusion processes; metrics which arise from infinite-dimensional spaces, for example Gromov–Hausdorf distance; matrices of distances in rigidity theory; separators between cluster centers, say $(. - z^{(i)})^T (. - z^{(i)})$ where $z^{(i)}$ is the ith cluster center; angular distance. In the last application mentioned, if vectors in a class cluster are in a particular direction, it is often convenient to measure the angular displacement between them. If vectors are scaled to have length one then $1 - x \cdot y$ may be one such measure. Here $x \cdot y$ is a suitable inner product.

12.2.4 Clustering algorithms

Clustering procedures fall into two main categories, supervised and unsupervised. In unsupervised procedures, the patterns are presented to the classifier with all the available information, for example, number of classes, targets and the classifier generates the pattern classes. In supervised procedures, the classifier is trained using a set of representative vectors whose classes are known. In both cases, supervised and unsupervised, it is very possible that the feature vectors may not be understood. We present examples of both categories.

Maxmin algorithm

This is a simple algorithm which illustrates well the idea of an unsupervised classifier. Pick your favorite metric and (*) assign $x^{(1)}$ as the first cluster center $z^{(1)}$. Now for each pattern $x^{(i)}$, compute its distance from each cluster center and choose the minimum distance to a cluster center. Next, select the pattern whose minimum distance is the largest, i.e., the one with the maximum of the minimum distances. If this distance is less than some pre-assigned threshold, assign all remaining patterns to the cluster with the closest center, if not call this the next cluster center and and go back to (*). Finally recompute the cluster centers as the average of the patterns in that cluster. We note that this algorithm may be very sensitive to the choice of threshold.

K-means algorithm

The K-means algorithm is a base unsupervised algorithm, used when we understand both the vector structure and the number of classes, and typically requires minimizing a performance criterion. Start with $k \geq 1$ initial cluster centers. Pick your favorite metric d. (*) After each iteration, distribute the pattern vectors amongst the k clusters S_i say according to the relationship:

$$x \in S_j \text{ if } d(x, z^{(j)}) \leq d(x, z^{(i)})$$

for all $i \neq j$. Now recompute the cluster centers in such a way as to minimize a performance criterion which for example in the case of the Euclidean metric, updates $z^{(j)}$ to be the mean of the vectors in S_j. If at the end of this step, the cluster centers are unchanged, the algorithm terminates. Otherwise repeat (*).

Supervised classifiers: neural networks

The two class Perceptron

Suppose we have a large amount of data in the form of real n-tuples $u \in \mathbb{R}^n$ where n is also large. Extracting information from this data or "data mining" is an important focus of modern mathematical and statistical research. Here we consider a case where the data is "clustered" into two subsets of points, Class 1 and Class 2. We describe a method to "train" a function to classify these points automatically. This is one of the simplest examples of a neural network, and illustrates what happens at the core of machine learning. The Perceptron is a function $\mathbf{R}(u) = (w, u) + w_0 = \sum_{j=1}^{n} w_j u_j + w_0$ taking any vector $u \in \mathbb{R}^n$ to a real number. Here the real n-tuple $w = (w_1, \ldots, w_n)$ is called a weight vector and is to be determined by the data. The number w_0 is called the scalar offset. For fixed w, w_0 the equation $\mathbf{R}(u) = 0$ defines a hyperplane in $\mathbb{R}^n$. (For $n = 2$

this is a line, for $n = 3$ a plane, and so on.) Suppose the data is **linearly separable**, i.e., suppose there exists some hyperplane $\mathbf{R}(u) = 0$ such that any Class 1 data point x satisfies $\mathbf{R}(x) > 0$ and any Class 2 data point y satisfies $\mathbf{R}(y) < 0$. For $n = 3$, for example, this would mean that there is a plane such that the Class 1 data points all lie above the plane and the Class 2 data points all lie below it. We will show that if the data is linearly separable then there is an efficient algorithm to determine a separating hyperplane.

To describe the algorithm and the verification of its validity, it is useful to embed the problem in $\mathbb{R}^{n+1}$. Thus a data point u is replaced by the $n + 1$-tuple $u' = (u, 1)$ and the weight vector becomes the $n + 1$-tuple $w' = (w, w_0)$. The equation for the hyperplane becomes $(w', u')' = 0$ where $(\cdot, \cdot)'$ is the dot product in R^{n+1}.

Suppose we have a finite set of training vectors $v^{(1)}, \ldots, v^{(m)}$, each of which belongs either to Class 1 or to Class 2, and we know which is which.

The Single Layer Perceptron Algorithm:

1. Initially, assign arbitrary values to w'. Call this $w'^{(1)}$. At the kth iteration the weights will be $w'^{(k)}$.
2. At iteration k we present a training vector v and augment it to v'.
   - If $v \in$ Class 1 and $(w'^{(k)}, v')' \leq 0$, set $w'^{(k+1)} = w'^{(k)} + v'$;
   - If $v \in$ Class 2 and $(w'^{(k)}, v')' \geq 0$, set $w'^{(k+1)} = w'^{(k)} - v'$;
   - Otherwise set $w'^{(k+1)} = w'^{(k)}$.
3. The process terminates when all the training vectors are correctly classified, i.e., when w' is unchanged over a complete training *epoch*.
4. If the process terminates then the final weight $w'^{(k)} = (w^{(k)}, w_0^{(k)})$ defines a separating hyperplane for the data.

Theorem 12.6 *If the classes (as represented by the training vectors) are linearly separable, then the algorithm always terminates with a separating hyperplane. Conversely, if the algorithm terminates then the training data is linearly separable.*

Proof Suppose the training vectors are linearly separable, so that there exists a linear hyperplane $\mathbf{S}(u) \equiv (z, u) + z_0 = 0$ such that $\mathbf{S}(x) > 0$ for all Class 1 training vectors x and $\mathbf{S}(y) < 0$ for all Class 2 training vectors y. We further simplify our analysis by writing $y' = (-y, -1)$ for all the Class 2 training vectors, so that the inequality $(z', x')' > 0$ will characterize the training vectors of both classes. (The distinction will be in the last component, +1 for Class 1 and −1 for Class 2.) Then step 2 of the algorithm can be rewritten as

– If v' is a data vector and $(w'^{(k)}, v')' \leq 0$, set $w'^{(k+1)} = w'^{(k)} + v'$; otherwise do nothing.

Note that for our proof we will not update w' unless we are presented with a training vector $x^{(k)}$ such that $(w'(k), x'^{(k)}) \leq 0$. Thus the algorithm generates a sequence of weight vectors

$$w'^{(k+1)} = w'^{(1)} + \sum_{j=1}^{k} x'^{(j)}, \quad k = 1, 2, \ldots,$$

where $(w'^{(j)}, x'^{(j)})' \leq 0$ for all j and the $x'^{(k)}$ form a sequence of training vectors with possible repetitions. Let $a > 0$ be the minimum value of the inner products $(z', x')'$ for all the (finite number of) training vectors x. Without loss of generality we can normalize z' to have length 1: $||z'||' = 1$. (This doesn't change the separating hyperplane.) Then

$$(z', w'^{(k+1)})' = (z', w'^{(1)})' + \sum_{j=1}^{k} (z', x'^{(j)})' \geq b + ka, \quad b = (z', w'^{(1)}).$$

Applying the Schwarz inequality to the left-hand side of this equation we obtain

$$||w'^{(k+1)}||' \geq b + ka. \tag{12.3}$$

Thus the norm squared of the weight vector $w'^{(k+1)}$ grows at least quadratically in k. On the other hand we have

$$||w'^{(k+1)}||'^2 = (w'^{(k)} + x'^{(k)}, w'^{(k)} + x'^{(k)})' = ||w'^{(k)}||'^2 + ||x'^{(k)}||'^2 +$$

$$2(w'^{(k)}, x'^{(k)})' \leq ||w'^{(k)}||'^2 + ||x'^{(k)}||'^2,$$

so, iterating,

$$||w'^{(k+1)}||'^2 \leq ||w'^{(1)}||'^2 + \sum_{j=1}^{k} ||x'^{(j)}||'^2 \leq ||w'^{(1)}||'^2 + ck \tag{12.4}$$

where c is the maximum value of $||x'||'^2$ as x ranges over the set of training vectors. This shows that the norm squared of the weight vector $w'^{(1)}$ grows at most linearly in k. The inequalities (12.3) and (12.4) can be reconciled only if for some k the inner products $(w'^{(k)}, x')'$ are strictly positive for all training vectors x. Then the algorithm terminates. Conversely, we see from our method of proof that if the algorithm terminates with weight vector $w'^{(k)}$ then it defines a separating hyperplane for the training vectors. □

For practical application, the training set needs to be sufficiently representative of the data that the Perceptron, once trained, can accurately classify data not in the training set. Also, of course, the data must be linearly separable and this is not always the case.

Exercise 12.9 Suppose the set of training vectors is such that $a = 1$, $c = 5$ in the proof of Theorem 12.6 and we start the Perceptron algorithm with $w'^{(1)}$ as the zero $n + 1$-tuple. Find the smallest k such that the algorithm is guaranteed to terminate in at most k steps.

Exercise 12.10 Find an example in two dimensions and an example in three dimensions of Class 1 and Class 2 training vectors that are not linearly separable.

Self-organizing memory (SOM)

The concept of SOM was first introduced by Kohonen in 1982. The basic idea is to keep high-dimensional distance accurately reflected in a given low-dimensional representation, i.e., clusters close in n space are close in the given network layout, or put another way, SOM proposes to do purely unsupervised clustering in high dimensions without any dimension reduction. The concept is to set up a grid of high-dimensional nodes which represent cluster centers in such a way that cluster centers close in the grid represent clusters centers close in high-dimensional space, i.e., close to each other as pattern vectors.

The algorithm proceeds as follows:

- Choose a set of nodes in advance and assign random initial cluster centers to the nodes;
- Present a pattern x to the network and find the closest cluster center say $c^{(j)}$;
- Assign the pattern to that cluster and recompute the cluster center $m^{(j)}$ as $m^{(j)} + \alpha(x - m^{(j)})$ where α is a parameter between 0 and 1;
- All "neighboring" cluster centers are also shifted towards x, $m^{(i)}$ is replaced by $m^{(i)} + \beta(x - m^{(i)})$ where β is a parameter between 0 and 1;
- Note that the values of α and β and the definition of "neighbors" are changed with each epoch so that the general form would be that $m^{(i)}$ is replaced by $m^{(i)} + h_{ij}(t)(x - m^{(i)})$ where $h_{ij}(t)$ is chosen to decrease to 0 in t (number of epochs) and eventually $h_{ij} = 0$ if $i \neq j$;

- Note that if we allow no neighbors (i.e., $h_{ij} = 0$ if $i \neq j$ for all epochs), then, in effect, SOM reduces to K-Means and so one may think of SOM as a spatially smooth form of K-Means.
- A modern generalization of SOM in several directions relates to prim diffusions.

12.3 Image representation of data

We are constantly flooded with information of all sorts and forms, and a common denominator of data analysis in many fields of current interest is the problem of dealing with large numbers of observations, each of which has high dimensionality.

Example 12.7 Suppose that a source X produces a high-dimensional data set

$$X := \left\{ x^{(1)}, \ldots, x^{(n)} \right\}, \, n \geq 1$$

that we wish to study. Typically, in a given problem of this kind, one is faced with a finite data set or a set of finite objects in a vector space X. For instance:

- Each member of X could be the frames of a movie produced by a digital camera.
- Pixels of a hyperspectral image in a computer vision problem, say face recognition.
- Objects from a statistical, computer science or machine learning model which need to be clustered.

The problem of finding mathematical methods in order to study meaningful structures and descriptions of **data sets** for learning tasks is an exploding area of research with applications as diverse as remote sensing, signal processing, critical infrastructure, complex networks, clustering, imaging, neural networks and sensor networks, wireless communications, financial marketing, and dynamic programming to name just a few. The advantage of X being linear is that we can now use the power of the linear algebra techniques we have developed earlier.

In the case of finite dimension, when dealing with these types of sets X, *high dimensionality* is often an obstacle for any efficient processing of the data. Indeed, many classical data processing algorithms have a computational complexity that grows exponentially with the dimension (the so-called *"curse of dimension"*).

On the other hand, the source of the data may only enjoy a limited number of degrees of freedom. In this case, the high-dimensional representation of the data is a natural function of the source and so the data has actually a low intrinsic dimensionality. Thus we can hope to find an accurate low-dimensional representation. We now describe a method for attacking this problem, based on probability and linear algebra theory.

12.3.1 The sample covariance matrix and principal component analysis

Suppose we have a multivariate distribution $\rho(t)$ in n random variables given by the n-vector t and we take m vector samples, $(T_1^{(k)}, \ldots, T_n^{(k)})$, $k = 1, \ldots, m$, independently. (Thus $T_i^{(k)}$ is the kth sample of the random variable $T_i \sim t_i$, a change of notation from that in Section 4.8.) The **sample covariance matrix** C^S is the $n \times n$ matrix with elements

$$C_{i,j}^S = \frac{1}{n-1} \sum_{k=1}^{m} (T_i^{(k)} - \bar{T}_i)(T_j^{(k)} - \bar{T}_j). \tag{12.5}$$

Here C_{ii}^S is the sample variance of the random variable t_i. Note that the sample data $T_i^{(k)}$ for fixed i is normalized by subtraction of the sample mean $\bar{T}_i$. The normalized data $T_i^{(k)} - \bar{T}_i$ has mean 0. The interpretation of the covariance $C_{i,j}^S$ between variables t_i and t_j for $i \neq j$ is that the variables are *correlated* if $C_{i,j}^S > 0$, *aniticorrelated* if $C_{i,j}^S < 0$ and *uncorrelated* if $C_{i,j}^S \approx 0$. An important application of the sample covariance matrix is its use in describing the relationship, or lack thereof, between data values of distinct random variables. The method for doing this is called Principal Component Analysis, which is none other than the application of a special case of the singular value decomposition to the sample covariance matrix.

From the sample data we form the $m \times n$ matrix X:

$$X_{k,i} = \frac{1}{\sqrt{n-1}}(T_i^{(k)} - \bar{T}_i) \quad k = 1, \ldots, m, \ i = 1, \ldots, n. \tag{12.6}$$

A special case of the Singular Value Theorem applied to X, see Subsection 1.7.1, says that there is a real orthogonal matrix O, $O^{\mathrm{tr}} = O^{-1}$, such that $C^S = O^{-1}DO$ where D is the $n \times n$ diagonal matrix

$$D = \begin{pmatrix} \sigma_1^2 & 0 & \cdots & \cdots & \cdots & 0 \\ 0 & \sigma_2^2 & \cdots & \cdots & \cdots & 0 \\ \vdots & \vdots & \ddots & & \vdots & \vdots \\ 0 & 0 & \cdots & \sigma_r^2 & 0 & \cdots \\ \vdots & \vdots & \vdots & & \vdots & \vdots \\ 0 & 0 & \cdots & \cdots & \cdots & 0 \end{pmatrix}.$$

Here, the nonzero eigenvalues of C^S are $\lambda_j = \sigma_j^2$ where $\sigma_1 \geq \sigma_2 \geq \cdots \geq \sigma_r > 0$, and r is the rank of C^S. Our interpretation of these results is that the orthogonal matrix O defines a transformation from our original coordinate axes, where data vectors took the form of n-component row vectors $t_i - \bar{t}_i$, to the principal axes. To understand this, note that $D = OC^S O^{-1} = OX^{\text{tr}} X O^{-1}$ so

$$(XO^{-1})^{\text{tr}}(XO^{-1}) = D.$$

With respect to the principal axes, the data matrix X takes the form XO^{-1} and the sample matrix C^S becomes the diagonal matrix $C'^S = D$. Thus with respect to the new axes the data variables are uncorrelated and there is no redundancy. The matrix D lists the principal variances in decreasing magnitude down the diagonal.

What we have done is find new orthogonal axes for the coordinates (new orthogonal basis for the vector space) such that:

- The axes explain the variance in the data in order.
- There is no redundancy between the coordinates.
- We can reduce the dimension in such a way that most of the information lost is the noise.

As we have shown above, amazingly, it is possible to do this under reasonable conditions (recall Example 1.85). We end this section by listing some assumptions and limitations of PCA: (1) the process is linear and orthogonal; (2) large variance implies information, small variance implies noise; (3) the statistics of the process is described typically by mean and variance.

Exercise 12.11 Show that $E(C^S) = C$, i.e., the expectation of C^S is C.

Exercise 12.12 Verify the following properties of the $n \times n$ sample covariance matrix C^S and the $n \times m$ matrix (12.6). (1) $C^S = X^{\text{tr}} X$, (2) $(C^S)^{\text{tr}} = C^S$. (3) The eigenvalues λ_j of C^S are nonnegative.

For more results on principal component analysis related to statistics see [87].

12.4 Image compression

There are three basic styles of compression. (1) Lossless: Here we have a recoding, no information is lost and the image can be perfectly reconstructed. (2) Threshold: Here, we remove certain aspects of information chosen to be least damaging to human perception. (3) Fractal: Here, we exploit the fractal nature of edges. There are also three basic criteria for compression so that (1)–(3) can be achieved. (a) Time for compression, (b) degree of compression achieved and (c) use to which the data is to be put. In what follows, we give a short description of image compression using wavelets.

12.4.1 Image compression using wavelets

A typical image consists of a rectangular array of 256×256 pixels, each pixel coded by 24 bits. In contrast to an audio signal, this signal has a fixed length. The pixels are transmitted one at a time, starting in the upper left-hand corner of the image and ending with the lower right. However for image processing purposes it is more convenient to take advantage of the $2D$ geometry of the situation and consider the image not as a linear time sequence of pixel values but as a geometrical array in which each pixel is assigned its proper location in the image. Thus the finite signal is two dimensional: x_{n_1,n_2} where $0 \leq n_i < 2^8$. We give a very brief introduction to subband coding for the processing of these images.

Since we are in two dimensions we need a $2D$ filter $\mathbf{H}$

$$\mathbf{H}x_{n_1,n_2} = y_{n_1,n_2} = \sum_{k_1,k_2} h_{k_1,k_2} x_{n_1-k_1,n_2-k_2}.$$

This is the $2D$ convolution $y = \mathbf{H}x$. In the frequency domain this reads

$$Y(\omega_1,\omega_2) = H(\omega_1,\omega_2)X(\omega_1,\omega_2)$$

$$= \left(\sum_{k_1,k_2} h_{k_1,k_2} e^{-i(k_1\omega_1+k_2\omega_2)} \right) \left(\sum_{n_1,n_2} \mathbf{x}_{n_1,n_2} e^{-i(n_1\omega_1+n_2\omega_2)} \right),$$

with a similar expression for the Z transform. The frequency response is 2π-periodic in each of the variables ω_j and the frequency domain is the square $-\pi \leq \omega_j < \pi$. We could develop a truly $2D$ filter bank to process this image. Instead we will take the easy way out and use *separable* filters, i.e., products of $1D$ filters. We want to decompose the image into low-frequency and high-frequency components in each variable n_1, n_2

separately, so we will use four separable filters, each constructed from one of our $1D$ pairs h^{low}, h^{high} associated with a wavelet family:

$$h_{n_1,n_2}^{(0)} = h_{n_1}^{low} h_{n_2}^{low}, \qquad h_{n_1,n_2}^{(2)} = h_{n_1}^{high} h_{n_2}^{low},$$

$$h_{n_1,n_2}^{(1)} = h_{n_1}^{low} h_{n_2}^{high}, \qquad h_{n_1,n_2}^{(3)} = h_{n_1}^{high} h_{n_2}^{high}.$$

The frequency responses of these filters also factor. Thus, $H^{(2)}(\omega_1, \omega_2) = H^{high}(\omega_1)H^{low}(\omega_2)$, etc. After obtaining the outputs $y_{n_1,n_2}^{(j)}$ from each of the four filters $\mathbf{H}^{(j)}$ we downsample to get $(\downarrow [2,2])y_{n_1,n_2} = y_{2n_1,2n_2}^{(j)}$. Thus we keep one sample out of four for each analysis filter. This means that we have exactly as many pixels as we started with (256×256), but now they are grouped into four (128×128) arrays. Thus the analyzed image is the same size as the original image but broken into four equally sized squares: LL (upper left), HL (upper right), LH (lower right), and HH (lower left). Here HL denotes the filter that is high pass on the n_1 index and low pass on the n_2 index, etc.

A straightforward Z transform analysis shows that this is a perfect reconstruction $2D$ filter bank provided the factors h^{low}, h^{high} define a $1D$ perfect reconstruction filter bank. The synthesis filters can be composed from the analogous synthesis filters for the factors. Upsampling is done in both indices simultaneously: $(\uparrow [2,2])y_{2n_1,2n_2} = y_{n_1,n_2}$ for the even–even indices. $(\uparrow [2,2])y_{m_1,m_2} = 0$ for m_1, m_2 even–odd, odd–even or odd–odd.

At this point the analysis filter bank has decomposed the image into four parts. LL is the analog of the low pass image. HL, LH, and HH each contain high-frequency (or difference) information and are analogs of the wavelet components. In analogy with the $1D$ wavelet transform, we can now leave the (128×128) wavelet subimages HL, LH, and HH unchanged, and apply our $2D$ filter bank to the (128×128) LL subimage. Then this block in the upper left-hand corner of the analysis image will be replaced by four 64×64 blocks $L'L'$, $H'L'$, $L'H'$, and $H'H'$, in the usual order. We could stop here, or we could apply the filter bank to $L'L'$ and divide it into four 32×32 pixel blocks $L''L''$, $H''L''$, $L''H''$, and $H''H''$. Each iteration adds a net three additional subbands to the analyzed image. Thus one pass through the filter bank gives 4 subbands, two passes give 7, three passes yield 10 and four yield 13. Four or five levels are common. For a typical analyzed image, most of the signal energy is in the low pass image in the small square in the upper left-hand corner. It appears as a bright but blurry miniature facsimile of the original image. The various wavelet subbands have little energy and are relatively dark.

If we run the analyzed image through the synthesis filter bank, iterating an appropriate number of times, we will reconstruct the original signal. However, the usual reason for going through this procedure is to process the image before reconstruction. The storage of images consumes a huge number of bits in storage devices; compression of the number of bits defining the image, say by a factor of 50, has a great impact on the amount of storage needed. Transmission of images over data networks is greatly speeded by image compression. The human visual system is very relevant here. One wants to compress the image in ways that are not apparent to the human eye. The notion of "barely perceptible difference" is important in multimedia, both for vision and sound. In the original image each pixel is assigned a certain number of bits, 24 in our example, and these bits determine the color and intensity of each pixel in discrete units. If we increase the size of the units in a given subband then fewer bits will be needed per pixel in that subband and fewer bits will need to be stored. This will result in a loss of detail but may not be apparent to the eye, particularly in subbands with low energy. This is called quantization. The compression level, say 20 to 1, is mandated in advance. Then a bit allocation algorithm decides how many bits to allocate to each subband to achieve that over-all compression while giving relatively more bits to high-energy parts of the image, minimizing distortion, etc. (This is a complicated subject.) Then the newly quantized system is *entropy coded*. After quantization there may be long sequences of bits that are identical, say 0. Entropy coding replaces that long string of 0s by the information that all of the bits from location a to location b are 0. The point is that this information can be coded in many fewer bits than were contained in the original sequence of 0s. Then the quantized file and coded file is stored or transmitted. Later the compressed file is processed by the synthesizing filter bank to produce an image.

There are many other uses of wavelet-based image processing, such as edge detection. For edge detection one is looking for regions of rapid change in the image and the wavelet subbands are excellent for this. Noise will also appear in the wavelet subbands and a noisy signal could lead to false positives by edge detection algorithms. To distinguish edges from noise one can use the criteria that an edge should show up at all wavelet levels.

For more information about image compression using wavelets see [42]. For the use of curvelets in image compression see [22, 23, 24]. For ridgelets in image compression see [20].

12.4.2 Thresholding and denoising

Suppose that we have analyzed a signal down several levels using the DWT. If the wavelets used are appropriate for the signal, most of the energy of the signal ($\sum_k |a_{j,k}|^2 = \sum_k (|a_{j-1,k}|^2 + |b_{j-1,k}|^2)$) at level j will be associated with just a few coefficients. The other coefficients will be small in absolute value. The basic idea behind thresholding is to zero out the small coefficients and to yield an economical representation of the signal by a few large coefficients. At each wavelet level one chooses a threshold $\delta > 0$. Suppose $x_j(t)$ is the projection of the signal at that level. There are two commonly used methods for thresholding. For *hard thresholding* we modify the signal according to

$$
y_j^{\text{hard}}(t) = \begin{cases} x_j(t), & \text{for } |x_j(t)| > \delta \\ 0, & \text{for } |x_j(t)| \leq \delta. \end{cases}
$$

Then we synthesize the modified signal. This method is very simple but does introduce discontinuities. For *soft thresholding* we modify the signal continuously according to

$$
y_j^{\text{soft}}(t) = \begin{cases} \text{sign}(x(t))(x_j(t) - \delta), & \text{for } |x_j(t)| > \delta \\ 0, & \text{for } |x_j(t)| \leq \delta. \end{cases}
$$

Then we synthesize the modified signal. This method doesn't introduce discontinuities. Note that both methods reduce the overall signal energy. It is necessary to know something about the characteristics of the desired signal in advance to make sure that thresholding doesn't distort the signal excessively.

Denoising is a procedure to recover a signal that has been corrupted by noise. (Mathematically, this could be Gaussian white noise $N(0,1)$.) It is assumed that the basic characteristics of the signal are known in advance and that the noise power is much smaller than the signal power.

The idea is that when the signal plus noise is analyzed via the DWT the essence of the basic signal shows up in the low pass channels. Most of the noise is captured in the differencing (wavelet) channels. In a channel where noise is evident we can remove much of it by soft thresholding. Then the reconstituted output will contain the signal with less noise corruption.

12.5 Additional exercises

Exercise 12.1 (1) Take each of the examples below, with 16 pixels and 3 bit dynamic range monochrome values and apply a contrast stretch and histogram normalization. What would you expect to see in real images of these types?

$$(A): \begin{bmatrix} 0 & 1 & 1 & 1 \\ 0 & 2 & 1 & 6 \\ 1 & 1 & 6 & 5 \\ 6 & 7 & 7 & 7 \end{bmatrix}, \quad (B): \begin{bmatrix} 7 & 6 & 5 & 4 \\ 6 & 7 & 5 & 5 \\ 5 & 5 & 6 & 7 \\ 4 & 5 & 7 & 6 \end{bmatrix},$$

$$(C): \begin{bmatrix} 2 & 3 & 2 & 3 \\ 3 & 2 & 3 & 5 \\ 3 & 6 & 5 & 6 \\ 6 & 5 & 6 & 5 \end{bmatrix}.$$

(2) Suppose that in a binary image you have k white pixels in a total of N pixels. Find the mean and variance of the image.

(3) (a) Show that the mean of an image with dynamic range of n bits is $1/2(2^n - 1)$ if histogram normalization has been applied to it. (b) Write down a formula for the variance of the images in (a). (c) What is the implication of (a)–(b) for contrast and brightness?

Exercise 12.2 (a) Consider each of the three images below. Each is the Fourier transform of an image. Roughly sketch the original image.

$$(A): \begin{bmatrix} \text{space} & \text{space} & \text{space} \\ . & . & . \\ \text{space} & \text{space} & \text{space} \end{bmatrix}, \quad (B): \begin{bmatrix} \text{space} & . & \text{space} & \text{space} \\ \text{space} & \text{space} & \text{space} & \text{space} \\ \text{space} & . & \text{space} & \text{space} \end{bmatrix},$$

$$(C): \begin{bmatrix} \text{space} & \text{space} & \text{space} & \text{space} & \text{space} \\ . & \text{space} & \text{space} & . & . \\ \text{space} & \text{space} & \text{space} & \text{space} & \text{space} \end{bmatrix},$$

(b) Suppose we take an image which consists of spaces everywhere except on the diagonal where it has a high pixel value. What will its Fourier transform look like? Explain your answer.

Try to work these out yourself from your understanding of Fourier transforms. Then review the paper with the web address given by: www.cs.unm.edu/~ brayer/vision/fourier.html

Exercise 12.13 Gray scale morphology operators: (a) Show that the following is true: if f is the intensity function for an image and k is a hemisphere structuring function then:

- $fdk = T[[f]dU[k]]$.
- $fek = T[U[f]eU[k]]$.
- $fok = T[U[f]oU[k]]$.
- $fck = T[U[f]cU[k]]$.

(b) Prove the Umbra Homomorphism Theorem:

- $U[fdg] = U[f]dU[g]$.
- $U[feg] = U[f]eU[g]$.
- $U[fog] = U[feg]dU[g] = U[f]oU[g]$.

Exercise 12.14 Find the opening and closing of the images (A)–(C) in Exercise 254 for suitable structuring elements of your choice.

Exercise 12.15 Prove that if A and B are umbras, then so is AdB.

Exercise 12.16 Consider an image with pixels over $[0, 4]$. The gray scale function for the image intensity is given by $f(x) = x - 1$ for $1 \leq x \leq 2$, $f(x) = 3 - x$ for $2 \leq x \leq 3$ and 0 elsewhere. The structuring function is a hemisphere of radius 0.25, i.e.,

$$k(x) = \sqrt{0.0625 - x^2} \quad -0.25 \leq x \leq 0.25.$$

Find the opening and closing of f by k and discuss what this should do to the visual aspects of the image.

Exercise 12.17 Write a program to implement the 3×3 skeletonization algorithm described above and try it out on some images.

Exercise 12.18 Consider the pattern classes $\{(0, 0), (0, 1)\}$ and $\{(1, 0), (1, 1)\}$. Use these patterns to train a perceptron algorithm to separate patterns into two classes around these. [Do not forget to augment the patterns and start with zero weights.]

Exercise 12.19 Suppose we have a collection of pattern classes and in each case the distribution of the patterns around the cluster center is Gaussian. Suppose also that the pattern components are uncorrelated. [This means the covariance matrix is the identity.] Find a formula for the probability that a given pattern vector x belongs to cluster w_j. In this case, given a pattern vector x, deduce that the decision as to which class it belongs to would be made simply on its distance to the nearest cluster center.

Exercise 12.20 Consider the following 20 two-dimensional vectors $x = \begin{pmatrix} x_1 \\ x_2 \end{pmatrix}$:

$(x_1 = 4.1, 2.8, 0.5, 3.1, 4.8, 1.5, 3.8, 2.4, 1.7, 2, 0.2, 3.5, 2.1, 0.8, 2.9, 4.4, 3.7,$
$5, 4, 1.3)$

$(x_2 = 9.7, 6.3, 2.4, 7.2, 10.2, 4.4, 9, 5.4, 4.7, 4.8, 1.5, 8, 5.7, 2.5, 6.7, 10, 8.8,$
$10.5, 9.3, 3.3)$

Without using a MATLAB program, find its principal components and comment on the results.

Exercise 12.21 In this chapter, we have given a self-contained introduction to some techniques dealing with the parsimonious representation of data. MATLAB offers several toolboxes which allow one to implement demos on data. Study the imaging and statistical pattern toolbox demos given in the links:

http://cmp.felk.cvut.cz/cmp/software/stprtool/examples.html
http://www-rohan.sdsu.edu/doc/matlab/toolbox/images/images.html

We have labored to produce the following reference list which allows the interested reader to follow applications and modern and current research in many topics in this area including our own research.

1. Compression, classification, clustering, neural networks, alignment, segmentation, registration, clustering, computer aided geometric design, subdivision schemes, PDE methods, inpainting, edge detection, shape detection, pattern detection and pattern recognition, shape spaces, resolution recovery. See [33, 34, 56, 57, 64, 65, 86] and the references cited therein.
2. Rigidity theory, matrix completions, image matching and shape matching data matching, shape and point cloud recognition, metrics, group invariance, symmetry. See [13] and the references cited therein.
3. Metrics, diffusion, percolation processes and linear, constrained, local and nonlinear dimension reduction of data, dimension estimation, PDE flows, prim diffusions and graph diffusions, self-organizing memory, information theory methods. See [8, 31, 32, 33, 44, 45, 46, 49, 64, 65, 86] and the references cited therein.
4. Optimal configurations, discretization, kernels, numerical integration, random matrix theory, minimal functionals. See [48, 51] and the references cited therein.

5. Compressive sensing, sparcity, wavelets, curvelets, edgelets, ridgelets, shearlets, sparse approximation, approximation using sparcity, sparse representations via dictionaries, robust principal component analysis, recovery of functions in high dimensions, learning, linear approximation and nonlinear approximation, the curse of dimensionality. See [6, 7, 9, 12, 14, 18, 20, 21, 22, 23, 24, 25, 27, 28, 29, 30, 35, 36, 37, 38, 39, 40, 41, 42, 43, 47, 52, 54, 58, 59, 60, 61] and the references cited therein.

References

[1] D. Achlioptas, Database-friendly random projections: Johnson-Lindenstrauss with binary coins, *J. Comp. Sys. Sci.*, **66** (4), (2003), 671–687.

[2] A. Akansu and R. Haddad, *Multiresolution Signal Decomposition*, Academic Press, 1993.

[3] G. E. Andrews, R. Askey and R. Roy, *Special Functions*, Encyclopedia of Mathematics and its Applications, Cambridge University Press, 1999.

[4] T. M. Apostol, *Calculus*, Blaisdell Publishing Co., 1967–1969.

[5] L. J. Bain and M. Engelhardt, *Introduction to Probability and Mathematical Statistics*, 2nd edn., PWS-Kent, 1992.

[6] R. Baraniuk, M. Davenport, R. DeVore and M. Wakin, A simple proof of the restricted isometry property for random matrices, *Const. Approx.*, **28**, (2008), 253–263.

[7] R. Baraniuk and M. B. Wakin, Random projections on smooth manifolds, *Found. Comput. Math.*, **9**, (2009), 51–77.

[8] M. Belkin and P. Niyogi, Laplacian eigenmaps for dimensionality reduction and data representation, *Neural Comput.*, **15** (6), (2003), 1373–1396.

[9] A. Berlinet and C. T. Agnan, *Reproducing Kernel Hilbert Spaces in Probability and Statistics*, Kluwer, 2004.

[10] J. J Benedetto and M. W. Frazier, *Wavelets: Mathematics and Applications*, CRC Press, 1994.

[11] J. J. Benedetto and A. I. Zayed, *Sampling, Wavelets and Tomography*, Birkhauser, 2004.

[12] P. Binev, A. Cohen, W. Dahmen and V. Temlyakov, Universal algorithms for learning theory part I: piecewise constant functions, *J. Mach. Learn. Res.*, **6**, (2005), 1297–1321.

[13] M. M. Bronstein and I. Kokkinos, *Scale Invariance in Local Heat Kernel Descriptors without Scale Selection and Normalization*, INRIA Research Report 7161, 2009.

[14] A. M. Bruckstein, D. L. Donoho and M. Elad, From sparse solutions of systems of equations to sparse modeling of signals and images, *SIAM Review*, **51**, February 2009, 34–81.

[15] A. Boggess and F. J. Narcowich, *A First Course in Wavelets with Fourier Analysis*, Prentice-Hall, 2001.

[16] S. Boyd and L. Vandenberghe, *Convex Optimization*, Cambridge University Press, 2004.

[17] C. S. Burrus, R. A. Gopinath and H. Guo, *Introduction to Wavelets and Wavelet Transforms. A Primer*, Prentice-Hall, 1998.

[18] A. Bultheel, Learning to swim in the sea of wavelets, *Bull. Belg. Math. Soc.*, **2**, (1995), 1–44.

[19] E. J. Candes, Compressive sampling, *Proceedings of the International Congress of Mathematicians, Madrid, Spain, (2006)*.

[20] E. J. Candes, Ridgelets and their derivatives: Representation of images with edges, in *Curves and Surfaces*, ed. L. L. Schumaker *et al.*, Vanderbilt University Press, 2000, pp. 1–10.

[21] E. J. Candes, The restricted isometry property and its implications for compressed sensing, *C. R. Acad, Sci. Paris Ser. I Math.*, **346**, (2008), 589–592.

[22] E. J. Candes and D. L. Donoho, Continuous curvelet transform: I. Resolution of the wavefront set, *Appl. Comput. Harmon. Anal.*, **19**, (2005), 162–197.

[23] E. J. Candes and E. L. Donoho, Continuous curvelet transform: II. Discretization and frames, *Appl. Comput. Harmon. Anal.*, **19**, (2005), 198–222.

[24] E. J. Candes and E. L. Donoho, New tight frames of curvelets and optimal representations of objects with piecewise-C2 singularities, *Comm. Pure Appl. Math*, **57**, (2006), 219–266.

[25] E. J. Candes, X. Li, Y. Ma and J. Wright, Robust principal component analysis?, *J. ACM*, **58** (1), 1–37.

[26] E. J. Candes, J. Romberg and T. Tao, Robust uncertainty principles: exact signal reconstruction from highly incomplete frequency information, *IEEE Trans. Inform. Theory*, **52**, (2006), 489–509.

[27] E. J. Candes, J. Romberg and T. Tao, Stable signal recovery from incomplete and inaccurate measurements, *Comm. Pure and Appl. Math.*, **59**, (2006), 1207–1223.

[28] E. J. Candes and T. Tao, Decoding by linear programing, *IEEE Trans. Inform. Theory*, **51**, (2005), 4203–4215.

[29] E. J. Candes and T. Tao, Near optimal signal recovery from random projections: universal encoding strategies, *IEEE Trans. Inform. Theory*, **52**, 5406–5425.

[30] E. J. Candes and T. Tao, The Dantzig selector: Statistical estimation when p is much larger than n, *Ann. Stat.*, **35**, (2007), 2313–2351.

[31] K. Carter, R. Raich and A. Hero, On local dimension estimation and its applications, *IEEE Trans. Signal Process.*, **58** (2), (2010), 762–780.

[32] K. Carter, R. Raich, W.G. Finn and A. O. Hero, FINE: Fisher information non-parametric embedding, *IEEE Trans. Pattern Analysis and Machine Intelligence (PAMI)*, **31** (3), (2009), 2093–2098.

[33] K. Causwe, M Sears, A. Robin, Using random matrix theory to determine the number of endmembers in a hyperspectral image, *Whispers Proceedings*, **3**, (2010), 32–40.

[34] T. F. Chan and J. Shen, *Image Processing and Analysis*, SIAM, 2005.

[35] C. Chui, *An Introduction to Wavelets*, Academic Press, 1992.

[36] C. Chui and H. N. Mhaskar, MRA contextual-recovery extension of smooth functions on manifolds, *App. Comput. Harmon. Anal.*, **28**, (2010), 104–113.

[37] A. Cohen, W. Dahmen, I. Daubechies and R. DeVore, Tree approximation and encoding, *Appl. Comput. Harmon. Anal.*, **11**, (2001), 192–226.

[38] A. Cohen, W. Dahmen and R. DeVore, Compressed sensing and k-term approximation, *Journal of the AMS*, **22**, (2009), 211–231.

[39] A. Cohen, I. Daubechies and J. C. Feauveau, Biorthogonal bases of compactly supported wavelets, *Comm. Pure Appl. Math.*, **45**, (1992), 485–560.

[40] A. Cohen and R. D Ryan, *Wavelets and Multiscale Signal Processing*, Chapman and Hall, 1995.

[41] A. Cohen, Wavelets and digital signal processing, in *Wavelets and their Applications*, ed. M. B. Ruskai *et al.*, Jones and Bartlett, 1992.

[42] A. Cohen and J. Froment, Image compression and multiscale approximation, in *Wavelets and Applications* (Marseille, Juin 89), ed. Y. Meyer, Masson, 1992.

[43] A. Cohen, Wavelets, approximation theory and subdivision schemes, in *Approximation Theory VII* (Austin, 1992), ed. E. W. Cheney *et al.*, Academic Press, 1993.

[44] R. R. Coifman and S. Lafon, Diffusion maps, *Appl. Comp. Harmon. Anal.*, **21**, (2006), 5–30.

[45] R. R. Coifman and M. Maggioni, Diffusion wavelets, *Appl. Comput. Harmon. Anal.*, **21** (1), (2006), 53–94.

[46] J. A. Costa and A. O. Hero, Geodesic entropic graphs for dimension and entropy estimation in manifold learning, *IEEE Trans. Signal Process.*, **52** (8), (2004), 2210–2221.

[47] F. Cucker and S. Smale, On the mathematical foundations of learning theory, *Bull. Amer. Math. Soc.*, **39**, (2002), 1–49.

[48] S. B. Damelin, A walk through energy, discrepancy, numerical integration and group invariant measures on measurable subsets of euclidean space, *Numer. Algorithms*, **48** (1), (2008), 213–235.

[49] S. B. Damelin, On bounds for diffusion, discrepancy and fill distance metrics, in *Principal Manifolds*, Springer Lecture Notes in Computational Science and Engineering, vol. 58, 2008, pp. 32–42.

[50] S. B. Damelin and A. J. Devaney, Local Paley Wiener theorems for analytic functions on the unit sphere, *Inverse Probl.*, **23**, (2007), 1–12.

[51] S. B. Damelin, F. Hickernell, D. Ragozin and X. Zeng, On energy, discrepancy and G invariant measures on measurable subsets of Euclidean space, *J. Fourier Anal. Appl.*, **16**, (2010), 813–839.

[52] I. Daubechies, Orthonormal bases of compactly supported wavelets, *Comm. Pure Appl. Math.*, **41**, (1988), 909–996.

[53] I. Daubechies, *Ten Lectures on Wavelets*, CBMS-NSF Regional Conf. Ser. Appl. Math, 61, SIAM, 1992.

[54] I. Daubechies, R. Devore, D. Donoho and M. Vetterli, Data compression and harmonic analysis, *IEEE Trans. Inform. Theory Numerica*, **44**, (1998), 2435–2467.

[55] K. R. Davidson and A. P. Donsig, *Real Analysis with Real Applications*, Prentice Hall, 2002.

[56] L. Demaret, N. Dyn and A. Iske, Compression by linear splines over adaptive triangulations, *Signal Process. J.*, **86** (7), July 2006, 1604–1616.

[57] L. Demaret, A. Iske and W. Khachabi, A contextual image compression from adaptive sparse data representations, *Proceedings of SPARS09, April 2009, Saint-Malo*.

[58] R. DeVore, Nonlinear approximation, *Acta Numer.*, **7**, (1998), 51–150.

[59] R. Devore, Optimal computation, *Proceedings of the International Congress of Mathematicians, Madrid, Spain, 2006*.

[60] R. Devore, Deterministic constructions of compressed sensing matrices, *J. Complexity*, **23**, (2007), 918–925.

[61] R. Devore and G. Lorentz, *Constructive Approximation*, Comprehensive Studies in Mathematics 303, Springer-Verlag, 1993.

[62] D. Donoho, Compressed sensing, *IEEE Trans. Inform. Theory*, **52** (4), (2006), 23–45.

[63] D. Donoho, M. Elad and V. Temlyakov, Stable recovery of sparse overcomplete representations in the presence of noise, *IEEE Trans. Inf. Theory*, **52** (1), (2006), 6–18.

[64] L. du Plessis, R. Xu, S. B. Damelin, M. Sears and D. Wunsch, Reducing dimensionality of hyperspectral data with diffusion maps and clustering with K means and fuzzy art, *Int. J. Sys. Control Comm.*, **3** (2011), 3–10.

[65] L. du Plessis, R. Xu, S. Damelin, M. Sears and D. Wunsch, Reducing dimensionality of hyperspectral data with diffusion maps and clustering with K-means and fuzzy art, *Proceedings of International Conference of Neural Networks, Atlanta, 2009*, pp. 32–36.

[66] D.T. Finkbeiner, *Introduction to Matrices and Linear Transformations*, 3rd edn., Freeman, 1978.

[67] M. Herman and T. Strohmer, Compressed sensing radar, ICASSP 2008, pp. 1509–1512.

[68] M. Herman and T. Strohmer, High resolution radar via compressed sensing, *IEEE Trans. Signal Process.*, November 2008, 1–10.

[69] E. Hernandez and G. Weiss, *A First Course on Wavelets*, CRC Press, 1996.

[70] W. Hoeffding, Probability inequalities for sums of bounded random variables, *J. Amer. Statist. Assoc.*, **58**, (1963), 13–30.

[71] B. B. Hubbard, *The World According to Wavelets*, A. K Peters, 1996.

[72] P. E. T. Jorgensen, *Analysis and Probability: Wavelets, Signals, Fractals*, Graduate Texts in Mathematics, 234 Springer, 2006.

[73] D. W. Kammler, *A First Course in Fourier Analysis*, Prentice-Hall, 2000.

[74] F. Keinert, *Wavelets and Multiwavelets*, Chapman and Hall/CRC, 2004.

[75] J. Korevaar, *Mathematical Methods (Linear algebra, Normed spaces, Distributions, Integration)*, Academic Press, 1968.

[76] A. M. Krall, *Applied Analysis*, D. Reidel Publishing Co., 1986.

[77] S. Kritchman and B. Nadler, Non-parametric detection of the number of signals, hypothesis testing and random matrix theory, *IEEE Trans. Signal Process.*, **57** (10), (2009), 3930–3941.

[78] H. J. Larson, *Introduction to Probability Theory and Statistical Inference*, Wiley, 1969.

[79] G. G. Lorentz, M. von Golitschek and M. Makovoz, *Constructive Approximation: Advanced Problems*, Grundlehren Math. Wiss, 304, Springer-Verlag, 1996.

[80] D. S. Lubinsky, A survey of mean convergence of orthogonal polynomial expansions, in *Proceedings of the Second Conference on Functions Spaces* (Illinois), ed. K. Jarosz, CRC press, 1995, pp. 281–310.

[81] B. Lucier and R. Devore, Wavelets, *Acta Numer.*, **1**, (1992), 1–56.

[82] S. Mallat, *A Wavelet Tour of Signal Processing*, 2nd edn., Academic Press, 2001.

[83] S. Mallat, A wavelet tour of signal processing - the Sparse way, Science-Direct (Online service), Elsevier/Academic Press, 2009.

[84] Y. Meyer, *Wavelets: Algorithms and Applications*, translated by R. D. Ryan, SIAM, 1993.

[85] W. Miller, Topics in harmonic analysis with applications to radar and sonar, in *Radar and Sonar, Part 1*, R. Blahut, W. Miller and C. Wilcox, IMA Volumes in Mathematics and its Applications, Springer-Verlag, 1991.

[86] M. Mitchley, M. Sears and S. B. Damelin, Target detection in hyperspectral mineral data using wavelet analysis, *Proceedings of the 2009 IEEE Geosciences and Remote Sensing Symposium, Cape Town*, pp. 23–45.

[87] B. Nadler, Finite sample approximation results for principal component analysis: A matrix perturbation approach, *Ann. Stat.*, **36** (6), (2008), 2791–2817.

[88] H. J. Nussbaumer, *Fast Fourier Transform and Convolution Algorithms*, Springer-Verlag, 1982.

[89] P. Olver and C. Shakiban, *Applied Linear Algebra*, Pearson, Prentice-Hall, 2006.

[90] J. Ramanathan, *Methods of Applied Fourier Analysis*, Birkhauser, 1998.

[91] H. L. Royden, *Real Analysis*, Macmillan Co., 1988.

[92] W. Rudin, *Real and Complex Analysis*, McGraw-Hill, 1987.

[93] W. Rudin, *Functional Analysis*, McGraw-Hill, 1973.

[94] G. Strang, *Linear Algebra and its Applications*, 3rd edn., Harcourt Brace Jovanovich, 1988.

[95] G. Strang and T. Nguyen, *Wavelets and Filter Banks*, Wellesley-Cambridge Press, 1996.

[96] R. Strichartz, *A Guide to Distribution Theory and Fourier Transforms*, CRC Press, 1994.

[97] J. Tropp, Just relax: convex programming methods for identifying sparse signals in noise, *IEEE Trans. Inf. Theory*, **52** (3), (2006), 1030–1051.

[98] B. L. Van Der Waerden, *Modern Algebra*, Volume 1 (English translation), Ungar, 1953.

[99] G. G. Walter and X. Shen, *Wavelets and other Orthogonal Systems*, 2nd ed., Chapman and Hall, 2001.

[100] E. T. Whittaker and G. M. Watson, *A Course in Modern Analysis*, Cambridge University Press, 1958.

[101] P. Wojtaszczyk, *A Mathematical Introduction to Wavelets*, London Mathematical Society Student Texts, 37, Cambridge University Press, 1997.

[102] R. J. Zimmer, *Essential Results of Functional Analysis*, University of Chicago Press, 1990.

[103] A. Zygmund, *Trignometrical Series*, Dover reprint, 1955.

Index

Made in the USA
Monee, IL
07 July 2026

56550170R00270